Handbook of OPTICAL PROPERTIES

Volume II
Optics of Small Particles, Interfaces, and Surfaces

Handbook of
OPTICAL PROPERTIES

Volume II
Optics of Small Particles, Interfaces, and Surfaces

Edited by

Rolf E. Hummel
P. Wißmann

CRC Press

Boca Raton   New York   London   Tokyo

Acquiring Editor:	Tim Pletscher
Project Editor:	Jennifer Richardson
Marketing Manager:	Susie Carlisle
Direct Marketing Manager:	Becky McEldowney
Cover design:	Dawn Boyd
PrePress:	Kevin Luong, Greg Cuciak, Carlos Esser
Manufacturing:	Sheri Schwartz

Library of Congress Cataloging-in-Publication Data

Thin films for optical coatings / edited by Rolf E. Hummel and Karl H.
   Guenther.
     p. cm.– – (Handbook of optical properties ; v. 1)
   Includes bibliographical references and index.
   ISBN 0-8493-2484-X
   1. Optical coatings. 2. Thin films. 3. Optical films.
 I. Hummel, Rolf E., 1934–   II. Guenther, Karl H. III. Series.
TS192.C36   1995
621.36– –dc20   94-32692
                                                            CIP

No claim to original U.S. Government works
International Standard Book Number 0-8493-2484-X
Library of Congress Card Number 94-32692
Printed in the United States of America 1 2 3 4 5 6 7 8 9 0
Printed on acid-free paper

EDITORS

Rolf E. Hummel is Professor of Materials Science at the University of Florida, Gainesville, Fl. He received his Ph.D. in 1963 from the University of Stuttgart, Germany and the Max-Planck-Institut for Materials Research, also in Stuttgart. His previous publications include the books *Opticals Properties of Metals and Alloys* (1971), *Electro- and Thermotransport in Metals and Alloys* (1977), *Electronic Properties of Materials* (1st ed. 1985, 2nd ed. 1992), *Handbook of Optical Properties, Volume I, Thin Films for Optical Coatings.*

Peter Wißmann completed his studies of Physics and Mathematics in Hannover and Munich, Germany. In 1965 he received the Dipl.-Phys., in 1967 he received his Dr. rer. nat. from the Technical University of Hannover, and in 1972 his Dr. rer. nat. habil. from the University Erlangen-Nürnberg. Since 1978 he has been a Professor of Physical Chemistry at the University of Erlangen-Nürnberg, and from 1988 to 1991 he served as Chairman of the Thin Film Division of the German Physical Society.

CONTRIBUTORS

Gedalyah A. Berger
Department of Physics
Queens College of CUNY
Flushing, New York

Leah Bergman
Research Associate
Department of Physics
North Carolina State University
Raleigh, North Carolina

Richard K. Chang
Professor
Department of Applied Physics
Yale University
New Haven, Connecticut

Janice L. Cheung
Postdoctoral Research Assistant
Department of Chemical Engineering
Worcester Polytechnic Institute
Worcester, Massachusetts

Gottfried H. Döhler
Professor
Insitut für Technische Physik
University of Erlangen
Erlangen, Germany

Thomas Fischer
Physik - Fakultät für Chemie
Technische Universität München
Garching, Germany

Anton Fojtik
Physik - Fakultät für Chemie
Technische Universität Müchen
Garching, Germany

Gerd Frankowsky
HL SP PE DES
Siemens AG
Muenchen, Germany

M.I. Freedhoff
Congressional Science Fellow
c/o Rep. Edward Markey
Washington, D.C

Azriel Z. Genack
Professor
Department of Physics
Queens College of CUNY
Flushing, New York

Richard F. Haglund, Jr.
Professor
Department of Physics and Astronomy
Vanderbilt University
Nashville, Tennessee

Andreas Hangleiter
4 Physikal Institut
Universität Stuttgart
Stuttgart, Germany

Volker Härle
Bereich Halbleiter
Siemens AG
Regensburg, Germany

Justin M. Hartings
Graduate Student
Department of Applied Physics
Yale University
New Haven, Connecticut

Rolf E. Hummel
Professor
Department of Materials Science and
Engineering
University of Florida
Gainesville, Florida

Michael Kempe
Research Associate
Department of Physics
Queens College of CUNY
Flushing, New York

Peter Kiesel
Physicist
Institut für Technische Physik
Universität Erlangen
Erlangen, Germany

Wolfgang Knoll
Director
Department of Materials Science
MPI for Polymer Research
Mainz, Germany

Uwe Kreibig
Professor
I. Physikalisches Institut
RWTH Aachen
Aachen, Germany

Matthias H. Ludwig
Adjunct Professor
Department of Materials Science and
Engineering
University of Florida
Gainesville, Florida

A.P. Marchetti
Senior Research Associate
Eastman Kodak Company
and
Center for Photoinduced Charge Transfer
University of Rochester
Rochester, New York

Peter Michler
Institut für Festkörperphysik
Universität Breman
Breman, Germany

Thomas Muschik
Physik - Fakuetät für Chemie
Technische Universität München
Garching, Germany

Robert J. Nemanich
Professor
Department of Physics
North Carolina State University
Raleigh, North Carolina

Dietmar Ottenwälder
4 Physikal Institut
Universität Stuttgart
Stuttgart, Germany

Vesselinka Petrova-Koch
Physik - Fakultät für Chemie
Technische Universität München
Garching, Germany

Ferdinand Scholz
4 Physikalisches Institut
Universität Stuttgart
Stuttgart, Germany

Rajiv K. Singh
Associate Professor
Department of Materials Science and Engineering
University of Florida
Gainesville, Florida

H. Stolz
Professor of Physics
Department of Physics
Universität Rostock
Rostock, Germany

Klaus Streubel
Department of Electronics, HMA
Royal Institute of Technology
Kista, Sweden

Chen Yao Tsai
Department of Electrical Engineering
National Cheng Kung University
Tainan, Taiwan

Stan Vepřek
Professor
Institut for Chemistry of Inorganic Materials
Technische Univesität München
Garching, Germany

Hartmut Vogelsang
Department of Physics
Universität-GH
Paderborn, Germany

Wolf von der Osten
Professor of Physics
Department of Physics
Universität-GH
Paderborn, Germany

Peter Wißmann
Institut für Physikalische und Theoretische
Chemie
Universität Erlangen-Nürnberg
Erlangen, Germany

Thomas Wirschem
Technische Universität München
Institut CAM
Garching, Germany

REVIEWERS

PREFACE

Optical Properties have been perceived in the past mostly as a bulk quality of materials. Eventually thin film optics became a science in its own right, particularly due to the interest of industry which applied thin films for optical coatings. Thus, the first volume in this series was devoted to the topics of "Thin Films for Optical Coatings". Volume II proceeds one step further in "miniturization". It summarizes our present understanding with respect to how small particles (mostly nanometer size) interact with light. Moreover, interfaces of only a few atomic diameters in size or surfaces may have completely different optical properties than the bulk or even thin films. This volume authoritatively addresses these topics. Fifteen experts in their respective fields present their experience, understanding, and views on a variety of pertinent optics-related subjects. Among them are quantum confinement in semiconductor nanocrystals, AgBr quantum crystallites, GaInAs-InP quantum wells, porous silicon, spark-processed light emitting silicon, nanosized metals, micro droplets, nipi superlattices, diamond particles, polymeric thin films and interfaces, etc. Thus, the covered topics relate to emerging technologies at a point at which industrial products based on this knowledge are often not yet available and promise to have an edge in future markets. In short, this unique book provides a state of the art, comprehensive coverage of various aspects of the optical properties of nanosized particles or layers. It is beneficial in the field as well as for researchers and graduate students who desire a dependable and thorough introduction into the field.

R.E. Hummel
University of Florida

P. Wißmann
Universität Erlangen-Nürnberg

CONTENTS

Chapter 1

Quantum Confinement in Semiconductor Nanocrystals

M. I. Freedhoff and A. P. Marchetti

Reviewed by D. S. Tinti

TABLE OF CONTENTS

ABSTRACT

Quantum confinement has been observed and studied in numerous semiconductor nanocrystals. Size restriction can result in shifts of the absorption edge to higher energy, in changes in the radiative properties, and in impurity exclusion effects. This review will discuss the typical phenomena of quantum confinement in semiconductor nanocrystals. A comparison of the photophysics of semiconductors with allowed lowest energy band gap transitions with indirect and forbidden gap semiconductors will be made. The direct band gap materials CdSe, PbS, and AgI will be compared with the indirect gap materials

Si and AgBr, as well as with the direct forbidden gap semiconductor Cu_2O. The synthesis, surface effects, radiative properties, and impurity effects associated with quantum confinement will be discussed.

1. INTRODUCTION

The fabrication of semiconductor nanocrystals that display quantum confinement effects is an area of current interest in materials research. As with any new technology, there is a drive to develop an understanding of the materials and their properties. There is also a drive to find commercial applications. Nowhere is that more evident than in research on porous silicon, where it is hoped that quantum confinement effects can result in the increase of the luminescence yield and a shift of the emission wavelength into the visible region of the spectrum. The goal of such research might be to produce optically tunable electroluminescent displays with high quantum yields from a material with a large, known technology base. While the struggle continues for applications employing quantum particle technology, it appears that quantum confinement effects have been used unknowingly for years, even centuries, to produce the aesthetic beauty of "stained" glass [Ekimov, 1994]. Thus nanoparticles have found a technological application long before they were known as nanoparticles.

This review will detail the general effects of size restriction found in nanocrystals of six different semiconductors. These materials were chosen because they represent a wide range of material properties, and several are of specific interest to the authors. Three of the semiconductors are direct band gap with allowed lowest energy band-to-band transitions. Two of the materials are indirect band gap semiconductors with momentum-forbidden lowest energy band-to-band transitions. The final material has a direct band gap but a forbidden lowest energy transition.

This review is meant to cover the basic properties of nanocrystals of several different but representative classes of materials. The intent is to give the reader a sense of the scope and character of quantum confinement in semiconductors. We have tried, where possible, to use the same relatively simple models for the shift to higher energies with decreasing particle size of the band edge absorption, exciton emission, etc. This was done to provide a consistent basis for comparison with experimental results. There are certainly more-complex models that can predict the blue shift with decreasing size that is one of the signatures of quantum confinement, but the basic physics is often lost as these models involve extensive parameterization and fitting techniques. With these comparisons it is hoped that the reader can come to some general conclusions about the nature and magnitude of quantum confinement effects and can develop a sense of the size regimes where the models are appropriate. The breakdown of the basic model of a nanocrystal as a perturbed semiconductor occurs in very small nanocrystals, where there are too few atoms or molecules (unit cells) in the cluster to exhibit semiconductor properties, and it is hoped that this will become clear through an examination of the characteristics and behavior of the six materials when subjected to size restriction.

The references cited are not meant to be exhaustive. Many references were chosen for their presentation and clarity of exposition. Some of the more recent references have been cited, particularly for the very busy field of research in porous Si. Here the collected work presented at one of the more recent conferences is referenced.

The size-related effects in nanocrystals (quantum dots) have much in common with the behavior of quantum well devices and superlattices. The most general model for a quantum well is the simple particle in a one-dimensional box. Nanocrystals generally require that the model include the three-dimensional nature of these crystals. Because of this three-dimensional confinement, electrostatic interactions between electrons and holes become more important. In one-dimensional quantum well structures electrons and holes are not confined in the other two dimensions, and entropy provides a driving force for them to remain distant from one another. In fact, in certain structures it is possible to confine the electron in one well and the hole in an adjacent well. Excitons can be manifest prominently in quantum wells, but, again, the confinement effects are one dimensional in nature.

The sides of the well in a quantum well structure are generally fabricated out of materials with similar lattice structures and material properties to the well material. This restriction occurs because of a need to have some matching of lattice constants and mechanical properties so that the structural transition will be continuous. This means that the barrier is often very finite with well depths of less than 1 eV. Nanocrystals that terminate to a liquid like hexane or to a polymer-like polyvinyl alcohol have well depths that are a few electron volts in magnitude, and this makes the infinite well approximation more appropriate.

Because of the precise nature of growth techniques for quantum well structures (i.e., molecular beam epitaxy), superlattices can be grown with repeating layers that have little or no size dispersity. This provides materials that are at times easier to study than nanocrystals, because the variable of inhomogeneous size distributions has been removed. However, because of the lattice-matching constraints mentioned earlier, as well as a lag in the technology where II-VI semiconductor growth is concerned, there are only a very finite number of superlattice materials available. The diversity of materials amenable to nanocrystal fabrication more than makes up for the annoyance of polydispersity in size that is often encountered.

1.1. Materials and Crystallographic Data

The materials to be covered in this review are: AgI, CdSe, PbS, AgBr, Cu_2O, and silicon. They were chosen because they represent a broad spectrum of semiconductor properties. The crystallographic data for these materials is given in Table 1.

Table 1 Crystallographic data for the materials discussed

Material	Crystal structure	Space group	Band gap	Lowest Abs	Ref.
β-AgI	Wurtzite	$P6_3mc$-C_{6v}^4	Direct	Γ_9-Γ_7	[von der Osten, 1982]
γ-AgI	Zincblende	$F\bar{4}3m$-T_d^2	Direct	Γ_8-Γ_6	[von der Osten, 1982]
CdSe	Wurtzite	$P6_3mc$-C_{6v}^4	Direct	Γ_9-Γ_7	[Broser et al., 1982]
	Zincblende	$F\bar{4}3m$-T_d^2	Direct	Γ_{15}-Γ_1	and [Madelung, 1992]
PbS	Cubic (fcc)	$P\bar{m}$-D_{2h}^{16}	Direct	L_6^+-L_6^-	[Madelung, 1992]
AgBr	Cubic (fcc)	$F\bar{m}3m$-O_h^5	Indirect	L_4^-,L_5^--Γ_6^+	[von der Osten, 1982]
Cu_2O	Cubic	$P\bar{n}3m$-O_h^4	Dir. Forb.	Γ_7^+-Γ_6^+	[Madelung, 1992] and [Goltzene and Schwab, 1982]
Si	Diamond	$F\bar{d}3m$-O_h^7	Indirect	Γ_{25}'-X_1	[Madelung, 1991]

AgI is an ionic semiconductor; that is, its room temperature transport properties are determined by the Frenkel disorder. It is a direct gap semiconductor with an allowed lowest energy band-to-band transition. It displays both exciton absorption and resonance emission. CdSe is a material with a direct band gap and an allowed lowest energy transition. The more covalent nature of CdSe produces surfaces with dangling bonds and/or reconstruction. These introduce midgap states that can trap carriers. It is a material that has been heavily investigated and that has been fabricated into small, very monodispersed nanocrystals. PbS is a material with a small band gap energy and a large static dielectric constant. In this material, quantum confinement effects will occur in large nanocrystals or microcrystals. The band gap in PbS is not at the zone center, but the lowest energy band-to-band transition is allowed.

AgBr, like AgI, is an ionic semiconductor. It is an indirect gap material, and thus the lowest energy band-to-band transition is forbidden by momentum selection rules. AgBr was the first material to exhibit impurity exclusion effects. It also is of interest, as are all the materials with forbidden or indirect transitions, because of the possible effects that quantum confinement might have on their selection rules. Cu_2O is an ionic semiconductor with a direct band gap, but a symmetry-forbidden lowest energy band-to-band transition. It is a material that exhibits easily observable Wannier exciton absorption. Si is a covalent indirect gap semiconductor with a momentum-forbidden lowest energy band-to-band transition. This material was one of the first to demonstrate surface reconstruction. Si is one of the most widely studied semiconductor materials, and it is of enormous economic importance. Both Si and CdSe nanocrystals seem to be likely candidates for early commercial applications.

2. PHYSICAL BASIS AND THEORY OF QUANTUM CONFINEMENT

Quantum confinement or size effects are the changes in the measurable properties of a semiconductor when the crystal size approaches the spatial extent of an electron and/or hole. This spatial extent or characteristic length for a particular material has been defined in several similar ways. A fictitious Bohr radius can be defined for the electron and hole by scaling the hydrogen atom radius by the effective mass of the carrier and the dielectric constant of the semiconductor [Efros and Efros, 1982]. The exciton radius can be calculated by treating the exciton as a scaled hydrogen atom in the effective mass approximation (see below) using the effective masses of the electron and hole and the dielectric constant [Brus, 1984]. Both of these metrics give a feeling for the size regime in which the spatial extent of the carrier or exciton will start to "notice" the crystal boundaries.

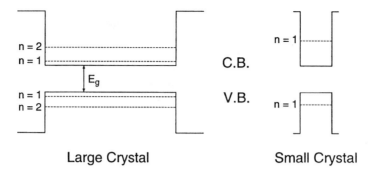

Figure 1 The energy level for a thick and thin semiconductor crystal modeled as levels in a one-dimensional box. In the thick crystal the band gap energy is only weakly blue shifted from the bulk values, E_g, while in the thin crystal the blue shift is substantial.

Quantum confinement can be understood at a simple level, although numerous complexities have been found and elucidated. There are two categories that need to be detailed. One is the case where the electron–hole pair is uncorrelated, and the other is when it is correlated or bound in some way [Brus, 1984]. In the first case, the quantities studied are changes in the material properties, such as the band gap energy when an uncorrelated electron–hole pair is created. The second case examines absorption to exciton states, exciton emission, and other electron–hole recombination processes that account for the Coulomb attraction between the electron and hole.

2.1. Particle-in-a-Box

An elementary picture of quantum confinement is obtained if one imagines that there is a semiconductor layer of variable thickness supported on a noninteracting transparent insulator. With the positions of the lowest conduction band and highest valence band known, the perturbation of these bands is given by the energy of a particle in a one-dimensional box (see Equation 1). What is assumed is that the bulk semiconductor properties (band gap energy, etc.) are perturbed by "adding the box;" that is, by making the layer thin. This is illustrated in Figure 1, which shows that the energy levels for a particle in a one-dimensional box depend on the size (thickness) of the box. The conduction band and the valence band are perturbed by the "box" and shifted to higher energies as the slab becomes thinner. In addition, it is known that the electron and hole masses differ from each other and from that of a free electron; that is, the effective mass approximation (EMA) is used [Kohn, 1957]. In this case, the masses of the electron and hole that are used in the calculation are those found from experimental measurements (such as cyclotron resonance). In applying Equation 1 the hole energies are the negative of electron energies when plotted on a normal electron energy plot, meaning that the conduction band will go up in energy and the valence band will be moved down.

The energy of a particle in a one-dimensional box with infinite walls is

$$E(n) = \frac{n^2 \hbar^2 \pi^2}{2m^* a^2} \tag{1}$$

where n is the quantum number (= 1, 2, 3, ...), h is Planck's constant, m^* is the effective mass, and a is the layer thickness [Kauzmann, 1957, or any elementary quantum mechanics textbook]. For the $n = 1$ levels the conduction and valence bands of the semiconductor are not changed with large layer thicknesses, and the existence of the box makes no measurable difference. But as the layer thickness decreases (usually into the tens of nanometers region), the box begins to cause the conduction band to move up in energy and the valence band to move down in energy. Thus, for some given thickness an absorption measurement would show that the lowest energy transition was larger than E_g, which is the value for the bulk material. The increase in the band gap energy is obtained by summing the shifts indicated by Equation 1 for the conduction and valence bands. This is qualitatively one of the simplest manifestations of quantum confinement.

2.2. Three-Dimensional Confinement and Coulomb Interactions

In a nanocrystal the confinement is three dimensional, and Equation 1 could be replaced by a similar expression for a particle in a three-dimensional box or sphere. As confinement in nanocrystals is considered in more detail, it becomes clear that there are more interactions than just the well barriers. The electron and hole, when near one another, have a coulombic attraction that must be taken into account. In addition, there can be interactions with the lattice which cause a relaxation around the particles that resemble "solvent" effects in nature. This approach has been worked out and well documented [Brus, 1984]. A simple expression for the change in the band gap energy when an uncorrelated electron–hole pair is created is given by

$$\Delta E = \frac{h^2 \pi^2}{2R^2} \left[\frac{1}{m_e^*} + \frac{1}{m_h^*} \right] - \frac{1.8 \, e^2}{\varepsilon R} + pol. \tag{2}$$

where R is the radius of the cluster, the m^* terms are the electron and hole effective masses, e is the charge of an electron, ε is the dielectric constant, and $pol.$ represents the polarization terms. This approach generally predicts an increase in the band gap energy with decreasing particle size. This expression will be used throughout this chapter for comparison with experimental data.

2.3. Confined Scaled Hydrogen Atom

When light absorption creates an exciton, the final state must include the Coulomb interaction in a specific form. That is, the electron and hole are correlated. In considering a Wannier or large-radius exciton, the EMA can be used and the semiconductor can be treated as a continuous medium with a dielectric constant, ε. Expressions have been derived to predict the energy for this case [Schmidt and Weller, 1986; Mohan and Anderson, 1989; Zicovich-Wilson et al., 1994].

A simple way to treat this situation is to take the exact calculations (or very good approximate calculations) that have been performed for a hydrogen atom in a spherical box and scale them using the reduced effective mass and the dielectric constant [Marin and Cruz, 1991a,b]. This approach should work well as long as the effective Bohr radius is large compared with a primitive lattice spacing. This means that there would be a generalized table or curve with which scaled experimental exciton data could be compared. The case of trapped excitons must be considered carefully when nanocrystal radii become comparable with the effective size of the complex [Nomura and Kobayashi, 1994]. Also, if trapping centers tend to be more prevalent near the surface, the surface can affect the dynamics as well as the energetics of carrier recombination.

2.4. Donor–Acceptor Recombination

Donor–acceptor recombination can occur in a semiconductor when an electron is trapped at an ionized donor and a hole is trapped at an ionized acceptor. An ionized donor is often a substitutional impurity with a net positive charge with respect to the lattice, and an ionized acceptor is an impurity or defect with a net negative charge with respect to the lattice. The electron bound at the donor can usually be modeled using the EMA, and thus the electron will take up a large "hydrogenic" orbit about the positive center. Donor–acceptor recombination occurs when the donor electron tunnels to the acceptor [Pankove, 1971]. The treatment of donor–acceptor recombination will require some knowledge of the nature of the donor and acceptor. If it can be assumed that the hole is deeply bound at the acceptor (and thus of small spatial extent) and that the electron is weakly bound in an extended hydrogenic orbit about the donor, then confinement would affect the donor much more than it would the acceptor with decreasing crystal size. The donor is treated as a confined hydrogenic system, and confinement effects on the acceptor are ignored until the cluster size becomes very small. The Coulomb interaction is specified for a given donor–acceptor pair, with each pair at a known pair separation in the lattice. The difficulty lies in the fact that the donors and acceptors are not free to move, and thus their distances from each other are fixed. For an ensemble of donors and acceptors, these distances will be distributed and can be averaged over.

2.5. Free Carrier–Trapped Carrier Recombination

This case might be one which is common in covalent semiconductors. Here the hole is trapped at the surface at broken or dangling bonds while the electron moves about the nanoparticle. This situation is very similar to the uncorrelated electron and hole case. Equation 2 will apply if we drop the hole kinetic energy term. The hole is treated as being deeply trapped and is therefore ignored. The electron is confined to the box, and the Coulomb interaction will have to be parameterized, which may yield a constant which is different from the one found from Equation 2. As in the case of donor–acceptor recombination, the trapped hole binding energy must be accounted for when comparing experimental data with a model. This is also a situation in which confinement effects can be observed in large particles, because in the strict application of the model, recombination will only occur at the surface.

2.6. Other Manifestations of Quantum Confinement

Quantum confinement effects are most easily observed as a blue shift in the absorption edge or the emission maximum, but movement of conduction and valence bands and changes in the volume of high-dielectric-constant material with confinement should change the ionization potential (work function) and the electron affinity of the material. It is expected that both the ionization potential and the electron affinity will decrease as the particle radius decreases [Brus, 1983; Makov et al., 1988]. Corresponding changes in the redox properties of nanoclusters should also be observed.

Saturation and nonlinear optical effects should also be observed in semiconductor nanocrystals. In the strong confinement limit, where the system is well represented by a particle-in-a-box model and where the assumption that the hole mass is larger than the electron mass is valid, the creation of a second electron–hole pair should blue shift the absorption edge by a value of $\Delta E = 0.2e^2/\varepsilon R$. The creation of a third electron–hole pair will cause that electron to occupy the second ($n = 2$) particle-in-a-box level as a result of the Pauli principle [Ekimov and Efros, 1988]. This will shift the absorption by an amount equal to the separation between the first and second levels. With more than one electron–hole pair per particle, the lowest transition will begin to saturate. Changes in oscillator strength and increases in the third-order susceptibility are also predicted as the particle radius decreases [Schmitt-Rink et al., 1987; Takagahara, 1987; Bryant, 1988].

3. EFFECTIVE MASS APPROXIMATION

3.1. Which Dielectric Constant Is Appropriate?

The EMA is used to estimate the radius and binding energies of shallowly trapped electrons (holes) and excitons [Kohn, 1957]. In this approximation an effective mass is used instead of the free–electron (hole) mass to account for interactions with the lattice. The effective mass is related to the curvature of the conduction and valence bands. Within this approximation Coulomb interactions are also scaled by a dielectric constant. The question is, then, what is the magnitude of this dielectric constant? Is it the high-frequency (optical) value, is it the static value, or is it something else?

The answer depends on a number of factors. This problem arises as the shallow–deep instability of impurity levels in semiconductors [Mukhopadhyay, 1985; Grinberg et al., 1985] and is considered at length in discussions on the theory of excitons [Knox, 1957]. The qualitative answer for a given semiconductor is: when the electron has a small effective mass and the material dielectric constant is large, the mean distance of the electron from the trapping center (ionized donor or hole) is large (orbital frequency of the electron and hole pair about their center of mass will be slower than, or approximately equal to, the frequencies of the lattice vibrations), then the appropriate value is the static dielectric constant, ε_0. This is the situation for many semiconductors. If the orbital frequency is higher than the phonon frequencies, a situation that corresponds to a smaller orbital radius, then the optical dielectric constant, ε_∞, is appropriate. With large effective masses and a small dielectric constant, the EMA is not valid at all and electrons (holes) are deeply trapped or excitons will be tight binding in nature.

There are several ways to determine whether or not the EMA is appropriate. The total dielectric function can be obtained for a given material. This, along with the electronic dielectric function, can be used to calculate a binding energy [Grinberg et al., 1985]. If this binding energy is approximately equal to that found using the EMA, the EMA can be considered to be valid. From the results of this sort of exercise for a number of materials, a simple method of testing the validity of the EMA has been put forth. First, the EMA with the static dielectric constant is used to calculate a scaled binding energy:

$$E_b = \frac{R_y m^*}{\varepsilon_0^2} \qquad (3)$$

where R_y is the Rydberg energy (13.6 eV) of a hydrogen atom and $m*$ is the appropriate effective mass. If E_b is less than 100 meV, then the combination of a small effective mass and large dielectric constant makes the EMA using the static dielectric constant appropriate [Grinberg et al., 1985]. If the calculated binding energy is approximately equal to or somewhat larger than 100 meV, the EMA may be valid but only if the high-frequency dielectric constant is used. In this case, care must be exercised. It should be fairly obvious that when the radius, R, of the particle becomes smaller than the characteristic length scale for the material, the Bohr radius or the exciton radius, the kinetic energy caused by confinement will become large, as will the electron (hole) velocity. In this situation the high-frequency dielectric constant will be more appropriate.

3.2. Physical Constants

Table 2 contains the values of the dielectric constants, effective masses, lowest band gap energy (E_g), exciton binding energy (E_b(ex)), and exciton Bohr radius (r_b(ex)) for the six materials considered in this review. An examination of this table indicates that the EMA is valid for all materials except Cu_2O. In fact, if data from the known line positions in the absorption of the lowest energy transition (yellow exciton series) of Cu_2O are used to calculate an effective dielectric constant, one obtains a value of 7.2 for $n \geq 2$ but a value of 6 for the $n = 1$ exciton. The value of 7.2 is very close to the experimental value of ε_0, while 6 is close to the experimental value of ε_∞. This suggests that Cu_2O is in an intermediate region for a ground state exciton. Isotropic averages of $m*$ and ε were used in all calculations when the material constants were anisotropic.

Table 2 Dielectric constants, effective masses, exciton binding energies, and exciton radii for the six materials discussed in this review

Material	ε_0	ε_∞	m_e^*	m_h^*	E_g(eV)	E_b(ex) (meV)	r_0(ex) (Å)
β-AgI (w)	7.0	4.2	0.21 (av. red. exciton mass)		3.0247	58(79)	18(23)
γ-AgI (z)	—	—	—	—	2.91	—	—
CdSe (z)	—	—	0.11	0.44	1.9	—	—
(w)	9.29$_\parallel$ 9.15$_\perp$	6.0$_\parallel$ 6.1$_\perp$	0.13 —	≥1.0$_\parallel$ 0.45$_\perp$	1.841	15(16)	52
PbS	169	17.2	0.105$_\parallel$ 0.080$_\perp$	0.105$_\parallel$ 0.075$_\perp$	0.286	0.02	2000
AgBr	10.64	4.68	0.289	1.71$_\parallel$ 0.79$_\perp$	2.7125	25	25
Cu_2O	7.11	6.46	0.99	0.58	2.1725	97	10
Si	12.1	—	0.1905$_\parallel$ 0.9163$_\perp$	0.537hh 0.153lh	1.1700	16(14)	37

Note: Values for E_b and r_0 are calculated using isotropic average effective masses where necessary (within the EMA). Parenthetical values are experimental data. When available, data is for liquid helium temperatures. The references are the same as Table 1.

4. SYNTHETIC METHODS

Many techniques have been used to synthesize semiconductor nanocrystals. An attempt has been made to outline the common methods utilized in fabricating the materials discussed in this chapter, as well as to refer the reader to other techniques and materials where appropriate. Quoted crystal sizes are obtained from transmission electron microscopy (TEM).

4.1. AgX (X = Br, I, Cl)

Three different methods have been used to synthesize silver halide nanocrystals. Each lends itself particularly well to creating crystals within a specific size regime as a result of properties such as solubility, kinetic ripening, and thermodynamic stability of the clusters.

4.1.1. Inverse Micelles

Inverse micelles has been used to synthesize many different materials, including metal chalcogenides [Lianos and Thomas, 1986; Steigerwald et al., 1988] and silver halides [Chen et al., 1994]. This technique invokes the basic principle that oil and water do not mix. The metal ions and salts/chalcogenides are dissolved in aqueous solutions. When small quantities are added to an organic solution (i.e., heptane) containing a surfactant such as dioctylsulfosuccinate sodium salt (AOT), semiconductor nanocrystals

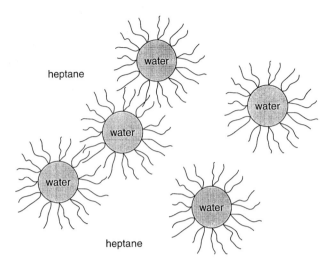

Figure 2 A schematic representation of reverse micelles. The surfactant molecules orient themselves such that the hydrophilic heads surround the aqueous solution while the hydrophobic tails extend into the oily hydrocarbon phase.

can be grown when small aqueous pools dispersed in the oily phase collide. Alternatively, the metal ion can be introduced in a micellar solution to an organic phase that already contains a small amount of salt/chalcogenide dissolved in it. The size of the water pools, and thus the size of the crystallites, is controlled to some extent by the ratio of the concentrations of water to AOT. A schematic representation of inverse micelles is given in Figure 2.

Silver bromide nanocrystals of 5 to 15 nm radius [Chen et al., 1994] in size have been prepared using this technique. One of the problems associated with micellar preparations is sample polydispersity, which results in the broadening of luminescence peaks arising from the recombination of excitons. This is due to the fact that crystal growth occurs when one water pool collides with another, and there is no way to keep the number of collisions uniform.

4.1.2. Nonaqueous Dispersions

This type of synthesis draws on the use of a restraining polymer as well as the decreased solubility of AgBr in nonaqueous media such as acetone. Silver trifluoroacetate and lithium bromide, both dissolved in acetone, are pumped into a solution of the polymer Butvar in acetone at a constant flow rate. Samples withdrawn at various stages during the preparation, which typically takes 25 min to complete, correspond to sizes of 20 to 50 nm in diameter and are more monodisperse than samples prepared from micelles [Freedhoff et al., 1995]. These "midsize" preparations of AgBr have been useful in quantum confinement studies since they are small enough to display the effects of size restriction, but monodisperse enough such that specific excitonic transitions can be resolved. This will be examined more closely in a later section.

4.1.3. Aqueous/Gelatin Preparations

Polymers and gelatin can be used as peptizing and restraining agents in aqueous syntheses of semiconductor nanocrystals, as they will prevent ripening of the clusters by physically separating them from one another. CdS [Fischer et al., 1989] and CdTe and ZnTe [Resch et al., 1989] have been prepared using the polymer hexametaphosphate, while AgBr nanocrystals that are in the 50-nm-diameter size range have been prepared using gelatin [Johannson et al., 1992] as a restraining agent. While these samples are quite monodisperse, they are too large to exhibit dramatic effects because of quantum confinement.

AgI has been prepared using the stabilizing agent poly(ethyleneimine) (PEI) [Mulvaney, 1993], and, unlike the AgBr preparations, these resulted in extremely small (3-nm-diameter) particles that exhibited significant quantum confinement effects. AgI nanocrystals in the ≥30-nm size regime have been prepared in gelatin [Berry, 1967].

4.2. CdE (where E = S, Se, Te)

These materials are typically prepared via organometallic routes. CdE [Murray et al., 1993] syntheses generally proceed by rapidly injecting the organometallic reagents into a hot solvent, slowly growing and annealing the crystallites, and then size selecting the samples. These syntheses are occasionally completed by capping the surface of the crystallite with stabilizing organic groups. These materials are often prepared in inverse micelles as well [Steigerwald et al., 1988]. Other materials prepared using organometallic syntheses include GaAs [Nozik et al., 1993] and GaP [Dougall et al., 1989]. The extreme monodispersity of CdSe samples synthesized via these techniques has allowed this material to be extremely well characterized.

CdE nanocrystals have also been created by dissolving small amounts of Cd^{2+} and chalcogenide in a borosilicate glass matrix at a high temperature followed by annealing [Borrelli et al., 1987]. Annealing techniques can also be used to embed nanocrystallites in a polymer matrix [Murray et al., 1993; Colvin et al., 1994]. This latter technique has been used to create a light-emitting device whose color is tunable through manipulation of the size of the CdSe nanocrystals.

4.3. Si

Silicon nanocrystals have been prepared by pyrolyzing disilane, thereby making a dilute aerosol of surface-oxidized Si nanoclusters [Wilson et al., 1993; Littau et al., 1993]. The high temperature aerosol is then bubbled through a solution of ethylene glycol to create a colloidal dispersion. The colloidal particles can then undergo a size-selective precipitation to obtain more monodisperse samples.

4.4. PbS

Lead sulfide nanocrystals have been synthesized through a variety of techniques, including colloidal preparation in aqueous solutions using polyethylene glycol [Nozik et al., 1985], polyvinyl alcohol [Gallardo et al., 1989], or polyphosphates [Machol, 1993] as restraining agents; growth in polymer films [Wang et al., 1987; Tassoni and Schrock, 1994]; growth in glasses [Borrelli and Smith, 1994]; growth in micelles and sol-gels [Ward et al., 1993; De Sanctis et al., 1995; Bliss et al., 1995; Gacoin et al., 1995]; and growth in alcohol and acetonitrile/methanol mixtures [Rossetti et al., 1985]. One particularly interesting method is nanocrystal growth in zeolite cavities.

Zeolites consist of networks of aluminate and silicate tetrahedra that are bridged by oxygen atoms, with internal cavities that can be used for growing extremely small semiconductor nanocrystallites. The zeolite must first be ion-exchanged with metal ions by stirring an aqueous solution of metal ions with the zeolite overnight. The powder must then be washed, heated, exposed to the desired chalcogenide, and then stored. Since each zeolite cavity size is identical, this synthesis results in extremely monodisperse size distributions of extremely small particles. Semiconductor nanocrystals successfully synthesized utilizing this technique include CdS [Wang and Herron, 1991] and PbS [Moller et al., 1989] in zeolite Y and mordenite.

4.5. Cu$_2$O

To the best of our knowledge, the only synthesis of Cu_2O nanocrystals has been carried out by mixing aqueous solutions of copper sulfate, tartaric acid, and sodium hydroxide and then adding an aqueous solution of ascorbic acid [Freedhoff et al., 1995]. Polyvinyl alcohol can be used as a restraining agent. The resultant nanocrystals are quite polydisperse, but do yield information on quantum confinement effects of this material.

5. DIRECT BAND GAP SEMICONDUCTORS WITH ALLOWED LOWEST ENERGY TRANSITIONS

5.1. CdSe

Cadmium selenide is a direct gap semiconductor whose nanocrystal properties have been studied using many techniques, including spectral hole burning [Gaponenko et al., 1994], electric field modulation [Colvin et al., 1994], X-ray characterization [Bawendi et al., 1989], and time-resolved photoluminescence and differential transmission spectroscopy [Ekimov and Kudryavtsev, 1992]. Organometallic syntheses have been developed [Murray et al., 1993, and references therein] that are capable of producing nearly monodisperse samples of CdSe crystallites ranging from 12 to 115 Å in diameter. These particles can then be capped with organic moieties in order to stabilize them against further growth. Similarly,

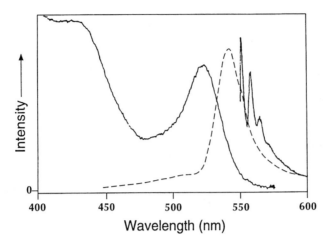

Figure 3 The absorption and luminescence spectra of 32 Å diameter CdSe nanocrystals at 15 K. Excitation at 440 nm (dashed). Excitation at 550 nm (solid). Reproduced from data in Bawendi et al. [1990; 1992].

monodisperse samples of CdSe have been grown through embedding in glass matrices [Borrelli et al., 1987, Bawendi et al., 1990, 1992]. In our earlier discussion of the theoretical basis of quantum confinement, we mentioned that these effects manifest themselves most dramatically when the size of the crystals approaches that of the effective excitonic radius. For CdSe, the excitonic radius is 52 Å, meaning that size restriction will be observable in comparatively large-sized crystals. This fact, as well as the ability to synthesize homogeneously sized samples, makes CdSe an ideal material to study.

One of the manifestations of quantum confinement is the onset of discrete transitions that are blue shifted in energy from the bulk, and these can be described equivalently as being either the allowed transitions of a particle-in-a-box or the allowed transitions between the altered molecular orbitals of the smaller samples. Figure 3 shows the low-temperature absorption and luminescence spectra of 32-Å-diameter CdSe clusters grown in organic glass [Bawendi et al., 1990, 1992]. Because there exists a distribution of sizes in the samples, these luminescence spectra are inhomogeneously broadened when the crystals are excited with 440-nm light and all crystal sizes can be resonantly excited. When the excitation is changed to 550 nm, however, only the larger crystals have resonant transitions available to them, and thus the luminescence spectrum consists of discrete lines. In order to remove the inhomogeneous line broadening, transient photophysical hole-burning [Alivasatos et al., 1988] experiments have been performed, which are capable of resolving vibronic contributions to the transitions.

Figure 4 shows a comparison of the theoretical change in peak position caused by quantum confinement as per Equation 2 [Brus, 1984] with data from literature references [Borrelli et al., 1987; Brus, 1990; Wilson et al., 1990; Nomura et al., 1991; Ekimov and Kudryavtsev, 1992; Brus, 1993; Persans et al., 1993]. There is a clear but qualitative agreement of experiment with theoretical predictions. There is also a clear trend to measure a larger crystal size than would be expected for a given blue shift in energy. This could be due to several factors: the growth of oxides or other impurities on the nanocrystal surface, which contribute to the size measurement but not the semiconductor properties; slight growth of the samples while being prepared for the sizing measurements; inhomogeneous broadening of samples and unequal branching of luminescence yields from size to size (i.e., there could be more larger crystallites, but the contribution to the luminescence could be greater from smaller crystallites in the sample); an alteration of the band structure caused by surface effects; and a size-dependent modification of the material properties of CdSe (i.e., the dielectric constant or effective masses) that would contribute an error to the theoretical results.

Researchers in this field hope that semiconductor nanocrystals will one day be useful in optical device fabrication and that the ability to change the nanocrystal size and hence the transition energy will lend itself to creating devices with a tunable range of emission wavelengths. In order to accomplish this, it is helpful to understand some of the nonlinear optical properties associated with these systems. Since CdSe nanocrystals are model materials to fabricate from a synthetic point of view, there has been a significant amount of work on their Stark and nonlinear optical effects.

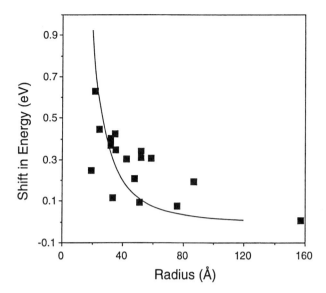

Figure 4 A comparison of the experimental blue shifts as a function of particle size for CdSe nanocrystals and the predictions from Equation 2. Data taken from Brus [1993]; Borrelli et al. [1987]; Ekimov and Kudryavtsev [1992]; Brus [1990]; Nomura and Kobayashi [1991]; Persans et al. [1993]; Wilson et al. [1990].

Nonlinear optical effects on semiconductor nanocrystals are thought to have potential technological interest because the discrete transitions allowed, as per the selection rules, can be both narrow and intense. In 1983 high third-order susceptibilities ($\chi^{(3)}$) [Jain and Lind, 1983] were reported for degenerate four-wave mixing signals in color filter glasses, and the origin of this was found to be small crystallites of CdS_xSe_{1-x} that were embedded in the glass. The crystallites were found to range in size from 100 to 1000 Å, and the $\chi^{(3)}$ measured was an order of magnitude larger than those for single-crystal CdS or CdSe. Theories to explain these effects draw on both the optical tunability of semiconductor nanocrystals as well as on the possibility that size restriction will enhance the nonlinear properties of the materials.

For nanocrystals in the weak confinement regime (sizes a few times the bulk Bohr radius), an excitonic description of the wave function is valid. The radiative lifetime of the lowest energy transition is given by Equation 4 and is predicted to shorten to the picosecond scale as the crystal volume increases [Takagahara, 1987; Hanamura, 1988; Brus, 1991]

$$\frac{1}{\tau_r} = \frac{64\,\pi\varepsilon}{3}\left(\frac{R}{\lambda}\right)^3 \frac{\Delta E_{LT}}{h} \tag{4}$$

where τ_r is the radiative lifetime, ε is the dielectric constant, R is the crystal radius, λ is the transition wavelength, and ΔE_{LT} is the longitudinal-transverse exciton splitting. This is only applicable at low temperatures and when the center of mass confinement energy is bigger than ΔE_{LT} [Brus, 1991]. Jain and Lind [1983] reported a nonlinear response time of less than 8 ns, the experimental resolution. This further suggests the potential for commercial applications requiring fast optical switching. It has been suggested [Rossmann et al., 1990] that these kinetic effects arise as a result of surface trapping of one of the carriers by impurities or intentionally added capping agents.

Quantum-size Stark effects have been observed in CdSe by several groups [Hache et al., 1989; Ekimov et al., 1990]. When an electric field F is applied to a crystallite of radius R, the shift in the lowest energy transition ΔE in the strong confinement limit as calculated using second-order perturbation theory [Rossmann et al., 1990] is

$$\Delta E(F) = \frac{-0.65(eFR)^2}{E_{low}} \tag{5}$$

where E_{low} is given by

$$E_{low} = \frac{h^2 \pi^2}{2 R^2} \left(\frac{1}{m_e^*} + \frac{1}{m_h^*} \right) \tag{6}$$

where m_e^* (m_h^*) is the effective electron (hole) mass. This theory will in fact overestimate the Stark shift, since it ignores the coulombic nature of the exciton. The applied field tends to drive the electrons and holes apart, but the Coulomb attraction between them will reduce their average separations. Thus, Stark shifts of nanocrystals are actually smaller than those of superlattices, and smaller than most theoretical predictions. It is thought that the most-pronounced effects will occur in larger crystallites where wave functions are more easily polarizable than smaller nanocrystals, but whose transitions are still discrete in nature. Experimental studies [Ekimov et al., 1990; Rossmann et al., 1990] confirm that a quadratic Stark shift does occur for larger crystallites and that field-induced broadening of the lowest energy transition also occurs. Since these effects could also be affected by the crystal surface or trapping, only a partial understanding of the confined Stark effect is possible at the current time.

The most surprising observation in studies of both the nonlinear optical properties and Stark effect on nanocrystals is that the largest effects do not occur in the smallest samples; rather, the most-pronounced differences from bulk spectra occur in samples which are on the order of several effective Bohr radii in size. These larger crystals are more easily polarizable than very small nanocrystals are, but they still undergo the discrete transitions that make nanocrystals a desirable material to use for optically tunable devices. In order for these materials to be used to develop actual technology, however, better control of the size distributions must be realized and a much more detailed understanding of the surface effects is necessary. Although CdSe is a material about which there is a large amount of information, there is still a long way to go before marketable-quality devices will be fabricated.

5.2. AgI

AgI, which is a direct band gap, ionic semiconductor, was one of the first materials in which it was recognized that quantum confinement had blue shifted the absorption of small particles [Berry, 1967]. Quantum-size effects on the absorption edge have been used to follow the nucleation processes in the study of AgI colloid formation. The colloid formation is initiated using an electron pulse [Schmidt et al., 1988]. In these very small nascent clusters (~2.3 nm diameter) the exciton absorption was shifted from 420 nm (2.94 eV) to about 315 nm (3.92 eV), although it is not clear that nanocrystals of AgI actually exhibit exciton absorption (see below). Other workers have reported similar band shifts at similar cluster sizes using a variety of materials to stabilize AgI nanocrystals [Tanaka et al., 1979; Micic et al., 1990; Mulvaney 1993]. Figure 5 shows the experimental shift in band gap energy as a function of particle size vs. the predicted shift found using Equation 2. The thresholds were determined from plots of $(\alpha h \nu)^{1/2}$ vs. $h \nu$, where α is the absorption coefficient. The particles were sized using light-scattering measurements [Schmidt et al., 1988].

The data given in Table 2 indicate that the EMA should be valid for AgI, and calculations using Equation 2 show that a discernible shift in the exciton absorption should occur with particles that have diameters between 20 and 40 nm. Figure 6 shows the absorption spectra at liquid helium temperature for several preparations of AgI nanocrystals in gelatin [Freedhoff et al., 1995]. The largest crystals have an average equivalent diameter of 320 nm, while the smaller samples are about 34 ± 10 nm and 97 ± 30 nm in diameter. There are several features evident from these spectra. First, there are three absorption peaks visible, two pronounced peaks at about 416 and 421 nm and a shoulder at 426 nm. The peaks at 416 and 421 nm are the two 1s exciton absorptions (W_1 and W_2) of the hexagonal (wurtzite) phase of AgI [von der Osten, 1982]. The shoulder at 426 nm is the exciton absorption of the cubic (zincblende) phase of AgI that often coexists with the wurtzite phase. Second, the large crystals show resonance exciton fluorescence coincident with the lowest exciton absorption of β-AgI. Finally, there is a blue shift of the exciton absorption in the preparation of the smallest nanocrystals.

The magnitude of the blue shift is 1.3 meV for the W_1 exciton (421 nm) in the 34-nm-sized material with respect to the 97-nm material. The blue shift is substantially larger for the W_2 exciton (416 nm) with respect to the larger-sized preparation. Here the shift is closer to 2.5 meV. A careful examination of the data available for AgI indicates that the parallel and perpendicular components of the reduced

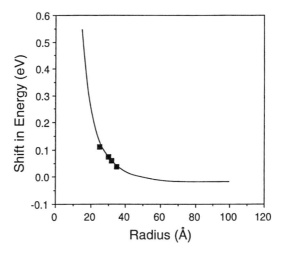

Figure 5 A comparison of the experimental band gap shifts for AgI and the prediction from Equation 2. Data taken from Schmidt et al. [1988].

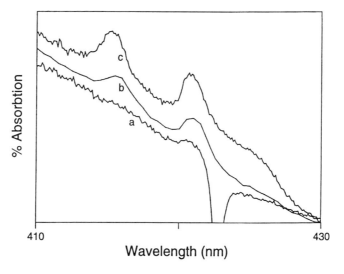

Figure 6 The absorption spectra for several gelatin preparations of AgI. (a) 320 nm diameter. (b) 97 nm diameter. (c) 34 nm diameter. Note the resonance exciton emission from the largest crystallites.

exciton mass differ for the W_1 and W_2 exciton. The isotropic average is, however, the same for both excitons, and to that level of approximation it is expected that both excitons should have the same shift caused by confinement. Comparison with the lowest exciton peak of the largest-sized sample cannot be made because resonance fluorescence distorts the peak position, and it is probable that a small amount of resonance fluorescence occurs in the 97-nm preparation. Thus, the best estimate for the shift is 2.5 ± 0.5 meV.

The low-temperature emission spectra (not shown) of all three preparations show what has been identified as donor–acceptor emission peaking at 445 nm [Marchetti and Eachus, 1992]. The low-temperature emission of the smallest nanocrystal preparation does not show any exciton luminescence. This could be because exciton scattering from the surface causes the excitons to decay and/or break up, or they may become trapped at donors and acceptors. There is some evidence that in an intermediate size range (>60 nm) the exciton emission intensity decreases with pump irradiation time for 325 nm excitation [Johannson and Marchetti, 1990]. This would suggest that some photochemistry (photography?) was occurring even at liquid helium temperatures. The obvious but speculative conclusion that can be reached is that silver clusters were formed and these quenched the exciton emission.

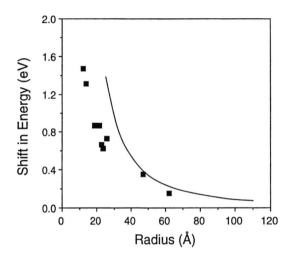

Figure 7 A comparison of the experimental blue shifts as a function of particle size for PbS nanocrystals and the predictions from Equation 2. Data taken from Wang et al. [1987].

5.3. PbS

PbS is a semiconducting material with a small band gap energy, small and almost equal effective electron and hole masses, and very large static and high-frequency dielectric constants (see Table 2). The EMA estimates the Bohr radius for the exciton of this material to be almost 200 nm, and the binding energy is very small. This means that quantum-size effects should be discernible in micrometer-sized crystallites. That is, the exciton absorption and emission (if they exist) should show measurable quantum confinement effects in crystals whose radii are five times the Bohr radius, although these changes from the bulk properties will be small in an absolute sense. In crystals where the crystal radius is smaller than the Bohr radius ($R_{cryst} < R_{Bohr}$), the exciton binding energy will be small or very small compared with the confinement energy, and it can be viewed as a perturbation. Thus, in monodispersed preparations of PbS nanocrystals the particle-in-the-box transitions should be observable.

There have been numerous preparations of PbS nanocrystals (see the synthesis section). Many of these preparations have produced, under the right conditions, a very characteristic spectrum with pronounced peaks at about 600, 400, and 300 nm.

A size series of PbS nanocrystals has been prepared in ethylene-15% methacrylic acid copolymer [Wang et al., 1987]. The absorption data were analyzed in the usual manner for a direct gap semiconductor [$\sigma h\nu = A(h\nu - E_g)^{1/2}$]. These data are shown in Figure 7 along with the shift in band edge predicted by Equation 2. This figure indicates that the general trend predicted by Equation 2 is followed, but that there are deviations which become larger as the particle size decreases. In this small-size region it is expected that some of the deviations may be due to uncertainties in sizing. Further, it is expected that with smaller nanocrystals the approximations used in deriving Equation 2 will break down. This is clearly evident with the smallest particles where the deviation is the largest.

A number of PbS nanocrystal preparations have led to very similar absorption spectra. One example is shown in Figure 8. The spectrum exhibits a lowest energy peak at 600 nm and stronger, higher energy peaks at about 400 and 300 nm. These PbS nanocrystals have been characterized as spheres and as elongated crystals with two-dimensional confinement [Rossetti et al., 1985; Gacoin et al., 1995]. The energy levels for a particle in a three-dimensional spherical or rectangular box are shown in Figure 9 [Kauzmann, 1957]. The levels for a spherical box are characterized by a principal quantum number, n ($= 1, 2, 3, \ldots$) and an angular quantum number, l ($= 0, 1, 2, 3, \ldots$). In the solutions for a particle in a spherical cavity, each value of n has an infinite number of l roots, thus for each n level there is an unlimited but discrete set of energy levels. For $n = 1$ there are $l = 0$ (1s), $l = 1$ (2p), $l = 2$ (3d), etc. levels, for $n = 2$ there are $l = 0$ (2s), $l = 1$ (3p), etc. levels. The energy ordering for the levels as a function of (n,l) is: 1s (1,0) < 2p (1,1) < 3d (1,2) < 2s (2,0) < 4f (1,3) < 3p (2,1), and so on. For a particle in a rectangular box the solutions are the sum of the one-dimensional solutions and are characterized by quantum numbers n_1, n_2, and n_3 ($= 1, 2, 3, \ldots$). These energy diagrams shown in Figure 9 are scaled to match the lowest transition energy for the preparations that give the nanocrystals of PbS with absorptions at 600, 400, and 300 nm.

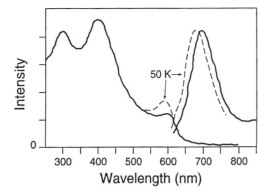

Figure 8 The absorption and emission spectra of PbS nanocrystals at room temperature and 50 K. The crystals are thought to have a diameter of 4.0 nm. Reproduced from data in Machol et al. [1993]; Machol [1993].

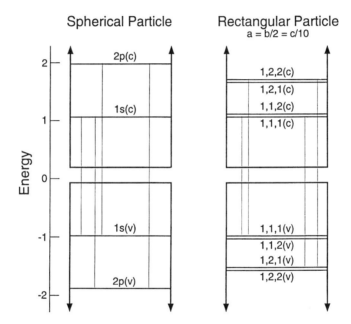

Figure 9 The energy levels for a semiconductor nanocrystal modeled as spherical and as a rectangular box. The levels are approximately scaled for the transitions found in the smallest nanocrystals of PbS.

For spherical particles, the three observed transitions at 600, 400, and 300 nm would be assigned to a 1s(v) to 1s(c) transition, to the degenerate 1s(v) to 2p(c) and 2p(v) to 1s(c) transitions and to a 2p(v) to 2p(c) transition, respectively. The notation used is taken from Kauzmann [1957]. The reason for the apparent symmetry between the conduction and valence band level scheme is the fact that the hole and electron effective masses are almost equal in magnitude. If the particles responsible for quantum confinement are rectangular, as proposed in several investigations [Gacoin et al., 1995], then the other level scheme in Figure 9 will be operative. In a simple rectangular box with dimensions $a = b/2 = c/10$, a reasonable fit to the observed spectrum is obtained when the energy levels are scaled to fit the lowest energy transition. The fit could be improved with a small adjustment in the box dimensions. This set of dimensions is almost certainly not unique, in contrast to the situation for a spherical box. The lowest energy transitions would be assigned to 111(v) and 112(v) to 111(c) and 112(c) transitions, the next to 111(v) and 112(v) to 121(c) and 122(c) as well as a 121(v) and 122(v) to 111(c) and 112(c) transitions, and the highest energy peak to 121(v) and 122(v) to 121(c) and 122(c) transitions.

In one investigation of PbS the spectrum, exhibiting 600-, 400-, and 300-nm peaks, was decomposed by curve fitting in to a spectrum containing five absorptions [Machol, 1993]. This fit was very good and gave peaks at 603 nm (2.05 eV), 572 nm (2.16 eV), 505 nm (2.45 eV), 403 nm (3.07 eV), and 294 nm

(4.21 eV). It is interesting to note that these transitions can be reasonably well fit by a particle-in-a-rectangular-box model with edge lengths in the ratio of 1:3:6 or 7, although it is probable that this fit is not unique. It is certainly possible that the multipeak spectrum could also be made up of several size distributions experiencing only one-dimensional confinement.

The dynamics of excitons in PbS nanocrystals have been studied in a size regime where the particle radius is much smaller than the Bohr radius (strong confinement). Standard time-resolved nonlinear absorption techniques were used. The transient absorption of the lowest exciton level exhibited quantum beats in the terahertz frequency region. The quantum beats are vibronic in nature and are assigned to confined transverse optical phonons [Machol, 1993; Machol et al., 1993].

Nanocrystals of PbS were one of several quantum-sized particles that were used to sensitize nano-porous TiO_2, ZnO, SnO_2, Nb_2O_5, and Ta_2O_5 [Vogel et al., 1994]. It was found that the relative positions of the energy levels at the interface could be optimized for efficient charge separation. Photocurrent yields approached 80%, and open circuit voltages were on the order of 1 V. It was found that the photostability of these electrodes was enhanced by surface modification.

6. DIRECT BAND GAP SEMICONDUCTOR WITH A FORBIDDEN LOWEST ENERGY TRANSITION

6.1. Cu_2O

Copper oxide represents a class of semiconductors different from those already discussed in this chapter. Although the band structure [Dahl and Swittendick, 1966] shows the lowest energy transition to be a direct transition, it is nevertheless forbidden. As it turns out, while the momentum selection rules are preserved for these materials, it is the parity selection rules that are violated. The valence bands of Cu_2O largely comprise copper d orbitals, while the conduction band is s-like in character. Since electronic transitions with $\Delta n = 2$ (n is the principal quantum number) are parity forbidden, copper oxide is thus classified as a direct forbidden semiconductor material. Since all quantum confinement studies have focused on either direct allowed or indirect transitions, a study of size restriction effects on this different type of semiconductor is of fundamental interest.

There has been a tremendous amount of work done on single-crystal copper oxide. Studies include absorption measurements of strain split excitons [Gross and Kaplyanskii, 1961], absorption by single crystals [Baumeister, 1961], Raman scattering and luminescence studies [Compaan and Cummins, 1972], Stark effect measurements [Deiss et al., 1970], and Zeeman effect measurements [Gross, 1962, and references therein]. The reason for this interest is that copper oxide is the first material to have established the existence of excitons [Hayashi and Katsuki, 1952].

As discussed in the theoretical section of this chapter, excitons can be described by using a scaled hydrogenic model. When the coulombic interaction between the electron and the hole is taken into effect, the solution for the allowed energies E_n is

$$E_n = E_g - \frac{2\mu e^4 \pi^2}{h^2 \varepsilon^2 n^2} \tag{7}$$

where E_g is the band gap (2.17 eV) [Agekyan, 1977], μ is the reduced exciton mass, e is the electron charge, h is Planck's constant, ε is the dielectric constant, and n is the principal quantum number. It has been shown that for $n \geq 2$ [Baumeister, 1961; Forman et al., 1971], this model is a good fit when the value of 7.2 for the static dielectric constant is used, but fails for $n = 1$ transition unless the high-frequency dielectric constant of 6.46 is used.

The transition dipole matrix element for the $n = 1$ transition is identically zero [Agekyan, 1977], although the transition is quadrupole allowed and therefore sometimes observable. The failure of effective mass theory for the $n = 1$ transition has been attributed to the small exciton radius, which makes the use of the static dielectric constant in the description of the coulombic interaction between the carriers an inaccurate one [Agekyan, 1977].

Figure 10 shows a simplified band structure for Cu_2O and shows the existence of four different energy transitions, all of which have hydrogenic series associated with them. While the "yellow" and "green" series are direct forbidden, the "blue" and "indigo" transitions are allowed. This section will focus on the lowest energy yellow series, particularly on the quantum confinement effects on the lowest forbidden $n = 1$ excitonic transition.

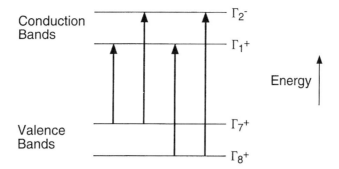

Figure 10 A simplified band structure for Cu_2O, showing the four lowest energy transitions. These transitions are the yellow, the green, and the two violet Wannier exciton absorption series. Data taken from Agekyan [1977].

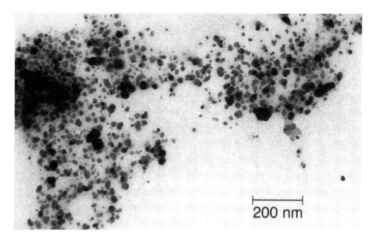

Figure 11 TEM micrograph of Cu_2O grown using an aqueous synthesis. The mean size of the crystallites were 7.0 ± 4.2 nm in radius.

The synthesis used to fabricate Cu_2O nanocrystals has been described in a previous section. A TEM micrograph of a typical preparation is given in Figure 11 and clearly indicates that although small clusters are present, a large size distribution exists [Freedhoff et al., 1995]. Quantifying any size restriction effects was therefore anticipated to be difficult, if not impossible, since each size of crystallite will undergo transitions only slightly different in energy from one another, which will result in an inhomogeneously broadened peak in either absorption or luminescence.

Figure 12 shows low-temperature luminescence spectra of both bulk (Aldrich, then annealed for 15 h under N_2 at 500 to 600°C) and nanocrystalline Cu_2O. The sharp peaks in the bulk spectrum in the 600 to 640-nm region correspond to previously observed and cataloged phonon-assisted transitions [Petroff, 1974]. The broad peaks in both the larger and smaller samples are probably due to recombination of carriers through donors/acceptors or other lattice imperfections. Unlike the case of AgBr, there was no enhancement of the luminescence intensity arising from direct recombination of the carriers in quantum-confined copper oxide; in fact, these excitonic transitions were not visible in luminescence at all. We attribute this to the incorporation of many lattice irregularities into the crystallites, resulting in a high density of donor and/or acceptor states available to the carriers upon excitation. This is in contrast to the elimination of carrier binding sites via impurity exclusion effects that occurs in the synthesis of quantum-confined AgBr.

Although it was not possible to observe band-edge exciton luminescence in nanocrystalline Cu_2O, band-edge excitation spectra could be taken using an argon ion-pumped tunable dye laser operating at 30 mW as the excitation source. Figure 13 shows an excitation spectra (monitored wavelengths were the luminescence peaks) of both the bulk single-crystal and nanocrystalline copper oxide samples [Freedhoff et al., 1995]. There are several features of the bulk spectrum that should be pointed out. First of all, the band edge is located at around 606 nm, in good agreement with literature values. The second feature is the two sets of narrow triplets, centered at 606 and 610 nm, respectively. The peaks centered

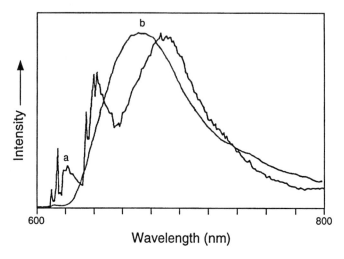

Figure 12 The low-temperature emission spectra from Cu_2O. (a) Powder, annealed under N_2 at 500 to 600°C and (b) aqueous synthesis of Cu_2O nanocrystals.

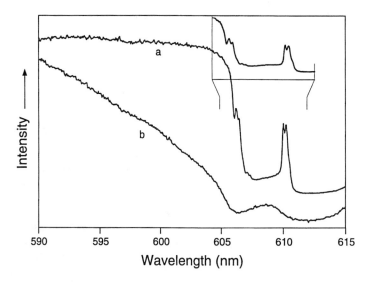

Figure 13 The excitation spectrum of bulk (a) and nanocrystalline (b) Cu_2O. The peak of the luminescence was monitored in each case. The inset shows a high resolution spectrum in the region the $n = 1$ transition of the bulk powder.

at 610 nm are thought to be due to the $n = 1$ exciton of the yellow series, with a phonon-assisted replica showing up at 606 nm. The two triplets are separated from each other by 103 cm^{-1}, matching up fairly well with the observed values of phonon frequencies of 105 to 110 cm^{-1} given in the literature [Elliott, 1961; Compaan and Cummins, 1972; Petroff, 1974]. The multiplicity of the lines is attributed to the removal of the degeneracy of the $n = 1$ transition that occurs as a result of crystal stress and deformation [Gross and Kaplyanskii, 1961].

These features are not as clearly visible in the smaller samples, although the band edge and $n = 1$ excitonic transition are easy to identify. The reason these features are broader and less well defined, as previously mentioned, is that this sample is extremely inhomogeneous in size, and what would be a steep absorption edge and narrow excitonic transition is inhomogeneously broadened as a result. It is, however, obvious that both the absorption edge and the $n = 1$ excitonic transition have shifted to higher energies in these smaller samples as per the quantum confinement effects discussed earlier. In order to attempt any quantitative modeling, a method of preparation yielding more-monodispersed samples must be developed.

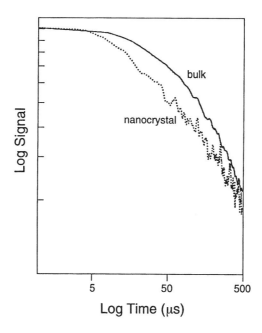

Figure 14 A log–log plot of the observed luminescence decay of bulk and nanocrystalline Cu_2O.

The final size restriction effect that has been observed in some preparations is that of a shorter observed emission lifetime in the smaller samples. Figure 14 shows a log–log plot of the observed intensity (maximum luminescence peak monitored) for both the bulk and smaller samples, and it is clear that the nanocrystals exhibit a significantly shorter lifetime [Freedhoff et al., 1995]. While this could be explained by the selection rules becoming more allowed because of a modification of the nanocrystal band structure through surface bond incorporation, it is also probable that surface-assisted recombination and changes in nonradiative pathways are also factors. See the section on AgBr for further examples and discussion of this phenomenon.

7. INDIRECT GAP SEMICONDUCTORS WITH FORBIDDEN LOWEST ENERGY TRANSITIONS

7.1. AgBr

Silver bromide is a material essential to the photographic process and, as such, is of practical interest. Since it is an indirect gap semiconductor, meaning that the lowest energy transition is forbidden according to momentum selection rules, the effect of size restriction on this material is of fundamental interest. It has been postulated that, as the crystals are made smaller, incorporation of the wave functions of the surface atoms into the band structure will result in a relaxation of these selection rules and a trend for the transition to be more allowed in nature [Rossetti et al., 1985].

A relatively simple calculation based on the Heisenberg uncertainty principle illustrates that although strong confinement of AgBr will ultimately result in an alteration of the band gap transition from forbidden to allowed, this change will only take place in extremely small nanocrystals.

Consider the band structure of AgBr, given in Figure 15. The valence band maximum is located at the L point, which is π/a away in k-space from the zone center at Γ, the conduction band minimum, where a is the appropriate lattice distance of AgBr [von der Osten, 1982]. The uncertainty principle can be expressed as:

$$\Delta \times \Delta k \geq h \tag{8}$$

In order for the conduction and valence band extrema to significantly overlap in k-space (i.e., $\Delta k = \pi/a$) the size of the nanocrystals must be smaller than 1 nm in size. It must therefore be concluded that other factors are contributing to the observation of size restriction effects on AgBr.

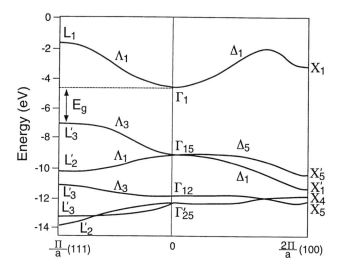

Figure 15 A simplified band diagram for AgBr. The lowest energy indirect transition is from L_3' to Γ_1. Data taken from Marchetti and Eachus [1992].

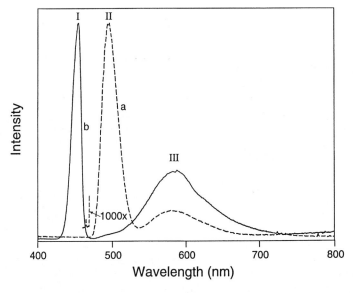

Figure 16 The low-temperature luminescence spectrum of bulk and nanocrystalline AgBr. (a) Bulk single crystal of AgBr and (b) a 15-nm-diameter micellar preparation. The exciton luminescence is blue shifted and enhanced in the preparation with nanocrystals. The emission spectra consists of three regions: (I) exciton emission, (II) iodide-bound exciton emission, and (III) donor–acceptor radiative recombination.

Quantum confinement manifests itself in three ways in AgBr. Johansson et al. [1991] showed a dramatic enhancement of the intensity of the recombination of the free exciton in 10 to 40 nm micellar samples as compared with 100 nm samples prepared in gelatin or bulk AgBr. A comparison of these low-temperature luminescence spectra is given in Figure 16. Region I (463 nm) is the radiative recombination of free and shallowly bound excitons. This region, which will henceforth be referred to as the free exciton luminescence (FEL) region, is the region generally examined under higher resolution. This region is quite intense in the micellar sample as per the quantum confinement effects discussed earlier, but is weaker in the larger sample. Region II (500 nm) corresponds to the recombination of excitons trapped at impurity iodide ions [Marchetti and Burberry, 1983] and will be referred to as the bound exciton luminescence (BEL) region. Intrinsic iodide impurities exist in all bromide sources at a concentration of approximately 1 ppm. Absorption to and luminescence from excitons trapped at iodide ions

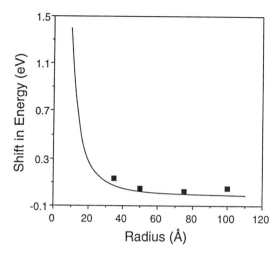

Figure 17 A comparison of the experimental band gap shifts for AgBr and the prediction from Equation 2. Data taken from Chen et al. [1994]; Marchetti et al. [1993].

are much more allowed processes than direct exciton absorption and recombination, and it is for this reason that this band is the dominant spectral feature in samples with larger crystallites. Region III (595 nm) arises from the pair recombination of excitons and charge carriers trapped by donors and acceptors [Burberry and Marchetti, 1985] and will be referred to as donor–acceptor (D-A) luminescence.

The second manifestation of quantum confinement in AgBr is best illustrated by comparing the FEL from different sizes of micellar AgBr [Marchetti et al., 1993]. As the particle size is made smaller, the peak of the luminescence shifts to higher energy as per the simple particle-in-a-box model. A plot of experimental data vs. the values predicted by Equation 2 is shown in Figure 17. The final observation of quantum confinement in AgBr is the change in the observed radiative lifetimes of some of the excitonic transitions from those of the bulk [Chen et al., 1994]. While some of these effects have been thought to be due to a breakdown in the momentum selection rules as the transition becomes less forbidden [Kanzaki and Tadakuma, 1991], it is probable that impurity exclusion and surface-assisted recombination are also factors.

In addition to the band-to-band transition of the indirect free exciton [von der Osten and Stolz, 1990], there exists a set of sharp transitions lying close to the band gap in energy in AgBr. The origin of these transitions has been extensively investigated [von der Osten, 1987; Sliwczuk and von der Osten, 1988; von der Osten and Stolz, 1990]. Nine of these transitions have been cataloged as the "BX_N" excitons and are attributable to shallow trapping by ionized donors [Sliwczuk and von der Osten, 1988], impurities such as Cd^{2+} [Sliwczuk and von der Osten, 1988] or Pb^{2+} [Sliwczuk and von der Osten, 1988], interstitial silver ions [von der Osten and Stolz, 1990], or vacancies and deformations of the lattice that occur during growth and fabrication [Sliwczuk and von der Osten, 1988]. Since the binding energies of these excitons all lie within a few meV of the free exciton, their wave functions so closely resemble that of the free exciton that they undergo transitions utilizing the same momentum-conserving phonons as the free exciton does.

While the micellar syntheses were successful in fabricating small AgBr samples that displayed quantum confinement effects, the size distributions were so inhomogeneous that any individual excitonic transitions were unresolvable from one another. It was this inability to observe size restriction effects on specific transitions that was the motivator for designing a synthesis of midsize AgBr particles (see synthesis section for details on this nonaqueous preparation).

Figure 18 shows the nonaqueous dispersion growth evolution of the FEL examined under higher resolution. At small sizes, there is a broad peak located around 461.5 nm that becomes narrower, moves to lower energies, and eventually resolves into two shifting peaks as the particles grow. By the end of the synthesis, these peaks have all but disappeared. The second feature is the set of peaks at 464.12 nm and 464.8 nm, that continue to grow throughout the synthesis but never change their peak positions.

Peak A (as labeled in Figure 18) was assigned to be the 0,0 transition of the BX_8 exciton, which is thought to be associated with trapping by ionized donors [Sliwczuk and von der Osten, 1988]. The peak position of this transition converges to 2.6800 eV in the larger crystals, corresponding to the accepted

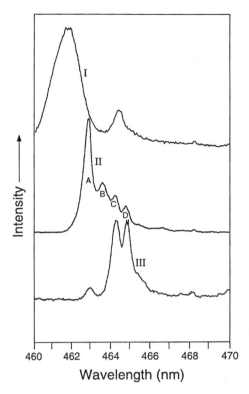

Figure 18 The radiative recombination of excitons in nonaqueous preparations of AgBr examined at higher resolution, (I) 1.0 min into the synthesis (16 nm diameter), (II) 6 min into the synthesis (31 nm), and (III) at the end (25 min) of the synthesis (52 nm).

value for the bulk BX_8 exciton. BX_8 is also known to have a weak TO phonon-assisted transition [Sliwczuk and von der Osten, 1988], which explains why only the 0,0 transition is visible.

Peak B was more difficult to assign since it was observed only over a narrow size range. Energetically it corresponds to the TO phonon-assisted transition of the free exciton, which is known to be a strong transition.

Peaks C and D are more enigmatic. While they appear to be exciton peaks in the energy region where exciton peaks are expected, there was no shift in energy with size, and the peaks do not correspond energetically to any previously observed bound excitons. The positions of the peaks do not change with excitation energy, nor do they have the same temperature dependence as the other excitonic transitions do. Measurements of the fluorescence lifetimes of these peaks do not vary with crystal size and are orders of magnitude different from lifetime measurements made on the known excitonic transitions. Since they do not appear in any other measurements of AgBr microcrystals made using different synthetic techniques, the conclusion reached as to the identity of these peaks is that they are probably a synthetically related impurity or trap state and that their location in the FEL region is purely coincidental.

Figure 19 shows a schematic representation of our overall understanding of the dynamics that occur upon excitation of AgBr. Since we are exciting with energy that is about 1 eV larger than the band gap energy, we are not creating excitons directly but rather free electrons and holes. These carriers then thermalize rapidly (in picoseconds) [Seeger, 1989] to the conduction and valence band edges, where excitons can form because of the coulombic attraction between the electrons and holes. Since the absorption of light must be a momentum-conserving process, the heavier holes will possess less kinetic energy than the electrons upon excitation. In addition, the holes couple to the lattice more effectively than the electrons [Marchetti et al., 1993] and will therefore thermalize to the valence band edge before the electrons reach the conduction band edge. Since some carriers will become trapped before forming excitons, the quantum yield of exciton formation will be less than unity.

Once an exciton is formed it can follow several different routes that result in its elimination.

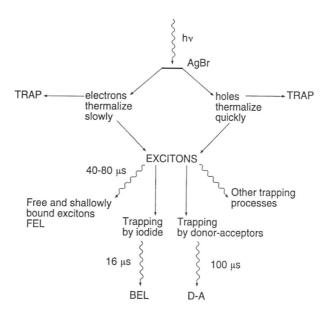

Figure 19 The pathways available to carriers in AgBr after excitation with above band gap light.

7.1.1. Recombination of the Free or Shallowly Bound Exciton

The electron and hole pair of the exciton can recombine radiatively either with or without the assistance of a momentum-conserving phonon resulting in FEL. Shallow trapping of excitons at impurities or other lattice irregularities also results in recombination of this variety. Lifetime measurements of the 0,0 transition of the BX_8 exciton (peak A in Figure 18) indicated that the measured fluorescence lifetime of 40 to 80 µs was in between that of the bulk (nanoseconds) [von der Osten and Stolz, 1990] and that of the micellar samples (100 to 700 µs) [Chen et al., 1994]. This can be explained by the fact that the very small micellar samples have fewer vacancies, interstitial ions, impurities, and defects in their lattices and thus fewer relaxation pathways open to the exciton other than direct recombination. As the particles grow, other more favorable processes become available to the excitons, and thus the observed lifetimes become much shorter.

7.1.2. Recombination of Carriers Trapped at Iodide Impurities

A second process the exciton can undergo is trapping by an iodide impurity and subsequent BEL. The lifetime of this luminescence has been found to be 16 µs in the bulk [Marchetti and Burberry, 1983] with a faster component that begins to contribute to the dynamics as the crystal size is decreased [Marchetti et al., 1993], but so far the trapping rate has not been modeled or measured.

In the simplest model, free excitons are able to move freely through a crystal with a root mean square (RMS) velocity determined by the crystal temperature,

$$V_{RMS} = \left(\frac{3kT}{M*}\right)^{1/2} \tag{9}$$

where k is Boltzmann's constant, T is the temperature, and M^* is the translational mass of the exciton. At 6 K, the RMS velocity is 1.40×10^4 m/s. The iodide-trapping rate would be determined by comparing the volume swept out by a traveling exciton in its lifetime with the probability of finding an iodide impurity within that volume.

The volume swept out by the exciton is the cross-sectional area (Equation 10) times the RMS velocity (Equation 9)

$$\sigma = \pi R_{ex}^2 \tag{10}$$

where R_{ex} is assumed to be the free exciton radius of 25 Å [Knox, 1963] (as calculated using a scaled hydrogen atom approximation). The cross-sectional area is found to be 19.6 nm², leading to

$$\sigma v_{RMS} = 2.75 \times 10^{14} \text{ nm}^3/\text{s} = \text{volume/time} \tag{11}$$

If we assume that the probability of trapping upon encounter is unity, then the excitons in the largest (50-nm) crystals will have swept out a volume equivalent to that of the crystal volume in about 0.5 ns. It can be safely concluded that even if that probability is not unity, the time scales in question are so much faster than the radiative decay of the FEL in this crystal size regime that if the exciton encounters an iodide impurity, it will become trapped prior to direct recombination. It must be pointed out that this model not only assumes a trapping probability of unity upon encounter, but also neglects to take other trapping processes into account.

When Poisson statistics were used to model the probability of finding an iodide impurity in a particularly sized sample [Freedhoff et al., 1995], it was found that even for the largest particle sizes of 50 nm (cube length), 40% of the sample should have no iodide impurities in it, and for the 20-nm crystals, that number increases to over 90%. It then becomes necessary to try to explain why the BEL is still such a dominant spectral feature even when 90% of a sample contains no iodide ions. This can be rationalized by recalling that the iodide-bound exciton transition is a more allowed transition than the free and shallowly bound exciton transitions are [Czaja and Baldereschi, 1979], and even though quantum size effects relax the momentum selection rules for the indirect transition somewhat, it still remains quite forbidden. Thus, the oscillator strength for the iodide-bound transition is so much stronger than that for the indirect transition that BEL dominates the spectrum even when the number of iodide impurities is very small.

Single photon–counting experiments on several different samples were performed in order to test the validity of the iodide-trapping model. The iodide-bound luminescence was monitored at 500 nm. A measure of the rise time of this luminescence would give a direct estimate of the iodide-trapping time. The results indicated that the trapping at iodide ions was taking place on a time scale limited by the picosecond laser pulse width, and no size dependence was observed. This is in clear disagreement with the results obtained via Poisson statistics.

There are several possible explanations for this. One is that the method utilized to calculate the iodide ion concentration was based on one used for bulk and melt-grown samples [Moser and Lyu, 1971]. It may be that for this dispersion synthesis, this is not a viable technique.

A more likely explanation draws on what has always been a generalized proposal for how bound exciton formation proceeds. Instead of exciton formation occurring prior to the binding to iodide, perhaps the free carriers themselves are sequentially trapped. The hole would first become bound to the negative iodide, and then the electron would become bound to it. That process would occur on a picosecond time scale [Seeger, 1989].

It is, however, an established fact that excitons that have already been created do get trapped at iodide ions. By exciting the samples right at the band gap energy, excitons are being created exclusively, and iodide-bound luminescence is still observed [Freedhoff et al., 1995]. Thus, it is not sufficient to postulate that the discrepancy between theory and experiment is due solely to the measurement of a sequential trapping lifetime, as a slower component arising from the trapping of formed excitons is expected as well.

A third possibility for the failure of the theory is that the trapping times quoted are in fact maximum trapping times; that is, the theoretical assumption made is that the excitons are formed as far away from the iodide impurity as possible and that they then have to traverse the length of the microcrystal before even encountering the trapping species. Since radiative recombination from the iodide ions is a far more allowed process than indirect recombination of the free exciton, it stands to reason that absorption taking place close to the impurity is also a less-forbidden process. It is possible that the absorption of light is taking place preferentially close to the trapping species, and thus the excitons become trapped immediately rather than having to travel through the microcrystals at all.

The final point to consider in this discussion concerns the way the iodide ions are incorporated into the crystal lattice. It has been shown [Marchetti et al., 1993] that as the crystal size decreases, the observed fluorescence lifetime of the iodide-bound exciton luminescence changes from a single exponential decay to a biexponential decay. This second lifetime is much faster (nanoseconds) than the bulk value of 16 μs [Marchetti and Burberry, 1983], and the extent of its contribution increases as the crystal size decreases. This has been postulated to result from the trend of iodide ions to aggregate close to the surface during nanocrystal fabrication. The surface then plays a role in the recombination process once the carriers are trapped at the impurity, resulting in faster carrier-assisted recombination. There is no information about the way the iodide is incorporated into the crystal lattice in these nonaqueous preparations, but it is possible that the fast measured trapping times come about as a result of surface-assisted processes.

In conclusion, it should be reiterated that AgBr is an indirect gap material for which many crystal size regimes have been studied extensively. Since the pathways available to free carriers upon excitation are numerous and varied, studies of this material access information on the photophysics of both free and trapped excitons.

7.2. Si

Si is potentially one of the most interesting materials for nanocrystal fabrication. The properties of crystalline silicon have been extensively studied, and there is a tremendous amount of information about this material. The hope for nanocrystals of silicon is to create a material with a tunable band gap energy, which is robust, and which has among other things a high luminescence yield. Much of the work on silicon has involved the fabrication of porous silicon through etching techniques. This produces a material that has many of the properties attributed to nanocrystalline silicon: higher luminescence yield and shorter wavelength emission. It is not universally accepted that the properties of porous silicon are entirely due to quantum confinement effects in what are postulated to be small, isolated silicon particles on the etched surface of the crystal.

The fabrication of quantifiable silicon nanocrystals is difficult. Isolated nanocrystals have been synthesized and examined, and it was determined that there was a relationship between the particle size and the optical absorption [Furukawa and Miyasato, 1988]. Recent work has focused on Si nanocrystals that were created in a microwave discharge of SiH_4 and H_2 or SiH_4, H_2, and Ar [Takagi et al., 1990]. Here the material was collected and examined by TEM, X-ray diffraction, Fourier transform infrared absorption, and photoluminescence. It was found that the Si nanocrystals which were presumably hydride-capped (SiH) were easily oxidized (60°C and 80% RH) in such a way that the crystals became covered with SiO_2. In these materials the room temperature photoluminescence yield increased and blue shifted with treatment up to a point. The photoluminescence peak energy was found to be a function of particle size, with the bluest emission belonging to the smallest particles. The size vs. photoluminescence energy presented in this work extrapolates to a luminescence maximum at 1.24 eV for bulk material. This is higher than the known band gap energy of 1.12 eV at room temperature. The optical band gap energies were measured but not given.

More-recent work has characterized the nanocrystals of silicon produced by pyrolysis of disilane [Littau et al., 1993; Brus et al., 1995]. The just-formed clusters are diluted with O_2 and He and are trapped in ethylene glycol. The oxygen treatment appears to cap the nanocrystals with SiO_2. These materials have been characterized using a number of standard techniques, including high pressure, liquid phase, and size exclusion chromatography. Nanocrystals were produced with mean sizes of 7.5, 4.5, and 3.2 nm, but it was felt from various analytical results that the Si clusters inside the SiO_2 shell were actually only 5, 2, and about 1 nm in diameter with photoluminescence peaks at 970, 770, and 660 nm, respectively.

Quantification of the quantum confinement-related shifts in the emission is difficult in silicon. The indirect band gap energy in silicon is 1.17 eV (see Table 2), and bulk silicon often has an emission band peaking at about 1.10 eV [King and Hall, 1994]. If this emission peak energy is used for a reference point, then the emission peaks at 970 (1.28 eV), 770 (1.61 eV), and 660 nm (1.88 eV) can be plotted with the predictions from Equation 2 using the heavy hole mass for silicon. This is shown in Figure 20 along with data taken from Takagi et al. [1990]. If the light hole effective mass is used in Equation 2, the fit to the experimental data seems not to be as good. However, uncertainties in the particle radii make any conclusion tentative at this point. What does seem clear from the experimental data [Takagi et al., 1990; Littau et al., 1993; Brus et al., 1995] is that decreasing the particle size further will not shift the emission from quantum-confined silicon out of the red region of the spectrum.

There has been a very large amount of work on porous silicon. Much of the work describes the production of materials with visible emission and characterization of the domains of silicon remaining after etching. A good recent reference to work in this area is the proceedings of the 1994 Materials Research Society Symposium [Collins et al., 1995]. There are strong parallels between the results of investigations of porous silicon and nanocrystalline silicon, but it is not universally accepted that all of the effects observed in porous silicon are due to size restriction.

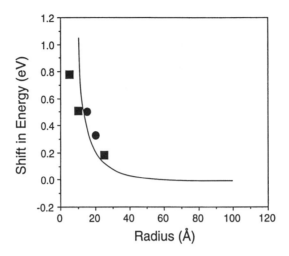

Figure 20 A comparison of the experimental blue shifts as a function of particle size for silicon nanocrystals and the predictions from Equation 2. Data taken from Takagi et al. [1990]; Brus et al. [1995].

8. PROBLEMS AND FUTURE OF NANOCRYSTALS

The materials discussed in this review are generally well behaved; that is, the blue shift of the absorption edge qualitatively follows the predictions of Equation 2. For some materials there are errors in sizing that are responsible for deviations from the model. For those materials where data are available there is a definite deviation from the energy predicted by the model in the small-size region. The predicted shifts in energy are much larger than those found through experiment. It may be that this deviation in the strong confinement size regime indicates the point at which materials can no longer be treated as confined semiconductors. Here the dominance of the surface and/or a transition to more molecular properties may be the cause of this breakdown.

A careful examination of the exciton emission data for AgBr in the weak confinement region suggests a small systematic deviation in the experimental data from the predicted shifts for a scaled confined hydrogen atom. A similar deviation may occur in the absorption data for AgI. It is possible that there is a systematic error in sizing. It is known that the peptizing agent gelatin used in the fabrication of silver halide microcrystals for photographic use leaves a residual layer that is 2 to 4 nm thick even after washing several times in warm water. If a similar effect occurs for the polymer used in the nonaqueous fabrication of AgBr nanocrystals, the observed deviation could be explained. This would suggest that more attention must be directed toward obtaining the size of the semiconductor core and not just the particle. This is the case for silicon nanocrystals which become covered with a layer of SiO_2. These problems point to a need for more-exacting techniques, such as lattice imaging, to be used in conjunction with standard TEM for particle sizing.

In addition to obtaining the correct size for particles, the shape now becomes important. In the case of PbS it appears that the latest measurements suggest that the material is rectangular and not spherical. The exact shape plays a role in the model used in the case of strong confinement. The detailed analysis of this material with a particle-in-a-box model will depend on what shape of box should be used. This situation also arises for AgI nanocrystals that are generally hexagonal bipyramids but which were analyzed as spheres. These are the details that can be incorporated to make the models more correct. Another detail to consider is that in almost all cases the anisotropy of the dielectric constant and effective masses were ignored and isotropic averages were used. These details may make some differences, but given the larger errors in sizing, they are probably relatively small.

An important change to the general model given by Equation 2 might be to incorporate a dependence of the dielectric constant on the kinetic energy of the particle in the box. That is, create a smooth transition from the static dielectric constant appropriate when the particle sizes are larger than the exciton diameter to the high-frequency dielectric constant when the diameters are smaller than the exciton diameter.

It is hard to predict where nanocrystal technology will find applications for devices or products until the actual features of some material make it the only viable candidate for that application. But some of the areas that have been targeted, such as luminescent displays, specialty glasses, and nonlinear optical components, seem to be appropriate. The materials that can be fabricated into nanocrystals seem to be limited only by the ingenuity of the scientists and engineers who are working on their fabrication. It seems probable that all manner of materials, not just semiconductors, will continue to be examined in the nanosize regime. Nanocrystals of all of these materials will continue to help us to understand the basic physics and chemistry of materials, and certainly the production and analysis of some of these materials will result in unique technological applications.

REFERENCES

Agekyan, V. T. 1977. Spectroscopic properties of semiconducting crystals with direct forbidden energy gap. *Phys. Status Solidi A*, 43:11–41.

Alivasatos, A. P., Harris, A. L., Levinos, N. J., Steigerwald, M. L., Brus, L. E. 1988. Electronic states of semiconductor clusters: homogeneous and inhomogeneous broadening of the optical spectrum. *J. Chem. Phys.*, 89:4001–4011.

Baumeister, P. W. 1961. Optical absorption of cuprous oxide. *Phys. Rev.*, 121:359–362.

Bawendi, M. G., Carroll, P. J., Wilson, W. L., Brus, L. E. 1992. Luminescence properties of CdSe quantum crystallites: resonance between interior and surface localized states. *J. Chem. Phys.*, 96:946–954.

Bawendi, M. G., Kortan, A. R., Steigerwald, M. L., Brus, L. E. 1989. X-ray structural characterization of larger CdSe semiconductor clusters. *J. Chem. Phys.*, 91:7282–7290.

Bawendi, M. G., Wilson, W. L., Rothberg, L., Carroll, P. J., Jedju, T. M., Steigerwald, M. L., Brus, L. E. 1990. Electronic structure and photoexcited carrier dynamics in nanometer sized CdSe clusters. *Phys. Rev. Lett.*, 65:1623–1626.

Berry, C. R. 1967. Structure and optical absorption of AgI microcrystals. *Phys. Rev.*, 161:848–851.

Bliss, D. E., Wilcoxon, J. P., Newcomer, P. P., Samara, G. A. 1995. Optical properties of lead sulfide nanoclusters: effects of size, stoichiometry and surface alloying. *Mater. Res. Soc. Symp. Proc.*, 358:265–269.

Borrelli, N. F., Smith, D. W. 1994. Quantum confinement of PbS microcrystals in glass. *J. Non-Cryst. Solids*, 180:25–31.

Borrelli, N. R., Hall, D. W., Holland, H. J., Smith, D. W. 1987. Quantum confinement effects of semiconductor microcrystallites in glass. *J. Appl. Phys.*, 61:5399–5409.

Broser, I., Broser, R., Hoffmann, A. 1982. II-VI compounds. In *Landolt-Bornstein*, New Series, Vol. 22a, pp. 202–224, Springer, Berlin.

Brus, L. E. 1983. A simple model for the ionization potential, electron affinity, and aqueous redox potentials of small semiconductor crystallites. *J. Chem. Phys.*, 79:5566–5571.

Brus, L. E. 1984. Electron-electron and electron-hole interactions in small semiconductor crystallites. *J. Chem. Phys.*, 80:4403–4409.

Brus, L. E. 1990. Structure, surface chemistry and electronic properties of quantum semiconductor crystallites. *J. Soc. Photogr. Sci. Technol. Jpn.*, 53:329–334.

Brus, L. E. 1991. Quantum crystallites and nonlinear optics. *Appl. Phys. A*, 53:465–474.

Brus, L. E. 1993. Radiationless transitions in CdSe quantum crystallites. *Isr. J. Chem.*, 33:9–13.

Brus, L. E., Szajowski, P. F., Wilson, W. L., Harris, T. D., Schuppler, S., Citrin, P. H. 1995. Electronic spectroscopy and photophysics of Si nanocrystals: relationship to bulk c-Si and porous Si. *J. Am. Chem. Soc.*, 117:2915–2922.

Bryant, G. W. 1988. Excitons in quantum boxes. *Phys. Rev. B*, 37:8763–8772.

Burberry, M. S., Marchetti, A. P. 1985. Low-temperature donor-acceptor recombination in silver halides. *Phys. Rev. B*, 32:1192–1195.

Chen, W., McLendon, G., Marchetti, A., Rehm, J. M., Freedhoff, M. I., Myers, C. 1994. Size dependence of radiative rates in the indirect band gap material AgBr. *J. Am. Chem. Soc.*, 116:1585–1586.

Collins, R. W., Tsai, C., Hirose, M., Koch, F., Brus, L. E. 1995. *Microcrystalline and Nanocrystalline Semiconductors, Materials Research Society Symposium Proceedings*, Materials Research Society, p. 358, Pittsburgh, PA.

Colvin, V. L., Cunningham, K. L., Alivasatos, A. P. 1994a. Electric field modulation studies of optical absorption in CdSe nanocrystals: dipole character of the excited state. *J. Chem. Phys.*, 101:7122–7138.

Colvin, V. L., Schlemp, N. C., Alivasatos, A. P. 1994b. Light emitting diodes made from cadmium selenide nanocrystals and a semiconducting polymer. *Nature*, 370:354–357.

Compaan, A., Cummins, H. Z. 1972. Raman scattering, luminescence and exciton–phonon coupling in Cu_2O. *Phys. Rev. B*, 6:4753–4757.

Czaja, W., Baldereschi, A. 1979. *J. Phys. C: Solid State Phys.*, 12:405–424.

Dahl, J. P., Swittendick, A. C. 1966. Energy bands in cuprous oxide. *J. Phys. Chem. Solids*, 28:931–942.

De Sanctis, O., Kadono, K., Tanaka, H., Sakaguchi, T. 1995. Synthesis of PbS semiconductor microcrystallites in situ in reverse micelles. *Mater. Res. Soc. Symp. Proc.*, 358:253–258.

Deiss, J. L., Daunois, A., Nikitine, S. 1970. Influence of an electric field on the quadropole line of Cu_2O. *Solid State Commun.*, 8:521–525.

Dougall, J. E., Eckert, H., Stucky, G. D., Herron, N., Wang, Y., Moller, K., Bein, T., Cox, D. 1989. Synthesis and characterization of III-V semiconductor clusters: GaP in zeolite Y. *J. Am. Chem. Soc.*, 111:8006–8007.

Efros, Al. L., Efros, A. L. 1982. Interband absorption of light in a semiconductor sphere. *Sov. Phys. Semicond.*, 16:772–775.

Ekimov, A. I. 1994. Optical properties of glass materials doped by semiconductor nanocrystals. *7th Europhysical Topical Conference: Lattice Defects in Ionic Materials*, Lyon, France.

Ekimov, A. I., Efros, A. L. 1988. Nonlinear optics of semiconductor-doped glasses. *Phys. Status Solidi B*, 150:627–633.

Ekimov, A. I., Efros, Al. L., Onushchenko, A. A. 1993. Quantum size effect in semiconductor microcrystals. *Solid State Commun.*, 88:947–950.

Ekimov, A. I., Efros, Al. L., Shubina, T. V., Skvortsov, A. P. 1990. Quantum size stark effect in semiconductor microcrystals. *J. Lumin.*, 46:97–100.

Ekimov, A. I., Kudryavtsev, I. A. 1992. Dimensional effects in luminescence spectra of zero-dimensional semiconductor structures. *Bull. Russ. Acad. Sci.*, 56:154–157.

Elliott, R. J. 1961. Symmetry of excitons in Cu_2O. *Phys. Rev.*, 124:340–345.

Fischer, Ch. H., Weller, H., Katsikas, L., Henglein, A. 1989. Photochemistry of colloidal semiconductors. 30. HPLC investigations of small CdS particles. *Langmuir*, 5:429–432.

Forman, R. A., Brower, W. S., Parker, H. S. 1971. Phonons and the green exciton series in cuprous oxide, Cu_2O. *Phys. Lett. A*, 36:395–396.

Freedhoff, M. I., Chen, W., Rehm, J. M., Meyers, C., Marchetti, A., McLendon, G. 1995a. Luminescence properties of silver bromide: from nanocrystals to microcrystals. *Proc. NATO Workshop on Semiconductor Nanocrystals*, (submitted).

Freedhoff, M. I., McLendon, G., Marchetti, A. 1995b. Quantum confinement effects on 100–400 Å silver bromide microcrystals. *Mater. Res. Soc. Symp. Proc.*, 358:301–306.

Freedhoff, M., Marchetti, A., McLendon, G. 1995c. Unpublished work.

Furukawa, S., Miyasato, T. 1988. Quantum size effects on the optical band gap of microcrystalline Si:H. *Phys. Rev. B*, 38:5726–5729.

Gacoin, T., Boilot, J. P., Gandais, M., Ricolloeau, C., Chamarro, M. 1995. Transparent sol-gel matrices doped with quantum sized PbS particles. *Mater. Res. Soc. Symp. Proc.*, 358:247–252.

Gallardo, S., Gutierrez, M., Henglein, A., Janata, E. 1989. Photochemistry and radiation chemistry of colloidal semiconductors. *Ber. Bunsenges. Phys. Chem.*, 93:1080–1090.

Gaponenko, S. V., Woggon, U., Uhrig, A., Langbein, W., Klingshirn, C. 1994. *J. Lumin.*, 60/61:302–307.

Goltzene, A., Schwab, C. 1982. Cuprous oxide. In *Landolt-Bornstein*, New Series, Vol. 17e, pp. 144–151, Springer, Berlin.

Grinberg, M., Legowski, S., Meczynska, H. 1985. The energy of shallow donor states in CdTe, GaAs, AgCl and AgBr semiconductors. *Phys. Status Solidi B*, 130:325–331.

Gross, E. F. 1962. Excitons and their motions in crystal lattices. *Sov. Phys. Usp.*, 5:195–218.

Gross, E. F., Kaplyanskii, A. A. 1961. Investigation of the influence of oriented deformations upon the spectrum of the basic absorption edge of single Cu_2O crystals. *Sov. Phys. Solid State*, 2:2637–2650.

Hache, F., Ricard, D., Flytzanis, C. 1989. Quantum-confined Stark effect in very small semiconductor crystallites. *Appl. Phys. Lett.*, 55:1504–1506.

Hanamura, E. 1988. Very large optical nonlinearity of semiconductor microcrystallites. *Phys. Rev. B*, 37:1273–1279.

Hayashi, M., Katsuki, K. 1952. Hydrogen-like absorption spectrum of cuprous oxide. *J. Phys. Soc. Jpn*, 7:599–603.

Jain, R. K., Lind, R. C. 1983. *J. Opt. Soc. Am.*, 73:647–653

Johannson, K. P., Marchetti, A. P. 1990. Unpublished work.

Johannson, K. P., McLendon, G. L., Marchetti, A. P. 1991. The effect of size restriction on silver bromide. A dramatic enhancement of free exciton luminescence. *Chem. Phys. Lett.*, 179:321–324.

Johansson, K. P., Marchetti, A. P., McLendon, G. 1992. Effect of size restriction on the static and dynamic emission behavior of silver bromide. *J. Phys. Chem.*, 96:2873–2879.

Kanzaki, H., Tadakuma, Y. 1991. Indirect exciton confinement and impurity isolation in ultrafine particles of silver bromide. *Solid State Commun.*, 80:33–36.

Kauzmann, W. 1957. *Quantum Chemistry.* pp. 183–188, Academic Press, New York.

King, O., Hall, D. 1994. Impurity related photoluminescence from silicon at room temperature. *Phys. Rev. B*, 50:10661–10665.

Knox, R. S. 1957. *Solid State Physics*, eds. F. Seitz and D. Turnbull, 5, pp. 37–59, Academic Press, New York.

Kohn, W. 1957. Shallow impurity states in silicon and germanium. *Solid State Physics*, eds. F. Seitz and D. Turnbull, 5, pp. 257–320, Academic Press, New York.

Lianos, P., Thomas, J. K. 1986. Cadmium sulfide of small dimensions produced in inverted micelles. *Chem. Phys. Lett.*, 125:299–302.

Littau, K. A., Szajowski, P. F., Muller, A. J., Kortan, R. F., Brus, L. E. 1993. Luminescent silicon nanocrystal colloids via a high temperature aerosol reaction. *J. Phys. Chem.*, 97:1224–1230.

Machol, J. L. 1993. Electronic and vibrational properties of lead sulfide nanocrystals. *Ph.D. Dissertation*, Cornell University, Ithaca, NY.

Machol, J. L., Wise, F. W., Patel, R. C., Tanner, D. B. 1993. Vibronic quantum beats in PbS microcrystallites. *Phys. Rev. B*, 48:2819–2822.

Madelung, O., Ed. 1991. *Semiconductors Group IV Elements and III-V Compounds.* pp. 11–28, Springer-Verlag, Berlin.

Madelung, O., Ed. 1992. *Semiconductors Other than Group IV Elements and III-V Compounds.* pp. 11–43, Springer-Verlag, Berlin.

Makov, G., Nitzan, A., Brus, L. E. 1988. On the ionization potential of small metal and dielectric particles. *J. Chem. Phys.*, 88:5076–5085.

Marchetti, A. P., Burberry, M. S. 1983. Optical and optically detected magnetic resonance studies of AgBr:I⁻. *Phys. Rev. B*, 28:2130–2134.

Marchetti, A. P., Eachus, R. S. 1992. The photochemistry and photophysics of the silver halides. *Advances in Photochemistry*, eds. D. Volman, G. Hammond, and D. Neckers, Vol. 17, pp. 145–216, John Wiley & Sons, New York.

Marchetti, A. P., Johannson, K. P., McLendon, G. L. 1993. AgBr photophysics from optical studies of quantum confined crystals. *Phys. Rev. B*, 47:4268–4275.

Marin, J. L., Cruz, S. A. 1991a. Enclosed quantum systems: use of the direct variational method. *J. Phys. B*, 24:2899–2907.

Marin, J. L., Cruz, S. A. 1991b. On the use of direct variational methods to study confined quantum systems. *Am. J. Phys.*, 59:931–935.

Micic, O. I., Meglic, M., Lawless, D., Sharma, D. K., Serpone, N. 1990. Semiconductor photophysics 5. Charge carrier trapping in ultrasmall silver iodide particles and kinetics of formation of silver atom clusters. *Langmuir*, 6:487–492.

Mohan, V., Anderson, J. B. 1989. Effect of crystallite shape on exciton energy: quantum Monte Carlo calculations. *Chem. Phys. Lett.*, 156:520–524.

Moller, L., Bein, T., Herron, N. 1989. Encapsulation of lead sulfide molecular clusters into solid matrices. Structural analysis with X-ray absorption spectroscopy. *Inorg. Chem.*, 28:2914–2919.

Moser, F., Lyu, S. 1971. Luminescence in pure and I-doped AgBr crystals. *J. Lumin.*, 3:447–458.

Mukhopadhyay, G. 1985. On shallow-deep instability of impurity levels in semiconductors. *Solid State Commun.*, 53:47–49.

Mulvaney, P. 1993. Nucleation and stabilization of quantized AgI clusters in aqueous solution. *Colloids Surf. A: Physicochem. Eng. Aspects*, 81:231–238.

Murray, C. B., Norris, D. J., Bawendi, M. G. 1993. Synthesis and characterization of nearly monodisperse CdE (E = S, Se, Te) semiconductor nanocrystals. *J. Am. Chem. Soc.*, 115:8706–8715.

Nomura, S., Kobayashi, T. 1992. Exciton-LO-phonon couplings in spherical semiconductor microcrystallites. *Phys. Rev. B*, 45:1305–1316.

Nomura, S., Kobayashi, T. 1994. An exciton with a massive hole in a quantum dot. *J. Appl. Phys.*, 75:382–387.

Nozik, A. J., Uchida, H., Kamat, P. V., Curtis, C. 1993. GaAs quantum dots. *Isr. J. Chem.*, 33:15–20.

Nozik, A. J., Williams, F., Nenadovic, M. T., Raj, T., Micic, O. I. 1985. Size quantization in small semiconductor particles. *J. Phys. Chem.*, 89:397–399.

Pankove, J. I. 1971. *Optical Processes in Semiconductors*. p. 143, Dover Publications, New York.

Persans, P. D., Silvestri, M., Mei, G., Lu, E., Yukselici, H., Schroeder, J. 1993. Size effects in II-VI semiconductor nanocrystals. *Braz. J. Phys.*, 23:144–150.

Petroff, Y. 1974. Luminescence et effet Raman résonant dans Cu_2O. *J. Phys.* (Paris), 35:C3-277–C3-285.

Resch, U., Weller, H., Henglein, A. 1989. Photochemistry and radiation chemistry of colloidal semiconductors. 33. Chemical changes and fluorescence in CdTe and ZnTe. *Langmuir*, 5:1015–1020.

Rossetti, R., Hull, R., Gibson, J. M., Brus, L. E. 1985. Hybrid electronic properties between the molecular and solid state limits: lead sulfide and silver halide crystallites. *J. Chem. Phys.*, 83:1406–1410.

Rossmann, H., Schålzgen, A., Henneberger, F., Måller, M. 1990. Quantum confined DC Stark effect in microcrystallites. *Phys. Status Solidi B*, 159:287–290.

Schmidt, H. M., Weller, H. 1986. Quantum size effects in semiconductor crystallites: calculation of the energy spectrum for the confined exciton. *Chem. Phys. Lett.*, 129:615–618.

Schmidt, K. H., Patel, R., Meisel, D. 1988. Growth of silver halides from the molecule to the crystal: a pulse radiolysis study. *J. Am. Chem. Soc.*, 110:4882–4884.

Schmitt-Rink, S., Miller, D. A. B., Chemla, D. S. 1987. Theory of the linear and nonlinear optical properties of semiconductor microcrystallites. *Phys. Rev. B*, 35:8113–8125.

Seeger, K. 1989. *Semiconductor Physics*, pp. 198–206, Springer-Verlag, Heidelberg.

Sliwczuk, U., von der Osten, W. 1988. Optical studies of weakly localized exciton states in silver halides. *J. Imaging Sci.*, 32:106–113.

Steigerwald, M. L., Alivasatos, A. P., Gibson, J. M., Harris, T. D., Kortan, R., Muller, A. J., Thayer, A. M., Duncan, T. M., Douglass, D. C., Brus, L. E. 1988. Surface derivatization and isolation of semiconductor cluster molecules. *J. Am. Chem. Soc.*, 110:3046–3050.

Takagahara, T. 1987. Excitonic optical nonlinearity and exciton dynamics in semiconductor quantum dots. *Phys. Rev. B*, 36:9293–9296.

Takagi, H., Ogawa, H., Yamazaki, Y., Ishizaki, A., Nakagiri, T. 1990. Quantum size effects on the photoluminescence in ultrafine Si particles. *Appl. Phys. Lett.*, 56:2379–2380.

Tanaka, T., Saijo, H., Matsuraba, T. 1979. Optical absorption studies of the growth of microcrystals in nascent suspensions III. Absorption spectra of nascent silver iodide hydrosols. *J. Photogr. Soc.*, 27:60–65.

Tassoni, R., Schrock, R. R. 1994. Synthesis of PbS nanoclusters within microphase-separated diblock copolymer films. *Chem. Mater.*, 6:744–749.

Vogel, R., Hoyer, P., Weller, H. 1994. Quantum sized PbS, CdS, Ag_2S, Ab_2S_3 and Bi_2S_3 particles as sensitizers for various nanoporous wide band gap semiconductors. *J. Phys.*, 98:3183–3188.

von der Osten, W. 1982. I-VII compounds. In *Landolt-Bornstein*, New Series, Vol. 17b, p. 273, Springer, Berlin.

von der Osten, W. 1987. Shallow bound excitons in silver bromide. *Physica*, 146B:240–255.

von der Osten, W., Stolz, H. 1990. Localized exciton states in silver halides. *J. Phys. Chem. Solids,* 51:765–791.

Wang, Y., Herron, N. 1991. Synthesis and characterization of nanometer-sized semiconductor clusters. *Res. Chem. Intermed.,* 15:17–29.

Wang, Y., Suna, A., Mahler, W., Kasowski, R. 1987. PbS in polymers. From molecules to bulk solids. *J. Chem. Phys.,* 87:7315–7322.

Ward, A. J. I., O'Sullivan, E. C., Rang, J., Nedeljkovic, J., Patel, R. C. 1993. The synthesis of quantum size lead sulfide particles in surfactant-based complex fluid media. *J. Colloid Interface Sci.,* 161:316–320.

Wilson, W. L., Bawendi, M. G., Rothberg, L., Jedju, T., Carroll, P. J., Brus, L., Steigerwald, M. L. 1990. Photophysics of capped CdSe microcrystals exhibiting quantum confinement. In *SPIE, Nonlinear Optical Materials and Devices for Photonic Switching,* 1216:84–87.

Wilson, W. L., Szajowski, P. F., Brus, L. E. 1993. Quantum confinement in size-selected, surface-oxidized silicon nanocrystals. *Science,* 262:1242–1247.

Zicovich-Wilson, C., Planelles, J. H., Jaskolski, W. 1994. Spatially confined simple quantum mechanical systems. *Int. J. Quantum Chem.,* 50:429–444.

Chapter 2

Spectroscopy of Confined Indirect Excitons in AgBr Quantum Crystallites

Heinrich Stolz, Hartmut Vogelsang, and Wolf von der Osten

Reviewed by S. W. Koch

TABLE OF CONTENTS

1. INTRODUCTION

Because of their fundamental and expected practical importance, quantum-size effects of excitons in nanocrystals have recently attracted increasing attention (for references see, e.g., Brus [1991]; Itoh et al. [1991]; Flytzanis and Hutter [1992]; Bányai and Koch [1993]). The small crystalline particles have sizes of the order of a few exciton Bohr radii so that the unit cell is already fully developed, as in the volume crystal, but the energy spectra are still discrete because of the spatial confinement. Such mesoscopic systems are frequently called *quantum dots.*

While the main body of theoretical and experimental work in this field refers to excitonic transitions in II-VI and I-VII semiconductors like, e.g., CdSe and CuCl, having direct energy gaps, it is only recently that indirect gap nanocrystals have been studied. The strong impact here came from the discovery of red and blue luminescence in porous silicon (for a recent review see, e.g., Lockwood [1994]); other systems recently under intense study are Ge, GaP, and AgBr. Compared with direct gap systems, a peculiar feature of indirect gap quantum dots is the expected strong size dependence of the radiative transition rate. This size dependence is due to the mixing of direct with indirect states and the breakdown in translational symmetry [Takagahara and Takeda, 1992], although different results were also reported [Tolbert et al., 1994].

Quantum-size effects in AgBr nanocrystals were first observed by groups at Kodak (USA) [Johansson et al., 1991; Johansson et al., 1992; Marchetti et al., 1993] and Fuji Photo (Japan) [Kanzaki and Tadakuma, 1991]. The crystallites with radii of a few nanometers are produced either in a gelatine matrix or in

0-8493-2485-8/97/$0.00+$.50
© 1997 by CRC Press, Inc.

reversed micelles in solution solidified at low temperature. In comparison with bulk AgBr, the main effect of the spatial confinement consists of a blue shift of the exciton emission. This shift is accompanied by strong enhancement of the emitted intensity which is believed to originate either from indirect–direct state mixing, resulting in relaxed wave vector selection rules or from the absence of the impurities (impurity exclusion effect) that is to be anticipated in the small particles from statistical considerations [Johansson et al., 1991; Kanzaki and Tadakuma, 1991].

An important discovery that makes AgBr quantum crystallites unique compared with all other indirect-gap materials is the occurrence of resonant Raman scattering (RRS) under excitation in the free exciton (FE) region [Scholle et al., 1993; Pawlik et al., 1993]. Actually this is very similar to bulk AgBr, where the lowest exciton state is situated at point L in the Brillouin zone and phonon-assisted absorption and emission associated with various momentum-conserving phonons, TA(L), TO(L), LA(L), LO(L), give rise to characteristic two-phonon scattering (see, e.g., von der Osten [1991]). In the nanocrystals inves-tigated here, additional one-phonon scattering appears. Exploiting the cross section permits the unam-biguous identification of the zero-phonon (ZP) transition due to direct–indirect state mixing and allows one to determine its strength in various confined excited states. Moreover, the narrow line Raman spectra reveal details of the exciton–phonon coupling as a function of particle size and renders possible well-defined size-selective investigations. In connection with time-resolving techniques, lifetimes as long as several hundreds of microseconds are obtained, in agreement with an earlier measurement by Masumoto et al. [1992].

The reported experimental results will be compared with two different theoretical models, the indi-vidual-particle confinement picture [Kayanuma, 1988] and the donorlike exciton model [Ekimov et al., 1989], the latter providing a good basis to describe energy states and radiative transition rates for AgBr quantum dots.

In 10 to 100 times larger microcrystals with rather uniform size and shape produced as photographic "emulsions" [Johansson et al., 1992] we measured luminescence and luminescence excitation spectra. These studies reveal the transition from bulk behavior to quantum confinement and simultaneously show the importance of surface trapping as an exciton relaxation mechanism. For the smallest microcrystals the appearance of high-energy exciton luminescence lines suggests the onset of quantum confinement of the exciton translational motion.

2. EXPERIMENTAL PROCEDURES

2.1. Preparation

Two types of crystallites different in preparation, size, and shape were investigated. The *microcrystals* consisted of nearly monodispersed small AgBr crystals of cubic shape with edge lengths of 50 and 450 nm, respectively. They were fabricated by Dr. A. P. Marchetti at the Kodak Research Laboratories in the form of photographic "emulsions" in gelatin, as described by Johansson et al. [1992]. Electron microscopy of correspondingly prepared material reveals that the standard deviation in size for the 450-nm crystallites is $\sigma_R = 30$ nm while the smaller ones have $\sigma_R = 9$ nm [Marchetti, 1993]. Also in this size regime, deviations from the cubic shape are beginning to occur and the particles tend to take on a more spherical shape (compare, e.g., Figure 8 in Johansson et al. [1992]). Subsequent to fabrication, the emulsions were freeze-dried (for reasons of transportation) and thereby partly degelled. Powder of this material was then carefully solved in cold water to give a finely dispersed mixture which was pipetted on a glass substrate. After drying, it formed a layer with the remainder of the gelatin serving as an adhesive medium between the crystallites (Figure 1).

The AgBr *nanocrystals* were prepared in reversed micelles starting with separate water solutions of KBr and AgNO$_3$ and following essentially the steps described previously [Johansson et al., 1991; Scholle et al., 1993]. These particles are spherical in shape with diameters as small as a few nanometers. As represented schematically in Figure 2, the essential step is the diffusion-controlled collisional reaction between micelles (AOT: dioctyl sulfosuccinate) containing the different solutions whereby the small AgBr crystallites precipitate. As a major change, instead of using n-heptane for most preparations we used 1,2,4-trimethylcyclohexane (TMCH) as a solvent because it remains more transparent in the solid state when cooled down to the measuring temperature. Usually the preparations were carried out at room temperature. In some exceptional cases the preparation temperature was lowered to 4°C. As a result of this measure, an increase of the (integrated) FE luminescence over the simultaneously present donor–acceptor (D–A) band (see Section 4) by a factor of five was found, compared with the room temperature preparations.

Figure 1 Scanning electron micrograph of a sample consisting of small monodispersed cubic AgBr microcrystals. The edge length of each cube is 450 nm. (From Timme, M. et al. *J. Luminesc,* 55, 79, 1993. With permission.)

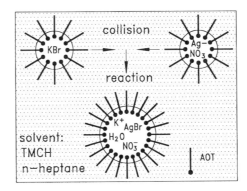

Figure 2 Schematic illustration of the preparation of AgBr nanocrystals in reversed micelles. AOT: dioctyl sulfosuccinate; TMCH: 1,2,4-trimethylcyclohexane.

Because of the light sensitivity of silver halides above 70 K, in any method of sample fabrication care was taken to keep the material in dark red light during all handling and to expose it to visible light only after it was cooled down to the low temperatures for the measurement ($T = 6$ K).

2.2. Spectroscopy

For the continuous wave (cw) luminescence and Raman measurements, we used standard optical equipment consisting of a tunable dye laser operated with Stilbene 3, a double monochromator ($f = 0.85$ m), and a photon counting system. To record the time-resolved data, two different setups were employed that cover a wide range of times from milli- to picoseconds. For the longer decay times, the cw setup was supplemented by a chopper and suitable boxcar electronics allowing readout of the photomultiplier signal in preselected time windows. With this setup, lifetimes and time-discriminated spectra could be measured in a range between 1 µs and 100 ms with a maximum time resolution of 200 ns. The short decay times were measured by means of a picosecond apparatus with 5 ps excitation pulses of a dye laser synchronously pumped by the third harmonic of a Nd^{3+}-YLF laser system (repetition rate 76 MHz). The emitted light, dispersed by a double monochromator ($f = 1$ m), was detected by a fast photomultiplier tube using time-resolved single photon counting with a time resolution of 40 ps (see, e.g., Stolz [1994]). All luminescence spectra were corrected for the spectral sensitivity of the detection system.

3. INDIRECT EXCITON STATES IN QUANTUM DOTS

The small crystallites investigated still have diameters large compared with the lattice constant ($a_0 \simeq$ 0.6 nm) and contain at least 10^3 Ag$^+$Br$^-$ ion pairs. In a first approximation we therefore expect that the electronic states in these crystallites can be described by effective mass theory. The finite size is represented by a confining potential, most simply with infinite high barriers. The exciton energies as a function of dot size can be calculated by means of the model of a particle moving in this potential well with radius R equal to the crystallite radius (for a review of the theory of quantum dots see Bányai and Koch [1993]). This model makes the analogy to the impurity problem obvious (see Bassani and Pastori Parravicini [1975]).

In this section, we will discuss the electronic states in AgBr dots from this point of view and derive the radiative transition rates of the lowest exciton states necessary to later interpret the experimental results.

3.1. Energy States

Considering at first the case of *weak confinement*, i.e., crystallites with particle radii R significantly larger than the exciton Bohr radius R_B ($R \gg R_B$), the exciton is confined as a whole. Its translational motion becomes quantized with approximate energies

$$E_{nl} = E_{gx}^i - Ry_x + \frac{\hbar^2 k_{nl}^2}{2M^* R_{\text{eff}}^2} \tag{1}$$

with quantum numbers $n = 1, 2, \dots$ and $l = 0, 1, 2 \dots$ (s,p,d ...) [Kayanuma, 1988]. E_{gx}^i and Ry_x are the energy of the indirect band gap and the bulk exciton binding energy, respectively, k_{nl} denote the nth zero of the spherical Bessel function, and M^* is the exciton translational mass. R_{eff} represents an effective quantum dot radius given by $R_{\text{eff}} = (R - 2\eta R_B)$ for the lowest 1s state. The factor η takes into account the exciton dead layer at the surface and amounts to $\eta = 1.4$ for a hole-to-electron effective mass ratio $\sigma = m_e/m_h = 4$ [Kayanuma, 1988] that applies approximately to AgBr (see below). As in AgBr the exciton Bohr radius is $R_B = 2.32$ nm the weak confinement situation is appropriate for our microcrystals. With $R = 225$ and 25 nm (see Table 1 for the other relevant quantities), the exciton gap position ($n = 1$) and, hence, the indirect absorption edge would occur blue shifted relative to that in an extended (bulk) sample by 0.005 and 0.46 meV, respectively.

Table 1 Parameters relevant in analyzing the optical properties of quantum dots in AgBr

m_e^*	0.288[a]	ε_0	12.45[a]	Ry_x	28 meV[b]
$m_{h\parallel}^*$	1.79[a]	ε_∞	4.5[a]	Ry_e	35 meV
$m_{h\perp}^*$	0.79[a]	ε_{eff}	11.0	Ry_h	137 meV
M^*	1.4[b]	E_{gx}^i	2684 meV[b]	a_0	0.573 nm[a]
R_B	2.32 nm	R_e	1.84 nm	R_h	0.48 nm
$\hbar\Omega(TO(\Gamma))$	11 meV[b]	$\hbar\Omega(TA(L))$	6.7 meV[b]	$\hbar\Omega(TA(X))$	3.9 meV[b]
$\hbar\Omega(LO(\Gamma))$	17 meV[b]	$\hbar\Omega(LA(L))$	11.8 meV[b]	$\hbar\Omega(LA(X))$	5.8 meV[b]
		$\hbar\Omega(TO(L))$	8.3 meV[b]	$\hbar\Omega(TO(X))$	13 meV[c]
		$\hbar\Omega(LO(L))$	16.5 meV[c]	$\hbar\Omega(LO(X))$	17 meV[c]

Note: Effective masses m^*, M^* in units of electron mass; indirect exciton gap energy (E_{gx}^i), phonon energies ($\hbar\Omega$), and lattice constant (a_0) at LHeT. For calculation of Bohr radii (R_B, R_e, R_h) and binding energies (Ry_x, Ry_e, Ry_h), see text.

[a] Values from *Landolt-Börnstein New Series III 17b: Semiconductors*, O. Madelung, Ed., Springer, Berlin 1982, p. 253.

[b] Values from W. von der Osten, in *Topics in Applied Physics, Light Scattering in Solids VI*, M. Cardona and W. Güntherodt, Eds., Springer, Berlin, 1991, 361.

[c] Values from Y. Fujii, S. Hoshino, S. Sakuragi, H. Kanzaki, J. W. Lynn, and G. Shirane, *Phys. Rev. B* 15, 358, 1977.

In the case of *strong confinement* ($R \leq R_B$ [Kayanuma, 1988]), which we favored in a previous analysis of AgBr nanocrystals [Pawlik et al., 1993], the electron and hole states are individually confined in space. Neglecting the Coulomb interaction and surface polarization effects, by solving Schrödinger's equation for spherical dots one obtains the wave functions of the electron (e) and hole (h) states

$$\phi_{n,l,m}^{e,h} = \sqrt{\frac{2}{a^3}} \frac{j_l(k_{nl}r/R)}{j_{l+1}(k_{nl})} Y_{l,m}(\vartheta,\varphi) \tag{2}$$

(j_l denotes the spherical Bessel function of order l and $Y_{l,m}$ the spherical harmonics) and the energies

$$E_{nl}^e(R) = E_g + \frac{\hbar^2 k_{nl}^2}{2m_e^* R^2} \tag{3}$$

$$E_{n'l'}^h(R) = -\frac{\hbar^2 k_{nl}^2}{2m_h^* R^2} \ . \tag{4}$$

As in the case of exciton confinement above, the energies of the electron and hole states depend on quantum numbers n, $n' = 1,2 \ldots$ and l, $l' = 0,1,2 \ldots$ (s,p,d ...) as well as on the effective masses, but not on magnetic quantum numbers m, m'. Neglecting the effective hole mass anisotropy, here and in the following an average value $m_h^* = (m_{h\parallel}^* + 2m_{h\perp}^*)/3$ is taken resulting in $\sigma \approx 4$.

Following Kayanuma's theory further, the exciton wave function is written as the product of the single particle states giving exciton energies

$$E_{nl,n'l'}(R) = E_{nl}^e(R) - E_{n'l'}^h(R) - CA \ . \tag{5}$$

CA represents the Coulomb interaction energy. While using the Coulomb term $CA = 1.786e^2/(4\pi\varepsilon_0\varepsilon_{eff}R)$ to calculate the 1s1s state energy, in case of the 1s1p and 1p1s states we took $CA = 1.835e^2/(4\pi\varepsilon_0\varepsilon_{eff}R)$ which was derived as the average between the 1s1s and 1p1p Coulomb energies. ε_{eff} denotes an effective relative dielectric constant (see Table 1) calculated from the exciton binding energy corresponding to an exciton Bohr radius $R_B = 2.32$ nm. In addition, an exciton correction term ($EX = 0.248Ry_x$) was taken into account which is of importance only for the lowest (1s1s) state.

Inserting values appropriate for the indirect exciton in AgBr (see Table 1), we calculated the energies of the various states as a function of particle size R (see dotted line in Figure 3 for the 1s1s state). For a dot of radius $R = R_B$, e.g., one obtains 220 and 50 meV for the electron and hole quantization energies, while the Coulomb energy amounts to about 100 meV.

For quantum dots with radii R somewhat larger than the exciton Bohr radius R_B, a case that actually applies better to the micelle type AgBr nanocrystals, the individual-particle confinement picture is no longer a good approximation, as the relative motion of the electron and the hole is influenced by their Coulomb attraction. To study the exciton states in this *intermediate* situation several groups [Hu et al., 1990; Takagahara and Takeda, 1992; Bányai and Koch, 1993] have performed elaborate numerical calculations. Instead, we apply the donorlike exciton model of Ekimov et al. [1989] that reproduces most of the essential features of quantum-confined excitons while keeping the single-particle picture (see Bányai and Koch [1993] for comments). This model is based on the fact that for most semiconductors the hole effective mass m_h^* is much larger than the electron effective mass m_e^*. Consequently, as reflected by the corresponding Bohr radius $R_{e,h} = 4\pi\varepsilon_{eff} \cdot \hbar^2/(m_{e,h}^* e_0^2)$, the motion of the hole subject to the Coulomb interaction spatially is much more restricted. As a first approximation the ground state electron wave function is calculated variationally with the hole fixed at the center of the dot, using the one-parameter trial function derived from the 1s single-particle state

$$\phi_{100}^e(r,\vartheta,\varphi) = \sqrt{N(\beta)} \, j_0(k_{00}r/a) \exp(-\pi\beta r/R) Y_{0,0}(\vartheta,\varphi) \tag{6}$$

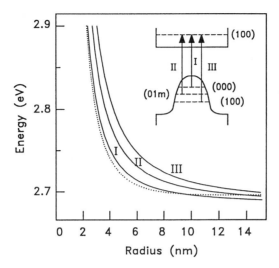

Figure 3 Radius dependence of energies for the transitions between the lowest electron state ($n = 1, l = 0$, $m = 0$) and the (000) (I), (01m) (II), and (100) (III) hole states in the donorlike exciton model (Equation 11). For comparison, the 1s1s transition energy is shown (dotted curve) calculated from the individual-particle confinement model Equation 5. The inset represents the confining potentials, the energy states, and the possible transitions in the donorlike exciton model.

Here, $N(\beta) = (1 + \beta^2)\beta/[R(1 - \exp(-2\pi\beta))]$ is a normalization constant and β the variational parameter, which is determined from minimizing the ground state energy

$$E_e(\beta) = Ry_e \frac{R_e}{R} 4\pi N(\beta) \left[\frac{\pi R_e}{4R\beta} \left(1 - \exp(-2\pi\beta)\right) - 2\int_0^R z j_0(z)^2 \exp(-2\beta z)\, dz \right]. \tag{7}$$

In a second step, the motion of the hole in the potential of the electron provided by the charge distribution $e_0|\phi^e|^2$ is considered. The simple classic picture of a positive charge in a homogeneously filled, negatively charged sphere shows that the interaction energy depends quadratically on the distance of the hole from the dot center. The hole, therefore, moves in the potential of a spherical harmonic oscillator. Taking into account the image potential of the charges due to the dielectric surrounding with dielectric constant ε_d (in our case a frozen aqueous solution of KNO_3), we can derive the oscillator frequency as [Ekimov et al., 1989]

$$\omega_0 = \sqrt{\frac{2e_0^2}{4\pi\varepsilon_0 \varepsilon_{\text{eff}} m_0 m_h^* R^3} \left(\frac{\pi^2}{3} + \frac{\varepsilon_{\text{eff}} - \varepsilon_d}{\varepsilon_{\text{eff}} + 2\varepsilon_d} \right)} \tag{8}$$

The energy levels of the hole are given by [Flügge, 1976]

$$E_h(n, l) = \hbar\omega_0(2n + l + 3/2) \tag{9}$$

where $n = 0, 1, \ldots$ and $l = 0, 1, \ldots$ represent the oscillatory and the angular momentum quantum numbers, respectively. The hole wave function is that of a spherical harmonic oscillator [Flügge, 1976]

$$\phi_{nlm}^h(\vec{r}_h) = \sqrt{N_{nlm}}\; r_h^l \exp(-\lambda r_h^2/2) F(-n, l+3/2; \lambda r_h^2) Y_{lm}(\vartheta_h, \varphi_h) \tag{10}$$

with $\lambda = m_0 m_h^* \omega_0/\hbar$, m the magnetic quantum number and F the confluent hypergeometric function. Of course, this approximation is valid for the lowest excited hole states only, the higher lying states becoming increasingly influenced by the size restriction of the dot.

By taking as an estimate $\varepsilon_d = 2.72$ derived from the index of refraction of liquid water ($n = 1.65$) at a wavelength of 20 μm [Hellwege, 1962], the hole quantization energy is increased to about 100 meV, demonstrating the strong additional confinement due to the electrostatic interaction, which is also found for other materials like GaAs (see, e.g., Bányai and Koch [1993]; Hu et al. [1990]).

From Equations 7 and 9 the energy of the electron–hole pair in a state $i = (nlm)_h \otimes (100)_e$ is given by the sum of the electron and hole energies

$$E_{ex}^i = E_e(\beta_{min}) + E_h(n,l) . \tag{11}$$

The results of a calculation of the size-dependent exciton energies in AgBr quantum dots based on the donorlike exciton model and using the parameters in Table 1 are shown in Figure 3. In the limit $R \to \infty$ the energy approaches $E_{gx}^i - Ry_e$, while the hole oscillation frequency goes to zero. However, from calculating the wave function for a dot of 25 nm in radius, we expect the hole to be still centered in the dot, the oscillation frequency amounting to about 2 meV in that case. Compared with the individual-particle confinement model (dotted line), the ground state energy is not very much different, demonstrating that this quantity alone is not very sensitive to model parameters. Very importantly, however, the next higher optically allowed excited state, the $n = 1$, $l = 0$ hole state (see Section 3.2), is quite close to the ground state. To give an example, for $R = 2R_B$ in the donorlike exciton model an energy difference of only 50 meV is calculated (see also line III in Figure 3) while 200 meV would result from the individual-particle confinement model.

In the foregoing discussion, the spins of the electron and hole have been neglected, but can be included by using the analogy to the impurity problem [Czaja and Baldereschi, 1979]. Accordingly, the hole states from the four inequivalent L points of the Brillouin zone have to be superimposed, thus forming a representation of the full O_h point group of the crystal. Taking only the uppermost $L_{4,5}^-$ states into account, two Γ_8^- states are formed, split by valley–orbit interaction. The lowest exciton state is then formed of the uppermost one of these hole states and the lowest electron state. As a result of exchange interaction, it splits into the optically forbidden $\Gamma_3^- + \Gamma_5^-$ triplet and the allowed Γ_4^- singlet states which are shifted to higher energies by the exchange energy Δ. In the following, we are only concerned with the allowed singlet states and can thus ignore the spin degeneracy.

3.2. Phonon Line Shape and Radiative Transition Rate

As in bulk samples, one expects the indirect transitions associated with the momentum conserving TO(L) and LA(L) phonons to be the relevant exciton–phonon processes. Moreover, due to mixing of the L with the Γ point valence band states, ZP transitions may become allowed as well.

To calculate the transition rates of the dominant TO(L) and the ZP process, we apply standard second-order perturbation theory [Dimmock, 1967; Bassani and Pastori Parravicini, 1975]. It is of advantage to express the rates in terms of momentum matrix elements $\langle \Psi_j | \vec{e} \cdot \vec{\pi} | \Psi_0 \rangle$, where $|\Psi_j\rangle$ denotes the electron–hole pair states in the dot, $\vec{e}$ the light polarization vector, and $\vec{\pi}$ the crystal momentum operator. From these, the absorption cross section $\sigma_A(\hbar\omega)$ of a single dot and the radiative transition rate W_{rad} can be calculated by the standard relations

$$\sigma_A(\hbar\omega) = \frac{\pi e^2}{\varepsilon_0 n^2 m_0^2 \omega v_{en}} \sum_j \left|\langle \Psi_j | \vec{e} \cdot \vec{\pi} | \Psi_0 \rangle_{av}\right|^2 \delta(E_j - \hbar\omega) \tag{12}$$

$$W_{rad} = \int \frac{e^2 \omega}{\pi \varepsilon_0 m_0^2 c^2 v_{en} \hbar} \sum_j \left|\langle \Psi_j | \vec{e} \cdot \vec{\pi} | \Psi_0 \rangle_{av}\right|^2 \delta(E_j - \hbar\omega) \, d\hbar\omega \tag{13}$$

with the subscript av indicating that a suitable average over the polarization states of the light has to be performed in the final expressions. n denotes the index of refraction and v_{en} the energy transport velocity in the medium which, neglecting dispersion, is given by $v_{en} = c/n$.

To calculate the TO(L) transition rate to an exciton state $(n,l,m)_h \otimes (n',l',m')_e$ we use the envelope function approximation [Bányai and Koch, 1993], assuming that the exciton wave function in the quantum dot is a product of the single particle states (in the missing electron scheme [Bassani and Pastori Parravicini, 1975])

$$\Psi = \phi^e_{n'l'm'}(\vec{r}_e)\phi^h_{nlm}(\vec{r}_h)u_c(\vec{r}_e)\otimes u_v(\vec{r}_h) \tag{14}$$

whereby $\phi^{e,h}$ are given by either Equation 2 or Equations 6 and 10 depending on the confinement. u_c, u_v denote the periodic part of the Bloch functions of the conduction and the valence bands at point Γ and L, respectively. This approximation is valid only in the limit of intermediate confinement, i.e., for $R < 5R_B$, while for larger particle sizes the correlation of the electron and hole motion due to the Coulomb interaction would become more and more important [Bányai and Koch, 1993].

Decomposition of the single particle states into Bloch states allows one to write the momentum matrix element as

$$\langle\Psi|\vec{e}\cdot\vec{\pi}|\Psi_0\rangle = gF_{nlm,n'l'm'}(\vec{q}-\vec{k}_L)p^{(2)}_{cv} \tag{15}$$

where g is a degeneracy factor due to averaging over the polarization, $p^{(2)}_{cv}$ is the second order k-vector independent band-to-band momentum matrix element, and $\vec{q}$ denotes the wave vector of the momentum-conserving phonon participating in the indirect transition. $\vec{k}_L$ denotes the wave vector of the L point.

In Equation 15 we introduced the function

$$F_{nlm,n'l'm'}(\vec{q}) = \frac{1}{(2\pi)^{3/2}}\int d^3r\,\phi^e_{nlm}(\vec{r})\phi^h_{n'l'm'}(\vec{r})\exp(-i\vec{q}\,\vec{r}) \tag{16}$$

as the Fourier transform of the overlap of electron and hole envelope functions in the dot. This relation shows that as long as the exciton correlations are ignored the line shape of an indirect transition in a quantum dot, via the dispersion of the momentum-conserving phonon, reflects the Fourier transform of the electron and hole wave functions. By utilizing this unique feature, indeed, information on the hole wave function could be obtained [Pawlik et al., 1995].

The analysis of the experimental bulk absorption and emission allows one to deduce the momentum matrix element between the valence and conduction band states. Using the expression for the bulk indirect TO(L) absorption strength at T = 0 K [Dimmock, 1967]

$$\alpha_{TO(L)}(\hbar\omega) = \frac{e^2}{4\pi\varepsilon_0 nm_0^2 c\omega}\left(\frac{2M^*}{\hbar^2}\right)^{3/2}\frac{\Omega}{\pi R_B^3}\left|g_{abs}\right|^2\left|p^{(2)}_{cv}\right|^2 \tag{17}$$

the TO(L)-assisted exciton radiative transition rate is obtained as

$$W^0_{rad} = \frac{ne^2\omega}{\pi\varepsilon_0 m_0^2 c^3\hbar}\cdot\frac{\Omega}{\pi R_B^3}\left|g_{em}\right|^2\left|p^{(2)}_{cv}\right|^2 \tag{18}$$

(Ω: volume of the crystal unit cell, $g_{abs,em}$: degeneracy factors). Inserting the values for AgBr from Sliwczuk et al. [1984], one deduces $W^0_{rad} \approx 1\cdot10^3\text{ s}^{-1}$. The radiative rate for the quantum dot is obtained by summing the contributions of all TO(L) phonons with different $\vec{q}$, which leads to the simple expression

$$W^{TO}_{rad}(R) = G^0_{nlm,n'l'm'}\frac{\pi R_B^3}{R^3}W^0_{rad} \tag{19}$$

with

$$G^0_{nlm,n'l'm'} = R^3\int d^3r\left(\phi^e_{nlm}(\vec{r})\phi^h_{n'l'm'}(\vec{r})\right)^2. \tag{20}$$

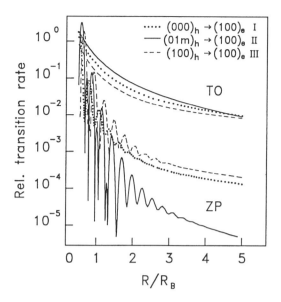

Figure 4 Dependence of the relative transition rates of the TO(L) and ZP processes in AgBr quantum dots calculated using the donorlike exciton model upon (normalized) particle radius. To obtain absolute values the TO(L) rates have to be multiplied by $W^0_{rad} = 1 \cdot 10^3$ s^{-1} and the ZP rates by $2 \cdot 10^5$ s^{-1}.

Applying at first the individual-particle confinement model, from the electron and hole functions (Equation 2) one obtains for the transition into the lowest pair state lsls

$$G^0_{\text{lsls}} = \frac{1}{2}\left(2\text{Si}(2\pi) - \text{Si}(4\pi)\right) = 0.672 \tag{21}$$

(Si(x): sine integral) showing that the radiative rate W^{TO}_{rad} increases with $1/R^3$ as the dot size becomes smaller. Qualitatively, this result can be understood by considering that the lowest quantized level collects the oscillator strength of all bulk states up to the next higher level. The correspondingly calculated TO(L) transition rate for the donorlike exciton model using the wave functions Equation 6 and 10 is plotted as the dotted line in Figure 4, showing a qualitatively similar dependence on dot radius.

For the excited electron–hole pair states, Equation 20 shows that transitions between *all* states become allowed, their relative strengths depending sensitively on the detailed nature of the wave function of the states and, thus, on the quantum dot radius. To demonstrate this radius dependence, we have also computed (see Figure 4) the relative rates $G^0_{nlm,n'l'm'}\ \pi R^3_B/R^3$ for TO(L) transitions from the first two excited hole states into the lowest electron state in the donorlike exciton model with parameters appropriate for AgBr (see Table 1). For all transitions the rates increase with decreasing dot radius. Most surprisingly, in the range of sizes shown, the $(01m)_h \rightarrow (100)_e$ transition is larger in strength than the dipole-allowed transition from the $(000)_h$ state but gets much smaller for larger dots. We note that even in the individual-particle confinement model the Δl, $\Delta m = 0$ selection rule for dipole transitions (see, e.g., Kayanuma [1988]; Bányai and Koch [1993]) is not valid for indirect gap materials.

The ZP transition rate originates from a mixing of the L point and Γ point hole states due to the confining potential. For dots still large in size compared with the lattice constant, the mixing is small and a perturbational treatment is sufficient. In complete analogy to the impurity problem [Bassani and Pastori Parravicini, 1975], the momentum matrix element can be calculated giving as dominant contribution [Takagahara and Takeda, 1992]

$$\left\langle \Psi \left| \vec{e} \cdot \vec{\pi} \right| \Psi_0 \right\rangle_{ZP} = F(\vec{k}_L) \int d^3r \left(u_c(\vec{r}) \vec{e} \cdot \vec{\pi} u_v(\vec{r}) \right) \tag{22}$$

whereby $F(\vec{k}_L)$ gives the strength of direct–indirect state mixing, and the last factor is the momentum matrix element of the direct transition. Inserting the individual-particle confinement wave functions (Equation 2) results in

$$F(\vec{k}_L) = \frac{1}{2}\left(\frac{2\text{Si}(k_L R) - \text{Si}(k_L R + 2\pi) - \text{Si}(k_L R - 2\pi)}{k_L R}\right) \qquad (23)$$

providing an explicit $1/R^2$ dependence of the transition rate (see Equation 13), which is modulated by the $\text{Si}(k_L R)$ behavior, resulting in an effective $1/R^6$ dependence. The results for the donorlike exciton model are also displayed in Figure 4. As the absolute magnitude of the required direct transition momentum matrix element is unknown, we scaled the transition rate that is proportional to $|F(\vec{k}_L)|^2$ to reproduce the experimental results (see Section 5).

Concerning the ZP processes two points should be stressed. First, the ZP rates increase with decreasing dot size as expected, but at a much smaller power law than in the individual-particle confinement case. At small dot sizes strong oscillations of the ZP rates are predicted that have their origin in the interference of the electron and hole wave functions at large wave vector. Second, consistent with the axial symmetry of the hole state, also the $(010)_h \rightarrow (100)_e$ transition is weakly allowed, but shows an increase with dot size which is steeper than that of the allowed transitions.

For dot sizes $R > 5R_B$, i.e., in the weak confinement regime, the size dependence is expected to change because of the increasing importance of exciton corrections. They give rise to an additional factor $\propto R^3$ due to the so-called giant oscillator strength effect [Kayanuma, 1988] leading to a size-independent transition rate of the indirect process as in the bulk.

4. LUMINESCENCE AND RESONANCE RAMAN SPECTRA

The most obvious signature of quantum confinement is a shift of the excitonic luminescence lines to higher energies relative to the bulk band gap (see Equation 5 or 11). For indirect excitons, because of momentum conservation, the luminescence always consists of a superposition of the various phonon-assisted processes, making the quantitative analysis difficult. Furthermore, as a consequence of the usual excitation high above the band gap, quantum dots of all sizes contribute to the emission and give rise to inhomogeneous broadening. Correspondingly, detailed information can be obtained only if the size distribution is narrow and the line broadening sufficiently small compared with the energy separation of the quantized states.

Different from that, under resonant excitation conditions well-defined crystallite sizes are excited [Scholle et al., 1993]. The resulting resonance fluorescence consists of sharp lines that reflect the radiation and phonon coupling of the excitons in the quantum dot.

4.1. Resonance Fluorescence of a Single Quantum Dot

Before presenting experimental results, the spectral features of resonance fluorescence of a single AgBr quantum dot with indirect gap will be derived applying the theory developed in Section 3.2. The exciton ground state, after being excited, essentially decays via radiative recombination. As shown in the inset of Figure 5 this consists of a ZP and a momentum-conserving TO(L) as well as a weaker LA(L) phonon process (see Table 1 for energies). In addition, phonon sidebands will occur (lower spectrum in Figure 5). These reflect coupling to near zone-center acoustic (LA(Γ)), and optical phonons (LO(Γ)) as well as X point phonons, all known to strongly interact with the indirect exciton in bulk AgBr [von der Osten, 1991]. The participation of X point phonons, clearly showing up in the experimental spectra, becomes possible because of intervalley scattering of the exciton state.

Accordingly, in the resonance fluorescence spectrum one expects a ZP transition (coincident with the exciting laser line at energy E_L), a 1TO(L) line associated with an additional ZP transition, a 2TO(L) line, and the corresponding LA(L) transitions (upper spectrum), each accompanied by a phonon sideband. In an ensemble of quantum dots with a broad size distribution all these lines have the character of resonance Raman scattering, as they shift in parallel to the exciting laser frequency.

Excitation of higher excitonic states (see Section 3) leads to the same phonon processes, albeit with different coupling strengths. In addition, relaxation into the ground state may occur, giving rise to characteristic fluorescence lines. The energy shift of these lines is expected to increase with excitation

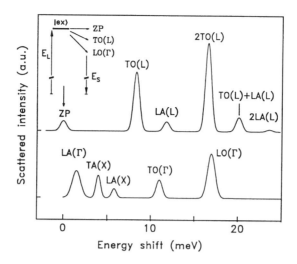

Figure 5 Principal resonance fluorescence spectrum of a single AgBr quantum dot. Upper curve: transitions involving ZP and momentum-conserving L point phonon processes; lower curve: sideband involving Γ and X point phonons (phonon energies from Table 1; peak of acoustic sideband assumed at 2 meV; coupling function $\tilde{g}(\Omega)$ and line widths chosen arbitrarily). To generate the spectra Gaussian line shapes and transition rate ratios $W^{ZP}/W^{TO} = 1/3$ and $W^{LA}/W^{TO} = 0.16$ [Sliwczuk et al., 1984] were assumed. The arrow marks the laser excitation photon energy. Inset: representation of the excitation and radiative decay processes.

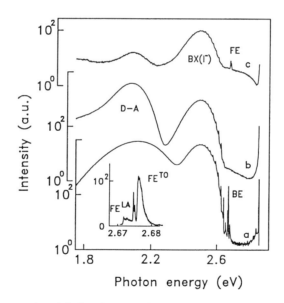

Figure 6 Luminescence spectra of AgBr microcrystals — edge lengths 450 nm (b) and 50 nm (c) — in comparison with a spectrum of a large crystal (a) excited at E_L = 2.85 eV. T = 8 K. The bands around 2.5 and 2.1 eV originate from the BX(Γ) and D–A-type recombinations. The narrow line structure at 2.68 eV is due to free and weakly bound exciton luminescence (FE, BE). The spectra are plotted in logarithmic scale and shifted relative to each other along the ordinate. Inset: near–band edge luminescence of the bulk sample showing TO(L)- and LA(L)-assisted FE emission. (From Timme, M. et al. *J. Luminesc.*, 55, 79, 1993. With permission.)

photon energy, because the energy difference between ground and excited states varies with dot size (see Figure 3).

For larger crystals, the strength of the ZP transition should decrease (see Figure 4) reflecting the gradual transition to the bulk behavior with Raman scattering involving momentum-conserving phonons only.

4.2. Luminescence Spectra

4.2.1. Microcrystals

Typical low-temperature luminescence spectra of two microcrystal samples together with a spectrum of a large silver bromide single crystal are displayed in Figure 6. The broad bands with maxima around 2.5 eV (495 nm) and 2.1 eV (590 nm) are due to the recombination of bound excitons (BX) at residual iodide (I⁻) (see, e.g., von der Osten and Stolz [1990] and to D–A pair recombination, respectively. These assignments are based upon the time evolution that was examined in the 450-nm microcrystal sample [Timme et al., 1993]. Like in large crystals, the BX (I⁻) luminescence, after subtraction of a long-lived background, decayed exponentially with decay time $\tau \approx 19$ μs (see von der Osten and Stolz [1990]), while for the D–A luminescence (measured at various photon energies across the band) a characteristic nonexponential decay was found with a shortest lifetime of about 28 μs. The intensity ratio of the two bands varied considerably for the three cases shown. Typically in the 450-nm sample, the D–A band was much more intense than the BX (I⁻) band, while in the 50-nm sample the reverse was true. Moreover, in the large single crystal both peak position and width of the D–A band are different from those in the microcrystal spectra, consistent with the suggested composite nature of this band [Marchetti et al., 1991].

A characteristic feature of the microcrystal luminescence is the broad background that extends over all measured spectral ranges below the excitation line (in Figure 6 at $E_L = 2.85$ eV). This background is strongest for the 50-nm-size crystals. We favor the idea that it originates from surface states. This is substantiated by excitation spectroscopy in these microcrystals, which revealed the increasing importance of surface states for exciton trapping becoming the dominant exciton annihilation process in the small microcrystals [Timme et al., 1993].

In the 50-nm samples, indications are found that size quantization already sets in. This is supported by the appearance of the near-edge luminescence for excitation above the band gap (Figure 7). In large AgBr single crystals, the TO(L) and LA(L) phonon-assisted FE recombination gives rise to two characteristic emission lines with Maxwellian line shape (see inset to Figure 6). Different from that, in the 50-nm samples another narrow emission component shows up (line marked by * in Figure 7) about 4 meV higher than the FE^{TO} replica. Assuming weak confinement, in a previous paper Timme et al. [1993] tentatively assigned this energy difference to emission from excited quantized exciton levels in cubic crystals for which, however, a substantially reduced side length had to be assumed. The donorlike exciton model, which already seems to be applicable in this size regime (see above), now would give even better agreement. For crystallites having an approximately spherical shape (as justified by the experimental observation, see Section 2) with radius $R = 25$ nm, it predicts an energy separation of the right magnitude between the two lowest hole oscillator states $(000)_h$ and $(100)_h$ which the observed transitions would correspond to. In particular, this model would explain quite naturally the increase in FE intensity for the 50-nm relative to the 450-nm sample by hole localization in the dot center, which inhibits the surface trapping of the holes. Also, this interpretation would be consistent with the excitation spectrum of the BX(I⁻) transition shown in Figure 7. Because of the onset of indirect absorption processes, the intensity here starts to increase by one TO(L) phonon energy above the band gap, its blue shift expected from exciton quantization, however, being too small to be observable especially because of inhomogeneous broadening.

4.2.2. Nanocrystals

The low-temperature luminescence of a typical AgBr nanocrystalline sample (estimated average dot radius 5 nm, see below) is substantially different from that of bulk AgBr and microcrystalline samples and essentially consists of two bands (Figure 8, left). The one at the low energy side, varying in strength and peak position with sample preparation, is ascribed to D–A-type transitions quite analogous to the bulk material. The other band is assigned to recombination of size-restricted free excitons and is found by orders of magnitude more intense than in the bulk or microcrystalline samples (see Figure 6). Compared with bulk AgBr, where the dominant TO(L)-assisted emission component of the indirect FE is situated at energy $E_{gx}^i - \hbar\Omega^{TO(L)}$, the emission (with peak energy 2.732 eV in Figure 8) in the nanocrystallites is shifted to high energy clearly indicative of the quantum confinement of states. Depending upon (average) particle size, this shift amounts to 40 to 60 meV. It is comparable or slightly larger than the indirect FE binding energy ($Ry_x = 28$ meV, see Table 1) in bulk material demonstrating a case of intermediate confinement [Kayanuma, 1988]. Besides being due to the superposition of transitions from various excited states and involving different phonons, the width of the FE band (60 meV in the example above) is largely inhomogeneous. This is attributed to the statistical distribution in particle size which could be demonstrated by the change in line shape and the occurrence of characteristic

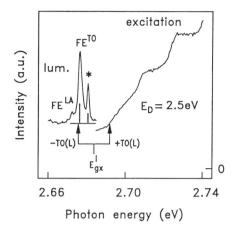

Figure 7 Near–band edge free exciton luminescence (lum.) spectrum and BX(I⁻) luminescence excitation profile (excitation) of the 50-nm microcrystals. FE^{TO,LA} denote the TO(L) and LA(L) replica of the indirect free exciton emission. E^i_{gx}: indirect exciton gap in bulk AgBr. The arrows mark the threshold for emission, –TO(L), and absorption, +TO(L), with phonon creation. Also shown are calculated transitions from quantized exciton states (see text). The excitation profile is detected at energy E_D = 2.5 eV. The pronounced modulation is attributed to LO(Γ) phonon scattering. (From Timme, M. et al. *J. Luminesc.*, 55, 79, 1993. With permission.)

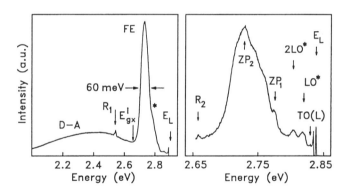

Figure 8 Left: Luminescence spectrum of AgBr nanocrystals excited at E_L = 2.90 eV. FE and D–A denote recombination luminescence from free excitons and donor–acceptor-like states. E^i_{gx}: indirect exciton gap energy in bulk AgBr. The * marks the $(100)_h \rightarrow (100)_e$ transition. Spectral bandwidth: 0.6 meV. Right: FE spectrum obtained in a nanocrystal sample upon resonant excitation at E_L = 2.84 eV. TO(L), LO*, and 2LO* mark Raman processes, ZP$_n$ denotes $(n00)_h \rightarrow (000)_h$ (n = 1,2) relaxation as described in the text. Spectral bandwidth: 0.5 meV. R$_1$, R$_2$ mark Raman lines of AOT and solvent. T = 6 K. (From Pawlik, S. et al. *Mater. Res. Soc. Symp. Proc.*, 358, 289, 1995. With permission.)

line narrowing under resonant (size-selective) laser excitation [Pawlik et al., 1993]. At the high-energy side of the FE band, a shoulder emerges (marked by * in Figure 8). From the energetic position it can be attributed to the luminescence from unrelaxed excited excitons in the $(100)_h \otimes (100)_e$ state (see Figure 3). We note that, different from the volume crystal (see Figure 6), no luminescence of excitons bound to isoelectronic I⁻ is seen in our nanocrystals. This decay channel appears strongly suppressed because of the impurity exclusion effect [Johansson et al., 1991; Kanzaki and Tadakuma, 1991].

Tuning the excitation toward lower energies (Figure 8, right), the spectral shape of the FE luminescence changes drastically. While near the laser line sharp features show up, which due to their linear shift with E_L can be identified as resonance Raman processes (see below), the FE band itself becomes highly structured. Quite pronounced are two lines (ZP$_{1,2}$) that are accompanied by additional sidebands. Although these lines shift when E_L is tuned, they clearly do not behave like Raman scattering. Consistent with the energy position relative to the laser line, these features are attributed to emission from the exciton ground state in dots of different sizes involving ZP and various phonon-assisted processes after excitation of an excited dot state. These processes will be discussed in some detail in Section 5.

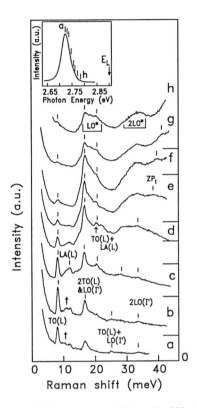

Figure 9 RRS spectra in AgBr nanocrystals (average crystallite radius $\langle R \rangle$ = 5.3 nm). Excitation photon energies E_L: 2.723 eV (a); 2.731 eV (b); 2.738 eV (c); 2.749 eV (d); 2.757 eV (e); 2.762 eV (f); 2.772 eV (g); 2.784 eV (h). T = 6 K. Spectral bandwidth: 0.4 meV. The lines marked by † are due to LA(Γ) scattering, ZP$_1$ denotes a ZP fluorescence line. Inset: exciton luminescence excited at E_L = 2.9 eV with E_L for the Raman spectra marked.

4.3. Resonance Raman Scattering

Valuable information on the AgBr quantum dot states and their interaction with phonons can be obtained from the study of Raman scattering that occurs upon resonant excitation in the exciton. Because of the dependence of exciton energy on particle size actually these processes probe size-selective properties.

In Figure 9, a series of spectra is shown taken at various excitation energies around the FE resonance. By comparison with the known bulk phonon energies (see Table 1), most of the scattering lines can be identified unambiguously. For excitation at lower exciton energies (spectra a, b, c) the dominant peaks are attributed to TO(L), 2TO(L), and weaker LA(L) scattering in complete analogy with the coupling of L point phonons in the bulk [Sliwczuk et al., 1984]. Further scattering processes are due to acoustic and optical phonons near the Γ point, LO(Γ) and LA(Γ). A detailed line shape analysis as a function of E_L confirms that actually two different lines are hidden under the structure around 17 meV, namely, the lines 2TO(L) and LO(Γ), with energies 16.5 and 17.0 meV. Upon excitation more to the high-energy side of the FE luminescence, the Raman spectrum (spectra d to h) changes gradually. While the TO(L) and 2TO(L) features are reduced in intensity, relatively intense and broad bands evolve around 16 meV (LO*) and 33 meV (2LO*). As will be discussed in detail below, we attribute them to strongly coupling optical phonons.

In order to understand the RRS spectra quantitatively, one has to consider the electron–phonon coupling of the resonant intermediate states. Thereby, one should note that by changing the laser energy not only quantum dots of a different size, but also different exciton states are excited, resulting in general in a complex superposition of different RRS spectra.

For an appropriate description of the spectra the transition probabilities for the fundamental ZP, TO(L), and LA(L) processes must be specified. While in principle these could be calculated using the formalism of Section 3.2, we prefer to take them as adjustable parameters in fitting the measured spectra. The coupling of the transitions to phonons can be formulated in analogy to the impurity problem

[Maradudin, 1966; Giesecke et al., 1972] by introducing a spectral coupling function $\tilde{g}(\Omega)$ which allows us to express the spectral line shape of the phonon band as

$$K(\omega) = A\omega \int_{-\infty}^{\infty} e^{i(\omega-\omega_0)t} e^{-S+g(t)} \, dt \; . \tag{24}$$

Here $g(t)$ is the Fourier transform of the coupling function

$$g(t) = \int \tilde{g}(\Omega) e^{-i\Omega t} \, d\Omega \tag{25}$$

and ω, ω_0, and Ω are the photon frequency, the electronic transition (or excitation) frequency, and the phonon frequency, respectively. S denotes the Huang–Rhys coupling strength which is given by

$$S = \int \tilde{g}(\Omega) d\Omega = g(0) \tag{26}$$

and represents the average number of phonons that are involved and give rise to the sideband. The ZP line, represented by a δ-function, is given by $K_0(\omega) = 2\pi A e^{-S} \delta(\omega - \omega_0)$. To obtain the overall Raman spectrum, the phonon sideband $K(\omega)$ (Figure 5, lower curve) has to be convoluted with itself to account for the process of absorption followed by emission via the intermediate exciton state. In trying to fit the measured spectra, however, we found that the L point phonon scattering has to be treated separately so that $K(\omega)$ is formed by all except the L point phonons. These couple only through the ZP component to $K(\omega)$ accounting for the fact that, while in principle the translational symmetry is broken, the wave vector at the L point is still a good quantum number, i.e., indirect–direct state mixing is weak, corresponding to $F(\vec{k}_L)$ in Equation 23 being small. Accordingly, the sideband only reflects coupling of the states to Γ point phonons and, because of the obviously still existing k-star degeneracy, X point phonons.

Fits of two Raman spectra calculated by means of this procedure are reproduced in Figure 10 and found to nicely agree with experiment. In these fits the TO(L)/LA(L) intensity ratio and the line widths were taken from the bulk [Sliwczuk et al., 1984] while the ZP and TO(L) phonon transition probabilities were varied. As to the excitation photon energies we have chosen simple situations (spectra a and h in Figure 9) in which the $(000)_h \otimes (100)_e$ exciton ground state and the $(100)_h \otimes (100)_e$ excited state, respectively, are the dominant intermediate states (see Section 5.3, Figure 15). From the fits the coupling functions $\tilde{g}(\Omega)$ are derived and displayed in the inset of Figure 10. Inspection of $\tilde{g}(\Omega)$ shows that the resulting coupling strength S is considerably larger at higher than at lower excitation photon energy ($S = 3.0$ and 0.75, respectively). While coupling to acoustic phonons increases in proportion to S, the line shape for the optical phonons undergoes considerable changes, the most obvious effect being the relatively stronger growth of a 15-meV line (labeled SP in Figure 10). This line is revealed as a well-defined and reproducible shoulder in the RRS spectra in all our samples (see, e.g., Figure 12, below). Based on the energetic position, we believe that surface phonons (SP) are responsible. This type of phonon was reported in quantum dots of GaP, too [Efros et al., 1991], and is anticipated to be especially strong in highly polar materials when the electronic charge distribution is concentrated near the surface. Using the donorlike exciton model, our calculations show that the hole wave function in the ground excitonic state is restricted to the center of the dot so that SP coupling is weak. In contrast, the wave functions of the excited hole states extend up to the surface of the dot, resulting in enhanced coupling, qualitatively consistent with our observation. According to Klein et al. [1990], the frequencies of the surface modes depend on the dielectric environment of the dots and are given by

$$\Omega_l = \sqrt{\frac{\Omega_{TO}^2(l+1)/l + \Omega_{LO}^2 \varepsilon_\infty / \varepsilon_d}{(l+1)/l + \varepsilon_\infty / \varepsilon_d}} \tag{27}$$

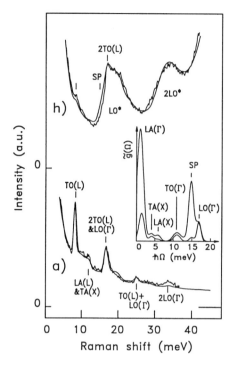

Figure 10 Experimental resonant Raman spectra (noisy lines; spectra a, h from Figure 9) in comparison with fits by the linear coupling model (smooth lines). Inset: one-phonon-coupling functions $\bar{g}(\Omega)$ resulting from the fits of the spectra and demonstrating the change in electron–phonon interaction with excitation energy. The coupling functions are scaled relative to each other according to Equation 26. The derived Huang–Rhys coupling strengths are $S = 3.0$ and $S = 0.75$ for spectra h and a, respectively. SP denotes the surface phonon. (From Vogelsang, H. et al., *J. Luminesc.*, (in print), 1996. With permission.)

with $l = 1, 2, \ldots$. Assuming water as the surrounding dielectric ($\varepsilon_d = 2.72$), we calculate the lowest mode to occur at 14.2 meV, the series limit of surface mode states giving rise to the observed structure at 15.0 meV.

It is evident from Figure 10 (spectrum h) that our calculation also reproduces the prominent features LO* and 2LO*. In previous investigations [Scholle et al., 1993; Pawlik et al., 1995], these bands, which occur in all AgBr quantum dot samples upon excitation on the high-energy side of the FE luminescence, have been assigned to forbidden L point phonon scattering. Our line shape analysis now suggests that they originate from multiphonon processes (average phonon number $S = 3$) caused by the strong coupling of the excited exciton states with various acoustic and optical phonons.

4.4. Time-Resolved Measurements

While RRS provides knowledge on exciton energies and phonon coupling, time-resolved investigations are needed to uncover the dynamic relaxation behavior of the indirect FE states [Pawlik et al., 1993]. As a first step, the time dependence of the emission intensity monitored at various energies across the FE band was measured and found to consist of two major components with lifetimes $\tau_1 \approx 3$ ns and $\tau_2 \approx 500$ μs. Subsequent detailed measurements under various conditions of excitation and detection (see Figure 11 for three examples) together with a quantitative analysis revealed that these times must be ascribed to the D–A and FE states, respectively. Besides being unexpectedly long in all investigated samples, the FE decay times exhibit an about 10% increase with decreasing E_L across the FE band (in Figure 11 from 446 to 536 μs). Different from the nonexponential time behavior that is found for confined direct excitons like in CuCl (see, e.g., Itoh et al., [1991]), in AgBr the FE decays monoexponentially over one to two orders of magnitude. Moreover, the lifetime is considerably longer than that known from typical AgBr bulk samples in which it is usually of the order of several nanoseconds and determined by nonradiative decay (see, e.g., von der Osten [1991]). The measured lifetimes are only a factor of two smaller than the bulk radiative lifetime of the indirect transition (see Section 3.2) and have the right order of magnitude expected theoretically in quantum dots (see Equation 19). From this, one has to draw

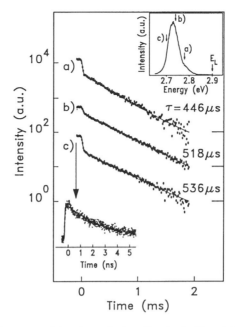

Figure 11 Decay of the luminescence intensity excited at E_L = 2.90 eV and detected within the FE band at photon energies E_S = 2.78 eV (a), 2.74 eV (b), 2.70 eV (c) (see inset). The intensity is plotted in a logarithmic scale with the curves shifted relative to each other along the ordinate. Points: experimental data; full lines: fits by a sum of two exponentials plus a small constant background contribution with the resulting FE decay times indicated. T = 6 K. Time window: 50 μs. The extra curve shows the fast decay component (lifetime 2.5 ns) measured after picosecond pulse excitation using time-resolved single-photon counting.

the conclusion that nonradiative processes are not very important and the quantum efficiency is close to one. This is in contrast to materials with a direct optical transition where usually nonradiative processes determine the relaxation behavior of the confined excitons in quantum dots [Brus, 1991].

With this knowledge, by using the time-resolving techniques described in Section 2, we were able to record time-discriminated FE emission and RRS spectra. Representative results are displayed in Figure 12, which once again demonstrate that the FE transition is long-lived. The long lifetime becomes very much apparent from the Raman spectra (right side, middle curve). Like in bulk AgBr [von der Osten, 1991], these lines can be thought of as hot luminescence processes, with their time behavior reflecting the lifetime of the intermediate exciton state.

As seen from Figure 12 (left), the D–A band which partly overlaps the FE peak is shorter lived than in the bulk and exhibits an appreciable range of decay times which depend upon (detection) photon energy. Even though the detailed nature of this band is still unknown, the shift of its maximum to lower energies with time seems to be consistent with D–A-type states as origin (compare with the lower two spectra in Figure 6).

5. PARTICLE-SIZE EFFECTS

An implication of quantum confinement very promising for future applications of nanocrystals is the size dependence of energy eigenstates and radiative transition rates, as this opens up the possibility of tailoring the optical properties. Because we did not have the facilities to determine the particle size directly by electron microscopy and also principal difficulties exist in exactly relating measured optical properties to the broad distribution of particle sizes, we have tried to deduce size-dependent information by comparing the results of various spectroscopic methods, like resonance fluorescence, Raman scattering, and lifetime measurements, with theory. Especially helpful here again is the indirect nature of the exciton transition in AgBr, since ZP and L point phonon transitions depend differently on the electron–hole wave functions. As in this way both experiment and theory are tested with respect to their overall consistency, reliable information on the size distribution and on size-dependent properties can be anticipated. In fact, the sizes determined in this way are found to reasonably well agree with electron microscopic measurements in correspondingly produced samples [Johansson et al., 1992], the compar-

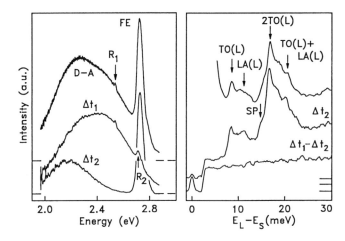

Figure 12 Left: Time-discriminated luminescence spectrum of AgBr nanocrystals measured with a digital lock-in. The fast (Δt_1) and slow (Δt_2) components contain lifetime contributions $\tau \le 20$ μs and $\tau \ge 100$ μs, respectively. Upper curve: time-integrated spectrum. R_1, R_2 are (fast) Raman lines of AOT and solvent. $E_L = 2.90$ eV; $T = 6$ K; spectral bandwidth: 0.5 meV. Right: Time-discriminated Raman scattering resonantly excited at $E_L = 2.75$ eV with a 5-ps laser pulse. The slow component Δt_2 is measured with a 5-ns time window set at 5-ns after excitation; the fast component $\Delta t_1 - \Delta t_2$ is calculated, with Δt_1 representing a 5-ns window right after the excitation pulse. Upper curve: time-integrated spectrum with the Raman processes denoted. $T = 6$ K. Spectral bandwidth: 0.6 meV. SP marks the surface phonon discussed in Section 4.3. (From Pawlik, S. et al. *Mater. Res. Soc. Symp. Proc.*, 358, 289, 1995. With permission.)

ison showing, e.g., that an average size of $\langle R \rangle = 5.5$ nm would correspond to the FE luminescence line position in Figure 8.

5.1. Size Dependence of Exciton Energies

A primary goal in studying the size-dependent properties is to determine the variation of exciton energies with quantum dot radius. In AgBr this can be done by monitoring the ZP transitions that occur in the course of relaxation following excitation of the quantum dot states (lines $ZP_{1,2}$ in Figure 8, right side). By analyzing resonantly excited time-gated fluorescence spectra (see, e.g., Figure 3 of Pawlik et al. [1993]), the energy positions relative to the laser line (at E_L) of two series of these ZP lines can be extracted. They are plotted in Figure 13 vs. excitation photon energy, the increase of the ZP shift with increasing E_L suggesting that these transitions are due to relaxation of different excited states and subsequent exciton annihilation in dots of two different sizes. As the ZP transitions have to be optically allowed, only $(n00)_h \otimes (100)_e$ excited states with $n = 1,2$ are to be expected in the range of excitation energies investigated. By calculating the energies of these states using the donorlike exciton model (for $n = 1$, compare curve III in Figure 3), the full lines are obtained, showing good agreement with the experimental data. As these calculations involve no free parameters at all, we take this as strong support that the donorlike exciton model provides a suitable description of the electron–hole pair states in AgBr providing in particular the dot radius as a function of exciton energy (see Figure 3).

5.2. Exciton Lifetime

Further information on the size dependence of exciton properties is provided by the radiative decay rates, which can be measured in AgBr quantum dots by the decay of resonance fluorescence. In a sequence of lifetime measurements E_L was varied and the emitted signal detected each at the 1TO(L) resonant Raman line. As the dominant excitation process here is absorption via the ZP and 1TO(L) phonon components followed by 1TO(L)-assisted and ZP emissions, respectively, crystallites of only two different sizes are excited and probed providing especially "clean" data.

Similar to the already discussed lifetime measurements in the FE luminescence (see Figures 11 and 12), besides a fast (3-ns) nonresonant background, we find in all decay curves the long-lived FE component having decay times of several hundreds of microseconds. Its amplitude scales with the TO(L) Raman line intensity which reflects the FE strength. The corresponding FE decay rates deduced from the analysis vs. E_L are displayed in Figure 14, together with a typical decay curve. According to the donorlike exciton model, the range of excitation energies displayed corresponds to a variation in particle

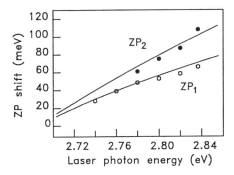

Figure 13 Energy positions of ZP lines ZP_1 and ZP_2 relative to the excitation photon energy E_L ("ZP shift") vs. E_L. Points: experimental data deduced from time-gated fluorescence spectra. Lines: calculation based on the donorlike exciton model (see text).

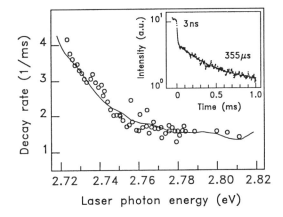

Figure 14 Decay rate of the TO(L) scattering intensity vs. excitation photon energy across the FE band. $T =$ 6 K. Circles: experimental data obtained by fitting the decay curves using two exponentials. Full lines: model calculation (see text). Inset: example of a decay curve showing the two time components.

radius between 3 and 6 nm (see Figure 3). While at higher energies (between 2.76 and 2.82 eV), i.e., smaller radii, the rates are almost constant, they increase by a factor of two at lower laser energies. This result seems to be inconsistent with the decay measurements of the FE luminescence, which rather show an increase in *lifetime* at lower energies (see Figure 11). Apparently, the FE decay after high-energy excitation involves additional relaxation processes leading to the observed long decay times far off the excitation photon energy. Therefore, these times do not represent the actual lifetime of the confined exciton states.

As to the change of the *exciton lifetime* in the nanocrystals relative to that in bulk AgBr, sources for both radiative and nonradiative decay have to be considered:

1. As already pointed out, an essential effect in quantum dots is mixing of L and Γ point valence band states. It results in the occurrence of a ZP transition and, depending on particle size, makes the optical transition more allowed. Furthermore, as discussed in Section 3.2, the indirect transition probability becomes size dependent according to Equation 20. Both these effects give rise to *radiative* lifetimes shorter than in the bulk.
2. The *nonradiative* processes are determined by two counteracting effects. One is that due to the exclusion of impurities from the volume of the crystallites [Johansson et al., 1991; Kanzaki and Tadakuma, 1991; Masumoto et al., 1992] nonradiative decay channels might largely be suppressed. On the other hand, due to the large increase of the surface-to-volume ratio, trapping and annihilation at surface states becomes much more probable. Actually such a behavior could be observed in the "large" microcrystals, where trapping at the surface becomes the dominant exciton relaxation channel (see Section 4.2.1).

The trend in the experimental data is that the decay rate of excitons becomes smaller in the smaller dots (Figure 14). From this, one has to conclude that the contribution of radiative decay is either negligible or almost constant in the energy range investigated. Therefore, the increase in decay rate with dot size is caused by nonradiative decay, implying that nonradiative decay channels due to surface trapping become less important the smaller the dots are. This surprising result can be explained straightforwardly in the donorlike exciton model by being aware of the fact that in silver halides hole trapping is the dominant nonradiative relaxation channel of excitons. The strong hole localization at the dot center, present according to this model, inhibits hole trapping at the surface in the smaller dots.

To quantitatively explain the measurements one has to take into account that for a given size distribution one actually excites not only the ground state, but also higher excited states (compare Figure 3). The measured decay rates therefore are a weighted average over the various states of which, due to the time resolution of the applied technique, only the slow component is registered (see inset to Figure 14). From the theory of Section 3.2 the total decay rate for each quantum dot state i may be written as a sum over the radiative TO(L) and ZP contributions plus the nonradiative part

$$\Gamma_{tot}^i = A_{TO} W_{rad}^{i,TO} + A_{ZP} \left| F^i\left(\vec{k}_L\right) \right|^2 + \Gamma_{nrad}^i . \tag{28}$$

A_{TO} is a parameter that takes into account processes not explicitly included in the theory, e.g., the LA(L) process (from absorption measurements one expects $A_{TO} = 1.16$ [Sliwczuk et al., 1984]), as well as an uncertainty in the value of the bulk radiative lifetime. A_{ZP} contains the unknown radiative lifetime of the direct transition in bulk AgBr and has to be considered as a scaling parameter. Taking the Gaussian size distribution determined from the analysis of RRS (see below), we calculated the weighted average of the rates over the hole states $(000)_h$, $(01m)_h$, and $(100)_h$, which is shown as the full line in Figure 14. For the nonradiative decay, besides a contribution representing surface trapping that varies proportional to $(E_L - E_{gx}^i)^{-2}$, in the calculation, we assumed constant contributions $\Gamma_{nrad} = 1 \cdot 10^3$ s^{-1} for the $(000)_h$ and $(01m)_h$ transitions and $\Gamma_{nrad} = 1 \cdot 10^5$ s^{-1} for the (100) transition in order to obtain the best fit to the data. Moreover, $A_{ZP} = 2 \cdot 10^{12}$ s^{-1} and $A_{TO} = 1.5$ was used, which is in very good agreement with expectation, since no other adjustable parameters are involved. The weak modulation present in the calculated curve is caused by the oscillatory behavior of the ZP rate (compare Figure 4), but the experimental data have still to be improved to experimentally verify this effect. From the fit, the ratio of ZP to TO(L) radiative decay rate can be estimated in quantum dots with $R = 5$ nm to be about 3:1 for the $(000)_h \rightarrow (100)_e$ transition, suggesting that the indirect–direct mixing represents an appreciable effect.

5.3. Raman Scattering Cross Section

While the lifetime measurements enabled us to explore the transition rates as a function of particle size, they do not give information on the size distribution which, however, is contained in the RRS cross section. Using the model of *absorption followed by emission* applicable in AgBr [von der Osten, 1991], the cross section for scattering involving a ZP- or phonon-assisted process $\alpha + \beta$ via an intermediate state j for a dot with radius R may be written as

$$S_{\alpha+\beta}^j(R) \frac{W_\alpha^j \Gamma_{rad,\beta}^j}{\Gamma_{tot}^j} \rho(R) . \tag{29}$$

Here α, β = ZP, TO(L), LA(L), W_α^j denotes the absorption probability which is proportional to the radiative rate, the Γ terms are decay rates in obvious notation, and $\rho(R)$ stands for the size-distribution function. The dot radius R is determined by the laser excitation photon energy E_L and the phonon frequency Ω_α via energy conservation

$$E_L = \hbar\Omega_\alpha + E_{ex}^j(R) . \tag{30}$$

The total cross section is given by the sum over all possible processes.

Comparing the integrated 1TO(L) and 2TO(L) Raman line intensities, we can deduce the contributions of the various intermediate states, because the two processes exhibit a different resonance behavior

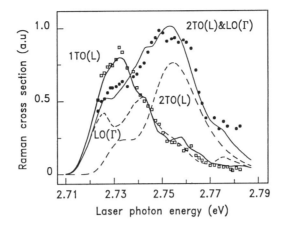

Figure 15 Cross section for TO(L) and combined 2TO(L) and LO(Γ) resonant Raman scattering at 8.3 and 17 meV, respectively. Full points and open squares represent the integrated intensities obtained from the analysis of the spectra in Figure 9. The full lines are the result of the model calculation discussed in the text. The dashed lines show the decomposition of the 17 meV line intensity into 2TO(L) and LO(Γ) components.

depending upon the type of state involved. In Figure 15, the measured integrated intensities of the 8.3- and 17.0-meV Raman structures are plotted vs. excitation photon energies as obtained from analyzing spectra like in Figure 9. The 8.3-meV line due to 1TO(L) scattering shows the maximum in cross section blue shifted relative to the peak of the FE luminescence (in this sample at 2.725 eV) by approximately one TO(L) phonon energy, as expected. The 17.0-meV line cross section exhibits a more complex behavior with the maximum around 2.755 eV, i.e., considerably higher than by one additional TO(L) phonon energy. From the discussion above, this line is known to be composed of a 2TO(L) and a 1LO(Γ) contribution of varying strengths.

To quantitatively calculate the various RRS cross sections the states $\{(000)_h, (100)_h, (01m)_h\} \otimes (100)_e$ have to be taken into account as these participate in the scattering. The calculations are based on Equation 29 using the energy-dependent radiative rates (or absorption probabilities) that followed from the fit of the measured lifetimes (Section 5.2). As size distribution we assumed a Gaussian, choosing the mean radius and standard deviation to be $\langle R \rangle = 5.3$ nm and $\sigma_R = 0.42$ nm that reproduced the experimental data best. The results are shown in Figure 15 as full lines, whereby we decomposed the 17-meV line into its 2TO(L) and LO(Γ) contributions (dashed lines). Very good overall agreement with experiment is achieved, and even finer details of the cross section are obviously reproduced within the accuracy of the data. Compared with the previous analysis of Raman and luminescence data by means of the individual-particle confinement model [Scholle et al., 1993], the resulting particle sizes are generally larger and agree much better with results of electron microscopy of similarly prepared AgBr nanocrystals [Johansson et al., 1992].

Along with the lifetime measurements in Figure 14, the results now allow one to understand in detail the measured RRS spectra:

1. At low energies around 2.72 eV, the scattering involves only the exciton ground state. This state couples strongly to light via a ZP transition, but the TO(L) intensity is reduced because of the short lifetime of the states. Also, the interaction with LO(Γ) phonons is strong.

2. Around 2.75 eV, the next higher state $(01m)_h \otimes (100)_e$ starts to become dominant. Strong interaction with TO(L) phonons leads to the maximum in the 2TO(L) cross section while the ZP transition, and thus the 1TO(L) line, is weak.

3. At energies higher than 2.77 eV, the $(100)_h \otimes (100)_e$ state takes over the scattering. This state is characterized by a rather strong coupling to the polar surface phonons. In the particular sample analyzed, the decay rate of this state may be large resulting in a very weak 1TO(L) line, consistent with the reduced intensity of the high-energy shoulder in the FE luminescence that we attribute to this state (compare Figure 8 and inset of Figure 9). In other samples, however, the intensities of both the 1TO(L) line and the high-energy luminescence band are found much stronger (compare, e.g., Figure 8, excited at $E_L = 2.84$ eV). We believe that this sample dependence is connected to the hole wave function of this state extending much more to the surface and, hence, being very sensitive to surface trapping.

6. CONCLUSIONS

The effect of quantum confinement upon the spectroscopic properties and the dynamic behavior of indirect excitons was investigated in small AgBr crystallites ranging in size from 450 nm down to about 5 nm.

In monodispersed *microcrystals* that have uniform cubic shape with 450-nm and 50-nm edge lengths, surface annihilation of excitons was found to be an important relaxation mechanism. Already in the 50-nm samples the near-edge exciton luminescence suggests the onset of quantum confinement of exciton translational motion.

Much smaller *nanocrystals* with radii comparable with the exciton Bohr radius but appreciable spread in size were produced in reversed micelles. In that case the exciton emission is inhomogeneously broadened and exhibits a substantial blue shift and enhanced intensity because of the spatial confinement. Besides luminescence, upon resonant size-selective excitation first- and second-order Raman scattering was observed, the participation of a ZP process clearly revealing mixing of L with Γ point states. The Raman cross section demonstrates that several quantum dot excited states are involved in the scattering. A line shape analysis of the spectra in terms of a linear-coupling model revealed a strong increase of coupling strength of LA(Γ) and surface optical phonons for the spatially more extended excited exciton states while the radius dependence of electron–phonon coupling in the investigated energy range is of minor importance.

From size-selective, time-resolved measurements exciton lifetimes on the order of 500 µs were found depending on the radius of the dot. Compared with the bulk radiative lifetime, they are smaller by a factor of about two, demonstrating that in these nanocrystals nonradiative processes are not as important as in the bulk.

The experimental results are consistently described in the donorlike exciton model in which the hole is localized at the dot center. Applying this model, exciton energies and radiative rates of direct and indirect transitions are deduced and the mixing of L and Γ point states is considered as a function of dot size.

ACKNOWLEDGMENTS

The authors thank Dr. A. P. Marchetti (Kodak, Rochester, NY) for supplying the microcrystalline samples and Prof. S. W. Koch (Marburg) for carefully reading the manuscript and for valuable suggestions. They are indebted to the Deutsche Forschungsgemeinschaft for financial support.

REFERENCES

Bányai, L. and Koch, S. W. 1993. *Semiconductor Quantum Dots,* World Scientific, Singapore.

Bassani, F. and Pastori Parravicini, G. 1975. *Electronic States and Optical Transitions in Solids,* Pergamon Press, Oxford.

Brus, L. 1991. Quantum crystallites and nonlinear optics. *Appl. Phys. A,* 53:465–474.

Czaja, W. and Baldereschi, A. 1979. The isoelectronic trap iodine in AgBr. *J. Phys. C. Solid State Phys.* 12:405–424.

Dimmock, J. O. 1967. Introduction to the theory of exciton states in semiconductors. In *Semiconductor and Semimetals, Volume 3 Optical Properties of III-V Compounds,* ed. R. K. Willardson and A. C. Beer, pp. 259–319, Academic Press, New York.

Efros, Al. L., Ekimov, A. I., Kozlowski, F., Petrova-Koch, V., Schmidbauer, H. and Shumilov, S. 1991. Resonance Raman spectroscopy of electron–hole pairs — polar phonon coupling in semiconductor quantum microcrystals. *Solid State Commun.* 78:853–856.

Ekimov, A. I., Efros, Al. L., Ivanov, M. G., Onushchenko, A. A. and Shumilov, S. 1989. Donor-like exciton in zero dimension semiconductor structures, *Solid State Commun.* 69:565–568.

Flügge, S. 1976. *Rechenmethoden der Quantentheorie,* 3rd ed., Springer-Verlag, Berlin.

Flytzanis, C. and Hutter, J. 1992. Nonlinear optics in quantum confined structures. In *Contemporary Nonlinear Optics,* eds. G. P. Agrawal and R. W. Boyd, pp. 297–365. Academic Press, London.

Giesecke, P., von der Osten, W. and Röder, U. 1972. Analysis of the R_2 absorption structures in LiF, KCl, and KBr. *Phys. Stat. Sol. (b)* 51:723–734.

Hellwege, A. M. 1962. *Landolt-Börnstein, Zahlenwerte und Funktionen aus Physik, Chemie, Astronomie, Geophysik und Technik, Band 8,* eds. J. Bartels, H. Borchers, H. Hausen, K.-H. Hellwege, Kl. Schäfer, and E. Schmidt, Vol. II(8), Springer, Berlin.

Hu, Y. Z., Lindberg, M. and Koch, S. W. 1990. Theory of optically excited intrinsic semiconductor quantum dots. *Phys. Rev. B* 42:1713–1723.

Itoh, T., Furumiya, M., Ikehara, T., Iwabuchi, Y., Kirihara, T. and Gourdon, C. 1991. Size-dependent optical properties of confined excitons in CuCl microcrystals. In *Proc. Taiwan–Jap. Workshop on Solid Opt. Spectr.,* World Scientific, Singapore.

Johansson, K. P., Marchetti, A. P. and McLendon, G. L. 1992. Effect of size restriction on the static and dynamic emission behavior of silver bromide. *J. Phys. Chem.* 96:2873–2879.

Johansson, K. P., McLendon, G. L. and Marchetti, A. P. 1991. The effect of size restriction on silver bromide. A dramatic enhancement of free exciton luminescence. *Chem. Phys. Lett.* 179:321–324.

Kanzaki, H. and Tadakuma, Y. 1991. Indirect-exciton confinement and impurity isolation in ultrafine particles of silver bromide. *Solid State Commun.* 80:33–36.

Kayanuma, Y. 1988. Quantum-size effects on interacting electrons and holes in semiconductor microcrystals with spherical shape. *Phys. Rev. B* 38:9797–9805.

Klein, M. C., Hache, F., Ricard, D. and Flytzanis, C. 1990. Size dependence of electron–phonon coupling in semiconductor nanspheres: the case of CdSe. *Phys. Rev. B* 42:11123–11132.

Lockwood, D. W. 1994. Optical properties of porous silicon. *Solid State Commun.* 92:101–112.

Maradudin, A. A. 1966. Theoretical and experimental aspects of the effects of point defects and disorder on the vibrations of crystals. *Solid State Phys.* 18:273–420.

Marchetti, A. P. 1993. Private communication.

Marchetti, A. P., Burberry, M. S. and Spoonhower, J. P. 1991. Characterization of an intermediate-case exciton in the 580 nm emission of Cd-doped and pure AgBr. *Phys. Rev. B* 43:2378–2383.

Marchetti, A. P., Johansson, K. P. and McLendon, G. L. 1993. AgBr photophysics from optical studies of quantum confined crystals. *Phys. Rev. B* 47:4268–4275.

Masumoto, Y., Kawamura, T., Ohzeki, T. and Urabe, S. 1992. Lifetime of indirect excitons in AgBr quantum dots. *Phys. Rev. B* 46:1827–1830.

Pawlik, S., Scholle, U., Weber, Th., Stolz, H. and von der Osten, W. 1993. Time resolved resonance raman scattering and luminescence from confined excitons in AgBr quantum dots. *J. Physique IV, Coll. C5, Suppl. J. Physique* II:151–154.

Pawlik, S., Stolz, H. and von der Osten, W. 1995. Luminescence and resonance Raman spectroscopy of indirect excitons in AgBr nanocrystals. *Mater. Res. Soc. Symp. Proc.* 358:289.

Scholle, U., Stolz, H. and von der Osten, W. 1993. Resonant Raman scattering and luminescence from size-quantized indirect exciton states in AgBr microcrystals. *Solid State Commun.* 86:657:661.

Sliwczuk, U., Stolz, H. and von der Osten, W. 1984. Indirect-forbidden exciton transitions in AgBr. *Phys. Stat. Sol. (b)* 122:203–209.

Stolz, H. 1994. *Time-Resolved Light Scattering from Excitons,* Springer Tracts in Modern Physics, 130, Springr, Berlin.

Takagahara, T. and Takeda, K. 1992. Theory of the quantum confinement effect on excitons in quantum dots of indirect-gap materials. *Phys. Rev. B* 46:15578–15581.

Timme, M., Schreiber, E., Stolz, H. and von der Osten, W. 1993. The effect of size restriction on exciton relaxation in AgBr microcrystals. *J. Luminesc.* 55:79–86.

Tolbert, S. H., Herhold, A. B., Johnson, Ch. S. and Alivisatos, A. P. 1994. Comparison of quantum confinement effects on the electronic absorption spectra of direct and indirect gap semiconductor nanocrystals. *Phys. Rev. Lett.* 73:3266–3269.

von der Osten, W. 1991. Light scattering in silver halides. In *Topics in Applied Physics, Light Scattering in Solids VI,* eds. M. Cardona and W. Güntherodt, pp. 361–422, Springer-Verlag, Berlin.

von der Osten, W. and Stolz, H. 1990. Localized exciton states in silver halides. *J. Phys. Chem. Solids* 51:765–791.

Chapter 3

Photoluminescence Studies on GaInAs/InP Quantum Wells

Ferdinand Scholz, Gerd Frankowsky, Andreas Hangleiter, Volker Härle, Peter Michler, Dietmar Ottenwälder, Klaus Streubel, and Chin-Yao Tsai

Reviewed by W. Seifert

TABLE OF CONTENTS

1. INTRODUCTION

For a wide field of modern applications in long- and short-distance communications, information storage, display, material processing, telemetry, etc., optoelectronic semiconductor devices are needed. These are made not from the elemental semiconductors like Si or Ge, which find their applications in conventional electronics and integrated circuits, but from **compound semiconductors**,* among which the III-V compounds (constituted from elements of the IIIrd and Vth column of the periodic system) have the greatest importance. These compound semiconductors have in most cases a direct band structure; i.e., the conduction band minimum and the valence band maximum are both situated in the center of the Brillouin zone. This property allows a direct interaction between the electrons and holes and the light without participation of further particles like, e.g., phonons.

Moreover, a wide variety of binary compounds (e.g., GaAs, InP) is available which additionally can be mixed to form ternary (e.g., GaInAs), quaternary (e.g., GaInAsP), or even more complex multinary

---

* See Section 9 for definitions of terms in bold type.

materials, thus allowing a coarse and fine tuning of most characteristic parameters as, e.g., band gap, lattice constant, refractive index, etc. Additionally, different semiconductor layers can be epitaxially combined, thus allowing a local variation of these properties. This opens a huge field for the design of particular device structures with very specific characteristics (see, e.g., Baets [1987]).

These compound semiconductor materials and device structures require specific preparation and characterization methods. Only modern epitaxial methods like **metalorganic vapor phase epitaxy** (MOVPE) or **molecular beam epitaxy** (MBE) enable the realization of such sophisticated **heterostructures** as are in use in modern **optoelectronic devices**. One key point is the implementation of extremely thin layers (in the range of a few nanometers) embedded in semiconductor material of wider band gap, so-called **quantum wells**. Many material properties now change from a three-dimensional to a two-dimensional mode, thus influencing essentially most device functional characteristics.

The purpose of this chapter is to shed light on the characteristics of such thin layers and the respective interfaces whose importance obviously increases with decreasing layer thickness. Besides microscopic methods or X-ray analysis, optical spectroscopy is a very powerful tool for acquiring detailed information about these quantum wells and interfaces. We demonstrate that, by these means, we get a deep insight into the microscopic structure of these interfaces thus enabling their optimization, on the one hand, and we can deduce new properties resulting from such materials artificially structured on a nanometer scale by the epitaxial growth process, on the other hand.

As an example, we will focus on one specific material family, GaInAs/InP heterostructures, which can be tuned to band gaps around 0.8 eV (corresponding to light wavelengths of about 1.5 μm), finding their applications in devices for modern far-distance optical telecommunications using glass fibers as "optical wires."

2. QUANTIZATION IN THIN FILMS

2.1. One-Dimensional Potential Wells

Let us consider a thin layer of semiconductor material F (F stands for "film") with low band gap (e.g., GaInAs) and thickness L_z embedded in a semiconductor material B (B for "barrier") with higher band gap (e.g., InP) as depicted in Figure 1. The free carriers, e.g., the electrons in the GaInAs conduction band (on which we will focus our considerations, although most statements hold for free holes in the valence band as well), which formerly could move in all three space directions, now are restricted to move only in the film plane because of the potential barriers formed by the high–band gap material. This localization in z-direction (perpendicular to the film plane) results in a quantization of the electronic states which can easily be calculated following well-established textbook methods for rectangular potential wells (see, e.g., Messiah [1965] and Schiff [1968]). The "ansatz" for solving the Schrödinger equation for a particle in the conduction band potential well consists of a free-running electron wave function Ψ_F in the central film, whereas exponentially damped wave functions Ψ_B are taken in the outer parts for particle energies lower than the InP band gap. These wave functions and the particle current densities (related to their first derivatives) have to be joined continuously at the interfaces while taking into consideration the different effective masses m_F and m_B in the film and the barrier material, respectively.

The resulting simple system of differential equations can be solved and leads to two transcendent equations (one symmetric and one antisymmetric) for the energy eigenvalues:

$$\omega \tan(\omega L_z / 2) = m_F / m_B \cdot \lambda$$

$$\omega \cot(\omega L_z / 2) = m_F / m_B \cdot \lambda \tag{1}$$

where

$$\omega^2 = 2 m_F E / \hbar^2 \tag{2}$$

and

$$\lambda^2 = 2 m_B (V - E) / \hbar^2 \tag{3}$$

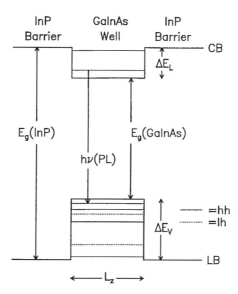

Figure 1 Conduction and valence band alignment of a quantum well structure (schematically). Electron and hole subbands are also indicated.

with E = free energy of particle. In our example (Figure 1) the potential V is just the conduction band offset between the barrier and the well. Taking the well-known material properties (conduction band offset, three-dimensional electron masses m_B and m_F of InP and GaInAs, respectively [Landolt-Börnstein, 1982; Skolnick et al., 1986; Gershoni and Temkin, 1989] into account, the eigenvalues for the electrons E_n^e can be calculated. We notice that the former GaInAs conduction band has changed into several subbands whose lowest energies increase for decreasing film width L_z. The number (n) of subbands decreases with decreasing width, but at least one remains even for the thinnest quantum wells.

2.2. Optical Transitions in Quantum Wells

We can find similar relations for the holes in the valence band. Owing to their different effective masses, the degeneracy of the heavy and light hole states at the center of the Brillouin zone in the three-dimensional bulk crystal now is canceled, and the holes form two different ladders of eigenstates (as suggested in Figure 1).

These features are directly observable in optical emission experiments where electrons are excited from the valence subbands (thus generating a hole) to the conduction subbands and recombine with the hole by emission of a photon following the selection rule

$$\Delta n = 0 \qquad (4)$$

Thus, the allowed transition energies are built by the quantization energies E_n^i of the electrons ($i = e$), heavy ($i = hh$) and light ($i = lh$) holes, and the GaInAs band gap E_g (see Figure 2):

$$E_n^h = E_g + E_n^e + E_n^{hh} \qquad (5)$$

for electron–heavy hole transitions and

$$E_n^l = E_g + E_n^e + E_n^{lh} \qquad (6)$$

for electron–light hole transitions.

Under normal low-excitation conditions, the emitted photon energy directly represents the energy difference between the lowest conduction band minimum and the highest valence band maximum (lowest solid line in Figure 2), because higher excited carriers relax very quickly into these extrema. Thus, the

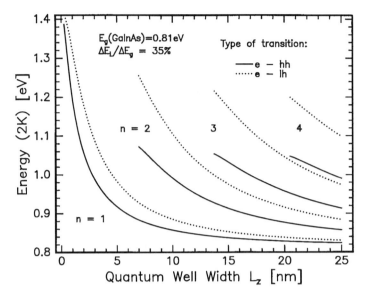

Figure 2 Calculated transition energies for the allowed ($\Delta n = 0$) *e–hh* and *e–lh* transitions in a GaInAs/InP quantum well at low temperature ($T = 2$ K).

effective band gap or the emission wavelength of a device containing such a thin film can be tuned just by the geometric width L_z of this quantum well. More generally, transitions between the electron and hole subbands are allowed if the above-mentioned selection rule is obeyed. These transition energies are sketched in Figure 2. They may be directly visible in absorption spectroscopy experiments.

Strictly speaking, the measured energetic position of the emitted photons resulting from these transitions has to be slightly corrected, because the electrons and holes form **excitons** before their recombination, which have a binding energy of about 3 to 9 meV, reducing the observed photon energy [Grundmann and Bimberg, 1988].

2.3. Photoluminescence on Quantum Wells: Basic Considerations

Obviously, there exists a direct relation between the quantization-induced band gap variations and the optical transitions. Therefore, optical spectroscopy is a key experiment to assess these features.

The simplest experiment may be **photoluminescence** spectroscopy; that is, the sample, which contains one or several quantum wells, is excited by light of rather high energy, in most cases fairly above the band gap of the barrier material. Thus, electron–hole pairs are generated which relax to the energetically most favorable position, i.e., the minimum of the lowest conduction band and the maximum of the uppermost valence band in the quantum well. Subsequently, they recombine under emission of a photon with energy equal to the "effective" band gap (see Figure 1). In the ideal case of a rectangular potential well with well-known composition, we can easily deduce the quantum well thickness L_z from the detected emission line. But in practice, the composition of such a thin layer may not be exactly known and the shape of the quantum well may not be perfectly rectangular. All this influences the energetic position of the emission line. Moreover, the quantum well properties (width, composition, interface abruptness) may vary laterally. Such effects will result in a broadening of the emission line.* Therefore, photoluminescence spectroscopy can be used to obtain detailed information about these material and interface properties.

A typical test structure contains several quantum wells of various thicknesses (Figure 3a) which give rise to the respective emission lines at different spectral positions simultaneously (Figure 3b). The experimental setup which was used to measure the spectra described here is schematically sketched in Figure 4. The semiconductor samples are excited by an Ar ion laser (wavelength: 514 nm). The laser light is filtered for spectral and intensity reasons. Then, it is focused on the sample surface. The emitted

---

* It is interesting to note that imperfections in the interface abruptness do not necessarily increase the emission line width, provided that they are laterally homogeneous. Even triangular or parabolic potential wells have ideally sharp eigenvalues.

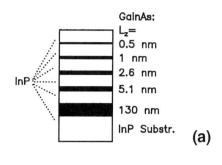

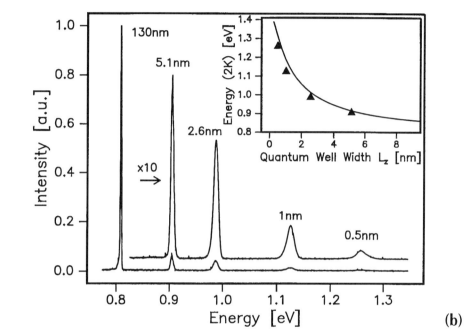

Figure 3 (a) Layer sequence of the GaInAs/InP multi-single-quantum-well (MSQW) structure used in these studies. (b) Photoluminescence spectrum of such an MSQW structure. The inset shows the comparison of the peak energies to the theoretical curve (compare Figure 2).

light is focused onto the entrance slit of a monochromator, where it is spectrally resolved and detected at the exit slit using a cooled Ge detector and conventional lock-in technique.

Besides, luminescence may be excited by other means. An elegant way to get locally resolved optical spectra is **cathodoluminescence**, where the electron–hole pairs are excited by the electron beam of a scanning electron microscope. Some examples for this method will also be presented in this chapter.

3. QUANTUM WELL PREPARATION

3.1. Metalorganic Vapor Phase Epitaxy

As mentioned above, such quantum well structures are fabricated today mainly by two epitaxial methods:

1. MBE, where the different elements are evaporated into an ultra-high vacuum (UHV) chamber, thus forming a molecular beam onto a given substrate, and
2. MOVPE, where the different elements are transported as gases to a tubelike reactor where they contribute to the growth process.

We would like to focus here more on the latter method, because it seems somewhat more flexible, in particular if heterostructures are to be grown that contain both arsenic and phosphorus. As the group III and group V elements are not gases per se, suitable precursors are used which contain the elements in larger molecules. Hydrides like AsH_3 and PH_3 are the simplest gaseous precursors for the group V

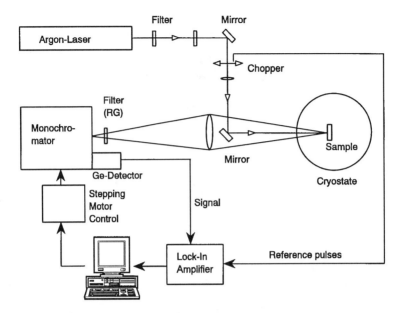

Figure 4 Setup for photoluminescence experiments (schematically).

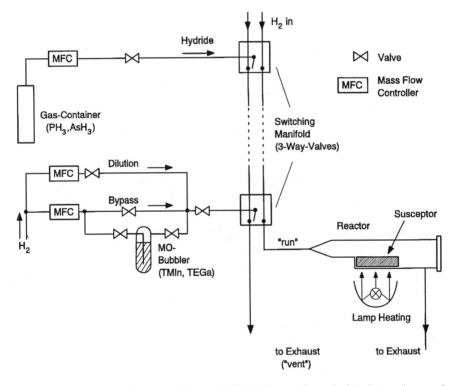

Figure 5 Setup of an MOVPE apparatus (schematically). Only one channel of each type (gas and metalor-ganics) is sketched. Present day systems typically have about four channels of either type.

elements, whereas metalorganic compounds like trimethyl-gallium [$(CH_3)_3Ga$, TMG] or trimethyl-indium (TMIn) have to be used for the group III elements, thus giving their name to this technique. These compounds are liquid or solid with rather large vapor pressure (typically some hectopascals) enabling their transport to the reactor via stainless steel tubes by an inert carrier gas (mostly used: H_2). A typical MOVPE system is sketched schematically in Figure 5. Within the reactor, typically a quartz tube with rectangular cross section to ensure lateral flow homogeneity, the gases are transported from the main gas stream to the substrate by gas phase diffusion [Ludowise, 1985, van de Ven et al., 1986]. Near the substrate, which is heated to elevated temperatures (around 600 to 700°C), and on its surface, the elements needed for the epitaxial growth are released from the molecules by pyrolytic reactions. Subsequently, they may diffuse on the surface to find their appropriate place in the growing layer. An incorporation at a surface step is preferred compared with a nucleation on a step-free surface. This results in a so-called two-dimensional growth mode (or a monolayer-by-monolayer growth) instead of a three-dimensional growth mode, an important prerequisite for the growth of large areas of homogeneous thin layers. In well-optimized MOVPE reactors, thickness homogeneities of better than 1% over seven 2-in. wafers are easily achievable [Frijlink et al., 1991].

Under normal growth conditions, the gas phase diffusion of the precursors is the limiting step in the growth process, and thus it is called *mass transport limited*. Therefore, simple linear relations hold among amount of supplied precursor gases, time, layer thickness, and composition, and the advantages of this vapor phase epitaxial technique are quite evident. The amount of every single element can easily and independently be controlled by the respective gas flow, thus enabling a simple control of the growth rate, thickness, composition, layer sequence, and even doping (the doping elements can be fed to the growing film in the same way as the main material elements). Under typical MOVPE conditions, growth rates of 1 to 3 µm/h can be established; i.e., the growth of quantum wells with a thickness of few nanometers needs some seconds and thus can be well controlled.

Today, MOVPE is applied to grow a wide variety of different materials, including most types of compound semiconductors, high-temperature superconductors, ferroelectric films, and much more. Many different elements of the periodic system can be deposited and combined to a wealth of compound materials, provided the key question is solved — the availability of suitable metalorganic compounds for the respective element, i.e., compounds which have the appropriate chemical stability and vapor pressure, which can be fairly easily synthesized and purified to a satisfying level.

The most important field for MOVPE is the growth of III-V-compound semiconductors. The excellent quality of single "bulk" layers, complex heterostructures, and sophisticated devices, not only for research, but also for mass production, is continuously proving the high standard which has been obtained in this epitaxial technique.

3.2. Interface Preparation

Of course, we are faced with some problems regarding the growth of thin films. Quantum wells just contain very few atomic monolayers (with a thickness of typically 0.3 nm), and thus any imperfections to the vertical abruptness or the lateral homogeneity will influence strongly the quantum well properties as mentioned earlier. This requires first an optimization of the "hardware." In order to keep the gas concentration transients as abrupt as possible when switching from one precursor to another, a so-called vent-run valve configuration is used at the inlet to the reactor (Figure 5). This enables the different gas flows to be stabilized while flushing them to the exhaust ("vent") before they are needed for the growth process. Then, they are abruptly switched to the reactor ("run") by a three-way valve without changing the gas velocity. A further important point is the optimization of the reactor tube geometry to ensure a laminar gas flow. Moreover, the gas velocity and, thus, the gas composition transients can be made faster by reducing the reactor pressure down to 10 to 100 hPa (low-pressure MOVPE).

Furthermore, the "software" has to be optimized, i.e., the switching sequence of the gases at the heterointerface. To this end, growth interruptions are introduced to ensure the optimal termination of the first layer, before the growth of the subsequent layer is initiated. For most III-V compounds, including InP and GaInAs, the growth rate is solely controlled by the transport of the group III elements to the growing surface, whereas the group V precursors have to be supplied to the reactor in excess (i.e., the so-called V/III ratio is much larger than unity, typically about 100 for InP and about 35 for GaInAs). On the other hand, the group V elements tend to desorb from the epitaxial surface at elevated temperatures, thus requiring a stabilization of the grown surface by an adequate hydride flow during the growth interruptions. Therefore, we have divided these interruptions into three parts (Figure 6). First, the growth of the underlying layer is stopped by switching the group III precursors(s) to vent (t_1, t_4). Then, all group V hydrides may be switched off for a short time (t_2, t_5). The growth of the next layer may be prepared by switching on the adequate hydride (t_3, t_6) before the growth is continued by opening the new group III precursor valves.

Many studies have been done to optimize the switching sequence at the heterointerface (see, e.g., Wang et al. [1989]; Grützmacher et al. [1990]; Streubel et al. [1992]; Samuelson and Seifert [1994]). In the next section, we would like to demonstrate how optical spectroscopy can be applied to improve

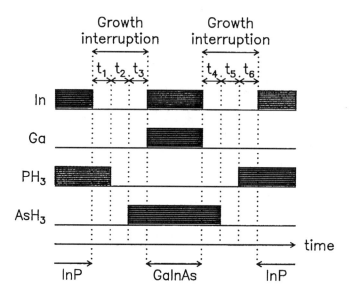

Figure 6 Time diagram of the growth interruption switching procedure

the growth process for the generation of perfect interfaces and quantum wells. It goes without saying that further characterization methods as, e.g., X-ray diffraction studies or microscopic studies (e.g., transmission electron microscopy, TEM), are necessary for a full and correct interpretation of the results obtained by optical spectroscopy. But these methods will only be mentioned briefly here.

3.3. Experimental Details

For these studies, we have grown GaInAs/InP quantum well structures on InP substrates oriented exactly to the ⟨001⟩ direction. We used a horizontal reaction chamber with rectangular cross section in order to increase the lateral homogeneity of the grown films. The substrates were placed on a molybdenum susceptor which was heated by halogen lamps to a typical growth temperature of 620°C. We used the conventional compounds TMIn, triethyl-gallium (TEGa), AsH$_3$, and PH$_3$ as precursors and palladium-diffused H$_2$ as carrier gas at a reactor pressure of 80 hPa. The quality of the grown materials was checked by standard methods. We obtained excellent low-temperature photoluminescence (LT-PL) spectra for both InP and GaInAs. The composition homogeneity of the GaInAs films was proved by narrow lines (full width at half maximum, FWHM < 1.5 meV) of the excitonic transition in LT-PL at 810 meV and by sharp diffraction peaks (FWHM < 25 arcsec) in high-resolution X-ray diffraction (HRXRD). Hall measurements (van der Pauw method) showed low background doping concentrations of $n \leq 10^{15}$ cm^{-3} and high electron mobilities at 77 K of $\mu \geq 80.000$ cm^2/Vs for both GaInAs and InP layers.

Besides the above-mentioned growth interruptions, the basic growth parameters for quantum wells were identical to those of thick bulk layers. Owing to the mass transport–limited growth mode, the growth rate is well established immediately after the respective switching point and thus is considered to be constant over time [Wang et al., 1990].

4. OPTIMIZATION OF GROWTH INTERRUPTIONS

4.1. Lower Interface

Of course, the switching times t$_1$ to t$_6$ have to be carefully optimized for any respective heterointerface. For this, optical spectroscopy on quantum well structures turns out to be an excellent tool. As explained in Section 2.3, the energetic position of the photoluminescence peak of a quantum well not only reflects its thickness L_z (compare Figure 2), but is also influenced by the interface abruptness. Thus, we have optimized these switching times by growing multiple single quantum well structures (MSQWs) (see Figure 3a) and evaluating them by low-temperature photoluminescence.

The growth of the lower InP barrier is stopped by switching the TMIn precursor to vent. In order to enable a perfect exchange of the group III gases, we intended to have a rather long growth interruption. This may help to minimize any memory effects which could result from TMIn deposits on the reactor

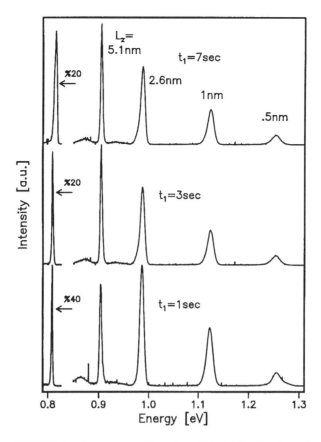

Figure 7 PL spectra of MSQW structures grown with various PH$_3$ stabilization times t_1.

walls or imperfect gas transients. But this may require a stabilization of the InP surface by still flushing PH$_3$ to the reactor (t_1). In fact, we observed that t_1 is less critical. For times between 0 and 7 sec (Figure 7) or even much longer (several minutes), we did not find significant differences in our PL spectra or the crystal morphology, which demonstrates that, on the one hand, group III memory effects seem not to play a significant role (compare also results from, e.g., Clawson et al. [1993] and Meyer et al. [1992]), and, on the other hand, the InP surface still is not affected by such long growth interruptions when stabilized by PH$_3$.

But this material combination obviously needs as well a perfect switching of the group V hydrides, thus requiring a total interruption where no precursor gas is fed into the reactor (t_2). Now, a compromise must be found between long interruption times t_2 to suppress any precursor carryover or memory effects and short times t_2 to limit the risk of a thermal degradation of the InP surface. In fact, the InP surface only remains perfect for times shorter than about 30 s. But again, much shorter interruption times are satisfactory to suppress any unwanted memory effects owing to our general growth conditions. Besides the optimized reactor geometry, the use of three-way valves with suppressed dead volume, and the reduced growth pressure, any memory effects stemming from deposits on the susceptor are effectively reduced by the use of a molybdenum susceptor having a much less porous surface than any conventional graphite susceptor. Best results have been obtained for a total interruption of about 5 s (t_2) without any stabilization of the InP surface ($t_1 = 0$).

It seems reasonable to prepare the growth of the subsequent GaInAs layer by switching first the AsH$_3$ to the reactor (t_3) before resuming the growth by opening the In and Ga valves. But, unfortunately, this drastically shifts the quantum well photoluminescence peaks to lower energies (Figure 8). This has been studied by many groups (e.g., Streubel et al. [1992]; Böhrer et al. [1992]; Seifert et al. [1994]) and can be explained by an exchange of the P in the InP surface by As. This process can be reduced by switching on the Ga and In gas flows simultaneously with the AsH$_3$ flow, but never can be avoided completely, as has been observed by many researchers, not only from photoluminescence results, but also by, e.g.,

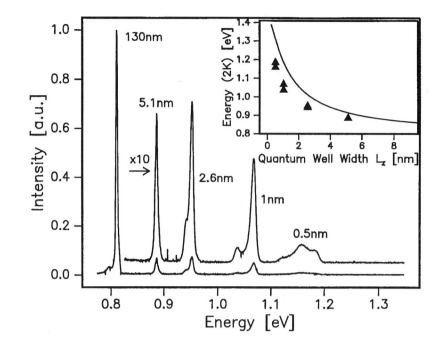

Figure 8 PL spectra of an MSQW structure grown with an AsH₃ purging period t_3 = 3 s.

X-ray diffraction (e.g., Kamei and Hayashi [1991]) or Raman spectroscopy (e.g., Mozume [1995]. Unfortunately, this As/P exchange takes place when the first As atoms come to the InP surface even if the GaInAs growth starts at the same moment. Detailed studies of Seifert et al. [1994] have shown that even for t_3 = 0 about 60% of P may be exchanged by As at conventional growth temperatures around 600°C because of the lower bond strength of P compared with As in the InP surface. That is why even for optimized growth conditions, the PL peaks of the narrowest quantum wells fall below the theoretical line (compare Figure 3b). Alternatively, this effect is limited to the uppermost P layer for conventional growth conditions [Kamei and Hayashi, 1991; Seifert et al., 1994], and it has only minor influence on the properties of wider quantum wells (L_z ~ 10 nm) which are conventionally used in modern optoelectronic devices. Therefore, this problem seems not to limit most optical and device characteristics (besides the quantization energy) and thus can be easily taken into account in the heterostructure design.

4.2. Upper Interface
The situation is somewhat different at the upper interface when the gases have to be switched from GaInAs to InP growth. Owing to the larger As bond strength, the GaInAs surface is much more resistant against thermal treatment. Thus, the stabilizing interruption t_4 can be avoided. In fact, it must be avoided, because As shows a very pronounced memory effect [Seifert et al., 1994; Jiang et al., 1995]. It may be deposited on the substrate, the susceptor, and the reactor walls during t_4 and then is easily incorporated into the subsequent InP in rather significant amounts. Even when we chose t_4 = 0, this may be observed, when the GaInAs layer is grown with rather high V/III ratio, depending strongly on further parameters, such as susceptor porosity, susceptor size, reactor geometry, etc. In current commonly used larger MOVPE systems, this problem is in general stronger than the P/As exchange at the lower interface. Therefore, a larger transition region may be observed at the upper interface [Carey et al., 1987; Samuelson and Seifert, 1994]. But being no "intrinsic" effect, it may be avoided by applying the appropriate "hard- and software" [Wang et al., 1989; Streubel et al., 1992], which may not be optimized for other requirements (e.g., large substrate size).

The preparation of the InP growth by a preflow of PH₃ results, similar as at the lower interface, in an exchange of P and As on the GaInAs surface [Streubel et al., 1992]. But, here, the effect is much less pronounced and can easily be suppressed by avoiding this preflow (t_6 = 0).

So we found, again, that a short total interruption of all carrier gases results in the best interface properties; i.e., we chose $t_2 = t_5$ = 5 s and $t_1 = t_3 = t_4 = t_6$ = 0 s.

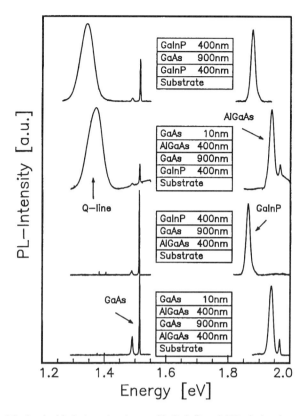

Figure 9 PL spectra of GaAs double heterostructures with GaInP and AlGaAs barriers, respectively. Only those structures show the Q-line which contains a GaAs on GaInP interface (see insets). (From Tsai, C. Y. et al. With permission.)

4.3. GaInP/GaAs Interfaces

The growth of GaInP/GaAs interfaces seems to be very similar to the InP/GaInAs case: at the hetero-interface, as well group III as group V elements change, the well material contains As and the barrier contains P. $Ga_xIn_{1-x}P$ can be grown lattice matched to GaAs for $x \sim 0.5$ and then has a rather large band gap of about 1.9 eV. Thus, this material system is attractive for optical devices (light-emitting diodes, lasers) which emit in the visible (red) part of the spectrum.

So we have intended to grow quantum wells with GaInP as barrier material and GaAs (having a band gap of 1.42 eV at 300 K) as well material at similar growth conditions as mentioned above with somewhat higher growth temperatures around 700°C, which are better suited for GaAs and GaInP single layers. But the photoluminescence results demonstrated that these experiments totally failed. Obviously, the abruptness of at least one interface was totally destroyed. This was even detectable on double hetero-structures with a rather thick GaAs layer of several 100 nm [Guimaraes et al., 1992], where a broad emission line below the GaAs band gap (further on called "Q-line") was detectable in LT-PL (upper trace in Figure 9) [Tsai et al., 1995].

Another material with larger band gap is $Al_xGa_{1-x}As$. This material can be easily grown on GaAs because its lattice mismatch to GaAs is very small for all compositions x. Moreover, only Al has to be switched on and off to produce a heterointerface, and these interfaces can therefore easily be grown with excellent quality. This opened the possibility of studying both interfaces independently [Tsai et al., 1995] by comparing structures where either the lower or the upper GaInP was replaced by AlGaAs with approximately the same band gap (see Figure 9). And, in fact, the Q-line only appears for those double heterostructures which contain the lower interface (transition from GaInP to GaAs, Figure 9). Obviously, we are again faced with the problem of P/As exchange [Bhat et al., 1992]. But this parasitic effect seems to be much stronger in this material system than in the InP/GaInAs system described above. The higher growth temperature can only partly explain this drastic effect: the Q-line remains visible for growth

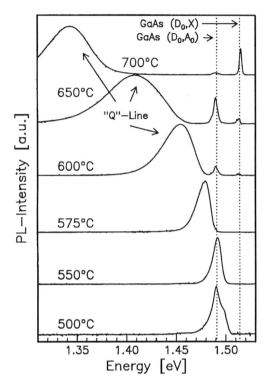

Figure 10 PL spectra of GaAs/GaInP structures grown at different temperatures. Note the development of the Q-line only for temperatures above ~575°C

temperatures down to about 600°C. Only for even lower temperatures (Figure 10) it vanishes. Therefore, some groups have optimized their growth conditions for such heterostructures to temperatures below 550°C. Consequently, they achieved excellent results when growing GaAs/GaInP quantum well structures (see, e.g., Omnes and Razeghi [1991]. But it seems as if optimum growth conditions have not been found so far [Samuelson and Seifert, 1994]. We found that the Q-line can also be suppressed by lowering the temperature just after the growth of the GaInP layer so that the first part of the GaAs layer is grown at 550°C or lower, whereas the remaining GaAs and the covering GaInP layer are again grown at high temperatures. This proves that the deterioration of the interface abruptness takes place during growth and is not a result of atomic intermixing by solid phase diffusion afterward.

This P/As exchange can also be found at InP/AlInAs interfaces [Vignaud et al., 1994]. It seems to be a common feature of interfaces, where the group V elements are changed from P to As, and reflects the strong differences in chemical bond strength between these elements.

5. MONOLAYER SPLITTING

Under optimized growth conditions, quantum well structures can be grown whose PL spectra match fairly well the expected peak positions (see Figure 3), except for the thinnest quantum wells, where the unavoidable P/As exchange at the lower interface significantly contributes to a downshift of the PL peaks.

But the spectra of some MSQW samples even show more peaks than expected from the number of quantum wells; in particular, the thinner well peaks may split in two, sometimes in three lines (Figure 11). A closer evaluation shows that the difference in energetic position of these split lines can be attributed to thickness differences of about one atomic monolayer (~0.29 nm) [Streubel et al., 1992]. Similar line splitting was observed for GaAs/AlGaAs quantum wells and interpreted as **monolayer splitting** [Tanaka and Sakai, 1987; Petroff et al., 1987; Bimberg et al., 1987]: the two-dimensional growth of the quantum wells is so perfect that only integer numbers of monolayers are formed. If the growth is interrupted so that nominally an atomic layer is not completed, then the atoms rearrange in a manner that they form islands where the thickness is exactly n monolayers, surrounded by other regions with thicknesses of exactly n + 1 monolayers. Consequently, the same interpretation was attributed to the line splitting in

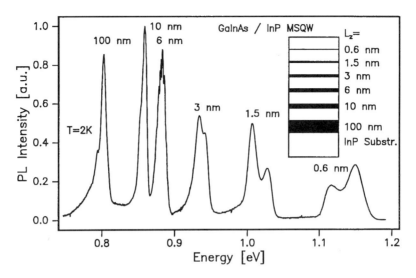

Figure 11 Low-temperature spectrum of an MSQW sample showing monolayer splitting for the 3, 1.5, and 0.6 nm quantum wells. This sample was grown with a rather large V/III ratio of 100 and an AsH₃ purge of 1 s (t_3).

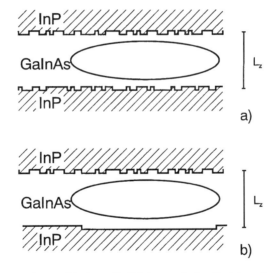

Figure 12 An exciton in a quantum well (schematically) demonstrating the interaction with interfaces of different smoothness.

GaInAs/InP quantum wells [Nilsson et al., 1991; Streubel et al., 1992]. Having in mind that these emission lines stem from excitonic recombinations, we can estimate that these monolayer flat islands have dimensions at least somewhat larger than the excitonic radius (about 10 nm, see Figure 12). If the islands are smaller, only a single broadened peak averaging over the excitonic radius is expected. In fact, these islands may even extend over some micrometers, as could be observed by locally resolved cathodoluminescence [Nilsson et al., 1990].

Obviously, these monolayer islands are formed during the epitaxial growth by an enhanced surface diffusion of at least the group III atoms. They build up only on substrates oriented fairly well to (001), whereas no islands can be found on substrates slightly misoriented by about 2° [Thijs et al., 1988]. Moreover, the growth conditions and, in particular, interruptions play a major role in the formation process; they build up more clearly when the atoms find more time for diffusion in a growth interruption [Grützmacher et al., 1990] or on a slowly growing surface. In most cases, only doublet lines can be observed, indicating that the monolayer terraces are only formed at one of the two interfaces. Although we have found some indications that the upper interface (change from GaInAs to InP) is responsible for their development [Streubel et al., 1992], most results published so far (e.g., Carey et al. [1987]; Samuelson and Seifert [1994]) point at the lower interface being more abrupt than the upper. Additionally, we found that the probability for monolayer peaks increases if we flush some AsH_3 over the underlying InP (i.e., $t_3 = 1$ to 5 s) or if at least the V/III ratio, i.e., the AsH_3 flux during the subsequent GaInAs layer growth, is rather high. But this results in a stronger P/As exchange as described above. It may be hypothetically assumed that the group III atoms (which define the growth rate) form atomically sharp interfaces, whereas the group V elements smoothen the vertical composition abruptness at the very same interface. High-resolution TEM studies, including a detailed analysis by the vector pattern recognition method [Ourmazd, 1989], support this hypothesis; on neither interface, could an atomically abrupt transition be detected, even on samples which show highly resolved monolayer peaks, but similar transition regions of about 0.3 nm (about one monolayer) have been evaluated [Streubel et al., 1992]. It was not possible to measure which sublattice (group III or group V) is responsible for these slightly smooth transitions. The upper interface may not contribute to the monolayer splitting because of the more or less pronounced As carryover effect.

6. MONOLAYER TERRACES IN SELECTIVE AREA EPITAXY

Further insight into the monolayer terrace formation process may be gained by a growth mode, where the growth rate varies intentionally on the wafer. One example for such a growth mode is **selective area epitaxy**, which allows epitaxial growth on locally well-defined areas. It is being currently investigated in many groups as a method for the monolithic integration of different III-V semiconductor devices on a single chip.

To this end, the substrate is covered by a dielectric film like SiO_2 or Si_xN_y. By conventional lithography, windows are opened in this film. In the subsequent epitaxy process, growth only occurs in the openings, where the semiconductor surface is in contact with the precursor gases, whereas no growth takes place on the masked parts of the wafer. This selectivity is strongly effective in MOVPE, reflecting that the semiconductor surface plays a major role in the chemical pyrolysis of the precursors. But because of this severe change of surface properties, the growth mode and, in particular, the growth rate change drastically in the open areas. This is caused by the precursor gases arriving on the masked areas where they cannot be used for growth. A large lateral concentration gradient in the gas phase builds up between the unmasked and the masked areas of the wafer, giving rise to a strong precursor gas diffusion from the latter to the former regions. Therefore, the growth rate increases in the unmasked areas and changes locally. The largest growth rates can be found at the opening edges, whereas the layers grow thinner in the center (Figure 13). Obviously, the growth rate now depends on the local position and the geometric arrangement of the masks.

When a ternary or multinary semiconductor layer is grown, several different precursor gases are subjected to this diffusion process. Owing to their different chemical and physical properties, they contribute differently to these growth rate modifications, which results in a local change of the composition [Thrush et al., 1992, Scholz et al., 1995]. When growing, e.g., $Ga_xIn_{1-x}As$, an In enrichment of the epitaxial film can be observed from the opening center to the edge (Figure 13). These features may be applied for the design of novel devices, because they allow the artificial control of the local band gap and thickness of a growing layer by the mask design. A well-known example is the integration of a semiconductor laser working on a given wavelength and a modulator whose band gap has to be somewhat larger [Aoki et al., 1991].

Besides the mask geometry, the growth temperature and the reactor pressure have a strong influence on the local layer properties. This motivated us to do systematic investigations on these topics. GaInAs/InP quantum well structures are well-suited test structures because thickness and composition modulations can be evaluated quite easily on the same layers. Besides scanning electron microscopy and surface profilometry for local thickness determination, cathodoluminescence turned out to be an excellent tool for these studies, allowing a locally well-defined excitation of electron–hole pairs whose luminescence reflects their local composition and their quantum well width simultaneously.

But besides the expected decrease of the luminescence peak energy when measuring over a selectively grown quantum well stripe due to increased In incorporation and growth rate at the opening edges, we observed a steplike change of the luminescence lines for the thinner wells (Figure 14). Taking into account the local composition variations as measured on bulky reference layers, the steps were shown

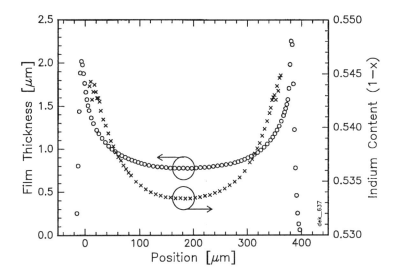

Figure 13 Growth rate enhancement (open circles) and In enrichment (crosses) in selective area epitaxially grown Ga$_x$In$_{1-x}$As. The former is measured by surface profilometry; the latter is evaluated from the band gap shift measured by cathodoluminescence.

to represent exactly single atomic layers. It turned out that much larger monolayer terraces extending over several 10 μm were formed than in planar growth, probably driven by the larger growth rate inhomogeneity. This opened an alternative chance to study the above-mentioned interface problems.

The influence of AsH$_3$ supplied to the lower interface was even more evident (Figure 15). Although the large red shift of all lines indicates clearly the As/P exchange for the structure where the GaInAs quantum well was grown with an AsH$_3$ purge at the lower interface ($t_3 = 5$ s), this sample shows well-developed monolayer terraces, whereas the sample grown with a comparatively low V/III ratio shows higher peak energies, indicating more abrupt interfaces, but no terraces. When flushing the upper GaInAs surface ($t_4 = 5$ s), no terraces could be found, and the cathodoluminescence signals are only weakly red-shifted owing to a slight As memory effect.

Here, again, only single, but no double, atomic layer steps could be detected, proving that these terraces are only formed at one interface. Moreover, the peak position on a single terrace is not constant, but changes slightly. This cannot be attributed to a local composition variation which was measured to be much less pronounced. So it may be explained by a smooth thickness variation taking place at the second (upper) interface, and the strongly different character of the lower and upper interface becomes again evident.

7. STRAINED QUANTUM WELLS

Up to now, we have only considered Ga$_x$In$_{1-x}$As films grown lattice matched to InP, i.e., $x \sim 0.47$. Over many years, the lattice-matching condition seemed to be a prerequisite for high-quality and long-term stable devices. But theoretical calculations suggested some fundamental advantages of **strained layers** as, e.g., reduced hole masses. Therefore, investigations on strained layers were initiated, and in fact many modern devices take advantage of strained layers.

One simple possibility to produce strained layers is the growth of lattice mismatched structures. The growing layer tends to hold at least the lateral lattice constant $a_\parallel$ of the substrate (Figure 16), which results in a biaxial strain

$$\varepsilon_\parallel = \left(\Delta a / a\right)_\parallel = \frac{a_\parallel - a_r}{a_r} \qquad (7)$$

where a_r represents the lattice constant of the unstrained material. This gives rise to a change of the vertical lattice constant $a_\perp$; the layer gets vertically strained as well. For the epitaxial surface oriented vs. [100] we find:

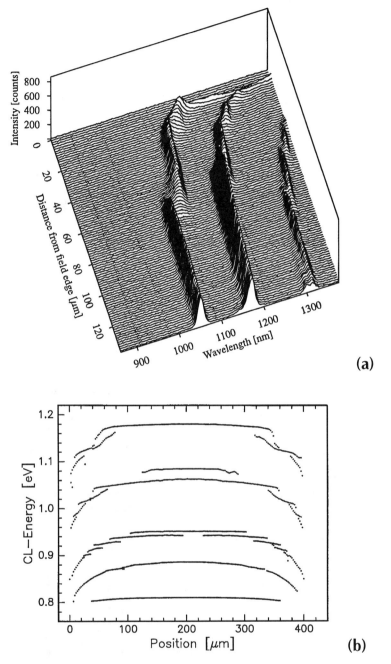

Figure 14 Locally resolved cathodoluminescence of an MSQW structure grown by selective area epitaxy in an open area being about 400 µm wide. (a) Original spectra for a region near the edge of the unmasked field; (b) peak maxima plotted over the local position (field edges at 0 and 400 µm).

$$\varepsilon_\perp = (\Delta a / a)_\perp = -\frac{2C_{12}}{C_{11}} \varepsilon_{\#} \qquad (8)$$

with the elastic constants C_{ij}. This, and the resulting symmetry break, has drastic consequences on many properties, in particular on the semiconductor band structure which motivated our work on strained layers.

The strain described in Equations 7 and 8 can only be accommodated within certain boundaries. If it becomes too large, the growing layer may relax by forming a huge number of dislocations. Matthews

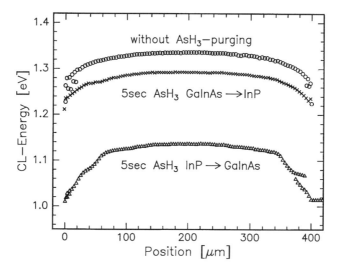

Figure 15 Peak positions measured from spectra as in Figure 14, plotted over local position. These samples have been grown with a rather low V/III ratio of 7 to suppress the monolayer terrace formation. In fact, no terraces are observed for the normal switching process (circles) and for an AsH₃ purge at the upper interface (crosses). But the terrace formation can be reestablished by purging the lower interface with AsH₃ (triangles).

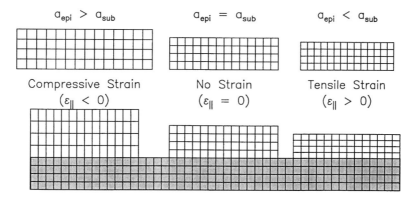

Figure 16 Formation of strain in a lattice mismatched epitaxial layer (schematically).

and Blakeslee [1974] calculated the critical thickness, up to which a fully strained layer may grow, under the assumption of a balance between strain and dislocation movement forces. These and similar calculations from others making use of an energy balance model [People and Bean, 1985] predict a critical thickness l_c depending on the particular strain. This critical thickness is in the range of some 10 nm for the compound semiconductors described here when strain levels of about 1% are present. This means that strain effects mostly are effective in quantum wells and that both strain and quantization effects will influence the band structure simultaneously. This may even result in unexpected phenomena, as described below.

7.1. Strain Effects and Band Structure

In the description of the strain effects, we follow mainly publications by Krijn [1991] and van de Walle [1989]. The influence of strain on the semiconductor band structure can be divided into hydrostatic and shear contributions. The former lead to a shift of the average valence band energy $E_{v,av} = E_{hh} - 1/3\Delta_0$ (with E_{hh} = valence band edge and Δ_0 = spin-orbit split-off energy of the unstrained layer), i.e., the average of the heavy hole, light hole, and spin-orbit split-off band:

$$\Delta E_{v,av} = a_v \left(2\varepsilon_\parallel + \varepsilon_\perp\right) \tag{9}$$

and of the conduction band energy

$$\Delta E_c = a_c \left(2\varepsilon_\parallel + \varepsilon_\perp \right) \tag{10}$$

where a_v and a_c represent the hydrostatic deformation potentials of the conduction and valence band, respectively.

The shear contribution causes a symmetry break in the lattice, giving rise to a cancellation of the heavy hole–light hole degeneracy at $k = 0$, and the valence bands shift by

$$\Delta E_{hh} = \frac{1}{3}\Delta_0 - \frac{1}{2}\,\delta E \tag{11}$$

for the heavy holes and by

$$\Delta E_{lh} = -\frac{1}{6}\Delta_0 + \frac{1}{4}\,\delta E + \frac{1}{2}\left[\left(\Delta_0 \right)^2 + \Delta_0 \delta E + \frac{9}{4}\left(\delta E \right)^2 \right]^{1/2} \tag{12}$$

for the light holes relative to the above-mentioned average $E_{v,av}$. Here,

$$\delta E = 2b\left(\varepsilon_\perp - \varepsilon_\parallel \right) \tag{13}$$

describes the strain-dependent shift with the shear deformation potential b. The influence of strain on the split-off band is not further considered in this chapter. Moreover, we will restrict our calculations on biaxial strain in the (001) plane which has the largest importance in practice because of the conventional substrate orientation.

With these relations, we can calculate the transition energies of band-to-band recombination of electrons in the conduction band and holes in the heavy ($i = h$) and light ($i = l$) hole valence bands:

$$\Delta E_{e,ih} = E_g + \frac{1}{3}\Delta_0 + \Delta E_c - \Delta E_{v,av} - \Delta E_{ih} \tag{14}$$

which is plotted in Figure 17 for the material system $Ga_x In_{1-x} As/InP$, using the material parameters collected by Krijn [1991]. We notice that in compressively strained layers ($x < 0.47$) the heavy hole band forms the uppermost valence band where the electrons recombine preferentially, whereas this is true for the light hole band in the case of tensile strain ($x > 0.47$). Moreover, the band gap is increased on the compressive side and decreased on the tensile side, compared with the unstrained GaInAs of the same composition. (Taking the composition-dependent band gap changes as evaluated by Götz et al. [1983] into account, we find the band gaps are smaller on the compressive side and slightly larger on the tensile side compared with the unstrained case of $Ga_{0.47}In_{0.53}As$ lattice matched to InP.)

It is evident that besides the band edges at $k = 0$, i.e., in the center of the Brillouin zone, which have been considered up to now, the complete band structure and, in particular, the effective carrier masses are changed by strain effects. Whereas the influence on the electron masses are quite small because of the s-orbital character of the conduction band, drastic changes are expected for the holes because of the p-orbital character of the valence bands. In fact, the valence band dispersion relations become anisotropic. The hole masses perpendicular to the epitaxial surface still remain unchanged (if we neglect the spin-orbit coupling) [Pikus and Bir, 1959; O'Reilly, 1989]:

$$m_{hh,\perp} = \frac{1}{\gamma_1 - 2\gamma_2} = m_{hh} \tag{15}$$

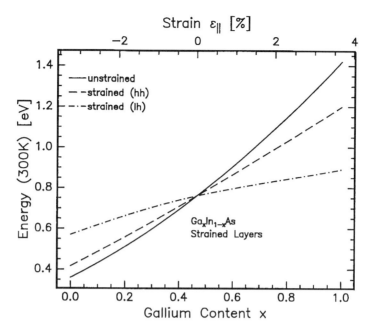

Figure 17 Conduction band–valence band transition energies for biaxially strained GaInAs layers. Solid line: unstrained case.

$$m_{lh,\perp} = \frac{1}{\gamma_1 + 2\gamma_2} = m_{lh} \tag{16}$$

where the γ_i are the Luttinger valence band parameters [Lawaetz, 1971].

But the masses parallel to the film plane change for small values of k quite obviously:

$$m_{hh,\parallel} = \frac{1}{\gamma_1 + \gamma_2} \tag{17}$$

$$m_{hh,\parallel} = \frac{1}{\gamma_1 - \gamma_2} \tag{18}$$

When we analyze these relations, we notice that in the film plane the formerly heavy holes get lighter and the formerly light holes get heavier. But the uppermost holes, which now form the valence band edge, are lighter in compressively as well as in tensile strained layers than the heavy holes in the unstrained case. This fact, and its consequences on further parameters (e.g., the density of states), is the main reason for the use of strained films in devices. In particular, semiconductor lasers show significantly reduced threshold currents and improved high-frequency modulation characteristics because of the reduced hole masses.

Now let us include quantization effects. Again, we have to consider mainly the more complex valence band. The quantization of the heavy and light holes is different because of their different masses (here, the mass components in vertical direction, $m_\perp$ have to be taken into consideration). Now, the reference level (i.e., the potential level of the well, see Figure 1) is represented by the strain-shifted heavy hole and light hole band edges. On the compressive side ($x < 0.47$, $\varepsilon_{xx} < 0$) the heavy hole remains the uppermost hole state. The situation is different on the tensile side: here, the uppermost hole state without quantization is a light hole. But in a quantum well, the light holes are more strongly downshifted than the heavy holes because of their smaller mass (compare Figure 2). Thus, the light holes may pass the heavy holes for quite thin quantum wells or for small strain values, and we come back to the "normal"

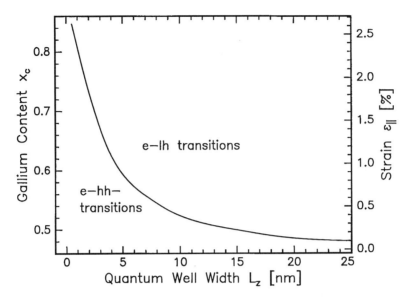

Figure 18 Critical composition x_c where the heavy and light hole subbands cross each other (solid line) as a function of quantum well width L_z.

situation. We can define a crossover point x_c depending on strain and quantum well thickness, where the heavy and light hole states change their positions. This point is always on the tensile side (Figure 18), being at $x = 0.47$ ($\varepsilon = 0$) for a well with infinite thickness.

7.2. Strained Quantum Wells: Photoluminescence Results

In order to investigate these strain-induced band gap variations, we have grown various quantum well structures with different strain and well width by simply changing the Ga/In precursor gas ratio [Härle et al., 1994]. In the conventional mass transport–limited growth regime, where the growth rate depends only on the group III precursor flow, the Ga to In ratio in the gas phase equals (more or less) the Ga to In ratio in the solid. Moreover, we have confirmed the composition and thickness by X-ray diffraction experiments on multi-quantum-well structures. Because of the rather small period of such a structure, additional satellite peaks are visible in the high-resolution X-ray spectrum whose peak distance is directly related to the period of the multi-quantum-well layers (Figure 19).

For the photoluminescence experiments, single quantum well structures have been grown in order not to exceed the critical thickness. On these samples, the strain-induced band gap variations depicted in Figure 17 could be easily confirmed (Figure 20, open symbols): on the compressive side, the measured points follow the steeper line of the electron–heavy hole transitions, whereas on the tensile side, the peak energies shift less according to the light hole transitions. The crossover of the thinner quantum well is shifted to larger tensile strain, as expected.

When we analyzed the original spectra (Figure 21), we realized that on some spectra additional low-energy peaks can be found. Obviously, they only appear for tensile strained structures, and their positions have been plotted in Figure 20 (filled symbols). They first were suspected to be impurity related. But PL spectra over a wide range of excitation intensities disproved this hypothesis [Härle et al., 1994].

7.3. Indirect Valence Band Structure

In order to explain this additional peak, we reconsidered our calculations. Up to now, we have tried to calculate the valence bands quite independently, without any interaction. But, as is evident at the critical heavy hole–light hole crossover point x_c, this consideration may fail. In fact, we are faced with "anti-crossing effects" at this point because of the strong interaction of the different valence bands. Such effects are well known from many similar problems in quantum mechanics. This can be calculated by perturbation theory, and six-band $\mathbf{k} \cdot \mathbf{p}$ calculations show [O'Reilly, 1989; Nido et al., 1993; Härle et al., 1994] that the uppermost valence bands become strongly nonparabolic because of these coupling effects even far away from the critical point. Near the critical composition x_c, we are faced with the

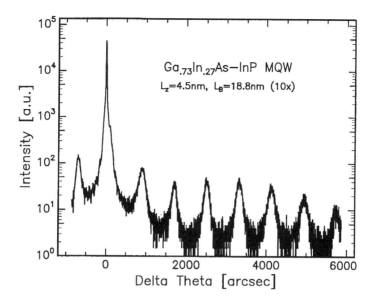

Figure 19 High-resolution X-ray diffractogram of a strained multi-quantum-well structure. The sharp satellite peaks on both sides of the main peak enable a very accurate evaluation of the superlattice period.

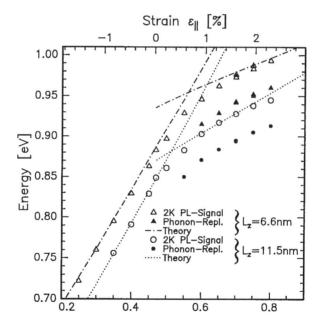

Figure 20 PL peak positions of strained GaInAs single quantum wells, compared with our calculated transitions. The results plotted in Figure 17 are corrected with respect to low temperature (band gap increases) and quantization of electrons and both types of holes. (From Härle, V. et al., *J. Appl. Phys.,* 75(10), 5067, 1994. With permission.)

formation of an absolute valence band maximum at $k \neq 0$, and an indirect band structure is formed (Figure 22).*

---

* As proposed for the first time by Gershoni and Temkin [1989], GaInAs/InP quantum wells form a type II interface (where transitions from the InP conduction band into the GaInAs valence band may be allowed) for large values of the Ga content x ($x > 0.8$). Such transitions may also be taken into consideration for the observed peaks. But obviously, they cannot account for our findings, because these are measured in structures with much lower Ga content.

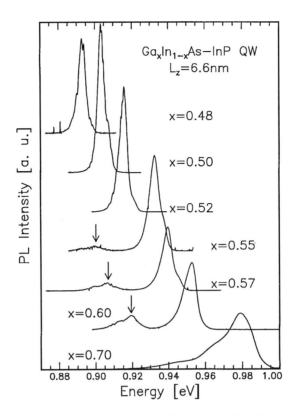

Figure 21 Original PL spectra of some data points of Figure 20. An additional peak at lower energy is clearly visible in tensile strained samples. (From Härle, V. et al., *J. Appl. Phys.,* 75(10), 5067, 1994. With permission.)

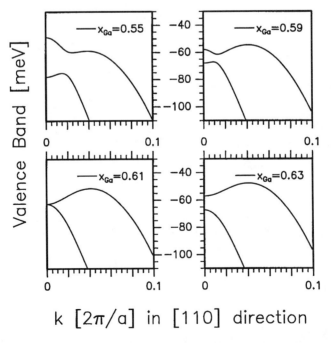

Figure 22 Dispersion relation of the uppermost valence subbands of strained quantum wells calculated by **k · p** perturbation theory. The build-up of an indirect maximum of the uppermost valence band can be seen for a composition *x* ~ 0.6. (From Härle, V. et al., *J. Appl. Phys.,* 75(10), 5067, 1994. With permission.)

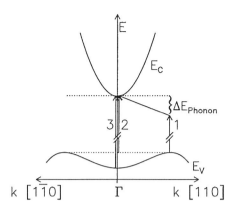

Figure 23 Indirect band structure (schematically) with possible optical transitions indicated (see text). (From Härle, V. et al., *J. Appl. Phys.*, 75(10), 5067, 1994. With permission.)

Now, the spectrum can be easily explained: as is well known from the indirect semiconductors Si and Ge, an additional particle, e.g., a phonon, is needed for the optical transition process in order to fulfill energy and momentum conservation simultaneously. The energy of the emitted photon is reduced by the energy which the phonon takes away (Figure 23, transition 1). That is why we find an additional PL line about 30 meV below the expected band gap. As evaluated by Nash and Mowbray [1989], longitudinal optical (LO) phonons couple more efficiently to excitons than transverse optical phonons. Moreover, the LO phonon energies are known to be 29 and 34 meV for InAs- and GaAs-type phonons, respectively [Landolt-Börnstein, 1982], in excellent agreement to our energetic difference of about 30 meV. So we conclude that this line is LO phonon assisted.

The main peak obviously is due to direct transitions without the assistance of further particles. This is not surprising holding in mind that the indirect maximum is still rather close to the Brillouin zone center ($\Delta k \sim 0.05 \cdot [2\pi/a]$). From direct semiconductors it is known that the excitons are not strongly localized in k space because of interface roughness scattering, alloy scattering, etc. Therefore, excitons localized at small k values have a finite probability to be found at $k = 0$. The phenomena observed here are similar to impurity-induced zero-phonon transitions in conventional semiconductors [Hopfield et al., 1966]. As a result of the localization, this transition may be even higher in intensity than the phonon-assisted transitions.

At elevated temperatures of about 70 K and more, the main peak jumps to higher energies. Now, we observe a direct transition at the Γ point (transition 3 in Figure 23) because of thermal activation of holes to $k = 0$.

7.4. Time-Resolved Measurements

This indirect band structure should have strong influence on the decay time of the PL signal as is known from other indirect semiconductors, as a result of the reduced recombination probability. Therefore, we have performed time-resolved PL experiments at low temperature [Michler et al., 1993; Härle et al., 1994]. The samples were excited by 5-ps pulses generated by the mixing of an Nd:YAG laser and a synchronously pumped pulse from a cavity-dumped dye laser. The signals were detected by photon-counting methods (for details see Michler et al. [1993a]).

As expected for such quantum wells [Brener et al., 1991; Michler et al., 1993a], we found decay times of a few nanoseconds for the unstrained and compressively strained samples (Figure 24). But on the tensile side, decay times in the range of several 100 ns have been observed. The position of the step-wise increase coincides perfectly with the critical point x_c (see Figure 18) being at lower strain values for the thicker quantum wells. These results additionally prove our assumptions of an indirect band structure in these samples which are, moreover, in perfect agreement to the theoretical calculations (see above).

7.5. Device Consequences

These findings are of major importance for the design of modern optoelectronic devices. In fact, several groups have already investigated the relation between the threshold current density of a quantum well semiconductor laser and the strain in the quantum well experimentally [van der Poel et al., 1993] and

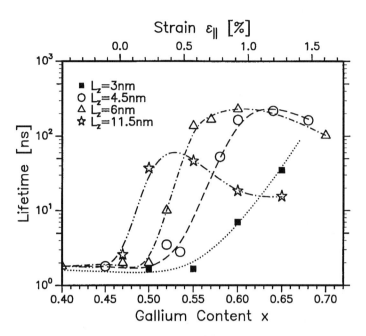

Figure 24 Photoluminescence decay times of differently strained single quantum wells of various thickness L_z. (From Härle, V. et al., *J. Appl. Phys.*, 75(10), 5067, 1994. With permission.)

theoretically [Seki and Yokoyama, 1994]. As expected, they find a strong improvement on the compressive side, which is directly attributed to the reduced hole masses. The worst threshold current is not measured at zero strain, but in slightly tensile strained structures at the critical composition x_c in good agreement with our findings. If tensile strained wells have to be used in order to increase the band gap as much as possible as, e.g., in short-wavelength GaInP lasers, then a strain level beyond the critical point should be chosen.

More generally spoken, strain is an additional degree of freedom when designing modern sophisticated optoelectronic devices. Another substantial fact — besides the minimized hole masses and density of states — is the exchange of the heavy and light hole subbands at the upper valence band edge, as described above. The type of holes involved in absorption and emission processes changes the polarization of the related photons: laser emissions stemming from heavy hole transitions show transverse electric (TE) polarization, whereas they show transverse magnetic (TM) polarization if light holes are involved. Such effects can be used for polarization-controllable devices which are needed in modern optoelectronic systems. A particular example may be a polarization-insensitive amplifier which makes use of the heavy hole–light hole crossover for slight tensile strain [Joma et al., 1993]. Moreover, the extension of the heavy and light hole wave functions — the latter is less localized than the former — in asymmetric quantum wells may enable novel or improved device functions [Tütken et al., 1994].

8. SUMMARY

This chapter has described some examples of ways photoluminescence can be used to investigate and optimize interface properties and strain effects of heterostructures and quantum wells. We have focused on the preparation of InP/GaInAs interfaces by MOVPE. The interface abruptness can be influenced by the design of growth interruptions. But, in particular, an exchange reaction of P and As at the lower interface (switching from InP to GaInAs) cannot be totally avoided resulting in a slight downshift of the photoluminescence signals of very thin quantum wells. This effect is even more severe when GaAs is grown on GaInP.

For an optimized growth procedure, very sharp interfaces can be grown which give rise to a splitting of the photoluminescence lines ("monolayer splitting"). This is even more obvious on quantum wells grown by selective area epitaxy, where large "monolayer" terraces can be found in locally resolved cathodoluminescence.

By detailed photoluminescence studies on strained GaInAs quantum wells, we found direct experimental evidence for an indirect band structure in tensile strained wells, which was predicted by theoretical calculations using $\mathbf{k} \cdot \mathbf{p}$ perturbation theory. It shows up in an additional low-energy line which can be attributed to a phonon-assisted optical transition. Furthermore, this was confirmed by time-resolved measurements, where a drastic increase of the photoluminescence decay time was observed.

These examples demonstrate that photoluminescence not only results in rather qualitative statements, but also helps to find quantitative and microscopic descriptions of sophisticated semiconductor heterostructure characteristics which are prerequisites for the development of modern optoelectronic devices.

9. DEFINING TERMS

Cathodoluminescence: Method of optical spectroscopy, where an electron beam (mostly the regular beam of a scanning electron microscope) is used to excite energetic states in crystals (or other materials), which then relax on a fairly short time scale by emission of light of characteristic wavelength (similar to photoluminescence, except for the excitation process).

Compound semiconductors: Crystals formed by at least two elements of two different columns of the periodic system, which have a band gap allowing thermal excitation of carriers at room temperature, as in elemental semiconductors (like Si or Ge). Most popular examples are III-V compounds (e.g., GaAs, InP) and II-VI compounds (e.g., CdS, ZnSe), composed by elements of IIIrd and Vth or IInd and VIth column, respectively.

Exciton: A quasi particle formed in a solid by the Coulomb interaction of an electron and a hole. Excitons in semiconductors are mostly weakly bonded with a rather large diameter extending over several lattice sites (Mott–Wannier exciton).

Heterostructures: Epitaxial structures that consist of several different materials which mostly have the same crystal structure or even the same lattice constant, but may differ in many other properties, e.g., band gap.

Metalorganic vapor phase epitaxy: Epitaxial method where all elements needed for the growing film are transported as gases to the reaction chamber. This simplifies drastically the control of the process. To overcome the problem that metal-containing gases do not exist, metalorganic compounds, like TMG, are used; these are liquids or solids, but they have a rather large vapor pressure, i.e., significant amounts of molecules in the gas phase at around room temperature allowing their gaseous transport by means of a carrier gas. Alternatively, this method is often called *metalorganic chemical vapor deposition* (MOCVD); the acronyms OMVPE and OMCVD (organometallic …) are also in use.

Molecular beam epitaxy: Epitaxial method where the elements needed for the epitaxial growth are thermally evaporated into a highly evacuated chamber and thus form a molecular beam directed onto a heated substrate.

Monolayer splitting: The main photoluminescence line of very narrow epitaxial films (quantum wells) may split into doublets or triplets. This is believed to be caused by thickness variations of the film of exactly one (or two) molecular monolayers at one or both interfaces.

Optoelectronic devices: Devices that enable the transfer of optical into electrical power or vice versa (like solar cells, light-emitting diodes) or that enable the control of light by electrical means or vice versa (e.g., electro-optical modulator).

Quantum wells: Very thin layers of a material with a given band gap surrounded by higher–band gap layers. The free carriers are confined within the low–band gap region. Thus, they obey quantization rules in one dimension as a "particle in a box." Consequently, they behave like two-dimensional particles.

Photoluminescence: Method of optical spectroscopy, where a light beam is used to excite energetic states in crystals (or other materials), which then relax on a fairly short time scale by emission of light of characteristic wavelength (similar to cathodoluminescence, except excitation process). The source of the exciting light is in most cases, but not necessarily, a laser with fairly high intensity.

Potential well: Description for the potential within a quantum well. If the transition from the lower-energy well to the higher-energy barrier is abrupt in real space, it is called a *rectangular potential well*.

Selective area epitaxy: Method where the epitaxial growth is locally inhibited on a substrate by, e.g., dielectric masks (SiO_2,), but enabled in mask openings, thus allowing definition of the lateral size of a device already at the epitaxial process.

Strained layers: When a heterointerface is grown where the upper material has a slightly different lattice constant than the underlying layer, the growing film still tends to keep the lateral lattice constant of the underlying material in order not to form dislocations. This results in a biaxial strain and a simultaneous change of the vertical lattice constant of the growing layer influencing the band structure and thus many fundamental properties.

Threshold current density: A semiconductor diode laser will emit laser light only if the internal gain is larger than the cavity losses. This is only true for a drive current (density) larger than some threshold value depending on the material and on the quality of the laser diode. Thus, this parameter is one of the most important diode laser characteristics.

ACKNOWLEDGMENT

This paper is the result of investigations done in the research groups of the authors during the last years. Many additional people have partly contributed to these studies, and their assistance and engagement is gratefully acknowledged.

REFERENCES

Aoki, M., Sano, H., Suzuki, M., Takahashi, M., Uomi, K., and Takai, A. 1991. Novel structure MQW electroabsorption modulator/DFB-Laser integrated device fabricated by selective area MOCVD growth. *Electron. Lett.* 27(23): 2138–2140.

Baets, R. 1987. Heterostructures in III-V optoelectronic devices. *Solid State Electron.* 30(11): 1175–1182.

Bhat, R., Koza, M. A., Brasil, M. J. S. P., Nahory, R. E., Palmstrom, C. J., and Wilkens, B. J. 1992. Interface control in GaAs/GaInP superlattices grown by OMCVD. *J. Cryst. Growth* 124: 576–582.

Bimberg, D., Christen, J., Fukunaga, T., Nakashima, H., Mars, D. E., and Miller, J. N. 1987. Cathodoluminescence atomic scale images of monolayer islands at GaAs/GaAlAs interfaces. *J. Vac. Sci. Technol.* B5(4): 1191–1197.

Böhrer, J. Krost, A., and Bimberg, D. 1992. InAsP islands at lower interface of InGaAs/InP quantum wells grown by metalorganic chemical vapor deposition. *Appl. Phys. Lett.* 60(18): 2258–2260.

Brener, I., Gershoni, D., Ritter, D., Panish, M. B., and Hamm, R. A. 1991. Decay times of excitons in lattice-matched InGaAs/InP single quantum wells. *Appl. Phys. Lett.* 58(9): 965–967.

Carey, K. W., Hull, R., Fouquet, J. E., Kellert, F. G, and Trott, G. R. 1987. Structural and photoluminescent properties of GaInAs quantum wells with InP barriers grown by organometallic vapor phase epitaxy. *Appl. Phys. Lett.* 51(12): 910–912.

Clawson, A. R., Jiang, X., Yu, P. K. L., Hanson, C. M., and Vu, T. T. 1993. Interface strain in organometallic vapor phase epitaxy grown InGaAs/InP superlattices. *J. Electron. Mater.* 22(2): 155–160.

Frijlink, P. M., Nicolas, J. L., and Suchet, P. 1991. Layer uniformity in a multiwafer MOVPE reactor for III-V compounds. *J. Cryst. Growth* 107: 166–174.

Gershoni, D. and Temkin, H. 1989. Optical properties of III-V strained layer quantum wells. *J. Lumin.* 44: 381–398.

Göetz, K.-H., Bimberg, D., Jürgensen, H., Selders, J., Solomonov, A. V., Glinskii, G. F., and Razeghi, M. 1983. Optical and crystallographic properties and impurity incorporation of GaInAs grown by liquid phase epitaxy, vapor phase epitaxy and metalorganic chemical vapor deposition. *J. Appl. Phys.* 54(8): 4543–4552.

Grundmann, M. and Bimberg, D. 1988. Anisotropy effects on excitonic properties in realistic quantum wells. *Phys. Rev.* B 38(18): 13486–13489.

Grützmacher, D., Hergeth, J., Reinhardt, F., Wolter, K., and Balk, P. 1990. Mode of growth in LP-MOVPE deposition of GaInAs/InP quantum wells. *J. Electron. Mater.* 19(5): 471–479.

Guimaraes, F. E. G., Elsner, B., Westphalen, R., Spangenberg, B., Geelen, H. J., Balk, P., and Heime, K. 1992. LP-MOVPE growth and optical characterization of GaInP/GaAs heterostructures: interfaces, quantum wells and quantum wires. *J. Cryst. Growth* 124: 199–206.

Härle, V., Bolay, H., Lux, E., Michler, P., Moritz, A., Forner, T., Hangleiter, A., Scholz, F. 1994. Indirect-band gap transition in strained GaInAs/InP quantum-well structures. *J. Appl. Phys.* 75(10): 5067–5071.

Hopfield, J. J., Thomas, D. G., and Lynch, R. T. 1966. Isoelectronic donors and acceptors. *Phys. Rev. Lett.* 17(6): 312–315.

Jiang, X. S., Clawson, A. R., and Yu, P. K. L. 1995. InP-on-InGaAs interface with Ga and In coverage in metalorganic vapor phase epitaxy of InGaAs/InP superlattices. *J. Cryst. Growth* 147: 8–12.

Joma, M., Horikawa, H., Xu, C. Q., Yamada, K., Katoh, Y., and Tamijoh, T. 1993. Polarization insensitive semiconductor laser amplifiers with tensile strained InGaAsP/InGaAsP multiple quantum well structure. *Appl. Phys. Lett.* 62(2): 121–122.

Kamai, H. and Hayashi, H. 1991. OMVPE growth of GaInAs/InP and GaInAs/GaInAsP quantum wells. *J. Cryst. Growth* 107: 567–572.

Krijn, M. P. C. M. 1991. Heterojunction band offsets and effective masses in III-V quaternary alloys. *Semicond. Sci. Technol.* 6: 27–31.

Landolt-Börnstein 1989. *Numerical Data and Functional Relationships in Science and Technology,* Vol 17: *Semiconductors,* Subvol. a: *Physics of Group IV Elements and III-V Compounds,* O. Madelung, Ed. Springer-Verlag, New York, 1982.

Lawaetz, P. 1971. Valence band parameters in cubic semiconductors. *Phys. Rev. B* 4(10): 3460–3467.

Ludowise, M. J. 1985. Metalorganic chemical vapor deposition of III-V semiconductors. *J. Appl. Phys.* 58(8): R31–R55.

Matthews, J. W. and Blakeslee, A. E. 1974. Defects in epitaxial multilayers, 1. Misfit dislocations. *J. Cryst. Growth* 27: 118–125.

Messiah, A. 1965. *Quantum Mechanics*, Vol. I, North-Holland Publ. Comp., Amsterdam.

Meyer, R., Hollfelder, M., Hardtdegen, H., Lengeler, B., and Lüth, H. 1992. Characterization of interface structure in GaInAs/InP superlattices by means of X-ray diffraction. *J. Cryst. Growth* 124: 583–588.

Michler, P., Hangleiter, A., Moritz, A., Fuchs, G., Härle, V., and Scholz, F. 1993. Indirect excitons in strained GaInAs/InP quantum wells. *J. Phys. IV, Coll. C5 Suppl. J. Phys. II* 3: 269–272.

Michler, P., Hangleiter, A., Moritz, A., Härle, V., and Scholz, F. 1993a. Influence of exciton ionization on recombination dynamics in InGaAs/InP quantum wells. *Phys. Rev. B* 47(3): 1671–1674.

Mozume, T. 1995 Characterization of interfacial structure of InGaAs/InP short-period superlattices by high resolution X-ray differentiation and Raman scattering. *J. Appl. Phys.* 77(4), 1492–1497.

Nash, K. J. and Mowbray, D. J. 1989. Exciton-phonon interactions in quantum wells and superlattices. *J. Lumin.* 44: 315–346.

Nido, M., Naniwae, K., Shimizu, J., Murata, S., and Suzuki, A. 1993. Analysis of differential gain in InGaAs-InGaAsP compressive and tensile strained quantum-well lasers and its application for estimation of high-speed modulation limit. *IEEE J. Quantum Electron.* 29(3): 885–895.

Nilsson, S., Gustafsson, A., and Samuelson, L. 1990. Cathodoluminescence observation of extended monolayer-flat terraces at the heterointerface of GaInAs/InP quantum wells grown by metalorganic vapor phase epitaxy. *Appl. Phys. Lett.* 57(9): 878–880.

Nilsson, S., Gustafsson, A., Liu, X., Samuelson, L., Pistol, M.-E., Seifert, W., Fornel, J.-O., and Ledebo, L. 1991. A luminescence study of the interface quality of GaInAs/InP single quantum wells grown by metalorganic vapour phase epitaxy. *Superlattices Microstruct.* 9(1): 99–102.

Omnes, F. and Razeghi, M. 1991. Optical investigations of GaAs-GaInP quantum wells and superlattices grown by metalorganic chemical vapor deposition. *Appl. Phys. Lett.* 59: 1034–1036.

O'Reilly, E. P., 1989. Valence band engineering in strained-layer structures. *Semicond. Sci. Technol.* 4: 121–137.

Ourmazd, A. 1989. Semiconductor interfaces: abruptness, smoothness, and optical properties. *J. Cryst. Growth* 98: 72–81.

People, R. and Bean, J. C. 1985. Calculation of critical layer thickness versus lattice mismatch for GeSi/Si strained-layer heterostructures. *Appl. Phys. Lett.* 47(3): 322–324.

Petroff, P. M., Cibert, J., Gossard, A. C., Dolan, G. J., and Tu, C. W. 1987. Interface structure and optical properties of quantum wells and quantum boxes. *J. Vac. Sci. Technol.* B5(4): 1204–1208.

Pikus, G. E. and Bir, G. L. 1959. Effect of deformation on the hole energy spectrum of germanium and silicon. *Sov. Phys. Solid State* 1: 1502–1517.

Samuelson, L. and Seifert, W. 1994. Metalorganic vapour phase epitaxial growth of ultrathin quantum wells and heterostructures. In *Handbook of Crystal Growth*, Vol. 3, D. T. J. Hurle, Ed., pp. 746–783. Elsvier Science B.V., Amsterdam, the Netherlands.

Schiff, L. I. 1968. *Quantum Mechanics,* Int. Student Ed., MacGraw Hill, New York.

Scholz, F., Ottenwälder, D., Eckel, M., Wild, M., Frankowsky, G., Wacker, T., and Hangleiter, A. 1995. Selective area epitaxy of GaInAs using conventional and novel group III precursors. *J. Cryst. Growth* 145: 242–248.

Seifert, W., Hessmann, D., Liu, X., and Samuelson, L., 1994. Formation of interface layers in GaInAs/InP heterostructures: a re-evaluation using ultrathin quantum wells as a probe. *J. Appl. Phys.* 75(3): 1501–1510.

Seki, S. and Yokoyama, K. 1994. Basic design principles for InGaAsP/InP strained-layer single-quantum-well lasers. *Optoelectron. Devices Technol.* 9(2): 205–217.

Skolnick, M. S., Tapster, P. R., Bass, S. J., Pitt, A. D., Apsley, N., and Aldred, S. P. 1986. Investigation of InGaAs-InP quantum wells by optical spectroscopy. *Semicond. Sci. Technol.* 1: 29–40.

Streubel, K., Härle, V., Scholz, F., Bode, M., and Grundmann, M., 1992. Interfacial properties of very thin GaInAs/InP quantum well structures grown by metalorganic vapor phase epitaxy. *J. Appl. Phys.* 71(7): 3300–3306.

Tanaka, M. and Sakaki, H. 1987. Atomistic models of interface structures of GaAs-AlGaAs (x = 0.2–1) quantum wells grown by interrupted and uninterrupted MBE. *J. Cryst. Growth* 81: 153–158.

Thijs, P. J. A., Montie, E. A., van Kaesteren, H. W., and 't Hooft, G. W. 1988. Atomic abruptness in InGaAsP/InP quantum well heterointerfaces grown by low-pressure organometallic vapor phase epitaxy. *Appl. Phys. Lett.* 53(11): 971–973.

Thrush, E. J., Gibbon, M. A., Stagg, J. P., Cureton, C. G., Jones, C. J., Mallard, R. E, Norman, A. G., and Booker, G. R. 1992. Selective and non-planar epitaxy of InP, GaInAs and GaInAsP using low pressure MOCVD. *J. Cryst. Growth* 124: 249–254.

Tsai, C. Y., Moser, M., Geng, C., Härle, V., Forner, T., Michler, P., Hangleiter, A., and Scholz, F. 1995. Interface characteristics of GaInP/GaAs double heterostructures grown by metalorganic vapor phase epitaxy. *J. Cryst. Growth* 145: 786–791.

Tütken, T., Hawdon, B. J., Zimmermann, M., Queisser, I., Hangleiter, A., Härle, V., and Scholz, F. 1994. Improved modulator characteristics using tensile strain in long-wavelength InGaAs/InGaAsP multiple quantum wells. *Appl. Phys. Lett.* 64(4): 403–404.

van der Poel, C. J., Ambrosius, H. P. M. M., Linders, R. W. M., Peeters, R. M. L., Acket, G. A., and Krijn, M. P. C. M., 1993. Strained layer GaAsP-AlGaAs and InGaAs-AlGaAs quantum well diode lasers. *Appl. Phys. Lett.* 63(17): 2312–2314.

van de Ven, J., Rutten, G. M. J., Raaijmakers, M. J., and Giling, L. J. 1986. Gas phase depletion and flow dynamics in horizontal MOCVD reactors. *J. Cryst. Growth* 76: 352–372.

van de Walle, C. G. 1989. Band lineups and deformation potentials in the model-solid theory. *Phys. Rev. B* 39(3): 1871–1883.

Vignaud, D., Wallart, X., and Mollot, F. 1994. InAlAs/InP heterostructures: influence of a thin InAs layer at the interface. *J. Appl. Phys.* 76(4): 2324–2329.

Wang, T. Y., Reihlen, E. H., Jen, H. R., and Stringfellow, G. B. 1989. Systematic studies on the effect of growth interruptions for GaInAs/InP quantum wells grown by atmospheric pressure organometallic vapor-phase epitaxy. *J. Appl. Phys.* 66(11): 5376–5383.

Wang, T. Y., Jen, H. R., Chen, G. S., and Stringfellow, G. B. 1990. Structural characterization of very thin GaInAs/InP quantum wells grown by atmospheric pressure organometallic vapor-phase epitaxy. *J. Appl. Phys.* 67(1): 563–566.

FOR FURTHER INFORMATION

The growth and properties of similar quantum wells and heterostructures as mentioned here are nicely described by L. Samuelson and W. Seifert in Chapter 17 (Metalorganic vapour phase epitaxial growth of ultra-thin quantum wells and heterostructures) of the *Handbook of Crystal Growth*, Vol. 3, D. T. J. Hurle, Ed., Elsevier Science B.V., Amsterdam, the Netherlands, 1994, pp. 746-783.

M. A. Herman et al. [*J. Appl. Phys.* 70(2): R1 (1991)] have presented a detailed review on "Hetero-interfaces in quantum wells and epitaxial growth processes: evaluation by luminescence techniques" with hundreds of useful citations.

A good introduction to the experimental methods mentioned here, in particular the epitaxial growth methods MOVPE and MBE may be found in *III-V-Semiconductor Materials and Devices* by R. J. Malik, Vol. Ed. (Vol. 7 of *Materials Processing, Theory and Practices,* F. F. Y. Wang, Ser. Ed., North-Holland, Amsterdam, New York, 1989). More-specific information on MOVPE is presented in *Organometallic Vapor Phase Epitaxy* by G. B. Stringfellow, Academic Press, Boston, 1989, and on MBE in *Techniques and Physics of MBE* by E. H. C. Parker, Ed., Plenum Press, New York, 1985.

About application of these methods, the proceedings of the respective international conferences may be studied, i.e., *J. Cryst. Growth,* Vols. 55, 68, 77, 91, 107, 124, and 145 (MOVPE) and *J. Cryst. Growth* Vols. 105, 127, and 150 (MBE).

An excellent introduction to all questions of interaction of light and semiconductors is presented in *Optical Processes in Semiconductors* by J. I. Pankove, Dover, New York, 1971.

Chapter 4

Light-Emitting Properties of Porous Silicon, Si+-Implanted SiO₂, and Si Colloids

Vesselinka Petrova-Koch, Thomas Muschik, Thomas Fischer, and Anton Fojtik

Reviewed by F. Koch

TABLE OF CONTENTS

ABSTRACT

The light-emitting properties of Si nanocrystals are reviewed. Their luminescing behavior is studied as a function of both the crystallite size and the surface termination. We consider three types of samples: porous silicon, Si precipitates in Si+-implanted SiO₂, and Si colloids. The combination of these three preparation methods allows us to vary the particle size from 100 Å down to clusters containing only a few Si atoms and to bypass some of the difficulties characteristic for the etched samples, such as connected crystallites and irregular shape. Particles with hydride, oxyhydride, and oxide surfaces are investigated. While for the first two coverages we observe a blue shift of the photoluminescence peak position with reduction of size, the thermally oxidized samples show constant, size-independent photoluminescence. Comprehensive understanding of this complex behavior is achieved by studying the steady state, time-resolved, and temperature-dependent photoluminescence and the absorption and photoluminescence excitation spectra. A strong indication for the involvement of the surface electronic states and species is provided by the infrared multiphoton excitation of the visible photoluminescence. Electroluminescing devices are realized from porous layers with liquid and solid electrodes and from implanted layers with solid electrodes.

1. INTRODUCTION

Since the discovery of visible photoluminescence (PL) in porous silicon (PS) at room temperature (RT)[7] more than 2000 papers have been published on theoretical aspects, preparation, characterization, and application of a variety of Si-based light-emitting samples. The majority of publications focus attention on the understanding of the phenomena. Three different categories of hypotheses have been proposed

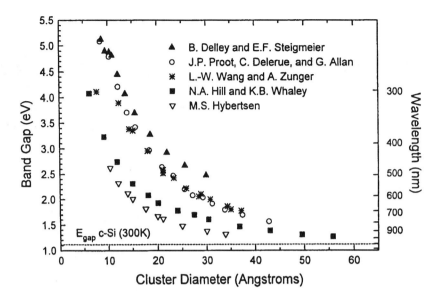

Figure 1 Different calculations of the band gap $E_{g(nc)}$ vs. cluster size.

(for review see Reference 23) — pure quantum-size effects,[7,47,58] pure surface/interface effects,[3,21,46] and surface-modified quantum-size effects.[13,24,29]

Because the Si nanocrystals (nc) are related to a bulk semiconductor with an indirect band gap, their behavior is expected to be more complex than the one of semiconductor nc with a direct band gap, like, for example, the CdSe nc.[4] Theoretically, most attention is paid to calculating the quantum-size effects, where only H-terminated Si nc have been considered.[20] A calculation of crystallites terminated with oxide will be of interest. Figure 1 summarizes several predictions for the shift of the band gap $E_{g(nc)}$ as a function of size.

The disagreement between the different calculations is of the order of 1 eV. There is a tendency to calculate lower $E_{g(nc)}$ values at a certain size in the more recent theoretical work. The predicted blue shifts exceed these of the CdSe nanocrystallites in the same size region. The matrix elements, however, remain almost unaffected when reducing the size.[20f] The Si nc are indirect semiconductors down to very small dimensions. Substantial advance is also achieved by the calculation of different Si-based molecules.[12] Only one work has been published on surface-modified quantum-size effects.[19] Experimentally, a number of Si-related particle systems that luminesce efficiently in the visible spectral region at RT are currently under investigation.

The electrochemically etched PS is the most popular because of the ease of preparation. As first noticed by Canham[7] and Lehmann and Gösele,[30] PS contains nanometer-sized Si crystallites and shows distinct quantum properties in its interaction with electromagnetic waves. The connectivity of the crystallites, their irregular shape, dispersed size, and imperfect surface cause additional problems in the understanding of the basically complicated Si nc behavior.

The second system under consideration is the Si colloid. Different methods for preparation have been suggested in Reference 17 and by Littau et al.[33] The colloid provides isolated particles with better-defined shape. There are problems with the size dispersion and imperfect surface termination. The nc powder also belongs in the category of isolated particles reported by Kanemitsu et al.,[22] and the compact CVD layers, obtained by subsequent oxidation of Si nanocrystalline grains, suggested by Veprek et al.[55]

More recently, yet another technique has been introduced to create the Si nanoparticles in a solid matrix, namely, the precipitation of excess Si by thermal treatment of Si⁺-implanted thermal oxide.[2,26,37,54] In contrast to PS, where bulk c-Si is used to produce the nanocrystals, here one starts by assembling the nanocrystals from individual atoms. Very small Si clusters can be achieved this way. The implanted layers are compact films and are processed and handled in a conventional manner for microelectronic technology. This gives them a decisive advantage in applications.

Investigating a variety of samples (PS, Si precipitates in thermal oxide, and Si colloids), we observe two principally different types of photoluminescing behavior: size-dependent PL for the hydride- and oxyhydride-terminated crystallites and constant, size-independent PL for the thermally oxidized ones. Here, we intend to provide a comprehensive view of this complex behavior. In Section 2 the preparation of the different samples is described. In Section 3 we deal with the photoluminescing and the absorption properties of the samples demonstrating a size effect, and Section 4 deals with the samples where no size effect on the PL is noticeable. The observation of visible light by infrared (IR) multiphoton excitation (MPE) represents a unique experimental demonstration of the involvement of surface electronic states and surface species in the light-emission process. In Section 5 results concerning the realization of electroluminescing devices out of the porous samples and of the implanted layers are shown.

2. SAMPLE PREPARATION AND STRUCTURE

2.1. Porous Silicon

PS is grown by electrochemical etching of B-doped p-type (100) c-Si substrates with a resistivity of approximately 1 Ω cm (p$^-$-substrates) or 5 mΩ cm (p$^+$-substrates) in a setup described in Reference 31. Ethanoic HF (1:1 in vol.) is used as an electrolyte. A porous skeleton is prepared this way containing crystallites with typical dimensions between 20 and 100 Å[31] and internal surface area of several hundred square meters per cubic centimeter, terminated by hydrogen in the fresh samples[38,59] or by oxihydride after aging in air.[13,38] Isotope substitution (deuteration of the crystallite surface) helps to assign the surface vibrational modes[59] and to study the compositional and structural disorder on the nanocrystalline surfaces. The Si–H$_x$ stretching modes at room temperature have typical FWHM of ~20 cm^{-1} or even more and do not narrow at low temperature. This inhomogeneous broadening is due to the compositional disorder and to perturbations of the Si–H and the Si–Si surface bonds. EPR measurements show in addition the existence of Si dangling bonds (DBs) with typical concentration of 10^{16} cm^{-3} for the efficiently luminescing samples.[34] As a result of these imperfections two types of electronic surface states appear in the forbidden gap of the nanocrystals, *tail states*, related to the perturbed bonds and *deep defects* due to the dangling bond, as previously discussed in Reference 24. Both affect the PL properties (see Section 3).

The size of the crystallites in PS is varied in three ways:

1. By the etch current density between 1 and 1000 mA/cm^2,[31]
2. By poststripping the samples in HF,[42] or
3. By thermal oxidation[40,41] at temperatures between 900 and 1000°C.

2.2. Si Precipitates in Si$^+$-Implanted SiO$_2$ Films

Samples with different oxide thicknesses implanted at different doses have been described in previous papers.[2,26,37,54] Here we restrict ourself to thermally grown SiO$_2$ with a thickness of 1000 Å on (100)-oriented Si substrates, implanted with Si$^+$ ions at doses 1, 2, and 5 × 10^{16} cm^{-2}. The SiO$_2$ matrix with the extra Si atoms represents an oversaturated solid solution where Si clusters are expected to form after heat treatment.[32] We annealed the samples at 900°C in forming gas (mixture of H$_2$ and N$_2$), varying the precipitation time between seconds and hours. Tunable PL was observed in the entire region from blue (3 eV) to red (1.5 eV), as we previously reported.[16] It will be discussed in Section 3. In contrast, a posttreatment of the layers in oxygen atmosphere leads to a constant, size-independent PL, as discussed in Section 4.

Figure 2 shows an HRTEM picture of a sample with Si precipitated in SiO$_2$. The Si crystallites are faceted. The annealing, which leads to the formation of precipitates, helps at the same time to reduce the paramagnetic defects caused by the implantation.[16]

2.3. Si Colloids

The Si colloids are prepared by combustion of SiH$_4$ in a syringe as described in References 17 and 18. This technology is simpler than the one suggested by Littau et al.[33] The powder deposited on the wall inside the syringe is suspended in a mixture of cyclohexane and 2-propanol. By adding HF, two phases form. The low phase consists of water, 2-propanol, HF, and particles; the upper phase mainly contains the nonpolar cyclohexane. After being etched by the HF in the low phase, the crystallites move into the upper phase. The luminescence develops with color being red in the beginning, turning to orange with time. No thermal oxidation is required in contrast to Reference 33. The preparation is performed in air. The slightly opalescent solution becomes nonopalescent after evacuation. When air is admitted, the PL quantum yield increases. Photoluminescing colloids with a quantum efficiency up to 20 to 25% are

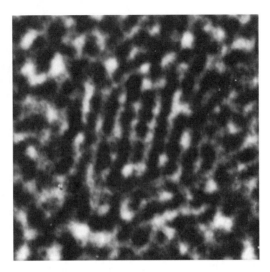

Figure 2   Faceted Si nanocrystal formed after annealing of Si⁺-implanted SiO₂.

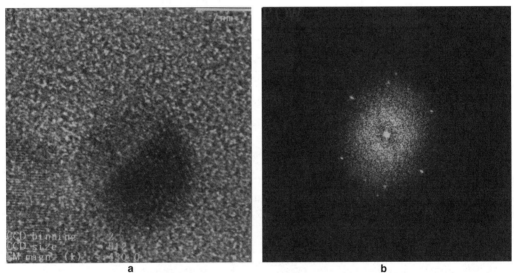

a   b

Figure 3   (a) HRTEM of a faceted colloidal Si nanocrystal and (b) the corresponding electron diffraction image.

obtained this way. Only the highly luminescing colloids are considered in this study. Figure 3 shows typical HRTEM of a colloidal particle and the corresponding electron diffraction picture. We see faceted Si crystallites, similar to those in the implanted layers.

In Figure 4 the quantum efficiency of different colloids is presented as a function of PL energy. The quantum yield decreases drastically for PL lower than 1.6 eV. There is a good qualitative agreement with theoretical predictions on quantum yield vs. size at RT.[20g]

3. PL OF HYDRIDE- AND OXYHYDRIDE-TERMINATED SAMPLES — THE CASE OF SURFACE-MODIFIED QUANTUM-SIZE EFFECTS

3.1. Tunability of the PL with Size

In Figure 5a and b two sets of tunable PL spectra are shown, obtained on PS and on Si⁺-implanted SiO₂ films, respectively. By combining the two types of samples, the spectral region between 1.1 and 3 eV was covered. With the etching technology the highest PL energy obtained so far for the continuously tunable PL is around 2 eV. The limitation in the latter case is most probably a result of the restricted possibility of making very small crystallites with the etching procedure. For sizes smaller than 20 Å the Si skeleton loses integrity. In contrast, in the Si⁺-implanted SiO₂ we start with very small clusters first.

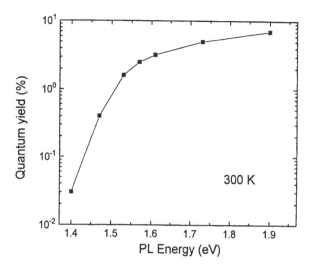

Figure 4 Quantum yield of colloids as a function of the PL energy.

In this sense the two techniques are complementary to each other. In the overlapping region their PL properties are similar, as shown later in this section.

For clusters containing several Si atoms only, the PL maximum shifts up to 3 eV. Although the last value is much higher than those observed in PS and the colloids,[18,49,58] it remains at least 1 eV lower than the theoretical prediction for very small crystallites (see Figure 1). The most likely reason for the systematically lower experimental PL value is the existence of surface tail states,[24] as will be discussed below. To what extent the characteristic Si–Si bond length remains preserved in the very small clusters is not established at the moment.

3.2. Superimposed Surface Effects

Many times it was pointed out in the literature that the crystallite size is not the only parameter that affects the PL energy position. Figure 6 demonstrates a typical red shift of the PL spectrum of PS when the degree of disorder on the surface is increased by partial desorption of the surface hydrogen. At the same time the PL intensity decreases while the FWHM increases. The quenching of the intensity of the RT-PL anticorrelates with the density of surface dangling bonds[34,36] (Figure 7). This is a clear demonstration that at RT the paramagnetic defects are nonradiative centers. Their density is seen to dominate the RT-PL decay time.[36] The increase of the PL FWHM after partial effusion of the surface hydrogen points to the creation of surface tail states involved in the radiative process, as discussed for amorphous alloys.[9]

Unique experimental evidence for the involvement of the surface electronic states and species in the luminescing process is provided by the IR MPE of the visible light, as reported recently by Chin et al.[11] and Diener et al.[14] for PS samples.

Using an optical parametric amplifier (OPA) pumped by active-passive mode-locked Nd-YAG laser,[11] generating 8-ps pulses tunable from 0.65 to 10 μm,[60] we excite the visible PL at different IR wavelengths: 1.064, 1.3, and 4.85 μm as shown in Figure 8. The spectra are very similar to each other and to the PL spectrum observed after excitation with visible light (532 nm in Figure 8). The most impressive indication, however, that the surface is involved in the light-emitting process is provided by the excitation of the visible light via resonances with the SiH_x-stretching vibrational ladder, as shown in Figure 9 for different pump intensities up to 2.86 GW/cm². At low intensities a monochromatic IR beam can resonantly excite the surface SiH_x bonds to the first ro-vibronic states, but subsequent resonant excitation to higher states is less likely, because they are rendered off-resonant by the existence of vibrational nonharmonicity. At sufficiently high beam intensities, direct MPE via near-resonant steps up the vibrational ladder becomes significant.[11]

3.3. Surface-Modified Quantum-Size-Effect Model

Based on the size- and surface-dependent PL observations, for the hydrogen- and oxyhydride-terminated crystallites, we have suggested the so-called surface-modified quantum-size model.[23,24] A

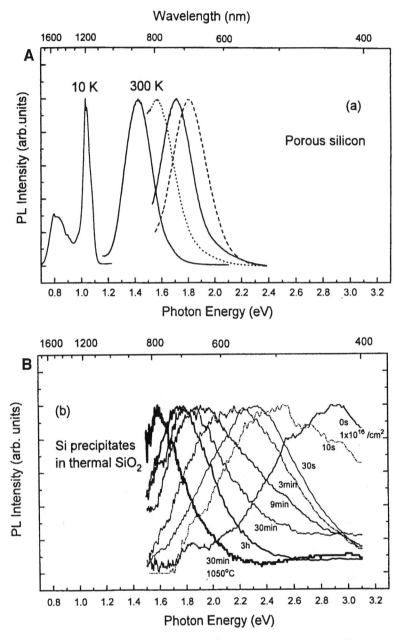

Figure 5　(a) Tunable PL in PS with different size after stripping. (b) Tunable PL in Si+-implanted SiO$_2$ layers (annealing temprature 900°C and implanted dose 5 × 10^{16} cm^{-2}).

qualitative sketch of this model is presented in Figure 10. The E$_c$ and E$_v$ are the conduction and valence band edges, the T$_c$ and T$_V$ are the conduction and valence band–related surface tail states, and the D are the deep defects (D+, neutral; D$^-$, singly occupied dangling bond). The reduction of the crystallite size leads to an increase of the gap E$_{g(nc-Si)}$ with respect to that of bulk Si E$_{g\,(c-Si)}$ (Figure 10b and c). The tail states follow the gap in a similar manner to the one discussed for amorphous alloys.[9] The increase of disorder leads to spreading of the tail states (Figure 10d). The DB-related defects, which are strongly localized electronic states, should not change their energy position.

3.4. PL Doublet

A simple way to determine the shift of ΔE$_c$ and ΔE$_v$ is to measure the PL of the samples at low temperature. Together with the visible band, a second band appears in the IR region, because the low-

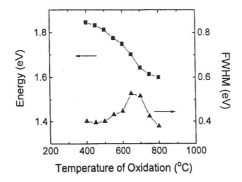

Figure 6 Red shift of the PL when desorbing H from the PS surface.

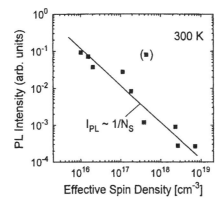

Figure 7 Anticorrelation between the PL intensity and the dangling bond density in PS.

temperature radiative recombination probability via an occupied DB increases exponentially.[35] Figure 11a demonstrates the PL pair for samples with different sizes of crystallites. When the size is reduced, the two bands shift into the blue spectral region by different amounts of energy.[23,41] Koch et al.[25] assigned the visible PL to recombination between conduction and valence band–related tail states and the IR band to recombinations between conduction band–related states and occupied dangling bonds. Therefore, the shift of the IR band with energy is a measure for the conduction band shift ΔE_c. The shift of the visible band then measures $\Delta E_v + \Delta E_c$. The experimental results show that the valence band in PS shifts twice as much as the conduction band ($\Delta E_v = 2\Delta E_c$). This "universal" correlation between the two bands is shown in Figure 11b and is in good agreement with theoretical predictions.[20b]

3.5. Temperature-Dependent and Time-Resolved PL Spectra

In Figure 12 a typical temperature dependence of the red PL in PS[36,61] is compared with that of a colloid[4] and an implanted layer. While there is a continuous increase of the PL intensity with decreasing temperature for the last two types of samples, the PS shows an anomalous temperature dependence, as discussed previously in the literature.[45,61] It is most likely that at low temperature in PS the carriers escape to regions containing dangling bonds and contribute either to the IR PL band or recombine nonradiatively.

Typical PL decay curves for implanted samples emitting the red spectral region (<2 eV) and for those with PL maximum in the green/blue region are presented in Figure 13a and b. While the red luminescence is microsecond-slow at RT like the one reported for PS[36,56] and colloids,[58] the tunable PL in the green/blue region is nanosecond-fast. There is no other Si-based material in which nanosecond-fast PL, tunable up to 3 eV is reported so far. These samples are currently under detailed investigation.

In Figure 14a and b the results of an extended study of the time-resolved PL for the slow red band in porous samples, measured at different temperatures, are presented[36]. Figure 14a shows the intensity vs. time for three different temperatures. In Figure 14b the maximum of the lifetime distribution as a function of temperature is depicted. For temperatures <60 K the lifetime is seen to increase over two orders of magnitude. The last result is very similar to the one reported for the red-emitting Si colloids.[58]

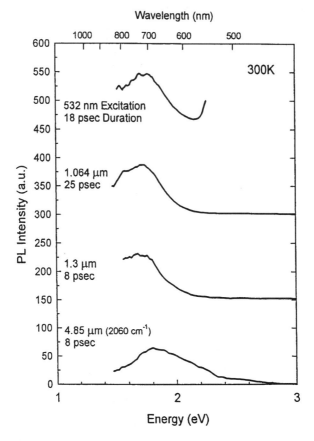

Figure 8 MPE spectra of PS for different IR PL wavelengths.

The radiative decay τ_r dominates the nonradiative decay τ_{nr} at T <60 K. The τ_r in the low-temperature region is temperature dependent. The typical τ_r of the red-emitting samples at low temperature is several microseconds. In agreement with other reports in the literature,[56,58] we find that the efficient red PL in the Si nanocrystals appears not because of significant increase of the radiative rate, but because of efficient suppression of the nonradiative processes.

3.6. Absorption and PL Excitation Spectra — Relation to Bulk Si

The absorption properties of different free-standing PS films,[23,24] of colloidal samples,[18] and of Si+-implanted SiO_2 layers[43] are studied by transmittance, reflectance, photothermal deflection, and PL excitation (PLE) spectroscopy. While the free-standing porous samples, which are usually optically thick, permit measurements up to ~2.5 eV, the diluted ones (colloids and the implanted layers) allow one to obtain the absorption spectra in the UV region up to 5.5 eV. The absorption spectra achieved with the free-standing porous layers are blue shifted with respect to that of c-Si,[20,24,30] and are featureless.[24] Besides this, a significant absorption is observed in the region below the gap of bulk c-Si,[23,24,57] which is caused by surface effects. In the UV part of the absorption spectrum (up to 5 eV), we observe a strong increase of the absorption for energies higher than 2.5 eV (Figures 15 and 16). In comparison with previous work,[18] the spectra obtained for the implanted samples and colloids with high quantum yields show better-pronounced absorption features in the region around the Γ'_{25}–Γ_{15} ~3.4eV, and the Γ'_{25}–Γ_2 ~4.5 eV transitions, characteristic for bulk Si.

In Figure 15 two absorption spectra measured on implanted samples are shown.[43] For comparison that of bulk c-Si is also presented. One clearly recognizes two features in this spectra, related to the ones observed in bulk (~3.4 and 4.5 eV), and an additional peak in the unannealed implanted layer around 3.8 eV, related most probably to defects.

In the absorption spectrum of the colloids presented in Figure 16, the features around 3.2 and 4.5 eV and the defect-related feature around 3.8 eV are apparent. For an unexplained reason the peaks in the

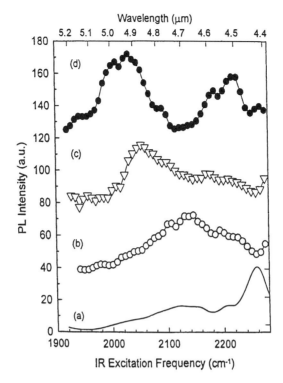

Figure 9 Excitation spectra of the visible PL in the region of the SiH-stretching vibrations different excitation powers: (a) FTIR spectra, (b) 0.4 GW/cm², (c) 0.8 GW/cm², and (d) 2.86 GW/cm².

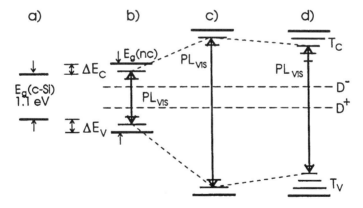

Figure 10 The surface-modified quantum-size-effect model for Si nanocrystals.

absorption spectrum of the implanted layers are better pronounced than those in the colloids. On the other side, in the colloids with higher quantum yields considered here the absorption peaks are better pronounced than in those reported previously for colloids with lower quantum yield.[4,18] In Figure 16 some differences between the absorption and the corresponding PLE spectra become apparent. The decrease of the PLE signal at energies higher than 4.5 eV, where the absorption is still increasing, could be due to ionization of the nanocrystals (the excited electrons escape into the liquid). The peak around 3.2 to 3.4 eV is better pronounced in the PLE spectrum. This could be due to the size selectivity provided by the PLE spectroscopy.

In Figure 17 PLE spectra are shown, detected at four different energies inside one PL spectrum. When detecting in the low-energy end of the PL spectrum arround 1.6 eV, we see a peak appear in the PLE spectrum around 2.6 eV (2.2 eV is known to be the energy of the Γ'_{25}–L_1 transitions in the c-Si). This feature vanishes in intensity for detection energies around 1.7 eV and disappears for those around

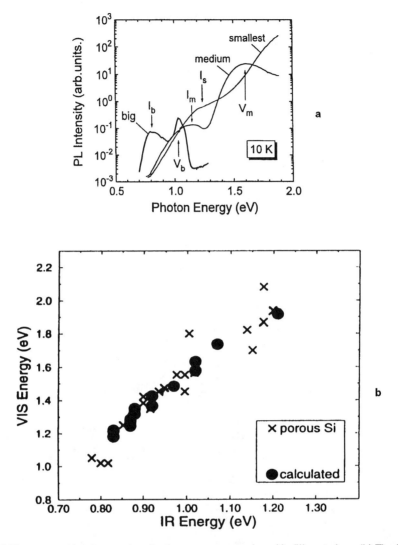

Figure 11 (a) PL spectra at low temperature for three porous samples with different sizes. (b) The "universal" correlation between the visible and the IR PL band. (From Hill, N. A. and Whaley, B. K., *MRS Proc.*, 358, 25, 1995. With permission.)

1.9 to 2 eV. For all detection energies a structure around 3.2 eV is observable. The shifts are much smaller compared with theoretical predictions like the one provided by Rama Krishna and Friesner.[48]

4. OXIDIZED NANOCRYSTALLINE Si AND THE EVIDENCE AGAINST THE SIZE EFFECT

It is established now that oxidation of PS[40,41] and of microcrystalline CVD layers[55] or CVD nanoparticles[22] yields a stably luminescing material that in many relevant aspects (intensity of the PL, FWHM, lifetime, etc.) equals the hydride- or oxyhydride-covered PS. However, there is one puzzling fact that categorically denies the existence of size effects; namely, in the substantially oxidized nanocrystals the PL energy remains nearly constant. Figure 18 demonstrates the convergence of the PL maxima of hydrided porous samples with different average sizes of the crystallites into the region around 1.55 to 1.65 eV after thermal oxidation, as observed by Steiner et al.[53]

The same fixed PL color (1.6 to 1.7 eV) is observed also in the oxidized CVD nc layers[55] and in the oxidized CVD powder.[22] In Figure 19a we see this to be also the case for the implanted samples after thermal postoxidation. For comparison, the evolution of the constant PL in PS at different oxidation

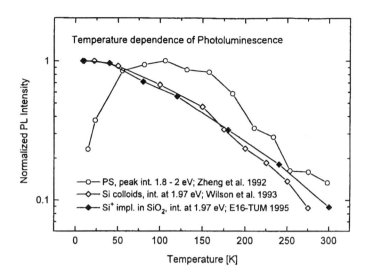

Figure 12 Temperature dependence of the red PL in PS,[44] Si colloid,[4] and Si+-implanted SiO$_2$.

temperatures is shown in Figure 19b.[36] There is remarkable similarity in the behavior of the two types of luminescing samples.

In all cases of oxidized Si nanocrystals another, blue/green PL band, constant in energy position, is also observed. Its intensity increases with oxidation temperature, while that of the red PL decreases. The last has been assigned to very small Si remnants,[33] to SiO$_2$-related defects,[15] to silanol groups in the films,[55] etc. The light-emitting mechanisms of the blue/green band with fixed energy position remain to be figured out.

For the red band Kanemitsu et al.[22] suggested a qualitative model, shown in Figure 20. It assigns the PL to the interface layers SiO$_x$ between Si nc and SiO$_2$ shell. However, the PL mechanism for this band is unresolved, too. In Reference 22 it is suggested that the PL color corresponds to the gap energy of the interface layer (substoichiometric oxide). The amorphous nature of this layer makes other explanations also possible.

In Figure 21a and b we show the time-resolved PL spectra at three different temperatures and the maxima of the lifetime distribution τ as a function of temperature between 4 and 300 K.[36] The dominant decay time τ increases very rapidly for T < 60 to 100 K. Compared with similar measurements for the tunable PL (see Figure 14), we see remarkable similarities between the temperature and the time dependence of the tunable and the size-independent red PL. One difference in the oxidized samples is the appearence of a second (much slower) decay time τ_T (hundreds of milliseconds at RT), which is most likely caused by trapping of photogenerated carriers in the oxide shell. After a certain dwell time they come back to the interface to recombine radiatively.

The implication of the surface states on other interesting properties, like the resonantly excited sideband PL[6] or the PL polarization[1] observed in the porous samples, is under investigation now.

5. ELECTROLUMINESCING DEVICES

A lot of effort has been devoted, since the discovery of PS, to realize electroluminescing (EL) devices. We have worked in several directions. One of these was to study the involvement of the surface states in the EL when the PS crystallites are in intimate contact with a liquid electrode. Usually, liquids containing strong redox species, for example, K$_2$(S$_2$O$_8$) in H$_2$SO$_4$, are used in a cathodically biased arrangement. Those electrolytes are known to inject holes in many wide gap semiconductors, where EL is observed.[44] Applying the same recipe on PS, visible EL with a quantum efficiency around 1% was reported,[8] which can be tuned over the entire visible range by varying the voltage applied on the device.[5] A pure quantum confinement model is suggested as an explanation for the tunability of the EL color.[5] We have searched for the IR band in the EL spectra,[50] which we expect to accompany the yellow/red emission in analogy with the observation of the PL pair discussed in Section 3. To be able to observe the IR EL we replaced the IR nontransparent electrolyte containing normal water by a heavy-water-

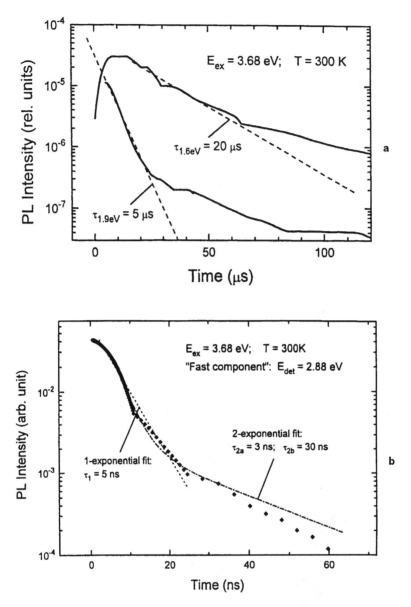

Figure 13 (a) Time-resolved PL of the slow decaying red PL band from SiO_2 layers implanted with 5×10^{16} cm^{-2} and annealed for 30 or 9 min at 900°C. (b) Time-resolved PL of the blue/green PL from an SiO_2 layer implanted with 1×10^{16} cm^{-2} and annealed for 3 min at 900°C.

based electrolyte ($D_2SO_4/K_2S_2O_8$). A typical spectrum containing the visible and the accompanying IR band is shown in Figure 22. By increasing the current beyond a certain level for some time the visible EL strongly degrades, as also reported by other authors,[5,8] while the IR EL survives the quenching.[50] One practical problem of the EL device with liquid electrodes is the liquid. A more serious problem is the very long decay time of this EL (several milliseconds at RT) as reported more recently.[51]

Another direction we followed more recently is the realization of a metal-insulator-semiconductor (MIS) structure using a photoluminescing implanted layer as an insulator.[28] At RT visible EL with a quantum efficiency of about 0.1% is achieved. A typical EL spectrum is demonstrated in Figure 23. After a deconvolution, three visible EL bands can be distinguished: one around 2.8 eV (blue color), the second around 2.3 eV (green), and the third around 1.8 eV (red). This EL is fixed in energy position; the EL spectra are identical with the cathodoluminescent spectra observed on such samples and are different from the corresponding PL spectra. The radiative centers for the EL are different from those responsible for the PL. The I-V characteristics allow us to conclude that Fowler-Nordheim (FN) tunneling

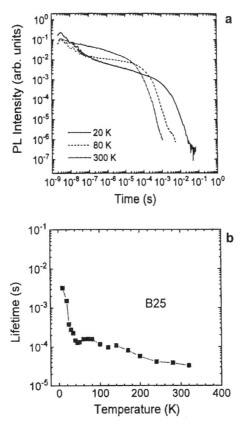

Figure 14 (a) PL intensity vs. time for three different temperatures, (b) maximum of the lifetime distribution vs. temperature in PS.

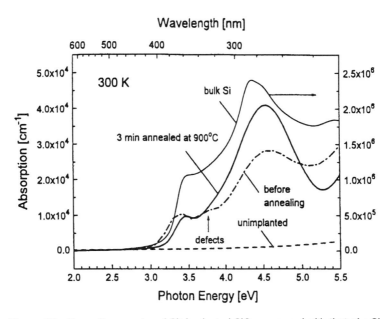

Figure 15 Absorption spectra of Si+-implanted SiO_2 compared with that of c-Si.

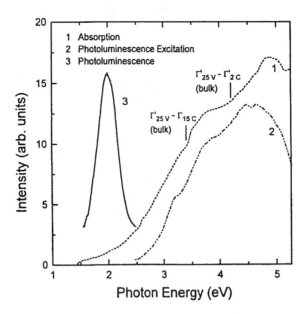

Figure 16 Comparison of the absorption and of the PLE spectrum of a red-emitting colloid.

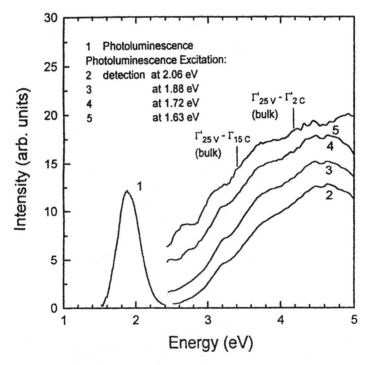

Figure 17 PLE spectra detected at four different PL colors in a colloid with 25% RT quantum yield; curves shifted and scaled for better clarity.

is the dominant mechanism of carrier injection. More work is in progress to optimize the preparation conditions and to understand the EL behavior of the samples.

6. CONCLUDING REMARKS

In this chapter an overview of the luminescing properties of PS, Si colloids, and Si-implanted SiO_2 layers is presented. Samples containing Si crystallites from big bulklike (100 Å or more) to very small

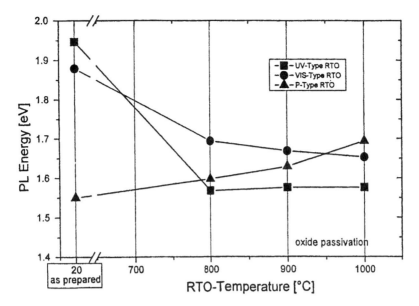

Figure 18 The convergence of the PL color in PS after thermal oxidation. (From Steiner, P., Doctoral thesis, Technische Universität, München, 1996. With permission.)

size (cluster containing only several Si atoms) have been prepared. Hydride, oxihydride, and pure oxide terminations of the surface have been considered. The PL is sensitive to both the quantum confinement effects and the surface chemistry. In this interplay of size and surface effects, one or the other can dominate. We found that the PL of the samples containing crystallites with hydride or oxihydride termination is dominated by size effects, while that of the oxidized samples is constant in energy position and is dominated by surface/interface effects. The experimental results can be understood by the surface-modified quantum-size model rather than by the pure confinement model. Two types of surface-related states have to be considered: deep defects due to the surface dangling bonds and surface tail states due to the reconfigured surface bonds. Because of the strongly localized nature of the dangling bond states, their behavior remains the same, like in amorphous Si and in degraded c-Si, and the knowledge established there can be adopted here. The deep defects have two implications for the luminescence behavior: they are seen to quench the RT visible PL or to cause an IR PL at low temperature. The peak energy position of the latter can be used as a measure of the conduction band shift vs. size. Because of the universal relation between the energy position of the IR and the visible PL, the shift of the visible band is a measure of the shift of the valence band–related states.

The experimental proof and understanding of the surface/interface-related tail states is more complicated. A unique demonstration for their involvement and for the involvement of surface species in the visible radiative process is the IR MPE of the visible PL in PS which was demonstrated recently. Concerning the understanding of the tail states, in contrast to the deep defects they can have a localization length over several lattice constants and can separate an electron–hole pair in the real space, as is known from studies on amorphous semiconductors. For the hydride- and oxihydride-terminated samples we have seen that tail states follow the shift of the band edges when the size of the crystallites is reduced. When the density of tail states in a sample containing certain sizes of crystallites increases, a red shift of the PL of the order of 200 to 300 meV and a broadening of its FWHM are observed as well. These are due to pure surface effects. For the oxidized samples the surface/interface tail states remain unchanged in energy position when the crystallite size is reduced. Comparing the temperature- and time-dependent characteristics of the tunable and the constant visible PL, we see remarkable similarities. In the latter case a separation of the e–h pairs in real space is responsible for the slow decaying PL. Because of the many similarities between the PL behavior of e–h pairs separated in the real or in the k-space, it is difficult to deconvolute these two contributions in the PL behavior of the Si nc systems.

We try to correlate the PL properties with the Si absorption features, performing absorption and PLE study. In the energy region below 2.5 eV we have seen featureless spectra, blue shifted with respect to that of bulk c-Si, and a substantial absorption in the region below 1.1 eV which is due to surface states.

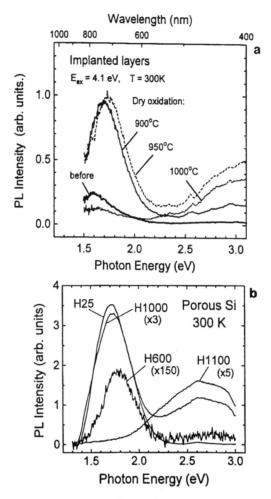

Figure 19 (a) The PL spectra of implanted samples after thermal postoxidation; (b) the PL spectra of similarly treated porous samples for comparison.

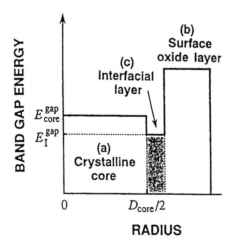

Figure 20 The model assigning the constant red PL in the oxidized Si nanocrystals. (From Kanemitsu, Y. et al., *Phys. Rev. B,* 48, 4883, 1993. With permission.)

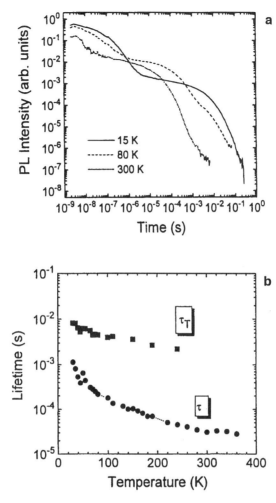

Figure 21 (a) PL intensity vs. time at three different temperatures in oxidized PS. (b) Maxima of the lifetime distributions τ and τ_T as a function of the temperature.

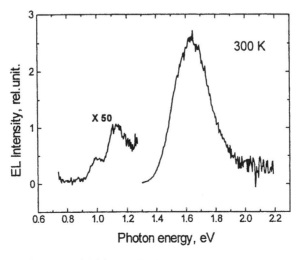

Figure 22 Visible and IR EL observed on n-type PS.

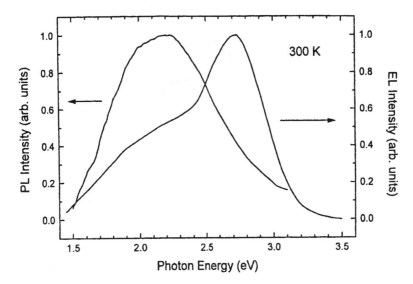

Figure 23 Typical EL spectrum of an Si-implanted Si layer in MIS structure. The corresponding PL spectrum is shown for comparison.

In the region higher than 2.5 eV several absorption bands related to those of bulk c-Si are recognized even for the samples with very small crystallites. The shifts of these bands vs. size, if any, are smaller than predicted by theory.

We worked on several projects dealing with the development of EL devices. PS with a liquid electrode was prepared and studied, which shows not only a visible, but also an IR EL band. We also succeeded in preparing EL MIS structures out of the Si-implanted SiO_2 layers with quantum efficiency of about 0.1%. Three EL bands with fixed color have been observed. The EL spectra differ from the PL and are very similar to the cathodoluminescence spectra for these films. The light-emitting centers in the case of PL and EL are different for the films prepared so far. Work in the direction of increasing the excess of Si in the SiO_2 and of improving the transport and injection properties without losing the PL in the films is in progress.

ACKNOWLEDGMENTS

We are pleased to acknowledge the many valuable discussions with Prof. F. Koch from the Technische Universität, München, and Prof. A. Henglein from Hahn-Meitner Institute, Berlin, and the close collaboration with Prof. Y. R. Shen and R. P. Chin at UC, Berkeley; F. Kozlowski and P. Steiner from IFT, München; D. Kovalev and G. Polisski at TU, München; and O. Sreseli from Ioffe Institute, St. Petersburg. We thank Prof. H. A. Atwater and K. Shcheglov from CALTECH, Pasadena, for the implanted samples at the beginning of our work and to M. Kranz, IFT, who performed the majority of the Si implantation experiments. Finally, we thank Prof. Y. Kanemitsu from Tsukuba University, Prof. R. Hummel from University of Florida, Prof. B. K. Whaley and N. Hill from UC, Berkeley, and many of the authors cited in this review for discussions.

One of us (VPK) is grateful to DFG for the Habilitations-Stipendium.

REFERENCES

1. **A. V. Andrianov, D. I. Kovalev, N. N. Zinovev, I. D. Yaroshetskii:** *JETP Lett.* 58 (1993) 427, **H. Koyama, N. Koshida:** *Phys. Rev. B* 52 (1995) 2649.
2. **H. A. Atwater, K. V. Shcheglov, S. S. Wang, K. J. Vahala, R. C. Flagan, M. L. Brongersma, A. Polman:** *MRS Proc.* 316 (1994) 409.
3. **M. S. Brandt, H. D. Fuchs, M. Stutzmann, J. Weber, M. Cardona:** *Solid State Comm.* 81 (1992) 257.
4. **L. E. Brus:** *Proc. of NATO ASI School on Nanophase Materials,* G. C. Hadjipanajis, Ed., Kluwer Acc. Publ., 1993.
5. **A. Bsiesy, M. A. Hory, F. Gaspard, R. Herino, M. Ligeon, F. Muller, R. Romestein, J. C. Vial:** *MRS Proc.* 358 (1995) 619.

6. P. Calcott, K. J. Nash, L. T. Canham, M. J. Kane, D. Brumhead: *Phys. Cond. Matter* 5 (1993) 191; *J. Lumin.* 57 (1993) 1257.

7. L. T. Canham: *Appl. Phys. Lett.* 57 (1990) 1046.

8. L. T. Canham, W. Y Leong, T. I. Cox, M. I. Beale, K. J. Nash, P. Calcott, D. Brumhed, L. L Taylor, K. J. Marsh: *ICPS 21*, P. Jiang and H.-Z. Zheng, Eds., World Scientific, Singapore, 1993, 1411.

9. R. Carrius, R. Fischer, E. Holzenkämpfer: *J. Lumin.* 24/25 (1981) 47.

10. R. P. Chin, V. Petrova-Koch, Y. R. Shen: 188 ECS Meeting (Chicago) 1995, in press.

11. R. P. Chin, Y. R. Shen, V. Petrova-Koch: *Science* 270 (1995) 776.

12. P. Deak, M. Rosenbauer, M. Stutzmann, J. Weber, M. S. Brandt: *Phys. Rev. Lett.* 69 (1992) 2531.

13. C. Diaz, K. Li, J. C. Campbell, C. Tsai: *Porous Silicon*, Z. C. Feng and R. Tsu, Eds., World Scientific, Singapore, 1994, 261.

14. J. Diener, M. Ben-Chorin, D. Kovalev, S. Ganichev: *Phys. Rev. B* 52 (1995) 8617.

15. P. M. Fauchet: *Porous Silicon*, Z. C. Feng and R. Tsu, Eds., World Scientific, Singapore, 1994, 449.

16. T. Fischer, V. Petrova-Koch, K. Shcheglov, M. S. Brandt, F. Koch: E-MRS Spring Meeting, Straßburg, *Thin Solid Films* 276 (1996) 100.

17. A. Fojtik, H. Weller, S. Fichter, A. Henglein: *Chem Phys. Lett.* 134 (1978) 677.

18. A. Fojtik, A. Henglein: *Chem Phys. Lett.* 221 (1994) 363.

19. V. Gavrilenko, F. Koch: *Appl. Phys.* 77 (1995) 3288.

20. (a) M. S Hybertsen: *Phys. Rev. Lett.* 72 (1994) 1514; (b) N. A. Hill, B. K. Whaley: *MRS Proc.* 358 (1995) 25; (c) L. W. Wang, A. Zunger: *J. Phys. Chem.* 98(1994) 2158; (d) J. P. Proot, C. Delerue, G. Allan: *Appl. Phys. Lett.* 61 (1992) 1949; (e) B. Delley, E. F. Steigmeier: *Phys. Rev. B* 47 (1993) 1397; (f) T. Uda, M. Hirao: Light emission from novel silicon materials, *J. Phys. Soc. Jpn.* 63 (1994) 97; (g) C. Delerue, G. Allan, M. Lanoo: *Phys. Rev. B* 47 (1993) 11024.

21. Y. Kanemitsu, H. Uto, Y. Masumoto, T. Futagi, H. Mimura: *Phys. Rev. B* 48 (1993) 2827.

22. Y. Kanemitsu, T. Ogawa, K. Shiraishi, K. Takeda: *Phys. Rev. B* 48 (1993) 4883.

23. F. Koch, V. Petrova-Koch: *Porous Silicon*, Z. C. Feng, R. Tsu, Eds., World Scientific, Singapore, 1994, 133.

24. F. Koch, V. Petrova-Koch, T. Muschik: *J. Lumin.* 57 (1993) 271.

25. F. Koch: *MRS Proc.* 298 (1993) 319.

26. T. Komoda, J. P. Kelly, A. Nejim, K. P. Homewood, P. L. F. Hemment, B. J. Sealy: *MRS Proc.* 358 (1995) 163.

27. D. Kovalev, I. D. Yaroshetzkii, T. Muschik, V. Petrova-Koch, F. Koch: *Appl. Phys. Lett.* 64 (1994) 214.

28. F. Kozlowski, P. Steiner, A. Wiedenhofer, V. Petrova-Koch, T. Fischer, F. Koch: submitted to *Appl. Phys. Lett.*

29. J. M. Lauerhaas, M. J. Sailor: *Science,* 261 (1993) 1567.

30. V. Lehmann, U. Gösele: *Appl. Phys. Lett.* 58 (1991) 856.

31. V. Lehmann, B. Jobst, T. Muschik, A. Kux, V. Petrova-Koch: *Jpn. J. Appl. Phys.* 32 (1993) 23.

32. I. M. Lifshitz, V. V. Slesov: *JETP* 35 (1959) 331.

33. K. A. Littau, P. F. Szajowski, A.-J. Müller, A. R. Kortan, L. E. Brus: *J. Phys. Chem.* 97 (1993) 1224.

34. B. K. Meyer, V. Petrova-Koch, T. Muschik, H. Linke, P. Omling, V. Lehmann: *Appl. Phys. Lett.* 63 (1993) 1930.

35. Y. Mochizuki, M. Mizuta, Y. Ochiai, N. Ohkubo: *Phys. Rev. B* 46 (1992) 12353.

36. T. Muschik: Doctoral thesis, Technische Universität München, 1995.

37. P. Mutti, G. Ghislotti, S. Bertoni, G. F. Cerfoloni, L.-Meda, E. Grilli, M. Guzzi: *Appl. Phys. Lett.* 66 (1995) 851.

38. V. Petrova-Koch, A. Kux, F. Müller, T. Muschik, F. Koch, V. Lehmann: *MRS Proc.* 256 (1992) 41.

39. V. Petrova-Koch, T. Muschik: *Thin Solid Films* 255 (1995) 246.

40. V. Petrova-Koch, T. Muschik, A. Kux, B. K. Meyer, F. Koch, V. Lehmann: *Appl. Phys. Lett.* 61 (1992) 943.

41. V. Petrova-Koch, T. Muschik, D. Kovalev, F. Koch: *Proc. of the ECS, Sol. Electr. Div.* 94(10) (1994) 523.

42. V. Petrova-Koch, T. Muschik, G. Polisski, D. Kovalev: *MRS Proc.* 358 (1994) 483.

43. V. Petrova-Koch, T. Fischer, K. Shcheglov, F. Koch: 188 ECS meeting, Chicago (1995) to be published.

44. B. Pettinger, M. R. Schöppel, H. Gerischer: *Ber. Bunsen Ges.* 80 (1976) 849.

45. C. H. Perry, F. Lu, F. Namavar, N. M. Kalkhoran, R. A. Soref: *Appl. Phys. Lett.* 60 (1992) 3117.

46. S. M. Prokes, O. J. Glembocki, V. M. Bermudez, R. Kaplan, L. E. Friedersdorf, P. C. Searson: *Phys. Rev. B* 45 (1992) 13788.

47. A. J. Read, R. J. Needs, R. J. Nash, L. T. Canham, P. D. J. Calcott, A. Qteish: *Phys. Rev. Lett.* 69 (1992) 1332.

48. M. V. Rama Krishna, R. A Friesner: *J. Chem. Phys.* 96 (1992) 873.

49. S. Schuppler, S. L. Friedman, M. A. Marcus, D. L. Adler, Y.-H. Xie, F. M. Ross, T. D. Harris, W. L. Brown, Y. J. Chabal, L. E. Brus, P. H. Citrin: *Phys. Rev. Lett.* 72 (1994) 2648.

50. O. Sreseli, V. Petrova-Koch, D. I. Kovalev, T. Muschik, S. Hofreiter, F. Koch: *ICPS 22*, D. J. Lockwood, Ed., World Scientific, Singapore, 1995, 2117.

51. O. M. Sreseli, G. Polisski, D. Kovalev, D. N. Goryachev, L. V. Belyakov, F. Koch: 188 ECS meeting, (Chicago) 1995, in press.

52. P. Steiner, F. Kozlowski, W. Lang: *MRS Proc.* 358 (1995) 665.

53. P. Steiner : Doctoral thesis, Technische Universität München, 1996.

54. T. Shimizu-Iwayama, S. Nuko, K. Saitok: *Appl. Phys. Lett.* 65 (1994) 1814.

55. S. Veprek, T. Wirschem, M. Rückschloß, H. Tamura, J. Oswald: *MRS Proc.* 358 (1995) 99.

56. J. C. Vial, A. Bsiesy, F. Gaspard, R. Herino, M. Ligeon, F. Müller, R. Romesten: *Phys. Rev. B* 45 (1992) 14171.
57. G. Vincent, F. Leblanc, I. Sagnes, P. A. Badoz, A. Halimaoui: *J. Luminesc.* (1993) 232.
58. W. L. Wilson, P. F. Szajowski L. E. Brus: *Science* 262 (1993) 1242.
59. B. Wurfel, V. Petrova-Koch, A. Nikolov, S. Hofreiter: unpublished.
60. J. Y. Zhang, J. Y. Huang, Y. R. Shen, C. Chen: *J. Opt. Soc. Am. B* 10 (1993) 1758.
61. L. Zheng, W. Wang, H. C. Chen: *Appl. Phys. Lett.* 60 (1992) 986.

Chapter 5

Luminescence of Nanocrystalline Spark-Processed Silicon

Matthias H. Ludwig

Reviewed by R. E. Hummel

TABLE OF CONTENTS

1. INTRODUCTION

Silicon is one of the most studied materials and, in its crystalline form, the basis of microelectronics. However, it does not emit light efficiently at room temperature because of its indirect band gap structure. Only weak emissions occur in the near-infrared at energies below the energy gap ($E_g \approx 1.14$ eV), and there is no visible luminescence. Therefore, silicon was not considered of much interest in the search for optically active components. The situation changed in the late 1980s with the observation of visible photoluminescence (PL) at room temperature from Si nanoparticles prepared by sputter deposition.[1,2] Furthermore, anodic etching turned out to be a different preparation technique to generate luminescing porous silicon (por-Si) with a high quantum efficiency.[3,4] Since then, a substantial number of papers have been published, mainly attracted by the prospects for making Si-based optoelectronic devices, as well as by the simplicity of the electrochemical preparation process. Compared with bulk silicon, morphological structure, as well as optical and electronic properties, of por-Si are drastically altered. The preparation is based on a series of electrochemical or chemical etchings involving hydrofluoric (HF) acid solutions. This technique yields a highly columnar Si skeleton and eventually forms an array of wires with diameters below 10 nm, which luminesce effectively at room temperature in the orange red part of the visible spectrum.[5,6,7] A similarly strong PL, but much more shifted into the blue and green spectral regions, was observed after spark processing of silicon.[8] It is the purpose of this chapter to describe in some detail what spark processing means and what results can be achieved by applying it to silicon.

Several theories have been proposed during the last few years to explain the unexpected behavior of Si to luminesce in the visible range at room temperature. The are falling principally into three distinct categories: (1) the quantum-confinement hypothesis,[3,9,10] (2) the molecular agents/defect state model,[11,12] and (3) the surface-modified quantum picture.[13,14] According to the pure quantum-size effect (model 1), the reduced dimensions of Si quantum wires or dots are believed to cause an increase of the optical band gap, thus leading to a blue shift of excitonic luminescence into the visible region. Model 3 includes the positional and compositional irregularities in the crystallite–passivant interface, which cause a tail

103

of energies in the band gap. Radiative transitions from interfacial tail states are also expected to be size sensitive, thus shifting to higher energies with decreasing particle diameters. Alternatively, the large surface/volume ratio of nanocrystalline Si can be of paramount importance. Oxides or remains from the chemical etch may form passivation layers and may generate recombination centers for excited carriers. Different recombination channels due to molecular agents (such as Si complexes with O and H or siloxene derivatives) were proposed as candidates for the observed PL (model 2).[15,16] Also, an oxide-related origin for the PL was suggested.[17,18] However, despite many theoretical and experimental studies, the mechanism(s) of the visible PL from por-Si or nanocrystalline Si (nc-Si) remain unclear.

Spark processing has been shown by Hummel and Chang[8] to be an alternative technique for preparing an Si structure, which strongly photoluminesces at room temperature. These results have in principle been confirmed by other authors.[19,20] The PL spectra of spark-processed Si (sp-Si) consist of two strong and some minor bands. The most intense emissions occur in the UV/blue peaking at 385 nm or in the green part of the spectrum around 525 nm, depending on the conditions during spark processing. The PL intensities are comparable with those of the most efficient electrochemically etched (red luminescing) por-Si specimens.[21,22] Spark processing is a dry technique and applicable in different gaseous environments. The involvement of OH groups, hydrogen, or contaminations like fluorine can therefore be excluded as major contributors for the origin(s) of the PL of sp-Si.[23,24] Furthermore, the PL of sp-Si also remains stable with respect to peak intensity and peak wavelength after etching in buffered HF acid up to 30 min, which should substantially diminish the SiO_x layers present. These results suggest that radiative recombinations via defect structures in SiO_x, as brought forward by others,[12,17,25] may not be considered as the principal origin for the PL of sp-Si. Moreover, PL is also observed when spark processing is applied to different indirect and direct band gap semiconductors, causing PL peak wavelengths, which are strongly blue-shifted compared with the band gap energies of the unprocessed samples.[26,27] It has been therefore proposed that the luminescence mechanism is related to the presence of nanocrystalline structures as found by transmission electron microscopy (TEM) or Raman measurements.[21,28]

This chapter provides an overview of the current knowledge about the optical and structural properties of spark-processed silicon. Section 2 is designated to describe the formation process and the main morphological implications for the Si substrate. In the following sections the optical properties are discussed. These are then used to develop (as far as is possible at present) a reasonably consistent picture of the luminescence from spark-processed Si.

2. PREPARATION AND STRUCTURE

2.1. Principle of Spark Processing

Spark processing utilizes a plasma-based discharge between two electrodes at a given pressure and gaseous environment. At low pressure ($<5 \times 10^2$ Pa) the mechanism is closely related to plasma processing, which is frequently applied in the fabrication of electronic circuits, taking advantage of the anisotropic etching behavior.[29] With increasing pressure, the type of discharge changes from glow to coronary and further to spark and arc discharges. Detailed studies about the mechanism of spark formation in air, which is utilized here, have been published by Sukhodrev,[30] and publications can be traced back to, e.g., Warburg,[31] who reported in 1909 about ionizing effects and ozone generation during sparking in air. Essentially, the applied voltage forces the cathode to emit electrons, which, on their way to the anode, ionize gas atoms/molecules. Thus, a plasma is generated. Within about 10^{-7} s after initiation the plasma channel reaches an equilibrium temperature of the order of 30,000 K.[29] The sharp increase in temperature leads to a radial expansion of the channel at supersonic speed and pressures near 1×10^5 torr occur. The ions in turn are accelerated toward the cathode where the ion bombardment liberates sufficient heat to melt and evaporate a volume of the cathode material to a certain depth. The vapor redeposits immediately after the end of the discharge, reacting thereby with the ions of the bombardment or incorporating them into a solid phase. It is interesting to note at this point that the condensation step is often accompanied by a clustering process, which results in the creation of small-scale elementary crystallites in the nanometer range.[32] Gas-phase nucleation or glow-discharge polymerization effects are also known to occur in discharge systems operating at high pressures.[33]

Spark processing of Si is commonly performed by using a unidirectional pulsed 15-kV source. As shown in Figure 1, unipolar high-voltage direct current (DC) pulses are applied between, for example, a tungsten tip (anode) and a silicon wafer (cathode). The polarity is decisive regarding the PL output from the processed Si. Only a negligible PL intensity is observed when the polarity is reversed. Furthermore, the luminescing properties do not depend on the material of the anode. Spark processing

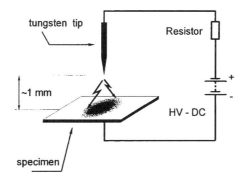

Figure 1 Scheme of the preparation technique. High-voltage DC pulses cause a discharge between the Si substrate (cathode) and a tip (anode).

between two Si electrodes or between Si and a metallic anode yields the same spectra and intensities, provided that no further conditions were changed. The electrodes are typically separated by a 1-mm gap. An increase in distance lowers the PL intensity but extends the processed area. Under low-pressure conditions ($\leq 1.5 \times 10^3$ Pa) a DC glow discharge in the milliampere range occurs.[19] At higher pressures ($\geq 10^5$ Pa) the increased negative differential resistance of the discharge plasma leads to a pulsed discharge behavior. The time constant of a single spark event or pulse length is estimated to be in the range of about 10 ns. With a repetition frequency of usually 16 kHz, a value determined by the voltage source, the off time between two consecutive spark events lasts about 60 μs. An estimation of the order of energy, which is set free during a discharge event, can be accomplished by taking the capacitance of the load into account. The latter ranges from 10 to 30 pF. Thus, an electrical charge of $q = 15$ kV * (10–30) pF = (0.15–0.45) mC is built up and moved to the load in about 10 ns, resulting in a peak current of $I = q/t = $ (15–45) A. This corresponds to an energy of approximately 5 mJ/pulse.

Macroscopically, spark processing of Si creates a circular pattern, the diameter of which depends on the processing time and the distance between the electrodes. (The microstructure will be discussed below.) In plan view, the center of the circular layer consists typically of a grayish-looking substance, surrounded by a brown halo. The thickness of the processed film, in cross-sectional view, is highest directly below the tungsten tip and has a formation rate of about 3 to 5 μm/min.

The luminescing properties of sp-films depend on a variety of parameters. Whereas a change of parameters, like the distance between electrodes, time of processing, or pulse frequency, mainly causes a linear variation of the PL intensity, modifications of the gaseous environment can substantially alter the spectral distribution of PL emission. For example, the specific spectra of sp-Si prepared in air under stagnant and under flowing conditions will be discussed in the next section. Also, a description of spectra obtained after spark processing in different gases and at various temperatures is presented below.

So far, most of the experiments have been performed employing Si where doping level and orientations of the single crystalline wafers were found to have no particular influence concerning luminescence distribution and intensity.[34]

2.2. Morphological Results

In the course of spark processing, the surface of the specimen becomes considerably roughened. The optical microimages in Figure 2 display the structure of the center area (directly below the anode). Whereas the image in Figure 2a was taken under white light illumination, the micrograph in Figure 2b depicts the PL of the structure when excited by ultraviolet (UV) laser light (325 nm, 3.81 eV). The bluish white PL clearly originates from the grainlike structures. Additionally, dark spots can be observed, which belong to holes having diameters of about 15 μm and extensions in depth up to 250 μm.

The scanning electron micrograph (SEM) in Figure 3a shows a similar area as above in plan view but with a higher resolution. The microstructure is distinguished by heavily indented and fine-shaped columns. Again, there are deep holes, which were inflicted by the spark discharge or, to be more precise, by the impact of the ion bombardment. Single crystalline particles in the nanometer range are revealed by high-resolution TEM as displayed in Figure 3b. It is noted that the {111} planes of the particles are rotated considerably with respect to each other. Consequently, the electron diffraction pattern of the crystalline area indicates polycrystallinity. The radii correspond to the d-spacing of silicon. The nano-

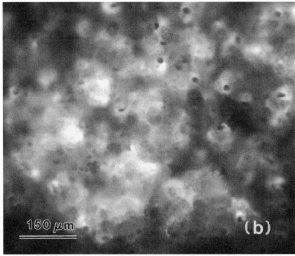

Figure 2 (a) Optical micrograph of sp-Si under white light illumination. (b) PL microimage from sp-Si excited by 325 nm UV light.

crystallites are surrounded by an amorphous phase, which consists of silicon oxide as identified by X-ray energy dispersion spectra.

Raman measurements essentially confirm the presence of nc-Si. Figure 4 compares the Raman spectra from bulk Si with those from sp-Si. The Raman shift for crystalline Si peaks at 520.9 cm^{-1} with a full width at half maximum (FWHM) of 4 cm^{-1}. For sp-Si Raman signals were found to shift to lower energies accompanied by peak broadening. The exact peak position and FWHM depend on sample preparation and spot position at the spark-processed area. Raman peak shifts can be caused by a phonon confinement in microstructures and correlate with the particle size. As described by Campbell and Fauchet[35] the peak position mainly depends on the size, while the width reflects the shape of the nanostructure. For the interpretation of the data from sp-Si, spherically shaped particles are assumed, which is consistent with the TEM results. From the spectra in Figure 4b and c, particle sizes of 3 to 5 nm can be inferred.[35,36]

In Figure 4c two types of fitting curves were used to model the experimental Raman data. The solid line represents a single Lorentzian fit. A far better description is achieved by a deconvolution on the basis of two Lorentzian curves (dotted line), which results in double peaks. Similarly shaped Raman spectra with a two-peak structure were also reported recently for por-Si and have been interpreted by a TO-LO phonon splitting.[37,38] Splitting of TO and LO phonons occurs when phonons at wave vectors

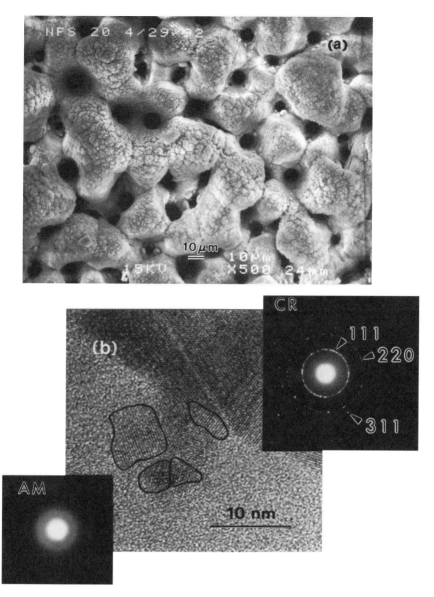

Figure 3 (a) SEM in plan view and (b) TEM of sp-Si. Some nanocrystallites are marked by circles. (CR = crystalline, AM = amorphous).

$q \neq 0$, which are nondegenerate in Si, contribute significantly to the Raman signal. A confinement of optical phonons in small crystallites may cause a relaxation of momentum conservation and, thus, the appearance of not only a peak shift but also a double-peak structure.

It should also be pointed out that no additional peaks near 480 and 464 cm^{-1} were observed, which indicates no substantial contribution to the Raman signal from a possible amorphous Si phase or from crystalline quartz, respectively.

The chemical composition of sp-Si has been evaluated by X-ray photoelectron spectroscopy (XPS) and Fourier transform infrared (FTIR) spectroscopy measurements.[34] Although both methods are integrating over a relatively large area (>100 μm for FTIR, ~5 mm for XPS in diameter) and are thus more sensitive to macroscopic properties, the results specify the presence and composition of certain species. From XPS measurements the core binding energies for Si 2p electrons in sp-Si were found to peak at 103.4 eV, which is close to the 103.7 eV value for Si^{4+}, that is, amorphous SiO$_2$ with silicon bonded to four nearest-neighbor oxygen atoms. Hence, XPS provides evidence for stoichiometric SiO$_2$ as a cover layer on sp-Si. A further study of depth-resolved X-ray emission spectra of sp-Si revealed that the strong

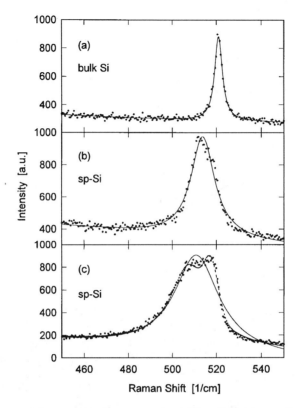

Figure 4   Room-temperature Raman shifts (λ_{exc} = 488 nm, 400 W cm^{-2}) for (a) crystalline Si, (b) and (c) different specimens of sp-Si. The peak maxima and FWHM are at (a) 520.9 and 4 cm^{-1}, (b) 513.8 and 6.5 cm^{-1}, (c) 510.9 and 24.0 cm^{-1} (solid line fit, deconvolution based on a single Lorentzian) or 507.7 and 517.9 cm^{-1} (dotted line, double Lorentzian).

oxidation extends up to 1000 nm.[39] The same degree of oxidation was found, however, for the luminescing center part of a spark-processed area, as well as for the nonluminescing halo which surrounds the optically active region.

FTIR spectra of sp-Si also favor those modes, which involve silicon–oxygen bonds in an SiO$_2$ stoichiometry. Figure 5 depicts a comparison of vibrational modes from fused SiO$_2$, sp-Si (prepared under air flow and under stagnant conditions), SiO, and Si in the range between 700 and 1400 cm^{-1}. It is evident that the spark-processed specimens are in close resemblance to the general absorption curve for fused SiO$_2$. Specifically, sp-Si displays a vibrational mode near to 1120 cm^{-1} with a shoulder at 1250 cm^{-1} which stands for asymmetric Si–O–Si stretchings and which are both characteristic for stoichiometric SiO$_2$. In contrast, the absorption modes for vapor-deposited SiO (curve d in Figure 5) and for Si with implanted O (curve e in Figure 5) are shifted or show no particular features, respectively. Additional absorption bands of sp-Si were observed[22] in the frequency ranges near 2250, 3350, and 3600 cm^{-1} which are commonly assigned to resonances from Si–H, N–H, and O–H bondings, respectively.[40,41] These bonds are surprisingly stable during annealings up to 900°C/3 h with only the O–H stretching modes being reduced significantly. However, when spark processing of Si is performed in a pure nitrogen atmosphere, PL intensity and distribution are very similar to those of a sample prepared under stagnant air conditions, but there are essentially no O–H bonding modes and a substantially diminished absorption due to Si–H vibrations. Therefore, analogies to the luminescence of siloxene (Si$_6$O$_3$H$_6$)[42] are not very plausible.

2.3. Formation Process

The above-given details allow a straightforward description of the way spark processing affects the silicon substrate. It can be assumed that the electric field, generated between the Si specimen (cathode) and a counterelectrode (anode) by a high-voltage source, forces electrons to leave the cathode, become accelerated, and ionize the surrounding gas molecules. The gas ions (such as oxygen or nitrogen), in turn, are accelerated toward the substrate and transfer a high momentum to the Si surface. Consequently,

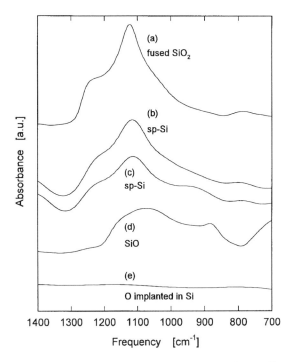

Figure 5 Absorption FTIR spectra of (a) fused SiO_2, (b) blue-luminescing sp-Si prepared under flowing air conditions, (c) green luminescing sp-Si prepared in stagnant air, (d) vapor-deposited SiO on Si, and (e) Si implanted by O (25 keV, 2×10^{16} cm^{-2}).

some areas of the silicon surface, which are impacted by the ion beam, undergo a localized flash evaporation.

The dimensions of the "worm" holes thereby created (diameters $d \approx 15$ μm and depths $l \approx 200$ μm) can be used to estimate the current density per discharge and the evaporation rate of silicon, which is vaporized in the time scale of a single spark event ($t \approx 10$ ns). As calculated above, peak currents of 15 to 45 A occur during a single discharge event, which have to leak through an impact spot on the sample surface, thereby creating the holes. Using the hole diameter to determine the spot area, the current density of a spark discharge can be estimated to be in the range of 10^7 A/cm^2. As a result of the impact, local flash evaporation of silicon occurs. The evaporation rate dm/dt can be determined by calculating the amount of silicon m, which filled a hole before evaporation:

$$\frac{dm}{dt} = \frac{d}{dt}\left(\frac{\pi}{4}\, d^2 l\rho\right) \tag{1}$$

With ρ (= 2.33 g cm^{-3}) as the density of Si, Equation 1 gives an evaporation rate of 8 g s^{-1}. This is too high a value to permit a dissipation of the Si vapor through the opening of the hole in the available time window. Thus, a high pressure forms within the hole. The vapor is mainly composed of Si atoms and ionized gaseous components as shown schematically in Figure 6. During the off times of the spark discharges the system cools down and the vapor phase immediately redeposits on the free surface. The resulting microstructure reflects the local, inhomogeneous composition of the quenched vapor phase. Several steps may be considered as decisive for the creation of luminescing centers. Among them are:

1. generation of localized structural defects by rapid quenching or ion implantation processes,
2. formation of wide gap amorphous silicon oxides (or nitrides or oxynitrides, depending on the gaseous environment) from well-intermixed vapor fractions, and
3. clustering of Si and eventual formation of nanocrystallites imbedded in amorphous Si oxide

Nucleation of Si is probably initiated in those parts of the vapor phase which are rich in Si. We have seen above (Figure 3) that randomly oriented Si nanocrystallites are imbedded within a noncrystalline

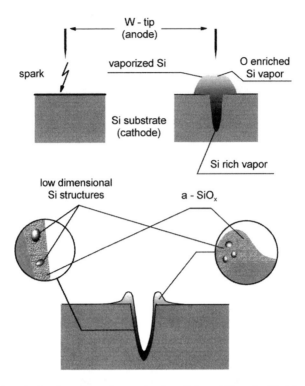

Figure 6 Principal steps in the formation of nanosized particles of Si embedded in silicon oxide by spark processing in air.

oxidic matrix. An epitaxial growth of Si nanoparticles on top of the substrate is probably hindered by the coexisting oxidic phase.

It is interesting to note in this context that a reversed polarity during spark processing (with the Si sample as anode) causes a different structure which photoluminesce only at a much lower intensity. The Si sample is then impacted predominantly by the electrons during the spark discharge and not by gaseous ions. Therefore, the transferred momentum is much lower and the generated heat is not sufficient for a comparable flash evaporation of Si. Additionally, the chances to create a specific microstructure, such as Si nanocrystallites embedded in an oxidic matrix, are rather low. An Si specimen processed in such a way exhibits a nearly flat surface without the structural peculiarities shown in Figures 2 and 3 and yields substantially no visible PL.

The observation of Si nanocrystallites imbedded within the matrix of sp-Si has raised much speculation about the possible role of quantum dots as origin of novel luminescing properties. The following sections focus therefore on relevant optical aspects of sp-Si to shed some more light on the possible mechanism(s) for luminescence. Additionally, comparisons to por-Si, amorphous Si, and silicon oxide are drawn.

3. OPTICAL PROPERTIES

3.1. Photoluminescence

Spark-processed Si typically displays strong PL at room temperature with maximum intensities in the blue, green, or red range, depending on the conditions during preparation. Figure 7 depicts several spectra, which were obtained by exciting the specimens with the 325-nm line (3.8 eV) of a continuous wave (cw) HeCd laser. This laser line was found to be high enough in energy to be sufficiently separated from the PL peak energies. A common feature for all PL bands of sp-Si consists in the linear variation of the PL intensity with excitation power when changed between 0.01 and 10 mW cm^{-2}. Also, the shapes of PL spectra are not affected by varying the excitation intensity. All spectra presented in Figure 7 were measured under identical conditions for wavelength and intensity comparisons. They were also corrected for the spectral response of the detection system.

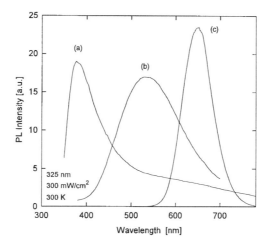

Figure 7 Room-temperature PL spectra (excited by the 325-nm line of an HeCd laser, 0.3 W cm^{-2}) of spark-processed Si, prepared under (a) flowing and (b) stagnant air conditions. All specimens were measured under identical conditions. For comparison, the red PL of por-Si is added (curve a).

When spark processing is performed in air, the PL peak position can be easily switched by a variation of process parameters. UV/blue PL (curve a in Figure 7) with peak maxima at 385 ± 10 nm (~3.22 eV) occurs after spark processing under flowing air conditions; that is, a stream of compressed, dry air is directed toward the discharges between the electrodes. The blue intensity increases with the velocity of air flow, but the peak position stays alike. An additional, secondary peak is sometimes observed at the low-energy side of the blue band, centered near 450 nm.

If, however, spark processing is conducted in stagnant air, PL occurs at higher wavelengths (curve b in Figure 7) having maxima around 525 ± 15 nm (~2.36 eV). The green PL intensity can be considerably enhanced when the specimen is heated during spark processing. Again, the peak position is essentially fixed, showing only a slight red shift at higher wafer temperatures. For certain conditions, both bands (blue and green peaks) can be observed for the same sample. It is, however, important to stress that no gradual peak shift or transition results.

A noticeable feature in the shape of the UV/blue emission spectrum (curve a of Figure 7) consists in its distinctive intensity tail into the red or toward lower energies. This tail can be defolded into a separate red peak near 650 nm (1.91 eV) when the specimen is exposed to the 514.5 nm (2.41 eV) laser line of an Argon ion laser.[8] There appears no shift in the peak position when the excitation wavelength is changed to 488 nm (2.54 eV). It is therefore reasonable to ascribe this third PL band to a specific radiative transition rather than to an extended distribution of states. A further increase of the excitation energy, however, triggers the UV/blue PL band and a superimposed spectrum as shown in Figure 7 (curve a) is eventually obtained.

Additional PL bands have been reported by Rüter et al.[19,43,44] after performing the plasma discharge treatment of Si in a low pressure system in the presence of H_2/CH_3OH, several alcohols, or $H_2/SiCl_4^!$. Emissions obtained under these conditions peak in the range from 430 to 550 nm. Furthermore, it has been shown by Steigmeier et al.[20] that the emission spectrum is widely independent of the gas species when the preparation is accomplished in nitorgen, oxygen, or hydrogen atmospheres. The resulting spectra display peaks at 620-660 nm. However, no PL response was found after spark-processing in helium and argon environments.

It is quite evident from Figure 7 that PL intensities from sp-Si are of similar magnitude as the orange/red luminescence from por-Si. An external quantum efficiency of up to 5% is observed at room temperature in both cases. Intensity differences appear, however, when the blue/green bands of sp-Si are compared with blue-luminescing por-Si. The latter PL emerges after rapid thermal annealings of por-Si in oxidizing atmospheres[45,46] and peaks between 460 and 480 nm. The PL intensity of blue-luminescing por-Si, however, was found to be several orders of magnitude smaller than that of sp-Si.[22]

In order to shed some more light on the origin(s) of PL in sp-Si one has to consider the three structures extracted in the foregoing sections nanoparticles, structural defects, and wide-gap alloys. Specific characteristics of clusters and defects will be addressed below within the sections dealing with temper-ature-dependent properties. The following discussion is focused on the matrix material which is provided mainly by silicon oxide.

It is a natural thought to compare the luminescing properties of sp-Si with those found in amorphous and fused silica. Despite the fact that silicon oxide is a wide–band gap material ($E_g \approx 9$ eV), there are various luminescence bands having peak energies at 1.9, 2.2, 2.7, 3.1, 4.3 to 4.7, and 6.7 eV (650, 563, 460, 400, 288, and 185 nm).[47,49] Most of these bands are assigned to different defect structures and found to increase in intensity after X-ray, gamma, or neutron irradiation; that is, after a deliberate raise of the defect density. Even though below–band gap optical excitation is sufficient to activate these radiative transitions, in all but one case (the 1.9-eV PL band) relatively large Stokes shifts occur. Typically, absorption bands are found at 7.6 eV (165 nm), 5.1 eV (240 nm), 4.8 eV (258 nm), or 2.0 eV (620 nm).[50]

In particular, an excitation energy of at least 5.0 eV is necessary to activate the bands at 2.7 and 4.3 to 4.7 eV. Both have been shown to result from triplet-to-singlet and singlet-to-singlet transitions, respectively, at Si–Si bonds located at an oxygen-vacancy site.[51,52] The 2.7-eV (460-nm) radiation is merely observed in oxygen-deficient silica with a low content of OH groups.[53] These aspects are of particular interest, as the PL spectra of sp-Si display no peak within that range (see Figures 7 and 10), in contrast to oxidized porous Si (see above). Stokes shifts of about 2 eV are characteristic for the 3.1 and 2.2 eV (400 and 563 nm) bands of silicon oxide which are normally extremely weak at room temperature. The presence, thus, is not very often reported and sometimes assigned to an unknown impurity.[48,54] For the 1.9-eV (650 nm) band, instead, a clear correlation to the nonbridging oxygen hole center (NBOHC) is drawn. The latter can be excited by two absorption bands, one at 4.8 eV and another one at 2.0 eV.[49,50]

Setting these considerations in context with the PL investigation of sp-Si described above, one can draw essentially two conclusions: (1) there is no direct match for the two main PL bands of sp-Si when compared to silica or damaged SiO_2, and (2) the here utilized laser line is to low in energy to achieve an efficient excitation of radiative transitions via known defects in silicon oxide (with the exception of the 1.9 eV (650 nm) band).

The first conclusion becomes even more obvious by comparing spectra of sp-Si with those of various types of silicon oxide as shown in Figure 8. Whereas essentially no PL emerges from fused silica (curve f in Figure 8), weak PL can be gathered from damaged silicon oxide, centering between 440 and 470 nm (2.82–2.64 eV), when the specimens are excited by 3.81 eV laser light at room temperature. The intensities, however, are at least three orders of magnitude smaller. The defect structures, which cause the radiative transitions in damaged silicon oxide shown in Figure 8, are related to oxygen deficiency centers. The same centers are probably responsible for the blue PL of oxidized, chemically etched por-Si. In other words, the PL of sp-Si does not only peak at different wavelengths, it is also much higher in intensity when compared to defect luminescence from silicon oxides.

As stated above, defolding of the tail PL of sp-Si in the red leads to a distinct band at 650 nm which is probably related to the presence of NBOHCs. Such centers are known, e.g., to appear in silica fibers when they were rapidly quenched from temperatures above 2000°C and result from the fission of strained Si-O bond.[49,50] Luminescence from the same center can be found again in sp-Si when excited with electron beams. (A supplementary discussion of the red luminescence around 2 eV is given in the section about cathodoluminescence, see below.)

3.1.1. PL Decay

Time-resolved PL measurements reveal that the luminescence of sp-Si decays very rapidly with lifetimes in the 5 to 10 ns range.[55] Figure 9 depicts the detailed decay characteristics of blue-luminescing sp-Si measured at various wavelengths. Similar decay characteristics were found for green-luminescing sp-Si (not shown). In all cases the PL decay does not depend significantly on the emission wavelengths. The decay dynamics is nonexponential, and the curves cannot be fitted by a bimolecular recombination law nor by a stretched exponential function. Specifically, an initially fast decay with a time constant of about 2 ns is superpositioned by a much slower decaying tail with time constants of about 10 to 30 ns. Nevertheless, these short lifetimes support the above-discussed model of efficiently competing green or blue channels, respectively, with transitions in the red.

Furthermore, the fast response suggests that the recombination is based on geminate carriers and thus on highly localized radiative centers, which provide a strong overlap of the wave functions. Taking these aspects into consideration, one would expect only a weak shift of the peak positions at low-temperature PL measurements.

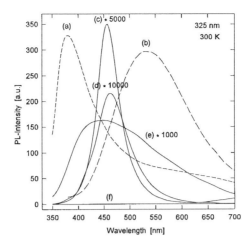

Figure 8 PL spectra of sp-Si and several Si–O compounds measured under identical conditions (325-nm cw laser, room temperature). (a) Blue-luminescing sp-Si after spark processing in flowing air, (b) green-luminescing sp-Si after spark processing in stagnant air at a temperature of 160°C, (c) X-ray damaged (40 keV) Borsilicate glass, (d) Si (IV) oxide powder (99.999% purity), (e) Si ion-implanted in fused SiO_2, (25 keV, 2×10^{11} cm^{-2}), (f) fused SiO_2.

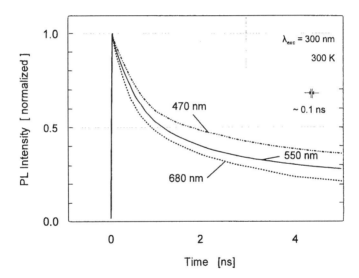

Figure 9 PL decays for sp-Si taken at several wavelengths as indicated. The excitation wavelength is 300 nm with pulse lengths of 2 to 3 ps.

The observed decay time constants of sp-Si are substantially shorter than the lifetimes of red-luminescing por-Si[56] or amorphous SiO_2[57] (typically 1 to 10 μs at room temperature).[57] However, an analogous fast PL decay in the range of 10 ns has been reported for the blue/green band of strongly oxidized por-Si.[14,58] The PL bands of SiO_2, which fall in the same wavelength range, are also much slower in their decay characteristics. For the 1.9, 2.2, 2.7, and 3.1 eV transitions in silicon oxide, time constants of 12 to 14 μs, 60 to 80 ns, 9.4 to 10 ms, and 110 μs, respectively, were measured.[49,53,59,60] It seems, therefore, unreasonable to consider transitions as they appear in SiO_2 as possible candidates to explain the blue and green PL of sp-Si.

3.1.2. Temperature-Dependent PL

The temperature-dependent PL spectra of blue- and green-luminescing sp-Si are shown in Figure 10. Although both bands display a broad and essentially featureless distribution, there are distinctive changes with temperature. As depicted in Figure 10a, the peak position of the blue PL stays fixed at about 385 nm and there is a concomitant second peak at slightly lower energy (~450 nm). An intensity tail into the

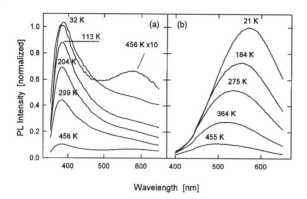

Figure 10 Temperature-dependent PL spectra of (a) blue- and (b) green-luminescing sp-Si, excited by the 325-nm line of a cw HeCd laser. Measurement temperatures are indicated.

yellow/red part of the spectrum is additionally discernible over the whole temperature range. At low temperatures the UV/blue band clearly dominates. With increasing temperatures the PL intensities generally decrease. However, the intensity decrease of the high-energy part of the spectrum is more pronounced, and at temperatures above 400 K the former tail luminescence emerges clearly as a separate band with a peak maximum near 570 nm (2.17 eV). To illustrate the changes in relative intensities between the blue and green/yellow PL contributions, Figure 10a also displays the spectrum taken at 456 K enhanced by a factor of ten. Quite evidently, the UV band is degrading stronger with increasing temperature than the tail luminescence.

The spectra of green-luminescing sp-Si obtained in the temperature range from 21 to 455 K are shown in Figure 10b. The most remarkable feature here is the shift of the center of the PL to lower wavelengths with increasing temperature accompanied by an overall decrease in intensity. The unusual red-shift is in strict contrast to the widening of bandgaps observed in c-Si, silica, or quantum dots. It resembles, however, very closely the thermal behavior of amorphous $SiO_x N_y$ or SiN_x excited by sub-bandgap energies as reported for a-SiN_x layers having high N contents in a temperature range up to 300 K by Austin et al.[61], and for annealing temperatures up to 800°C by Wakita et al.[62]

The fact, that both major PL bands of sp-Si do not follow the temperature-dependent relationships known for bandgap or excitonic transitions (either in bulk or quantum sized materials) is very decisive regarding possible mechanisms of PL. It is more reasonable to consider localized states playing a significant role in a similar way as proposed for amorphous Si:H where radiative transitions occur between band-tail states[62,63] Assuming such states to be embedded in an insulating matrix would prevent photoexcited electrons from diffusing away, which in turn suppresses nonradiative decay paths and may explain the high quantum efficiency. The stronger degradation effects for the high-energy transitions in Figure 10a can then be understood in terms of a thermal release of carriers from band-tail states: thermal excitation transfers weakly bound carriers to nonlocalized states where they recombine nonradiatively.

Figure 11a depicts the temperature dependence of the integrated PL intensity $I(T)$. At lower temperatures at about 20 K the PL intensities of sp-Si are generally enhanced by a factor of 10 to 20 when compared with those at room temperature. Above 290 K thermal quenching of $I(T)$ dominates, which can be described by an exponential dependence on $1/T$ with dissociation energies of 156 and 241 meV for the blue and green PL band, respectively. The lower dissociation energy for the blue luminescence confirms the above-described observation of a stronger degradation with temperature. However, these values are about three times larger than those found in por-Si[65,66] and indicate a stronger localization. Under the assumption of mainly coulombic attraction between an excited electron and a hole, the binding energy ΔE is

$$\Delta E = \frac{1}{4\pi\varepsilon\varepsilon_o} \frac{e^2}{d} \tag{2}$$

where d is the electron–hole distance and ε corresponds to the dielectric constant. By taking the average of the experimental activation energies of about 0.2 eV as equivalent to the binding energy, a distance

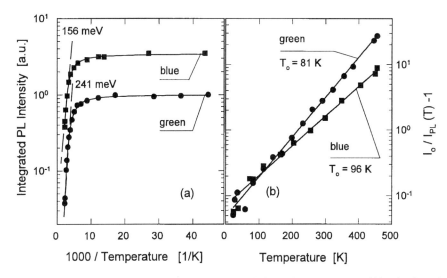

Figure 11 (a) Temperature dependence of the integrated PL intensity for green- and blue-luminescing sp-Si. (b) Logarithmic plot of $I_o/I(T) - 1$ vs. temperature for the same specimens.

of 0.6 nm results in Si ($\varepsilon = 12$) and of 1.9 nm in SiO_2 ($\varepsilon = 3.9$). Both values favor strong localization effects. Specifically, for Si the calculated electron–hole distance is close to the lattice constant (0.543 nm). The assumption of an electron–hole pair does not necessarily imply the presence of excitons (although it is a possibility). Trapping of carriers at the same center or at centers with states in the bandtail, which are close in distance, are alternative ways for pairing.

At lower temperatures ($T < 290$ K) the activation energies change continuously. A similar situation is known for amorphous materials. Collins et al.[67] have shown that the dependence of luminescence intensity on temperature $I(T)$ in amorphous Si:H can be expressed in the empirical form

$$I(T) = I_o \left/ \left[1 + c \exp\left(\frac{T}{T_o} \right) \right] \right.$$ (3)

where c is a constant independent of the temperature T. Equation 3 describes the temperature dependence of PL in disordered solids in terms of a characteristic temperature T_o, which is a measure of the system disorder and related to the tail width of the density of localized states into the gap region. The experimental data presented in Figure 10 were analyzed by using Equation 3, and the fitted results are shown by solid lines in Figure 11. In Figure 11b the logarithm of $I_o/I(T) - 1$ is plotted vs. temperature. The good fit by straight lines strongly indicates the presence of exponential band tails as assumed in Equation 3. Values for T_o are typically found in the range between 70 and 110 K. They are not dependent on the PL emission band nor the excitation power. For comparison, characteristic temperatures for por-Si vary between 40 and 70 K.[68,69] In conclusion, both materials show exponential band tails and considerable disorder. However, sp-Si is characterized by a higher density of tail states and higher degree of disorder.

It should be noted in passing that the PL intensity of sp-Si does not resemble the anomalous intensity behavior of por-Si. Specifically, for high-porosity samples[70] the intensity first increases as the temperature is lowered, but then decreases with further lowering of temperature. Maxima of the PL intensities in por-Si appear between 100 and 200 K.[71,72] These findings were discussed in terms of a transition in the electrical conduction mode from thermal activation to hopping-type.[73] The fact that the PL intensity of sp-Si shows no deterioration at lower temperatures may be interpreted by assuming that nonradiative transitions (via dangling bonds?) are efficiently blocked out, quite contrary to the situation with porous Si.

3.1.3. Degradation Characteristics

The PL of spark-processed Si is distinguished by a remarkable stability against laser radiation, HF etching, and heat treatment. The changes of intensity in time when exposed to UV laser light (325 nm,

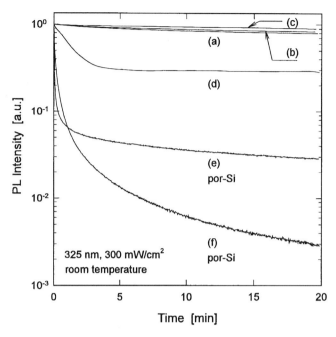

Figure 12 Decrease of PL intensity as a function of exposure to 325-nm laser light (0.3 W cm⁻²) for spark-processed Si prepared under (a) flowing and (b) stagnant air conditions, (c) air blown and subsequently annealed in air at 900°C for 3 h and (d) spark processed in pure nitrogen. For comparison, the intensity degradation of por-Si is added (e and f).

0.3 W/cm²) are shown in Figure 12 for differently prepared specimens. It is observed that, in the given time interval, spark processing in air provides a substance which emits essentially stable PL under UV laser excitation. Por-Si (Figure 12, curves e and f to clear) instead loses its PL intensity by about two orders of magnitude within the first 10 min when exposed to the same laser light. The difference between the stability of sp-Si and por-Si becomes even larger when the laser power density is increased. The rapid degradation of por-Si was shown to correlate with hydrogen loss due to high-energy illumination (threshold near 3.0 eV)[74] or with the release of SiH_3 due to heating effects.[75] It is therefore argued that other passivants, such as SiO_2, may stabilize the optical properties. Hence, the highest stability was found for Si spark processed under flowing air conditions and subsequently annealed in air at 900°C for 3 h (Figure 12, curve c). Spark processing in nitrogen atmosphere provides a somewhat less stable PL than that prepared in air (Figure 12, curve d).

The PL intensity of sp-Si and the wavelength of the PL maxima remain fairly stable during heat treatments in air or nitrogen, as Figure 13 demonstrates for the blue and the green band, respectively. Up to an annealing temperature of 600°C the fluctuation patterns in intensity for both bands are very similar. A temporary PL increase is observed after the 450°C anneal. This feature is commonly attributed[76,77] to the conversion of chemisorbed water to oxide-related silanol-type groups ($\equiv$Si–O–H), which photoluminesce in the blue/green. At temperatures above 450°C silanol desorbs or decomposes, eventually resulting in a dehydrogenated surface above 600°C. Concomitantly, the blue/green PL of silanol is essentially eliminated. In other words, the additional luminescence after a 450°C anneal can be traced to a superposition of PL from silanol on the PL from sp-Si. At lower temperatures (<350°C) the Si–OH groups are covered by water and therefore ineffective for PL, and after annealings at higher temperatures (>600°C) they are no longer present.

The PL of blue-luminescing sp-Si retains its intensity even for annealings up to 900°C, and there are no significant changes in the peak positions up to 1100°C. The sharp shift of the PL maximum after a 1200°C anneal can be correlated to the phase separation from SiO to SiO_2 and Si, which occurs at temperatures above 1050°C. It is interesting to note that after this heat treatment the PL peak shifts to 460 nm, which is in close resemblance to defect-induced PL in (defects containing) silicon oxides[78] and to the blue PL in strongly oxidized por-Si.[17]

The green band instead increases in intensity after annealings above 600°C, and the peak position stays essentially fixed. Very similar properties were found by Augustine et al.[79] for amorphous Si-rich

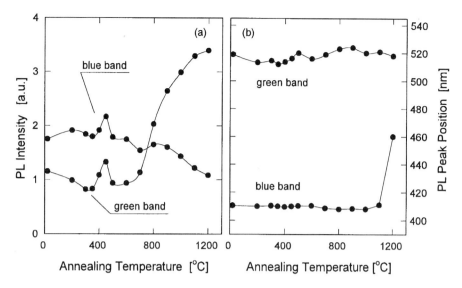

Figure 13 Relative intensities and peak positions of the blue and the green PL bands of sp-Si as a function of annealing temperature. Samples were annealed in a dry N₂ ambient for 30 min each. The data were obtained utilizing the 325-nm (3.82-eV) line of a cw HeCd laser at room temperature.

oxynitride (SiO$_x$N$_y$:H) films prepared by plasma-enhanced chemical vapor deposition. The room-temperature PL of SiO$_x$N$_y$:H films centers at 550 nm and increases by nearly three orders of magnitude after postdeposition annealing at temperatures larger than 850°C. The exact peak position depends on the stoichiometry of the composition and peak shifts in the visible range from 700 nm (1.8 eV) to 530 nm (2.34 eV) with increasing nitrogen content have been reported.[62,80] Furthermore, there were no phase segregation effects observed in silicon oxynitrides after heat treatments above 1100°C, which led to the conclusion that the presence of N or Si–N bonds prevent recrystallization and segregation effects by acting as a diffusion barrier for oxygen.[79]

As the annealing behavior of both bands is not affected by the annealing ambient (air or nitrogen), one might conclude that the observed luminescence is an intrinsic (but enhanced) property having distinct origins. Specifically, the blue band seems to be related to the presence of silicon oxides, and the green band displays similarities to silicon (oxy-)nitrides. (Still, the PL intensities of sp-Si exceed those of the "pure" Si compounds by several orders of magnitude.)

In contrast to the annealing behavior of sp-Si and SiO$_x$N$_y$:H the results for por-Si are strikingly different. The red PL of por-Si substantially decreases after heat treatments up to 700°C and completely vanishes between 900 and 1000°C. Concomitantly, a blue band emerges, centered near 460 nm.[58]

The effects of etching in buffered HF (NH₄F 40% 6 parts, HF 1 part) on peak positions and intensities of the UV/blue and green PL bands of sp-Si are shown in Figure 14 in a series of successive measurements. Buffered HF efficiently dissolves thermally grown SiO₂ with an etch rate of about 50 to 100 nm/min at room temperature. The etch rate for silicon nitride instead, when grown at a temperature of 1100°C, has been found to be as low as 14 nm/min.[81] (Therefore, silicon nitride is often used as a protection layer to cover silicon oxide.) For sp-Si an etch rate of 150 to 200 nm/min was found,[34] which is larger than both rates given above. The increase can be attributed to the rough structure and, therefore, to the increased surface area for sp-Si. Despite the rate enhancement the PL intensities, as well as the maxima positions, are remarkably stable. Within the first 30 min of consecutive etching periods there is no significant alteration of these properties. At further etching up to 3 h the intensities of both bands drop considerably by about two orders of magnitude. This decrease is accompanied by a complete loss of the initial peak positions, while the UV/blue band is red-shifted to a final value of 460 nm after 3 h of cumulative etching. The green emission, instead, shows first a strong blue shift, followed by a red shift to an equilibrium value of, again, 460 nm. In other words, both PL bands vanish completely during a 3-h etch, and the remaining luminescence shows essentially the same features: peak position at 460 nm and substantially reduced intensity. It should be emphasized at this point that the degraded PL at 460 nm is very similar in wavelength and lowered intensity to the PL found in Si (IV) oxide, irradiated fused silica, and in strongly oxidized por-Si (see Figure 8).

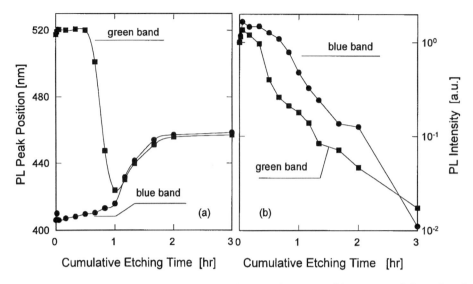

Figure 14   (a) PL maxima and (b) PL intensities of silicon spark processed in stagnant air (green) and under flowing air conditions (blue) as a function of the total etching time in buffered HF. The PL was measured at room temperature utilizing the 325-nm line of a cw HeCd laser.

In order to describe these features in a more comprehensive model, one may suppose the presence of three independent radiative bands, overlapping each other and centered at the following wavelengths: (1) at 385 nm (UV/blue), (2) at 460 nm ("oxide"-blue), and (3) at 520 nm (green). In the pre-etched stage the green and the UV/blue PL, respectively, are dominating in the corresponding specimens. After etching the green sample for 30 min, the green centers rapidly diminish leaving the UV/blue-luminescing states as the second in intensity, and the apparent blue shift results from the superposition. Eventually, the UV/blue centers are dissolved or washed out, too, and the weak PL at 460 nm can be detected. A similar process occurs for the originally UV/blue-luminescing sample with the only difference that there are no, or not enough, green centers to take over for an intermediate period. This model suggests a higher etch resistance for the blue centers, which is substantiated by the slower intensity decay over etching time as shown in Figure 14b.

3.2. Cathodoluminescence

Luminescence measurements using highly energetic electron beams as an excitation source provide a useful tool to investigate radiative transitions, which occur in silicon oxide. For the present study 10-keV electrons were used, which can easily create nonequilibrium carriers across the band gap of SiO_2 ($E_g \geq 8$ eV) and excite electrons into the absorption bands at 5.4 and 4.8 eV in amorphous silicon oxide. Thus, cathodoluminescence (CL) is a complementary method to PL (particularly, when a 3.81-eV light source is used, as in the present study) and may help to answer the question whether or not the PL from sp-Si stems from silicon oxide.

Figure 15a depicts some CL spectra taken on sp-Si, which display two emission bands, one centered at 480 nm and the second around 650 nm, respectively. It is evident from comparison with the PL response in Figure 7 that the distribution and the peak wavelengths of the CL emissions differ from those of PL spectra for the same material. Moreover, varied conditions during spark processing, which cause different PL maxima, have virtually no effect on the CL spectra. Specifically, both CL bands are independent of the preparation procedures of sp-Si.

As temperatures are lowered a competitive behavior between the blue and the red CL channels appears.[82] As shown in Figure 15, by increasing the temperature from 6 to 275 K the red CL loses about the same amount of intensity as the blue CL gains. The total of the integrated CL intensities essentially stays constant until above 275 K thermal quenching commences. The competitive behavior can be explained by assuming a smaller recombination rate (larger decay times) for the blue CL band compared with the red channel. Indeed, for sp-Si the CL decay constant for the blue sample was found to be about 50 ms at 5 K, whereas the time constant of the red CL is at least one order of magnitude shorter.[81] A similar blue CL is obtained in thermally grown SiO_2 having lifetimes considerably longer than 1 ms. The slow decaying, blue CL at wavelengths between 445 and 475 nm (2.8 and 2.6 eV) is an ubiquitous

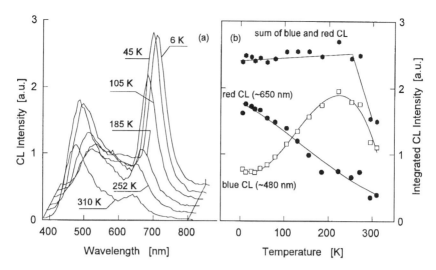

Figure 15 (a) CL spectra of sp-Si at various temperatures as indicated. (b) Temperature dependence of the integrated CL intensities of sp-Si for the red band (curve fitted according to Equation 3), the blue band (curve to guide the eye), and the total integrated CL intensity (curve to guide the eye). The specimens were excited by a 10-keV, 28-nA electron beam.

property in crystalline and amorphous SiO_2, often found to increase in Si–SiO_2 interface regions or in oxygen-deficient silicon oxide. It is attributed to transitions from triplet excited states of a twofold-coordinated silicon center.[83,84] Furthermore, time-delayed blue PL was reported for strongly oxidized por-Si by Kux et al.[64] with time constants in the order of 1 s when the excitation energy is above 4.3 eV.

However, all these results are in sharp contrast when compared with the blue PL band of sp-Si, which is not only blue-shifted to 385 nm (3.22 eV) but also much faster with decay times of 10 ns (see above). Additionally, sp-Si measured under identical conditions as described by Kux et al.[64] did not display a corresponding slow component of the blue PL.

Additionally, the blue CL band is characterized by a particular temperature dependence, not found during PL investigations of sp-Si. With increasing temperature, the CL intensity increases, peaking at about 275 K. At higher temperature the intensity decreases again. A fit by Equation 3 is not possible. A similar temperature behavior was found for the blue CL band at 468 nm (2.65 eV) in amorphous SiO_2 by Skuja et al.[59] Thus, the temperature dependence, the peak position, and the decay times of the blue CL from sp-Si are in close resemblance to the blue CL of silicon oxide and probably related to the presence of a twofold-coordinated silicon center.

The temperature dependence for the red CL band of sp-Si can be well described by Equation 3, and the solid line in Figure 15b represents the fitted solution. For the characteristic temperature T_o, values in the range between 97 and 106 K were found, which suggest disorder to a similar degree as detected by PL measurements. The fact that both CL and PL seem to have a similar disorder is not surprising, if one considers the same procedure of preparation. Spark processing generates comparable environments for all optically active centers.

As discussed earlier, the emission near 650 nm is commonly attributed to nonbridging oxygen centers (NBOC) having absorption bands near 4.8 eV and around 2 eV.[60,81] These centers are often found in radiation-damaged or electron-bombarded silica. As the spark discharge involves highly accelerated electrons and ions, the generation of a similar defect is very likely to occur in the proximity of spark-generated holes in silicon. NBOCs are therefore suggested to be responsible for the red CL band of sp-Si. (An alternative model ascribes the 1.9-eV CL to the presence of interstitial molecular ozone in a silica network,[85] which also seems to be applicable to sp-Si, as ozone is generated during the spark process.)

Another aspect of the red emission near 650 nm deserves attention. One of the linked absorption bands is low enough in energy to become excited by the laser lines used in this study. Indeed, a red PL is discernible when sp-Si is exposed to a low-energy light source ($2.0 < E_{exc} < 2.5$ eV). However, as shown above, the red PL disappears when the photon energy is raised (>3 eV), that is, because additional and fast recombination channels (blue and green PL bands) become activated. When sp-Si is exposed to highly energetic electrons, it is reacting in a different way. There are no other fast recombination paths accessible under CL conditions. The only second channel found in this study, the blue CL band at 480 nm, has a higher lifetime than the red CL band, and there is no resemblance to the blue PL band.

To explain these differences one should take into account that the penetration and thus activation depths for both methods are dissimilar. The depth distribution for the primary electrons of the CL study can be calculated with the help of the range–energy relation developed by Kenaya and Okayama[86] for 10-keV electrons the center of depth in SiO_2 lies at 2.5 to 3 μm with a maximum penetration up to 8 μm. (The distribution is not affected by diffusion of carriers as the diffusion length of electrons in SiO_2 has a maximum range of 3 to 5 nm.) On the other hand, amorphous silicon dioxide is almost transparent for light with an absorption coefficient of 10^{-3} cm^{-1} at 3.8 eV (325 nm). Even for $SiO_{1.5}$ the characteristic penetration depth for light of this wavelength still equals about 10 to 20 μm.[86] It is therefore reasonable to assume that CL and PL derive their light emissions from different depths and thus from probably dissimilar morphological configurations of sp-Si. The CL spectra are more surface related, and the PL spectra cover a larger depth range and may result from deeper-lying structures. Similar conclusions were also drawn for the relation of CL and PL in por-Si with the CL spectra mainly attributed to luminescing defects in silicon oxide.[87,88]

A further confirmation for the distinct origins of PL and CL in sp-Si was provided by an attempt to assign the light generation to certain positions in the microstructure.[89] Thereby, the PL was found to stem mainly from the granular structure (compare Figure 2b), whereas the CL dominantly originates from the walls of the holes, which were created during the preparation process. To illustrate this, Figure 16 depicts the surface topographies of the same area of sp-Si as observed by secondary electron (Figure 16a) and spectrally resolved CL (Figure 16b) scanning microimages. The secondary electron micrograph (Figure 16a) clearly displays black-appearing holes. The whitish areas result from charging effects, indicating the presence of electrically insulated regions (possibly SiO_2). The same black holes are the main source of luminescence in Figure 16b, which shows the red CL, taken at 644 nm and at room temperature. Finding the red CL in the proximity of the holes supports the above-discussed correlation with NBOCs, which should be generated by the ion beam impact during spark processing. Furthermore, the blue CL is often found in Si-rich regions or at Si–SiO_2 interfaces.[83,84] Both conditions occur inside the holes because of the rapid solidification of the flash-evaporated Si vapor partly intermixed with the surrounding gas atoms.

In essence, PL and CL of sp-Si are dissimilar in emission wavelengths, lifetimes, and temperature dependencies, and they derive their luminescence from different local positions of the microstructure. Specifically, the CL resembles a high degree of properties related to defects in silicon oxide.

4. SPARK PROCESSING OF OTHER MATERIALS

The observation of photoluminescence is not restricted to sp-Si. Visible PL or a substantial blue shift of the original PL appears also in spark-processed Ge, Sb, Bi, Sn, Te, As, and GaAs.[26,27] Figure 17 displays the normalized PL spectra of these materials. Each specimen was spark processed and measured under identical conditions. It is noteworthy that the principal occurrence of PL or a blue shift of PL after spark processing is only observed in materials with a relatively small band gap. When spark processing is applied to, e.g., carbon, no PL was detectable in the 350 to 850-nm range. Most of the intensities were found to be rather weak, that is, in a 1% range when compared with sp-Si. The only exception is sp-Ge, which provides a similarly intense luminescence. Comparable PL for Ge nanocrystals prepared by different methods has been shown in the literature.[91,92,93] Furthermore, the PL peak position in spark-processed, arsenic-containing substances (GaAs and As) shifts toward higher energies after storage in air. This blue shift is possibly related to an oxidation process. Although the specific origin of the PL in the spark-processed materials is not known at present, it seems to be evident that it is a more general property and not limited to a distinctive Si defect, molecule, or cluster.

5. SUMMARY AND CONCLUSIONS

It is apparent from the above-described results that spark processing of silicon generates Si nanoparticles which are embedded in an amorphous silicon oxide matrix. There is no evidence for a crystalline SiO_2 phase nor for amorphous silicon. Aside from silicon oxide, vibronic modes of Si–N, Si–H, and N–H are also present. Depending on the conditions during spark processing, intense PL occurs at bands near 385 or 525 nm when excited by 325-nm (3.8-eV) laser light. Toward lower temperatures the PL intensities increase continuously by nearly one order of magnitude (from 300 K to 20 K). Above 300 K the intensity

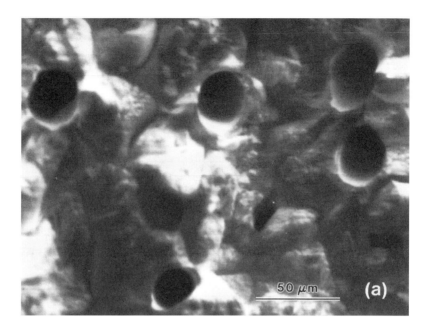

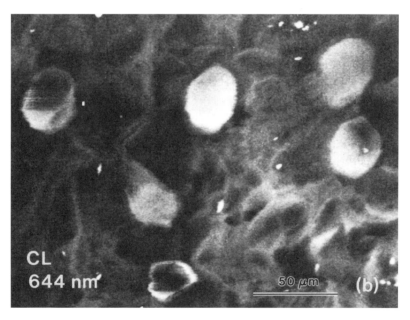

Figure 16 (a) Secondary electron microimage (1 kV primary electrons) of sp-Si in scanning mode. (b) Same area of sp-Si by a scanning CL micrograph taken at 644 nm and at room temperature (10 kV primary electrons).

loss follows an exponential law with further increasing temperature. There is no peak shift of the UV/blue emissions over the temperature range (20 K to 460 K), whereas the green band appears to shift to higher energies with rising temperature. The lifetime for excited carriers of both PL bands is shorter than in SiO_2 with decay times in the nanosecond range. Annealings up to 1100°C do not degrade radiative emissions either in intensity or in peak positions. A similar high stability is found against UV illumination (3.8 eV) and etching in HF acid. Excitation by 10-kV electrons causes different luminescing bands near 480 and 650 nm, which are typical for silicon oxide.

Three models need to be considered in an attempt to describe the luminescence of spark-processed silicon. Quantum confinement in nanostructures causes an increase of the band gap due to reduced dimensions. We have observed the presence of Si particles in sp-Si within a 3 to 5 nm range from TEM

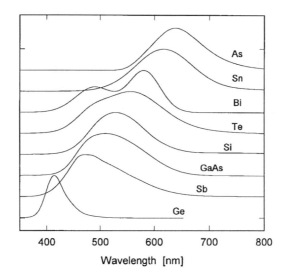

Figure 17 Normalized PL spectra of Ge, GaAs, Si, Te, Sb, Bi, Sn, and As after spark processing in air under stagnant conditions. The spectra are offset for clarity. (Excitation: 3.81-eV, 325-nm HeCd cw laser, room temperature).

and Raman measurements. Although they seem to be still too large to explain a blue shift in emission up to 3 eV (a recent study claims 1.3-nm clusters responsible for the 2-eV luminescence of por-Si[94]), the presence of even smaller and not yet resolved clusters in sp-Si cannot be excluded. These dimensions are hardly accessible by the above-utilized methods, but are very likely to be created during solidification of the spark-generated Si vapor phase. Specifically, as shown by a Monte Carlo simulation of high-temperature properties of small Si clusters,[95] the sizes of Si_7 and Si_{12} are particularly stable at melting temperatures. However, a quantum-dot-based luminescence in sp-Si should reflect very sensitively any changes of the size distribution in the course of modified conditions during spark processing (different temperatures, cooling rates, spark energies, etc.). Instead, the main PL bands at 385 and at 525 nm are very stable, and no transition in peak positions (e.g., after annealing in oxidizing atmospheres) occurs. Furthermore, equivalent short carrier lifetimes in the nanosecond range were found for both bands. In the quantum-confinement model they should change with the size of the crystallites.[96] Finally, at lower temperatures the energetic positions of the PL peaks are essentially fixed for the blue band or even shifted to lower energies for the green band. Both cases are not consistent with the properties of band–band or excitonic transitions in quantum wells. Thus, an explanation of the PL from sp-Si on the mere basis of quantum confinement is not suggested. For the same reasons, luminescence related to pure surface states of nanoparticles as proposed by Koch[13] for por-Si are also not very likely.

The strong localization effects in sp-Si in conjunction with the high degree of disorder, as well as the rapid decay data, seem to be more related to molecular or defectlike radiative transitions as found in amorphous or crystalline silica. Actually, there are several efficient PL bands in SiO_2 which are preferably attributed to broken Si bonds and the resulting undercorrelation of Si atoms. However, the excitation of these PL bands requires energies well above 4.8 eV (as the energy of the lowest absorption band). The blue CL band of sp-Si, observed by 10-keV electron beam excitation, shows clear signs of an SiO_2 origin (time constant > 50 ms, 470-nm peak, increasing intensity with increasing temperature up to 275 K). The PL excitation, however, with the 3.8-eV laser line used in the present experiments is not energetic enough to either stimulate one of the known, SiO_2-based radiative transitions or create nonequilibrium carriers above the 8- to 9-eV band gap of SiO_2. XPS and FTIR data for sp-Si prove the dominance of stoichiometric SiO_2. Transitions in the smaller band gaps of silicon suboxides are thus not very likely, though not excluded. Notwithstanding this possibility, radiative transitions in SiO_x should resemble the longer decay-time constants known for this material. Those are remarkably different from the short-lifetime data found for PL in sp-Si. In essence, SiO_2-related PL as a model to explain the PL from sp-Si is at best questionable.

A far better interpretation of the experimental data is possible by a combination of both models in the sense of an extended surface state model by including localized, disorder-induced defect states in the proximity of Si particles. A scheme of the phenomenological theory is shown in Figure 18: (1)

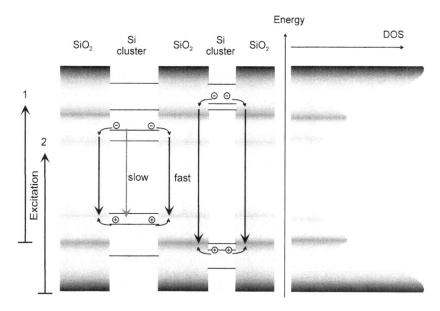

Figure 18 Schematic representation for a combined cluster/defect state model interpreting the visible luminescence from spark-processed Si. Photoexcited electron–hole pairs are generated in Si clusters with extended band gaps via excitation path 1. Tunneling of carriers to molecular defects provides a faster and much more efficient recombination path. With excitation path 2 the same energy is not sufficient to activate radiative channels.

Si clusters embedded in an amorphous SiO_2 matrix, which is rich in (2) molecular-like defects. Si nanostructures are generated in a wide distribution of sizes during spark processing and embedded in an amorphous SiO_2 matrix. The larger crystallites are directly detectable by TEM and Raman, and the smaller clusters are only accessible indirectly by their acting as efficient absorption points. In agreement with theoretical and experimental evidence, these Si clusters still have an indirect but extended band gap, thus shifting the absorption edge to higher energies in the visible part of the spectrum.[9,97] Consequently, nonequilibrium carriers can be created via "right-sized" clusters even by using an excitation source too low in energy for the absorption bands of SiO_2. However, radiative and nonradiative recombinations in Si clusters are still slow (in the millisecond range) because of the indirect band gap. Therefore, excited carriers may thermalize to interface states and further tunnel to low-dimensional defect states in the proximity of the clusters. As bulk, interface, and defect states are merging together, a pathway of states is accessible for excited carriers to move to nearby defect centers. Under these conditions radiative recombinations over molecular transitions are probable, which are much faster than recombinations within the clusters, permitting lifetimes in the nanosecond range. Moreover, superimposed electron–hole pairs and the strong overlap of the carrier wave functions suppress nonradiative channels and cause a high quantum yield. Both characteristics have been observed in sp-Si.

One can only speculate on the exact nature of the molecular defects at the present time. There seems to be no direct connection to simple defects like vacancies or for hydrogen to be involved in a similar way as in por-Si. This is proven by the stability during high-temperature annealings up to 1100°C for the blue and up to 1200°C for the green PL band. Already at lower temperatures a substantial amount of hydrogen is released, thereby generating dangling bonds, which should have quenched the PL intensity considerably when the luminescence mechanism is backed by those constituents.

The fractal-like network of sp-Si is characterized by a high degree of positional and compositional disorder, which makes it difficult to assign a definite type of bonding or arrangement as responsible for PL. Aditionally there is, however, a huge variety of silicon backbone polymers or Si-based macromolecules, most of which were found to luminesce in the UV or/and in the visible part of the spectrum.[98,99] Polysilanes are assumed to have a direct-type band structure with optical transitions allowed between σ–σ^* states formed by atomic orbitals of the Si skeleton.[100] The change of the luminescence intensity toward lower temperatures is very close to that of sp-Si with T_o values around 250 K.[101] However, Si-based polymers are very unstable against UV radiation, and a strong degradation of intensity occurs at temperatures above 77 K. The degradation is mainly caused by a dissociation of bonds between Si and substituents like H or organic molecules and resembles, thus, the behavior of hydrogen-terminated por-Si. For sp-Si, it might be more appropriate to consider Si-based macromolecules with substituents like oxygen or nitrogen.

ACKNOWLEDGMENTS

I wish to acknowledge the contributions of S.-S. Chang, J. Hack, M. Stora (University of Florida), A. Augustin (TU, München), and J. Menniger (Drude-Institut, Berlin), as well as the fruitful discussions with R. E. Hummel (University of Florida) and V. Petrova-Koch and F. Koch (TU München). This work was supported in part by a grant from the National Science Foundation, and Si wafers were provided by Wacker Chemitronic. The author gratefully acknowledges the financial support by a Habilitationsstipendium provided by the Deutsche Forschungsgemeinschaft.

REFERENCES

1. **Furukawa, S. and Miyasato, T.,** 1988. Three-dimensional quantum well effects in ultrafine silicon particles. *Jpn. J. Appl. Phys.* 27:L2207.
2. **Takagi, H., Ogawa, H., Yamazaki, Y., Ishizaki, A., and Nakagiri, T.,** 1990 Quantum size effects of photoluminescence in ultrafine Si particles. *Appl. Phys. Lett.* 56:2379.
3. **Canham, L. T.,** 1990. Silicon quantum wire array fabrication by electrochemical and chemical dissolution of wafers. *Appl Phys. Lett.* 57:1046.
4. **Lehmann, V. and Gösele, U.,** 1991. Porous silicon formation: a quantum wire effect. *Appl. Phys. Lett.* 58:856.
5. **Sarathy, J., Shij, S., Jung, K., Tsai, C., Li, K. H., Kwong, D.-L., Campbell, J. C., Yau, S.-L., and Bard, A. J.,** 1992. Demonstration of photoluminescence in nonanodized silicon. *Appl. Phys. Lett.* 60:1532.
6. **Fathauer, R. W., George, T., Ksendzov, A., and Vasquez, R. P.,** 1992. Visible luminescence from silicon wafers subjected to stain etches. *App. Phys. Lett.* 60:995.
7. **Herino, R., Bromchil, G., Barla, K., Bertrand, C., and Ginoux, J. L.,** 1987. Porosity and pore size distribution of porous silicon layers. *J. Electrochem. Soc.* 134:1994.
8. **Hummel, R. E. and Chang, S.-S.,** 1992. Novel technique for preparing porous silicon. *Appl. Phys. Lett.* 61:1965 and U.S. Patent No. 5,397,429/US005397429A.
9. **Brus, L.,** 1994. Luminescence of silicon materials: chains, sheets, nanocrystals, nanowires, microcrystals, and porous silicon. *J. Phys. Chem.* 98:3575.
10. **Calcott, P. D. J., Nash, K. J., Canham, L. T., and Kane, M. J.,** 1994. The spectroscopy of porous silicon. *Mater. Res. Soc. Symp. Proc.* 358:465.
11. **Rosenbauer, M., Stutzmann, M., Fuchs, H. D., Finkbeiner, S., and Weber, J.,** 1993. Temperature dependence of luminescence in porous silicon and related materials. *J. Lumin.* 57:153.
12. **Prokes, S. M. and Glembocki, O. J.,** 1994. Role of interfacial oxide-related defects in the red-light emission in porous silicon. *Phys. Rev. B* 49:2238.
13. **Koch, F., Petrova-Koch, V., Muschik, T., Nikolov, A., and Gavrilov, V.,** 1993. Some perspective on the luminescence mechanism via surface-confined states of porous silicon. *Mater. Res. Soc. Symp. Proc.* 283:197.
14. **Koch, F., Petrova-Koch, V., and Muschik, T.,** 1993. The luminescence of porous Si: the case for the surface state mechanism. *J. Lumin.* 57:271.
15. **Fuchs, H. D., Stutzmann, M., Brandt, M. S., Rosenbauer, M., Weber, J., Breitschwerdt, A., Deák, P., and Cardona, M.,** 1993. Porous silicon and siloxene: vibrational and structural properties. *Phys. Rev. B* 48:8172.
16. **Tamura, H., Rückschloss, M., Wirschen, T., and Veprek, S.,** 1994. Origin of the green/blue luminescence from nanocrystalline silicon. *Appl. Phys. Lett.* 65:1537.
17. **Tsybeskov, L., Vandshev, J., and Fauchet, P. M.,** 1994. Blue emission in porous silicon: oxygen-related photoluminescence. *Phys. Rev. B* 49:7821.
18. **Prokes, S.,** 1993. Light emission in thermally oxidized porous silicon: evidence for oxide-related luminescence. *Appl. Phys. Lett.* 62:3244.
19. **Rüter, D. and Bauhofer, W.,** 1993. Preparation of porous silicon by spark erosion. *J. Lumin.* 57:19.
20. **Steigmeier, E. F., Auderset, H., Delley, B., and Morf, R.,** Visible light emission from Si materials. *J. Lumin.* 57:9.
21. **Hummel, R. E., Ludwig, M. H., and Chang, S.-S.,** 1995. Strong, blue, room-temperature photoluminescence of spark-processed silicon. *Solid State Commun.* 93:237.
22. **Hummel, R. E., Ludwig, M. H., Chang, S.-S., and LaTorre, G.,** 1995. Comparison of anodically etched porous silicon with spark-processed silicon. *Thin Solid Films* 255:219.
23. **Hummel, R. E., Ludwig, M. H., Hack, J., and Chang, S.-S.,** 1995. Does the blue/violet photoluminescence of spark-processed silicon originate from hydroxyl groups? *Solid State Commun.* 96:683.
24. **Hummel, R. E., Ludwig, M. H., and Chang, S.-S.,** 1995. Possible mechanisms for photoluminescence in spark-processed Si. *Mater. Res. Soc. Symp. Proc.* 358:151.

25. **Nozaki, S., Sato, S., Ono, H., and Morisaki, H.**, 1994. Blue light emission from silicon ultrafine particles. *Mater. Res. Soc. Symp. Proc.* 351, 399.

26. **Ludwig, M. H., Hummel, R. E., and Chang, S.-S.**, 1994. Bright visible photoluminescence of spark-processed Ge, GaAs, and Si. *J. Vac. Sci. Technol. B* 12:3023.

27. **Ludwig, M. H., Hummel, R. E., and Stora, M.**, 1995. Luminescence of spark-processed materials. *Thin Solid Films* 255:103.

28. **Hummel, R. E., Morrone, A., Ludwig, M. H., and Chang, S.-S.**, 1993. On the origin of photoluminescence in spark-eroded (porous) silicon. *Appl. Phys. Lett.* 63:2771.

29. **Proud, J. M., Gottscho, R. A., Bondur, J., Garscadden, A., Heberlein, J., Herb, G. K., Kushner, M. J., Lawler, J., Lieberman, M., Mayer, T. M., Phelps, A. V., Roman, W., Sawin, H., Winters, H., Perepezko, J., Hazi, A. U., Kennel, C. F., and Gerardo, J.**, 1991. *Plasma Processing of Materials: Scientific Opportunities and Technological Challenges*. NRC Report, National Academy Press, Washington, D.C.

30. **Sukhodrev, N. K.**, 1982. On spectral excitation in a spark discharge. In Research on Spectroscopy and Luminescence, *Trans. P.N. Lebedev Phys.* Vol. XV, Part 3, Academy of Sciences of the USSR.

31. **Warburg, E.**, 1909. Über chemische Reaktionen, welche durch die stille Entladung in gasförmigen Körpern herbeigeführt werden. *Jahrb. Radioak. Elektron.* 6:181.

32. **Mandich, M. L. and Reents, W. D.**, 1992. How to grow large clusters from $Si_xD_y^+$ ions in silane or disilane: water them! *J. Chem. Phys.* 96:4233.

33. **Brodski, M. H.**, 1977. On the deposition of amorphous silicon films from glow discharge plasmas of silane. *Thin Solid Films* 40:L23.

34. **Chang, S.-S.**, 1993. Production and characterization of photoluminescing spark-processed Si. Ph.D. thesis, University of Florida, Department of Materials Science and Engineering.

35. **Campbell, I. H. and Fauchet, P. M.**, 1986. The effects of microcrystal size and shape on the one phonon Raman spectra of crystalline semiconductors. *Solid State Commun.* 58:739.

36. **Sui, Z., Leong, P. P., and Herman, I.P.**, 1992. Raman analysis of light-emitting porous silicon. *Appl. Phys. Lett.* 60:2086.

37. **Tsu, R., Shen, H., and Dutta, M.**, 1992. Correlation of Raman and photoluminescence spectra of porous silicon. *Appl. Phys. Lett.* 60:112.

38. **Wu, X.-L., Yan, F., Zhang, M.-S., and Feng, D.**, 1995. Influence of illumination of Raman spectra of porous silicon. *Phys. Lett. A* 205:117.

39. **Kurmaev, E. Z., Shamin, S. N., Galakhov, V. R., Sokolov, V. I., Ludwig, M. H., and Hummel, R. E.**, X-ray emission spectra and local structure of porous and spark-processed silicon. *J. Appl. Phys.* (submitted for publication).

40. **Lucovsky, G., Richard, P. D., Tsu, D. V., Lin, S. Y., and Markunas, R. J.**, 1986. Deposition of silicon dioxide and silicon nitride by remote plasma enhanced chemical vapor deposition. *J. Vac. Sci. Technol. A* 4:681.

41. **Gupta, P., Dillon, A. C., Bracker, A. S., and George, S. M.**, 1991. FTIR studies of H_2O and D_2O decomposition on porous silicon surfaces. *Surf. Sci.* 245:360.

42. **Brandt, M. S., Fuchs, H. D., Stutzmann, M., and Cardona, M.**, 1992. The origin of visible luminescence from "porous silicon": a new interpretation. *Solid State Commun.* 81:307.

43. **Rüter, D., Kunze, T., and Bauhofer, W.**, 1994. Blue light emission from silicon surfaces prepared by spark-erosion and related techniques. *Appl. Phys. Lett.* 64:3006.

44. **Rüter, D., Rolf, S., Bauhofer, W., Klar, P. J., and Wolverson, D.**, 1995. Luminescence properties of plasma treated silicon surfaces. *Thin Solid Films* 255:305.

45. **Petrova-Koch, V., Muschik, T., Kovalev, D. I., Koch, F., and Lehmann, V.**, 1993. Some perspectives on the luminescence mechanism via surface-confined states of porous Si. *Mater. Res. Soc. Symp. Proc.* 283:178.

46. **Kanemitsu, Y., Matsumoto, T., Futagi, T., and Mimura, H.**, 1993. Visible photoluminescence from rapid-thermal-oxidized porous silicon. *Mater Res. Soc. Symp. Proc.* 298:205.

47. **Friebele, E. J., Griscom, D. L., and Marrone, M. J.**, 1985. The optical absorption and luminescence bands near 2 eV in irradiated and drawn synthetic silica. *J. Non-Cryst. Solids* 71:133.

48. **Pio, F., Guzzi, M., Spinolo, G., and Martini, M.**, 1990. Intrinsic and impurity-related point defects in amorphous silica. *Phys. Status Solidi, B* 159:577.

49. **Stathis, J. H. and Kastner, M. A.**, 1987. Time-resolved photoluminescence in amorphous silicon dioxide. *Phys. Rev. B* 35:2972.

50. **Griscom, D. L.**, Optical properties and structure of defects in silica glass. *J. Ceram. Soc. Jpn.* 99:923.

51. **Tohmon, R., Shimogaichi, Y., Mizuno, H., Ohki, Y., Nagasawa, K., and Hama, Y.**, 1989. 2.7-eV luminescence in as-manufactured high-purity silica glass. *Phys. Rev. Lett.* 62:1388.

52. **Jones, C. E. and Embree D.**, 1976. Correlations of the 4.77–4.28 eV luminescence band in silicon dioxide with the oxygen vacancy. *J. Appl. Phys.* 47:5365.

53. **Nishikawa, H., Shiroyama, T., Nakamura, R., Ohki, Y., Nagasawa, K., and Hama, Y.**, 1992. Photoluminescence from defect centers in high-purity silica glasses observed under 7.9-eV excitation. *Phys. Rev. B* 45:586.

54. **Sakurai, Y., Nagasawa, K., Nishikawa, H., and Ohki, Y.**, 1994. Point defects in high purity silica induced by high-dose gamma irradiation. *J. Appl. Phys.* 75:1372.

55. **Hummel, R. E., Ludwig, M. H., Chang, S.-S., Fauchet, P. M., Vandyshev, Ju. V., and Tsybeskow, L.,** 1995. Time-resolved photoluminescence measurements in spark-processed blue and green emitting silicon. *Solid State Commun.* 95:553.

56. **Ookubo, N., Ono, H., Ochiai, Y., Mochizuki, Y., and Matsui, S.,** 1992. Effects of thermal annealing on porous silicon photoluminescence dynamics. *Appl Phys. Lett.* 61:940.

57. **Tohmon, R., Shimogaichi, Y., Munekuni, S., Ohki, Y., Hama, Y., and Nagasamwa, K.,** 1989. Relation between the 1.9 eV luminescence and 4.8 eV absorption bands in high-purity silicon glass. *Appl. Phys. Lett.* 54: 1650.

58. **Kovalev, D. I., Yaroshetzkii, I. D., Muschik, T., Petrova-Koch, V., and Koch, F.,** 1994. Fast and slow visible luminescence bands of oxidized porous Si. *Appl. Phys. Lett.* 64:214.

59. **Skuja, L. N., Streletsky, A. N., and Pakovich, A. B.,** 1984. A new intrinsic defect in amorphous SiO_2: twofold coordinated silicon. *Solid State Commun.* 50:1069.

60. **Skuja, L. N.,** 1992. Time resolved low temperature luminescence of non-bridging oxygen hole centers in silica glass. *Solid State Commun.* 84:613.

61. **Austin, I. G., Jackson, W. A., Searle, T. M., Bhat, P. K., and Gibson, R. H.,** 1985. Photoluminescence properties of a-SiN_x:H. *Philos. Mag.* B 52:271.

62. **Wakita, K., Makimura, S., and Nakayama, Y.,** 1995. Effect of annealing on photoluminescence spectra and film structure in a-SiN_x:H. *Jpn. J. Appl. Phys.* 34:1425.

63. **Street, R. A.,** 1981. Luminescence and recombination in hydrogenated amorphous silicon. *Adv. Phys.* 30:593.

64. **Fischer, R.,** 1985. Luminescence in amorphous semiconductors. In *Topics in Applied Physics* Vol. 36, M. H. Brodsky, Ed., pp. 159ff. Springer-Verlag, Berlin.

65. **Gardelis, S., Rimmer, J. S., Dawson, P., Hamilton, B., Kubiak, R. A., Whall, T. E., and Parker, E. H. C.,** 1991. Evidence for quantum confinement in the photoluminescence of porous Si and SiGe. *Appl. Phys. Lett.* 59:2118.

66. **Kux, A., Kovalev, D., and Koch, F.,** 1995. Slow luminescence from trapped charges in oxidized porous silicon. *Thin Solid Films* 255:143.

67. **Collins, R. W., Paesler, M. A., and Paul, W.,** 1980. The temperature dependence of photoluminescence in a-Si:H alloys. *Solid State Commun.* 34:833.

68. **Rosenbauer, M., Strutzmann, M., Fuchs, H. D., Finkbeiner, S., and Weber, J.,** 1993. Temperature dependence of luminescence in porous silicon and related materials. *J. Lumin.* 57:153.

69. **Muschik, Th.,** 1993. Lichtemission aus amorphen und mikroporösem Silizium. Dissertation, Technische Universität, München.

70. **Xu, Z. Y., Gal, M., and Gross, M.,** 1992. Photoluminescence studies on porous silicon. *Appl. Phys. Lett.* 60:1375.

71. **Zheng, X. L., Wang, W., and Chen, H. C.,** 1992. Anomalous temperature dependencies of photoluminescence for visible-light-emitting porous Si. *Appl. Phys. Lett.* 60:986.

72. **Perry, C. H. and Feng L., Namarav, F., Kalkhoran, N. M., and Soref, R. A.,** 1992. Photoluminescence spectra from porous silicon (111) microstructures: temperature and magnetic-field effects. *Appl. Phys. Lett.* 60:3117.

73. **Koyama, H. and Koshida, N.,** 1993. Electrical properties of luminescent porous silicon. *J. Lumin.* 57:293.

74. **Collins, R. T., Tischler, M. A., and Stathis, J. H.,** 1992. Photoinduced hydrogen loss from porous silicon. *Appl. Phys. Lett.* 61:1649.

75. **Zoubir, N. H., Vergnat, M., Delatour, T., Burneau, A., and deDonato, Ph.,** Interpretation of the luminescence quenching in chemically etched porous silicon by the desorption of SiH_3 species. *Appl Phys. Lett.* 65:82.

76. **Tamura, H., Rückschloss, M., Wirschem, T., and Vepcek, S.,** 1994. Origin of the green/blue luminescence from nanocrystalline silicon. *Appl. Phys. Lett.* 65:1537.

77. **Tanielian, M.,** 1982. Adsorbate effects on the electrical conductance of a-Si:H. *Philos. Mag.* B 45:435.

78. **Marrone, M. J.,** 1981. Radiation-induced luminescence in silica core optical fibers. *Appl. Phys. Lett.* 38:115.

79. **Augustine, B. H., Irene, E. A., He, Y. J., Price, K. J., McNeil, L. E., Christensen, K. N., and Maher, D. M.,** 1995. Visible light emission from thin films containing Si, O, N, and H. *J. Appl. Phys.* 78:4020.

80. **Boonkosum, W., Kruangam, D., and Panyakeow, S.,** 1993. Amorphous visible-light thin film light-emitting diode having a-SiN:H as a luminescent layer. *Jpn. J. Appl. Phys.* 32:1534.

81. **Ghandhi, S. K.,** 1983. *VLSI Fabrication Principles. Silicon and Gallium Arsenide.* J. Wiley & Sons, New York, pp. 494ff.

82. **Ludwig, M. H., Menniger, J., and Hummel, R. E.,** 1995. Cathodoluminescing properties of spark-processed silicon. *J. Phys. Condens. Matter* 7:9081.

83. **Skuja, L. N. and Entzian, W.,** 1986. Cathodoluminescence of intrinsic defects in glassy SiO_2, thermal SiO_2 films, and α-quartz. *Phys. Status Solidi, A* 96:191.

84. **Mitchell, J. P. and Denure, D. G.,** 1973. A study of SiO layers on Si using cathodoluminescence spectra. *Solid State Electron.* 16:825.

85. **Awazu, K. and Kawazoe, H.,** 1990. O_2 molecules dissolved in synthetic glasses and their photochemical reactions induced by ArF excimer laser radiation. *J. Appl. Phys.* 68:3584.

86. **Kanya, K. and Okayama, S.,** 1872. Penetration and energy-loss theory of electron in solid targets. *J. Phys. D: Appl. Phys.* 5:43.

87. **Philipp, H. R.,** 1971. Optical properties of non-crystalline Si, SiO, SiO_x and SiO_2. *J. Phys. Chem. Sol.* 32:1935.

88. **Suzuki, T., Sakai, T., Zhang, L., and Nishiyama, Y.,** 1995. Evidence for cathodoluminescence from SiO_x in porous Si. *Appl. Phys. Lett.* 66:215.

89. **Mitsui, T., Yamamoto, N., Takemoto, K., and Nittono, O.,** 1994. Cathodoluminescence and electron beam irradiation effect of porous silicon studied by transmission electron microscopy. *Jpn. J. Appl. Phys.* 33:L342.

90. **Ludwig, M. H., Menniger, J., Hummel, R. E., Augustin, A., and Hack, J.,** 1995. Position- and temperature-dependent optical properties of spark-processed silicon. *Appl. Phys. Lett.* 67:2542.

91. **Kanemitsu, Y., Uto, H., Masumoto, Y., and Maeda, Y.,** 1992. On the origin of visible photoluminescence in nanometer-size Ge crystallite. *Appl. Phys. Lett.* 61:2187.

92. **Liu, W. S., Chen, J. S., Nicolet, M.-A., Arbet-Engels, V., and Wang, K. L.,** 1993. Nanocrystalline Ge in SiO_2 by annealing of $Ge_xSi_{1-x}O_2$ in hydrogen. *Appl. Phys. Lett.* 62:3321.

93. **Sendova-Vassileva, M., Tzenov, N., Dimova-Malinovska, D., Rosenbauer, M., Stutzmann, M., and Josepovits, K. V.,** 1995. Structural and luminescence studies of stain-etched and electrochemically etched germanium. *Thin Solid Films* 255:282.

94. **Schuppler, S., Friedman, S. L., Marcus, M. A., Brown, W. L., Chabal, Y. J., Brus, L. E., and Citrin, P. H.,** 1995. Dimensions of luminescent oxidized and porous silicon structures. *Phys. Rev. Lett.* 72:2648.

95. **Dinda, P. T., Vlastou-Tsinganos, G., Flytzanis, N., and Mistriotis, A. D.,** 1994. The melting behavior of small silicon clusters. *Phys. Lett. A* 191:339.

96. **Proot, J. P., Delerue, C., and Allan, G.,** 1992. Electronic structure and optical properties of silicon crystallites: application to porous silicon. *Appl. Phys. Lett.* 61:1948.

97. **Fauchet, P. M., Tsybeskow, L., Peng, C., Dutagupta, S. P., von Behren, J., Kostoulas, V., Vonyshev, J. M. V., Hirschman, K. D.,** 1995. Light-emitting porous silicon: materials science, properties, and device applications. *J. Sel. Topics Quantum Elect.* 1:1126.

98. **Miller, R. D. and Michl, J.,** 1989. Polysilane high polymers. *Chem. Rev.* 89:1359.

99. **Watanabe, A., Miike, H., Tsutsumi, Y., and Matsuda, M.,** 1993. Photochemical properties of network and branched polysilanes. *Macromolecules* 26:2111.

100. **Takeda, K., Teramae, H., and Matsumoto, N.,** 1986. Electronic structure of chainlike polysilane. *J. Am. Chem. Soc.* 108:8187; **Takeda, K.,** 1994. Si skelton high-polymers: their electronic structures and characteristics. *J. Phys. Soc. Jpn. Suppl. B* 63:1.

101. **Kishida, H., Tachibana, H., Matsumoto, M., and Iokura, Y.,** 1995. Visible luminescence from branched silicon polymers. *J. Appl. Phys.* 78:3362.

Chapter 6

Photoluminescence from Nanocrystalline-Si/Amorphous-SiO₂ Composite Thin Films Prepared by Plasma Chemical Vapor Deposition

Stan Vepřek and Thomas Wirschem

Reviewed by J. Oswald

TABLE OF CONTENTS

ABSTRACT

The chapter summarizes recent results on the preparation and properties of thin films of nanocrystalline silicon (nc-Si)/amorphous SiO₂ grain boundaries (a-SiO₂) which show intense visible photoluminescence (PL). The films were prepared by plasma (CVD) of nc-Si with the desired crystallite size followed by oxidation and annealing in forming chemical vapor deposition gas. We present evidence for a large increase of the band gap due to the quantum localization in nc-Si embedded in a-SiO₂ matrix, which is in agreement with the original theoretical calculations. This, together with additional experimental data on the dependence of the PL intensity and decay time on the crystallite size, explains the large red shift between the onset of the excitation spectra and the PL. The data provide strong support for the mechanism of the PL to be due to the formation of photogenerated electron–hole pairs within the Si nanocrystals and their recombination on radiative centers either at the Si/SiO₂ interface or within the SiO₂ matrix. The strong decrease of the efficiency of the PL due to a decrease of the thickness of the a-SiO₂ grain boundaries is shown and its origin discussed. Delocalization of the photogenerated charge carriers due to ultrathin a-SiO₂ is excluded as the caurse of this effect. Microwave absorption is used to study the effect of the grain boundaries on the localization and delocalization of photogenerated charge carriers in pure nc-Si and nc-Si/a-SiO₂ composites. Finally, we show the strong decrease of the PL decay time to ≤ 500 ps due to molecular-like radiative centers which are formed in the nc-Si/a-SiO₂ composites by appropriate doping.

1. INTRODUCTION

Since the original reports on the efficient PL from porous silicon (PS)[1,2] the mechanism of the PL from PS and nc-Si has been subject to numerous investigations and discussions (see, e.g., Reference 3 to 7). Although experimental evidence has been given for the quantum-confinement model,[8] ample arguments and experimental facts have shown that, at least in silicon nanocrystals with surface passivated by oxide, the PL originates from radiative centers which are located either at the nc-Si/SiO₂ interface[6,7] or from some localized defects within the SiO₂ grain boundaries.[9] One of the apparently unresolved problems is the large red shift between the calculated band gap of small silicon nanocrystals[4,5,11-14] and the observed spectral distribution of the PL (see, e.g., References 3-5,9,10,15).

The first section of this chapter summarizes the preparation technique of nc-Si by plasma CVD and, in particular, the possibilities of the control of the crystallite size and mechanical stress in the films. The latter is important for the control of the average thickness of the grain boundaries both in nc-Si and in nc-Si/SiO$_2$ composite films.

In the next section we shall show that the original calculations of the increase of the band gap with decreasing crystallite size are in agreement with the measured blue shift in the PL excitation spectra. This, together with further experimental data to be summarized strongly supports the mechanism in which the electron–hole pairs are formed as a result of optical absorption within the Si nanocrystals followed by their trapping in radiative centers and finally emission of the "red" PL.

The third part of the chapter deals with the effect of the thickness of the grain boundaries and mechanical stress in the films on the localization of photogenerated charge carriers. The absorption of microwave energy at a frequency of about 10 GHz allows us to study the transport of the photogenerated carriers at a distance of several tens of nanometers, which is about a factor of ten larger than the typical size of the nanocrystals. The results enable us to estimate the different time scales for the trapping of the charge carriers and their radiative recombination.

The subsequent section discusses the origin of several PL features frequently observed from PS and sometimes erroneously attributed to size effect and quantum confinement. The relatively long decay time of the PL of the order of tens of microseconds makes the application of this material for fast optoelectronic communication (e.g., chip-to-chip) unlikely. Therefore, the final section of the chapter deals with the problem of how to achieve a less than 1 ns ($\geq$1 GHz) fast, efficient luminescence. Based on the information obtained regarding the mechanism of the "red" PL, we shall show how doping of the films with fast charge-transfer ions can lead to a $\leq$100 ps fast and efficient PL. The whole preparation procedure is fully compatible with standard silicon technology.

2. PREPARATION OF nc-Si AND nc-Si/a-SiO$_2$ FILMS

Compact nc-Si and nc-Si/a-SiO$_2$ thin films were prepared either by plasma CVD from silane or by chemical transport of silicon in glow discharge hydrogen plasma. Unlike PS prepared by anodic etching, our preparation technique allows us a precise control and determination of the crystallite size. The latter was also achieved by the technique of chemical precipitation (either in the gas,[16] liquid,[17] or solid[18] phase), but the plasma CVD allows us to prepare compact films with controlled thickness of the grain boundaries and the mechanical stress which, as it will be seen, are very important for the confinement phenomena and for the PL.

Depending on the exact goal of the given study, one of the following three alternative plasma CVD procedures was used (see, e.g., References 19 to 21 and references given there).

- Deposition of a-Si from pure monosilane at room temperature of the substrate, which yields a-Si with a high hydrogen content. After the deposition, the films are slowly recrystallized to give a small crystallite size in the range of a few nanometers followed by a light oxidation of their grain boundaries which, upon subsequent passivation steps, avoids the recrystallization to larger grains.
- Deposition of nc-Si/a-Si agglomerate films from silane diluted with hydrogen (about 1:99).
- Deposition of pure nc-Si without amorphous tissue (Figure 1) by means of chemical transport of silicon in hydrogen plasma.[21,22] This technique provides the best control of the crystallite size and of the mechanical stress in the films, which is documented by Figure 2.

Films deposited at floating potential have tensile stress within the grain boundaries which can be quantitatively measured by X-ray diffraction.[25] The concomitant dilatation of the Si–Si bonds results in potential barriers for the charge transport through the boundaries and, as it will be shown, in localization. Similarly this also applies to the phonons. Thus, the films deposited, e.g., at about 60°C and floating potential (Figure 2, left), and those deposited at 260°C and $V_b = -150$ V have the same average crystallite size, but they differ in the nature of the grain boundaries; there are dilatated Si–Si bonds and therefore a larger potential barrier for the transport through the grains in the former films, whereas the large compressive stress induced in films because of the ion bombardment during deposition at negative substrate bias (of about 20 to 30 kbar in the case under discussion[21,25]) removes the bond dilatation and decreases the potential barrier. Consequently, although both kind of films have the same average crystallite size, the charge carriers and phonons are more localized in the films deposited at floating potential and largely delocalized in those prepared under substrate bias. We shall show that these effects can be seen in the appropriate experiments with pure nc-Si.

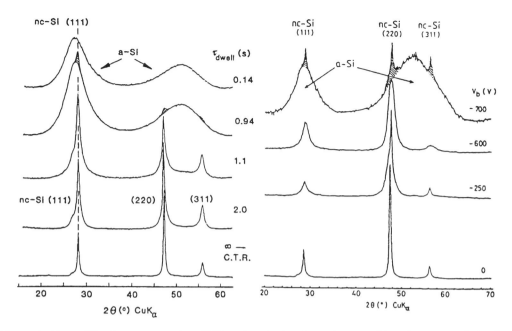

Figure 1 X-ray diffraction patterns of nc-Si films deposited under various conditions by means of chemical transport of silicon in hydrogen plasma.[23] The intensity of the diffuse scattering between the (220) and (311) Bragg reflections is a measure of the amount of a-Si component in the films (see Reference 24 for further details). It is seen that films prepared by chemical transport of Si are free of any a-Si up to a negative substrate bias (applied during the deposition) of −500 V.

As-deposited nc-Si shows very weak PL because of insufficient passivation of the surfaces of the crystallites. Interestingly, the concentration of dangling bonds as measured in these films by electron spin resonance (ESR) is very low (below the detection limit of $3 \cdot 10^{15}$ cm^{-3}).[26] Formation of at least 10 to 15 Å thick SiO_2 grain boundaries is necessary to obtain visible and efficient PL (Figure 3). A further improvement of the PL is achieved by annealing the nc-Si/a-SiO_2 films in forming gas (Figure 3). Films with a thicker a-Si tissue prepared, e.g., from diluted silane, enable us to obtain nc-Si/a-SiO_2 films with an even thicker silica tissue.

In order to optimize the PL efficiency one has to compromize between the sufficiently thick SiO_2 tissue, a high fraction of the crystallites, X_c, and small crystallite size, D. Because at a constant thickness of the SiO_2 tissue, X_c decreases with decreasing D, the preparation of such films requires large experience. Nevertheless, our standard reproducibility of the preparation of the samples with the desirable parameters is fairly satisfactory. This is very important. If one wants to study the dependence of some of the phenomena (PL efficiency, localization of electrons and phonons, etc.) on one of these parameters, all the other parameters must be kept constant. Neglecting this conditions led to much confusion in the literature, particularly regarding the calculation of the crystallite size and crystalline fraction from the Raman spectra.[27]

3. DEPENDENCE OF THE BAND GAP AND THE OPTICAL TRANSITION PROBABILITIES ON THE CRYSTALLITE SIZE

The measured spectral distribution of the PL in our compact films shows no shift with decreasing crystallite size (Figure 4a), and only a very small blue shift up to about 1.8 eV is found in isolated clusters with crystallite size of about 1.2 nm (Figure 4b). Theoretical calculations[12-14] (Figure 4) predict a much larger increase of the band gap of about 4 eV for this crystallite size. The shift experimentally found by Kanemitsu et al.[28] in organosilane chains and ladders (Figure 4b) agrees well with the calculations for Si chains by Delerue et al.[14]

The discrepancy between the calculated large increase of the band gap and the measured small blue shift of the PL is due to a large red shift of the PL as compared with the absorption, which is inherent in any mechanism of the PL where the energy is transferred from the crystallites to a radiative center (e.g., "surface-state-model"[6,7] or defects in the SiO_2 matrix[9,29,30]). In order to show this, we have measured

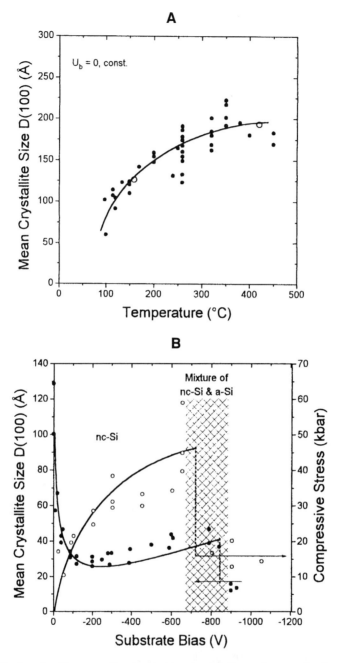

Figure 2 Crystallite size of nc-Si prepared by chemical transport. A: Dependence on the deposition temperature for $V_b = 0$ (wall potential, i.e., ion impact energy of $\leq$10 eV). B: Dependence on substrate bias for a constant temperature of 260°C. (From Vepřek, S. et al., *Phys. Rev. B.,* 36, 3344, 1987. With permission.)

the excitation spectra of the red PL for a series of samples with decreasing crystallite size. It is seen from Figure 5 that both the onset and the first maximum (which corresponds to the first direct transition $\Gamma'_{25} \rightarrow \Gamma_{15}$ in Si[31]) show a strong blue shift. The relatively broad onset of the excitation is due to the distribution of the crystallite sizes in the films and to the distribution of the states in the small nano-crystals.[5] Taking the photon energy which corresponds to 50% of the first maximum as the average value of the PL onset, we obtain excellent agreement with the calculated increase of the band gap (see filled

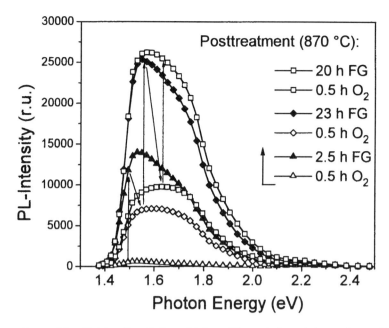

Figure 3 Dependence of the PL intensity with the oxidation of the grain boundaries and annealing of the films in forming gas.[19,24]

symbols in Figure 4a). Choosing a somewhat higher or lower value (e.g., 70%) influences the agreement insignificantly.

As mentioned above, theoretical calculations predict a strong increase of the rate of radiative recombination of electron–hole pairs (due to the progressive mixing of the quantum states in the k-space) with decreasing crystallite size.[11-14] Examples of such calculations are shown in Figure 6a together with our measurements of the efficiency of the red PL.[10,19,34] The agreement in the trend and magnitude is very good. However, the measured decay time is of the order of tens of microseconds (Figure 6b, upper curve), and it does not show any dependence on the crystallite size.

These results are compatible with the surface-statemodel of Koch et al.[6,7] which is schematically illustrated in Figure 7. Accordingly, the exciting UV light is efficiently absorbed by the small nanocrystals with the formation of an electron–hole pair. Because of the small size of the crystallites as compared with bulk exciton, the electron and/or hole are trapped in the surface states (possibly at the Si/SiO$_2$ interface), where they rest up to several tens of microseconds until the radiative recombination occurs. Of course, a large part of the photogenerated electron–hole pairs annihilates via nonradiative recombinations, thus giving the quantum efficiency of the red PL in the range of several percent. The energy of the emitted light and the PL decay time are determined by the nature of these states. Therefore, they do not show the strong dependence on the crystallite size which is predicted by the theories based on model of the electron–hole radiative recombination within the crystallite. Recent theoretical calculations of Allan et al.[36] have shown that exciton polaronic states can exist at the hydrogen-terminated surface of nc-Si and their energy (and possibly also the radiative recombination time) is largely independent of the crystallite size. Although it will be much more difficult to conduct such calculation for the nc-Si/SiO$_2$ interface, there is little doubt that similar states do also exist in this case. The apparent agreement of the calculated dependence of the radiative recombination time and of our measurements of the PL efficiency on the crystallite size can be easily understood. The theory considers the probability for the photon emission as a result of the transition from the excited state to the ground state. At least within the theoretical model of Delley and Steigmeier[12] this is equivalent to the inverse process, i.e., the photon absorption with the formation of electron–hole pair. The PL efficiency is proportional to the latter process, provided, of course, that the probability of nonradiative recombinations remains unchanged with decreasing crystallite size. The latter assumption is supported by the excellent agreement of the theoretical calculations and measurements in our films (see Figure 6a).

The fast blue PL induced by W^{n+} centers within the SiO$_2$ matrix will be discussed in the last section.

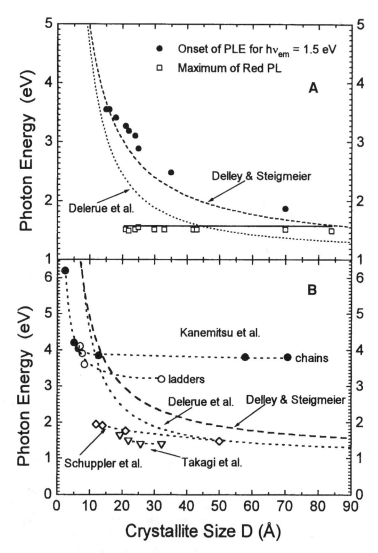

Figure 4 (A) Dependence of the band gap on the crystallite size calculated by Delley and Steigmeier[12] and Delerue et al.,[5,13,14] position of the red PL maximum[10] and the measured 50% onset of the excitation spectra of that PL.[27,32] (B) Blue shift of the maximum of the PL from isolated Si clusters measured by Schuppler et al.[15] and Tagaki[33] and that found in organosilane chains and ladders by Kanemitsu et al.[28]

4. EFFECT OF THE GRAIN BOUNDARIES ON THE EFFICIENCY OF THE PL AND THE LOCALIZATION PHENOMENA OF CHARGE CARRIERS AND PHONONS

The plasma CVD technique allows us to prepare films with the same crystallite size but with different thickness of the grain boundaries. It was shown that the decrease of the average SiO_2 thickness from about 10 to 15 Å to 2 to 3 Å results in a strong decrease of the PL efficiency by more than two orders of magnitude.[34] The question arises if this is due to the delocalization of the photogenerated carriers by tunneling through the SiO_2 interfaces or to other effects. Using the well known formula for the tunneling probability and the measured ratio of the PL intensity I_1/I_2, together with the corresponding SiO_2 thickness d_1 and d_2, one can calculate from Equation 1 the barrier height ΔV to be less than 1 eV. Assuming that the thinner barrier has a smaller ΔV_1 results in an even smaller value of ΔV_2. As such a value of the barrier height is too small for the thicker SiO_2, this clearly excludes the possibility that the observed decrease of the PL efficiency could be due to tunneling.

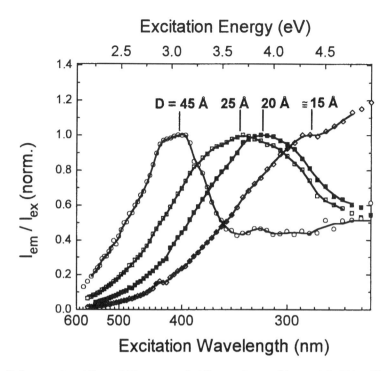

Figure 5 Excitation spectra of the red PL measured at the maximum of $h\nu_{PL} = 1.5$ eV for different crystallite sizes.[27,32]

$$\frac{I_1}{I_2} = \exp\left\{\frac{2}{\hbar}\sqrt{2m\Delta V}\left(d_1 - d_2\right)\right\} \tag{1}$$

It is well known that an interfacial layer of two to three lattice spacings is needed to relax the strain at an interface. Thus, the most probable reason for the decrease of the PL is the poor quality of the 2 to 3 Å thin oxide interface which is not sufficient for the necessary passivation of the grain surfaces. This conclusion is further supported by the fact that upon cooling the sample to 90 K only a small increase of the decay time by a factor of 2.4 is found.

In order to study the transport of photogenerated charge carriers through the grain boundaries we used a microwave (MW) apparatus described in Reference 34. It consists of a MW cavity with the necessary MW source, Schottky detector, and fast digital oscilloscope. The nc-Si sample deposited either on SiO_2 or on undoped Si substrate is placed within the cavity and illuminated by a 3-ns fast pulsed nitrogen laser. At the frequency of about 10 GHz the charge carriers can move within one half of the MW period to a distance of several hundred angstroms. Their scattering at the grain boundaries (and other possible defects) results in an absorption of the MW which can be measured as a transient signal following the laser pulse (Figure 8). If, however, the carriers are localized within the crystallites and, therefore, cannot move through the grain boundaries, no MW absorption is seen. This technique is much simpler and more sensitive than the absorption in infrared. The sensitivity of this technique is illustrated by Figure 8a, which shows no MW absorption for a sample with a crystallite size of 51 Å that was deposited on Si with floating substrate potential, resulting in a small tensile stress in the grain boundaries. If, however, negative substrate bias is applied during the deposition, compressive stress is induced within the film, which results in a better connectivity between the grains. As seen in Figure 8a such a sample shows MW absorption because the photogenerated carriers can move through the grain boundaries. A similar effect is found if nc-Si with a larger crystallite size is deposited at a higher temperature of 350°C on substrates with a different coefficients of thermal expansion (Figure 8b). Film deposited on silicon substrate shows MW absorption, but sample deposited on SiO_2 does not. Because of the higher deposition temperature, the nc-Si sample on silicon has better connectivity of the grains (higher conductivity), and,

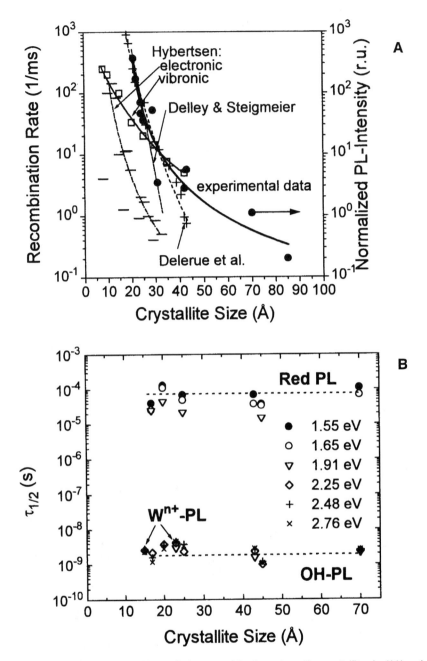

Figure 6 (A) Calculated dependence of the radiative recombination rate on the crystallite size[11-14] and measured efficiency of the red PL.[10,19,34] (B) Measured decay time of the red (upper curve) PL and those from silanol groups (OH-PL) and tungsten (W^{n+}-PL) in the nc-Si/SiO_2 films.[32,34,35]

therefore, the photogenerated carriers can move relatively easily through the boundaries. Due to the difference between the coefficients of thermal expansion between the film and the substrate, tensile stress is induced in the nc-Si film deposited on SiO_2 when the sample is cooled to room temperature. This results in an increased localization of the carriers within the crystallites and, consequently, much weaker MW absorption.

The localization of photogenerated charge carrier is necessary for efficient PL. Therefore, films deposited even at a small negative bias do not show any PL, but they do show MW absorption (Figure 8). However, after only a short oxidation where only a very weak PL can be measured, the MW absorption transient signal strongly decreases and changes its shape (Figure 9).

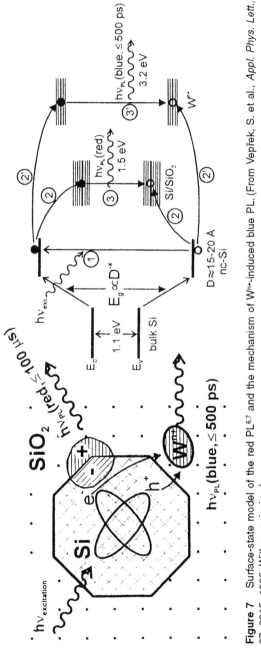

Figure 7 Surface-state model of the red PL[6,7] and the mechanism of W^{n+}-induced blue PL. (From Vepřek, S. et al., *Appl. Phys. Lett.*, 67, 2215, 1995. With permission.)

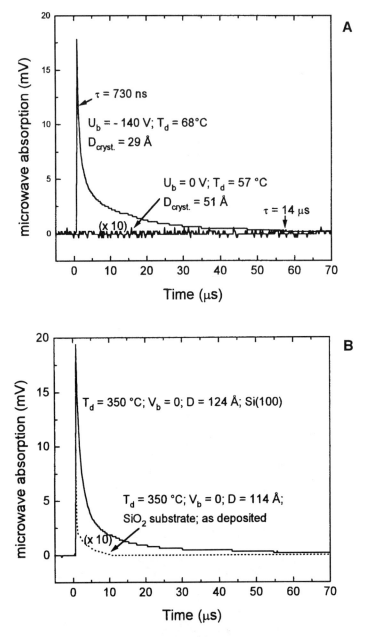

Figure 8 Transient microwave absorption due to photogenerated carriers. (A) Sample with a small crystallite size deposited at floating substrate potential (tensile stress) shows no MW absorption (lower trace), whereas sample deposited under negative bias of −140 V, which induces a large compressive stress of about 20 kbar, shows strong MW absorption (upper trace). (B) Similar effect is seen for a large-grain nc-Si deposited at 350°C on different substrates (see text).

The long decay time part of the MW absorption curve, which corresponds to the recombination of the photogenerated carriers, essentially vanishes, and only a short component remains. This corresponds to the trapping of the charge carriers within the surface states and possibly to an enhanced nonradiative recombination within the insufficiently passivated nc-Si/SiO$_2$ interface. With increasing oxidation the passivation is improved, as reflected in the increase of the PL efficiency.[10,19,34] Therefore, the contribution of the nonradiative recombination to the decrease of the MW absorption progressively decreases, and the decay time of the MW transient signal reflect more and more the trapping time of the photogenerated carriers within the surface states from which the PL occurs (see step 2 in Figure 7). Unfortunately, the

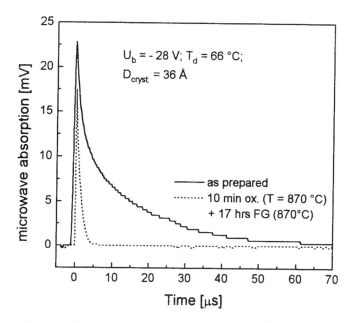

Figure 9 Transient MW absorption signal from a sample deposited at 66°C and substrate bias of −28 V (solid line) and after oxidation in pure oxygen (1 atm) for 10 min and annealing in forming gas for 17 h (both at 870°C). (From Vepřek, S. et al., *Mater. Res. Soc. Symp. Proc.*, 358, 99, 1995. With permission.)

MW transient signal is too weak in films which show PL with the highest efficiency and, therefore, the trapping time cannot be reliably measured. Nevertheless, the available data allow us to reliably separate the time scales for the various steps in Figure 7. The formation of the photogenerated carriers within the Si nanocrystals (step 1, Figure 7) occurs at the time scale of femtoseconds; their trapping at a time scale of much less than 1 μs may be even of the order of picoseconds. Finally, the radiative recombination and the PL emission from these states occur at a time scale of tens of microseconds. This separation of the various processes will be important for the discussion of the mechanism of the fast PL from tungsten-doped nc-Si/SiO$_2$ films.

The localization of carriers is evidently also the origin of an enhanced intensity of Raman spectra seen in such films. Figure 10 shows the dependence of the Raman intensity normalized to the Raman signal from CaF$_2$ on the substrate bias. The relative intensity of the X-ray scattering from the grain boundaries with expanded Si–Si bond distances is shown as well.[25] Grain boundaries with expanded bond distances represent barriers for the charge transport (weak or no MW absorption) resulting in localization. If, under a large compressive stress of 20 to 30 kbar (for $V_b = -150$ to -400 V, see Figure 2), the expansion decreases and finally vanishes (see Figure 10), the carriers can move easily through the grain boundaries and their localization vanishes as well. Since a similar effect was found also for the optical absorption above the band edge and for elastic light scattering,[25] as well as for the MW absorption (see Figure 8), there is little doubt that enhancement of the Raman and elastic scattering and of the optical absorption in nc-Si are due to the localization phenomena.

Relative intensity of the Γ'_{25} crystalline mode to the 480 cm^{-1} scattering from a-Si in the Raman spectrum is frequently used to evaluate the fraction of crystalline component in films consisting of a mixture of nc-Si and a-Si. The results presented here, as well as those published in our earlier paper,[25] show clearly that such a procedure can lead to large errors and, therefore, should not be used.

5. CHANGING THE COLOR AND DECAY TIME OF THE PHOTOLUMINESCENCE

During the last 2 years several papers reported the observation of a nanosecond fast green/blue PL from heavily oxidized PS or nc-Si which appears either after several weeks aging of the samples in air or immersion in water. We have shown that this PL is associated with silanol groups ≡Si–OH within the silica[37] and is observed also in other oxides, such as alumina, ZnO, and PbO$_2$.[20] It can be distinguished from other PL features in the green/blue/violet spectral range by the following characteristic properties: it appears independently of the red PL, and its appearance does not cause quenching of the red PL (see

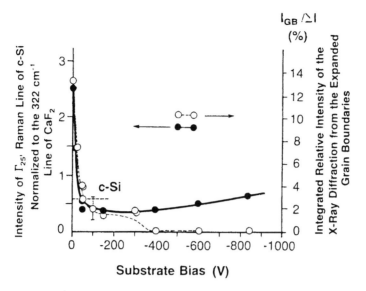

Figure 10 (a) Normalized intensity of Raman scattering and the relative intensity of X-ray scattering from the grain boundaries vs. substrate bias during deposition. $T_d = 260°C$, constant. (From Vepřek, S. et al., *Phys. Rev. B*, 36, 3344, 1987. With permission.)

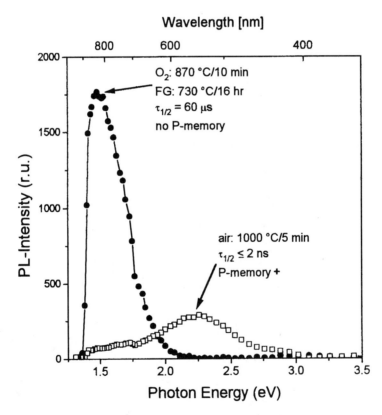

Figure 10 (b) Example of the standard red PL obtained by ultraclean processing and of a green one associated with defects in the SiO_2 which appears if such a the sample is, afterwards, annealed in air.

Reference 37); it shows a pronounced dependence on the temperature of isochronal annealing with a maximum intensity at about 400°C;[20,37] finally, it shows a strong polarization memory which is typical for a PL where the same dipole absorbs the excitation and emits the PL light.[20,37]

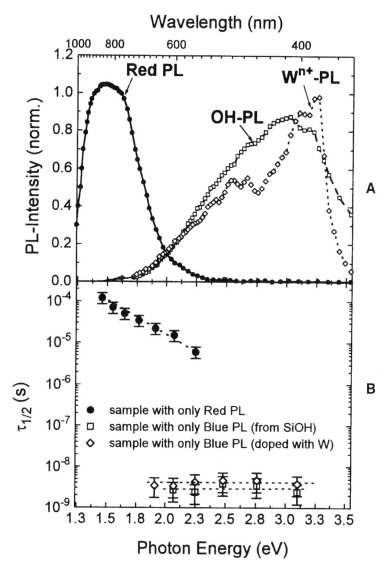

Figure 11 Comparison of the spectral distributions of the red, silanol-related, and tungsten-stimulated PL (A) and their decay times vs. the photon energy (B). (From Veprek, S. et al., *Appl. Phys. Lett.*, 67, 2215, 1995. With permission.)

Many other PL lines are occasionally observed which can be easily attributed to impurities and defects. It is well known that many organosilanes show efficient and fast PL in that spectral range, a typical example being "silicon vacuum grease"[38] which may easily contaminate the samples. In order to emphasize this point we show in Figure 11 the "red" PL obtained from our films if the oxidation is performed in ultrapure oxygen (no impurity found by neutron activation analysis[35]) and a "green" PL which is found if such a sample is afterwards annealed in air for 5 min. The latter PL is most probably from defects in SiO_2. Therefore, extreme care has to be taken to distinguish the various PL features in this spectral range. The spectral distribution, decay time, dependence on the thermal treatment, and the absence or presence of the polarization memory are useful, but not always sufficient criteria for attributing an observed PL to a given mechanism.

The slow decay rate of the red PL of the order of tens of microseconds is hardly of any interest for optoelectronic applications because the only niche for silicon appears to be fast optical communication from chip to chip which requires a clock frequency in the range of gigahertz. The calculated dependence of the radiative lifetime on the crystallite size in Figure 6a shows that only molecular-like centers of the

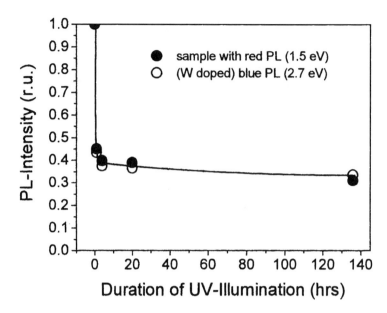

Figure 12   Photodegradation of the red and blue/violet PL under illumination with a strong UV lamp (see text).

size of few angstroms can reach such a fast PL. Since we know that the nc-Si/SiO$_2$ composite can efficiently form photogenerated electron–hole pairs and "store" them within the "surface-state" radiative centers for tens of microseconds, it was attractive to try to speed up the PL by an appropriate dye. Such a dye should facilitate the fast electron–hole recombination and light emission. The most likely candidate is a transition metal such as tungsten which, in the highest oxidation state, has an empty valence shell and can therefore provide fast charge-transfer transitions.

Although the preparation of such samples is rather difficult and still far from being optimized,[35] we have succeeded in obtaining from such W^{6+}-doped nc-Si/SiO$_2$ films efficient and about ≤100 ps fast PL.*Figure 12 shows the comparison of the spectral distribution and the dependence of the decay time on the PL energy for the three PL discussed here. The W^{6+}-related PL can be distinguished from the silanol-related one by the following characteristics: a clear, albeit relatively small difference in the spectral distribution and position of the maximum and, in particular, lack of any polarization memory because of the energy transfer from the crystallites (which absorbs the UV light) to the W^{6+} centers which emit the PL (see Figure 7). Furthermore, upon doping a sample, which initially showed a strong red PL, with tungsten, appearance of the blue/violet PL attributed to W^{6+} is accompanied by a concomitant decrease of the intensity of the red one (see Reference 35). The silanol-related green/blue PL does not cause such a quenching.

A further piece of evidence supporting the suggested mechanism is the identical photodegradation observed for both the red and the blue/violet PL shown in Figure 13. Upon illumination with a strong UV source (400 W Hg lamp used without any filter at a distance of 10 cm from the sample), the intensity of the PL measured under standard conditions (more than a factor of 100 Watt weaker intensity of the UV light) decreases and nearly saturates after about 15 h. The identical time dependence of the photo-degradation for both PL suggests that the same defect centers for nonradiative recombination are quenching both PL. As the photon energy of the exciting light of about 3.4 eV is less than the band gap of silica, formation of photogenerated defects at the Si/SiO$_2$ interface is the probable origin of the photodegradation.

So far, all the results seem to be in agreement with the suggested mechanism of the W^{6+}-stimulated PL. However, a mechanism can never be proved definitively, but only disproved if new results in disagreement with that mechanism are found. There is surely also a long way to go regarding the optimization of the doping and posttreatment of the samples, which requires simultaneous optimization of several parameters. A particularly difficult problem is the appropriate processing of the doped samples which makes sure that the energy transfer from the excited silicon nanocrystal will be transferred to the W^{n+} radiative center before

---

* The decay time of ≤2 to 3 ns reported in our original paper[35] was the resolution limit of the system.[35] More recently, we were able to repeat the measurements with another, more sophisticated apparatus and obtain a decay time $\tau_{1/2}$ of about ≤100 ps.

being trapped in the surface states that are responsible for the red PL (see Figure 7). Nevertheless, if the problems of reproducibility and technological optimization could be solved, the doping of nc-Si/SiO$_2$ films with fast dyes would be a very promising way toward fast optoelectronic devices based fully on silicon technology. We hope that the future work will bring further progress.

ACKNOWLEDGMENT

We should like to acknowledge the financial support of this work by the Federal Ministry for Research and Technology (BMFT) and by the Deutsche Forschungsgemeinschaft (DFG).

REFERENCES

1. L. T. Canham, *Appl. Phys. Lett.* 57, 1046, 1990.
2. V. Lehmann and U. Gösele, *Appl. Phys. Lett.* 58, 856, 1991.
3. L. Brus, *J. Phys. Chem.* 98, 3575, 1994.
4. M. S. Hybertsen, in *Porous Silicon Technology*, J.-C. Vidal and J. Derrien, Eds., Springer-Verlag, Berlin, 1995, 67.
5. C. Delerue, G. Allan, E. Martin and M. Lannoo, in *Porous Silicon Technology*, J.-C. Vidal and J. Derrien, Eds., Springer-Verlag, Berlin, 1995, 91.
6. F. Koch, V. Petrova-Koch, T. Muschik, A. Nikolov and V. Gavrilenko, *Mater. Res. Soc. Symp. Proc.* 283, 197, 1993.
7. F. Koch, V. Petrova-Koch and T. Muschik, *J. Lumin.* 57, 271, 1993.
8. P. D. J. Calcott, N. K. J. Nash, L. T. Canham, M. J. Kane and M. D. Brunhead, *J. Phys. Condens. Mater.* 5, L91, 1993; *J. Lumin.* 57, 257, 1993.
9. S. M. Prokes, *Appl. Phys.* 62, 3244, 1993; *J. Appl. Phys.* 73, 407, 1993.
10. M. Rückschloß, O. Ambacher and S. Veprek, *J. Lumin*, 57, 1, 1993.
11. M. S. Hybertsen, *Phys. Rev. Lett.* 72, 1514, 1994.
12. B. Delley and E. F. Steigmeier, *Phys. Rev. B* 47, 1397, 1993.
13. G. Allan, C. Delerue and M. Lannoo, *Phys. Rev. B* 48, 7951, 1993.
14. C. Delerue, M. Lannoo and G. Allan, *J. Lumin.* 57, 249, 1993.
15. S. Schuppler, S. L. Friedman, M. A. Marcus, D. L. Adler, Y.-H. Xie, F. M. Ross, T. D. Harris, W. L. Brown, Y. J. Chabal, L. E. Bruss and P. H. Citrin, *Phys. Rev. Lett.* 72, 2648, 1994.
16. W. L. Wilson, P. F. Szajowski and L. Brus, *Science* 262, 1242, 1993.
17. A. Fojtik, M. Giersig and A. Henglein, *Ber. Bunsenges. Phys. Chem.* 97, 1493, 1993; *J. Lumin.* 221, 363, 1994.
18. D. J. DiMaria, J. R. Kirtley, E. J. Pakulis, D. W. Domg, T. S. Kuan, F. L. Pesavenko, T. N. Theis, J. A. Cutro and S. D. Brorson, *J. Appl. Phys.* 56, 401, 1984.
19. M. Rückschloß, B. Landkammer and S. Veprek, *Appl. Phys. Lett.* 63, 1474, 1993.
20. M. Rückschloß, Th. Wirschem, H. Tamura, G. Ruhl, J. Oswald and S. Veprek, *J. Lumin.* 63, 279, 1995.
21. S. Veprek, in *Proc. Mater. Res. Soc. Symp. Europe,* Strasbourg, France, Eds. P. Pinard and S. Kalbitzer, Les Éditions de Physique, Les Ulis, 1984, 425.
22. S. Veprek, and V. Marecek, *Solid State Electron.* 11, 683, 1968.
23. M. Konuma, H. Curtins, F.-A. Sarott and S. Veprek, *Philos. Mag.* 55, 377, 1986; S. Veprek, M. Heintze, F.-A. Sarott, M. Jurcik-Rajman and P. Willmott, *Mater. Res. Soc. Symp. Proc.* 118, 3, 1988.
24. S. Veprek, Z. Iqbal and F.-A. Sarott, *Philos. Mag. B* B45, 137, 1982.
25. S. Veprek, F.-A. Sarott and Z. Iqbal, *Phys. Rev. B* 36, 3344, 1987.
26. S. Veprek, Z. Iqbal, R. O. Kühne, P. Capezzuto, F.-A. Sarott and J. K. Gimzewski, *J. Phys. C: Solid State Phys.* 16, 6241, 1983.
27. S. Veprek, T. Wirschem, M. Rückschloß, C. Ossadnik, J. Dian, S. Perna and I. Gregora, *Mater. Res. Soc. Symp. Proc.* 405, 1996 in press.
28. Y. Kanemitsu, K. Suzuki, Y. Nakayoshi and Y. Matsumoto, *Phys. Rev. B* 46, 3916, 1992.
29. W. E. Carlos and S. M. Prokes, *J. Appl. Phys.* 78, 2129, 1995.
30. S. M. Prokes and W. E. Carlos, *J. Appl. Phys.* 78, 2671, 1995.
31. M. L. Cohen and J. R. Chelikowsky, *Electronic Structure and Optical Properties of Semiconductors*, Springer-Verlag, Berlin, 1989.
32. Th. Wirschem, M. Rückschloß, J. Oswald and S. Veprek, *Appl. Phys. Lett.* 1995 submitted.
33. H. Takagi, H. Ogawa, Y. Yamazaki, A. Ishizaki and T. Nakagiri, *Appl. Phys. Lett.* 56, 2379, 1990.
34. S. Veprek, Th. Wirschem, M. Rückschloß, H. Tamura and J. Oswald, *Mater. Res. Soc. Symp. Proc.* 358, 99, 1995.
35. S. Veprek, M. Rückschloß, Th. Wirschem and B. Landkammer, *Appl. Phys. Lett.* 67, 2215, 1995.
36. D. Allan, C. Delerue and M. Lannoon, *Mater. Res. Soc. Symp. Proc.* 405, 1996 in press.
37. H. Tamura, M. Rückschloß, Th. Wirschem and S. Veprek, *Appl. Phys. Lett.* 65, 1537, 1994; *Thin Solid Films* 255, 92, 1995.
38. S. Veprek, Invited Talk at the Wacker Symp. 2nd Münchner Silicontage, August 1994, in: *Organosilicon Chemistry: From Molecules to Materials II*, N. Auner, Ed., VCH Verlagsgesellschaft, Weinheim, 1995.

Chapter 7

Optics of Nanosized Metals

Uwe Kreibig

Reviewed by M. Vollmer

TABLE OF CONTENTS

1. INTRODUCTION

Up to the mid 1970s the bulk solid and the planar surface were the favorite topics of matter physics and condensed matter science in general. Now, nanostructured metal has largely taken their position. Its variability appears to be so broad that limits are not yet in sight and technical applications are still dominated by promising proposals.

The topological structures of nanostructured metals may either be intentionally designed and tailored or be self-organized; they may be ordered or statistically disordered. The structurization on a "nanoscale" means, more precisely, sizes of structural units on the order of 10^0 to, say, 10^2 nanometers, where those units may differ from each other, both in their geometric and chemical properties. The size limitation of the units may include one, two, or three dimensions. This means topologies of layers, of wires or rods, and of clusters or nanoparticles, respectively.*

These scales lie, for many physical properties, in the "mesoscopic regime" where characteristic lengths like correlation lengths or mean free paths, l_j, of the bulk are larger than the sample: $l_j > D$. The wide variability of topological patterns appears to render it impossible to give a unified, general description of nanosized metals in the way that was so successfully applied to extended crystalline matter. Neither does it appear to be possible to give a general description of their optical properties.

---

* Here, we define *clusters* as aggregates of atoms in the order of 10^1 to 10^5 atoms, each. Synonymously, the term *nanoparticles* will be used. Typical molecular clusters consisting of less than, say, 10^1 atoms each are beyond the scope of this overview. They require theoretical approaches from molecule theories to describe their optical properties.

0-8493-2485-8/97/$0.00+$.50
© 1997 by CRC Press, Inc.

CHEMICAL COMPOSITION | STRUCTURAL COMPOSITION

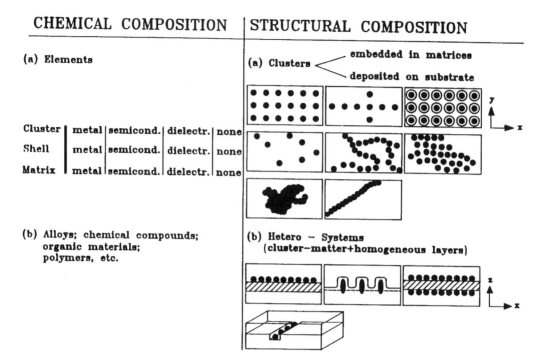

Figure 1 Compositions and topological structures of many-particle-systems/cluster matter.

Hence, for the following we reduce to several selected structural model elements, the optical properties of which will be treated both on the theoretical and the experimental stage. The extension to other structures gives, in general, additional problems of the numerics rather than of physics.

Excluded will be, by intention, the layer and wire geometries and the quantum dots, since these are all treated in other contributions to this book.

We will concentrate on systems based on single nanoparticles with more or less isometric dimensions. We start with the well-isolated single cluster of highly symmetric shape, like the sphere, ellipsoid, etc., treating dielectric theories with proper, realistic material properties for the cluster. The latter are introduced in Section 4 from various models of size effects. Here, we include the drastically important, yet usually neglected, additional influences of materials surrounding the cluster. Some information is also given concerning nonlinear optical properties of metallic nanoparticles and inelastic optical scattering (surface-enhanced Smekal-Raman scattering, SERS).

In the third section we change to many-nanoparticle-systems, i.e., to "cluster matter." Even for this particular kind of nanostructured matter, as a special kind of disordered, heterogeneous, or composite material, wide topological and chemical variability is obvious, as demonstrated in Figure 1. These systems are frequent in nature and widely used for practical and technological purposes as shown in Table 1.

We demonstrate various cluster–cluster interaction influences on the optical properties. After a short overview of the broad field of effective medium-models, we turn to the exact description of electromagnetic cluster–cluster interactions. The chapter ends with a discussion of the transition toward bulk material by coalescence and the compact nanophase material. Research on the optical properties of metal nanoparticles has been published since the beginning of this century with a steep increase in the last decades. Comprehensive citation is, in this chapter, replaced frequently by references to extended reviews concerning parts of the topics of this article. Among these are the books by van de Hulst [1957], Kerker [1969], Bohren and Huffmann [1983], and, as the most recent, one by Kreibig and Vollmer [1995].

2. ELECTRODYNAMIC EXCITATIONS OF THE SINGLE METAL CLUSTER

As indicated before, the lower size limit of clusters treated here is given by the postulate that the clusters should be describable by a dielectric function which is extrapolated from solid state models for metals:

$$\hat{\varepsilon}(\omega,\ldots) = \varepsilon_1(\omega,\ldots) + i\varepsilon_2(\omega,\ldots) = 1 + \hat{\chi}_\ell(\omega,\ldots) + \hat{\chi}_f(\omega,\ldots) + \hat{\chi}_{ib}(\omega,\ldots) \qquad (1)$$

Table 1 Examples of nanoparticles in nature and technology

Geocolloids (minerals, hydrosols, aerosols)	Quantum dots
Heterogeneous catalysts	Island films
Photography	Sol-gel systems
Color filters	Ferro fluids
Glass coloration (ruby, yellow, etc.)	Nanocrystalline materials
Solar absorbers	Sintered nanoceramics
Cermets, varistors	Cytochemical markers (medecin)
Recording tapes	Autoradiographic markers (medecin)

The indices ℓ, f, and *ib* of the respective susceptibility contributions stand for "lattice," "free electrons," and "interband transitions." The possibility is indicated that other variables beside the (circular) frequency ω influence $\hat{\varepsilon}$, such as temperature or cluster size or surface conditions, etc.

The classic electrodynamic theory of Mie [1908] and Gans and Happel [1909], compiled in the following, affords for $\hat{\varepsilon}$ gravid simplifications by assuming that the optical polarizability of the whole cluster (including its interface) is sufficiently described by one uniform quantity $\hat{\varepsilon}$ of the type of Equation 1. This means, in fact, that an average over the whole cluster of the local polarizabilities has to be performed, and this average $\hat{\varepsilon}$ no longer has the meaning given to the macroscopic dielectric function by its macroscopic definition. Instead, $\hat{\varepsilon}$ expresses microscopically the total linear polarization of the whole single cluster. As a consequence, the $\hat{\varepsilon}$ of the bulklike cluster material does, in general, not fit in a realistic description (as was frequently assumed). In order to keep the classic electrodynamics (in particular, the Maxwellian boundary conditions) applicable, $\hat{\varepsilon}$ has to be properly modified to include cluster-size and interface effects.

As a definition we may state: The proper $\hat{\varepsilon}(\omega)$ is found and modeled for some realistic cluster when, inserted into the formulae of the Mie theory, the calculated optical extinction spectra coincide with appropriate experimental spectra.

The aim is to keep Mie's and related electrodynamic theories applicable also to realistic clusters of small size by developing (quantum theoretical) microscopic models for the properties of a cluster which are cumulated in the introduction of proper correction or extension terms into $\hat{\varepsilon}$ of Equation 1. A justification of this rather indirect way to describe the optical response is that the electrodynamic description of the spherical cluster properties can be exactly performed, irrespective of cluster size (i.e., including retardation, multipole excitations, angle-dependent scattering, near-field properties, etc.) and, hence, quantitative comparison with experiments on metal clusters of arbitrary sizes is rendered possible.

2.1. Spherical Clusters

The exact (in the frame of Maxwell theory) electrodynamic description of the optical absorption and elastic scattering (usually expressed by the extinction which is defined as the sum of both) of the metallic sphere was first developed by Mie. *Elastic* means $\Delta\omega = 0$, i.e., luminescence and Smekal–Raman scattering are beyond this theory. In principle, Mie's theory is the analogon of Fresnel's formulae where the planar geometry of the former is changed to spherical symmetry. It is based on a spherical mode expansion method. These modes have, for the case of metallic clusters, later been interpreted as being — essentially — due to collective excitations of the electronic system, i.e., spherical surface plasmon polaritons (plasmons coupled to a macroscopic electromagnetic field) and eddy current modes, excited by the electric and magnetic fields, respectively. Besides these collective excitations of the conduction electrons, single electron–hole excitations also occur in clusters of realistic materials, described by Equation 1. While the latter are similar to the interband transitions known from the bulk metals, these collective modes are unique properties of the cluster state. In these modes the cylindrical symmetry of the field and the spherical one of the clusters are combined. The expansion set of modes is complete, their number being infinite. Nonmetallic clusters exhibit analogous modes of "cluster phonon polaritons."

The more or less sharp resonances, the Mie plasmon polaritons, are caused by the fact that collective oscillations of the conduction electrons are excited by the electromagnetic wave. Their dipole and higher multipole moments are due to surface charging, which is — compared with the planar geometry — extremely effective for spherical shapes (Figure 2). It has been shown (see Kreibig and Vollmer [1995]) that in free electron nanoparticles the total electronic oscillator strength extending in planar samples from the visible to zero frequency is compressed into these resonances. Since the electron densities of most metals do not differ strongly, the material-specific differences of the resonances are mainly due to

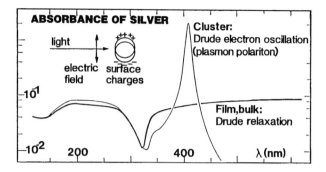

Figure 2 Computed optical absorbance spectra of silver films and silver clusters. Interband transitions below 325 nm.

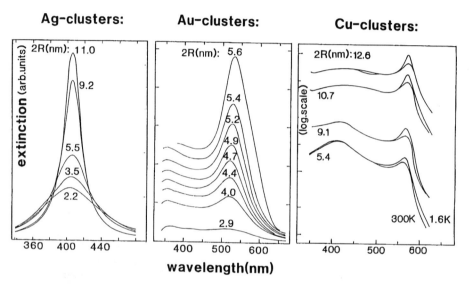

Figure 3 Measured extinction spectra of nanoparticles of silver, gold, and copper of various mean diameters 2R. Silver and gold at 300 K, copper at 300 and 1.6 K. The particles were prepared in photosensitive glasses (see Kreibig and Vollmer [1995]).

the band structure effects enclosed in Equation 1 in the term $\hat{\chi}_{ib}$ and also to material-specific kinds and amounts of excitation relaxation processes of the conduction electrons (Figure 3). This kind of collective excitation is unique for clusters and has no counterpart in the bulk matter or in thin films.

Only a few substantial extensions of Mie's theory were added later, like the inclusion of additional boundary conditions (see Kreibig and Vollmer [1995]), which allow the excitation of longitudinal volume plasmon modes in the particles.

The Mie theory was repeatedly reproduced and reformulated in various ways in the literature (in addition to the books cited before, see, e.g., Stratton [1941]; Born [1972]), so we restrict ourselves to a short summary of the original version.

For numerical purposes, the reformulation of Stratton [1941] has some advantages. An advanced FORTRAN computer program of this formulation of Mie's theory was recently published (see Haberland [1993]; Kreibig and Vollmer [1995]).

While Mie gave expressions for the extinction constant E, the scattering constant S, and the absorption constant A of macroscopic systems containing many (isolated, noninteracting!) clusters with a low volume filling factor $f = 10^{-6}$ for normalization, it is now common to derive, instead, the corresponding single-cluster cross sections.

In the following, an outline of the Mie theory for one particle including longitudinal excitations is given by listing the mathematical steps. (To obtain the experimentally accessible optical quantities, samples of typically 10^{10} or more particles are required. Many of the Mie clusters are therefore thought to be uniformly spread in a macroscopic volume with volume concentrations and sample thicknesses low enough to keep interactions among the particles and multiple light scattering events negligible.)

Calculation of the optical extinction of one cluster and extension to n clusters is as follows:

1. Introduction of spherical coordinates r, θ, ϕ, and particle radius R.
2. Plane, monochromatic incident electromagnetic wave (circular frequency ω, wavelength λ).
3. Solution of Maxwell's equations with proper boundary conditions for one arbitrary spherical particle of complex dielectric function $\hat{\varepsilon}(\omega)$, embedded in a dielectric medium with (real) dielectric constant ε_m.
   - Electric fields: $E = E$ (divergency free) + E (curl free);
   - Magnetic fields: $H = H$ (divergency free);
   - Boundary conditions at $r = R$:

$$E_o^{incident} = E_\Theta^{\iota\iota terior}; \quad E_\phi^{incident} = E_\phi^{\iota\iota terior} \tag{2}$$

   - current density: J_{total}^{normal} continuous (Sauter-Forstmann [1967] condition).

The linear response function of the particle material is $\hat{\varepsilon}\mu$, where $\hat{\varepsilon}(\omega) = \varepsilon_1 + i\varepsilon_2$ is averaged over the cluster volume as described above. The magnetic permeability μ is set 1 for the investigated high-frequency regions. (This means that magnetic response of the material, which may be effective at lower frequencies, is excluded throughout the following. However, the magnetic part of the electromagnetic wave contributes to the electronic excitations by creating eddy current modes.)

   - Separation of the transverse electromagnetic fields according to the radial field components: "electric partial waves" with $H_{r=R} \equiv 0$, $E_r \neq 0$; "magnetic partial waves" with $E_{r=R} \equiv 0$, $H_r \neq 0$. (Both are electromagnetic fields including Θ- and ϕ-components, too.)
   - Introduction of three scalar potentials, to separate the variables: Π_E, Π_M, Γ (indices E,M: electric and magnetic partial waves). They are solutions of the Helmholtz equations:

$$\Delta\Pi_{E,M} + K_{transverse}^2 \cdot \Pi_{E,M} = 0, \quad K_{transverse}^2 = \varepsilon_{trans}\omega^2/c^2 \tag{3a}$$

$$\Delta\Gamma + K_{longitudinal}^2 \cdot \Gamma = 0, \quad K_{longitudinal}^2 = \varepsilon_{long}\omega^2/c^2 \tag{3b}$$

   - Separation of the variables by a product ansatz, e.g., for Π_E holds

$$\Pi_E = F_1(r)F_2(\Theta)F_3(\phi) \tag{4}$$

Solutions are given in the following table:

Variable	Differential equation	Solution functions
r	Bessel	Cylinder functions (Bessel, Neumann); Index L
θ	Spherical harmonics	Legendre polynomials; Indices L, m
ϕ	Harmonic oscillation	cos/sin functions; Index m

   - Back-transformation of the potentials into fields of different polar and azimuthal order L, m. These are divided into 3×3 groups, i.e., each of the three electromagnetic waves — the incident wave, the wave in the interior of the particle, and the scattering wave — consists of three different contributions — the "electric" partial waves, the "magnetic" partial waves, and, if present, longitudinal waves.
4. Computation of intensities I (i.e., the Poynting vectors) from all field amplitudes and evaluation of the following experimentally accessible quantities for systems of n identical particles, packed loosely together in a macroscopic volume with volume concentration f. (This volume may be filled with some matrix material.) The corresponding (single-cluster) cross sections are for:
   - The extinction, i.e., the sum of absorption and scattering losses

$$\sigma_{ext} = \frac{2\pi}{|k|^2} \sum_{L=1}^{\infty} (2L+1) \operatorname{Re}\{a_L + b_L\} \tag{5}$$

- The elastic scattering (which can be calculated angle resolved or angle integrated, respectively)

$$\sigma_{sca} = \frac{2\pi}{|k|^2} \sum_{L=1}^{\infty} (2L+1) \left(|a_L|^2 + |b_L|^2 \right) \tag{6}$$

- The absorption, i.e., the production of heat and thermal radiation in the cluster, as the difference of Equations 5 and 6

$$\sigma_{abs} = \sigma_{ext} - \sigma_{scatt} \tag{7}$$

The "Mie coefficients," following from the boundary conditions, are

$$a_L = \frac{m\psi_L(mx)\psi'_L(x) - \psi'_L(mx)\psi_L(x)}{m\psi_L(mx)\eta'_L(x) - \psi'_L(mx)\eta_L(x)} \tag{8a}$$

$$b_L = \frac{\psi_L(mx)\psi'_L(x) - m\psi'_L(mx)\psi_L(x)}{\psi_L(mx)\eta'_L(x) - m\psi'_L(mx)\eta_L(x)} \tag{8b}$$

$m = \hat{n}/n_m$, where $\hat{n}$ denotes the complex index of refraction of the particle and n_m the real index of refraction of surrounding medium. k is the wave vector and $x = |k|R$ the "size parameter". $\psi_L(z)$ and $\eta_L(z)$ are Riccati–Bessel cylindrical functions (see, e.g., Born [1972]). The prime indicates differentiation with respect to the argument in parentheses. (In Mie's original notation, Equation 5 contains the imaginary instead of the real part because of different definitions of a_L and b_L.)

These cross sections are related to the absorption, scattering, and extinction constant $A(\omega)$, $S(\omega)$, and $E(\omega)$, respectively, and the according intensity losses ΔI_j of a system of number density Z of the (noninteracting) clusters by the Lambert–Beer law:

$$\Delta I_j = I_0 \left(1 - \exp\left(-Z \cdot \sigma_j \cdot d\right)\right), \quad \begin{array}{l} d = \text{sample thickness} \\ I_0 = \text{incident intensity} \end{array} \tag{9}$$

For details see, e.g., Mie (1908).

Both in Mie's theory and the Gans–Happel theory (see Equation 11) n_m and ε_m are restricted to be real. The embedding medium, thus, is assumed to be nonabsorbing. Recently, several attempts were made to extend to absorbing matrix material. This will be treated later.

It is a general rule that the importance of higher multipoles for A and S increases with particle size. However, the extinction of the dipolar mode always predominates, as shown in Figure 4 for the example of Ag clusters, where the respective maxima are plotted against the cluster radius. (It should be noted that these maxima lie at different frequencies which depend on R.)

If the nanoparticles are sufficiently small ($2R \ll \lambda$; "quasistatic regime") that scattering S is almost zero, then the extinction $E(\omega)$ contains merely dipolar plasmon absorption losses ($L = 1$) and the famous equation for absorbing spherical particles in a dielectric matrix is obtained

$$E(\omega) = A(\omega) = 6 f \varepsilon_m^{3/2} (\omega/c) \frac{\varepsilon_2(\omega)}{\left(\varepsilon_1(\omega) + 2\varepsilon_m\right)^2 + \varepsilon_2^2(\omega)} \tag{10}$$

c = vacuum velocity of light.

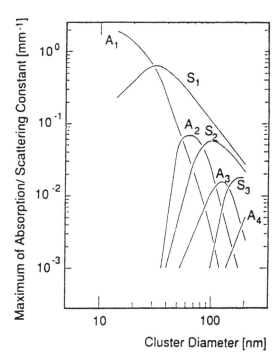

Figure 4 Mie resonances of silver nanoparticles: calculated size dependence of the maxima spectral of the modes which contribute most. A_1, S_1 = absorption, scattering. L = multipolar order.

Surprisingly, this equation holds quantitatively for, e.g., Ag and Au particles in the resonance region only if $2R \leq 10$ to 15 nm. For the purpose of quick, but only semiquantitative estimations, this quasistatic formula can be extended formally to include the higher multipolar modes. Then the "2" in the denominator has simply to be replaced by $((L + 1)/L)$, L = mode index. Physically, there is the restriction that, if higher multipolar modes are excited, the dipolar scattering is no longer to be neglected, and the quasistatic approximation is surpassed.

It should be noted that Equation 10 does not comprise explicitly the cluster size; i.e., identical optical spectra should be obtained for clusters of all sizes within the quasistatic regime. Yet, Mie has already mentioned that this independence is only formal and that, below some critical size, experimental spectra have to reflect the transition from the solid to the molecular state. In the frame of Equation 10 these ensuing intrinsic cluster-size effects have to be included by size-dependent dielectric functions $\hat{\varepsilon}(\omega,R)$.

Influences of $\hat{\varepsilon}(\omega)$ on the E spectra are shown schematically in Figure 5. The graphs A and B demonstrate that peak positions and also widths are essentially determined by $\varepsilon_1(\omega)$. Flat ε_1 spectra, i.e., small $d\varepsilon_1/d\omega$, yield broad bands. Hence, the band widths are not indicative only of damping mechanisms and energy dissipation! The graphs C and D show that, in principle, multipeak structures can be produced by proper ε_1 spectra (C), which, however, may be damped away when ε_2 is sufficiently large (D). So, band widths are also influenced by all damping mechanisms entering ε_2. In the special case of a Drude material, the plasmon polariton band width equals the relaxation frequency enclosed in the Drude–Sommerfeld- $\hat{\varepsilon}$. Both low $d\varepsilon_1/d\omega$ and high ε_2 in the vicinity of resonance frequencies ω_s cause the cluster plasmon polariton bands in most materials to be smeared out past recognition, and only few materials exhibit sharp and selective bands. These are, among others, the alkalis, noble metals, and aluminum.

While, e.g., in Ag and Na nanoparticles, the Mie resonances are well developed, they are suppressed for other materials, like Cu, Fe, Pt, etc., mainly because of lifetime broadening via strong plasmon decay channels. The plasmon lifetime may be limited either by strong conduction electron relaxation and/or radiation damping (in the larger particles) or by transformation of the collective excitation into electron–hole pair excitations. The latter takes place in Cu because of spectral overlap with interband transition excitations.

It should be pointed out that no assumptions whatsoever were made about the origin of the particular spectra: these can be ionic contributions as well as electronic ones. Well-developed Mie-phonon polariton bands are found, e.g., in MgO particles.

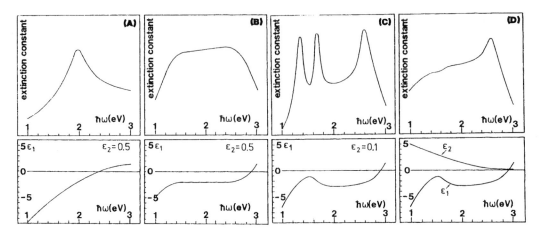

Figure 5 Schematics of Mie resonances of spherical metal clusters. The dielectric functions were arbitrarily chosen to demonstrate the relative influences of ε_1 and ε_2. Widths and positions and many-peak structures are mainly due to the spectrum of $\varepsilon_1(\omega)$.

For instance, one extinction peak of some semiconductor may be due to optical phonons while another may be due to the conduction electrons and a third may appear in the interband transition region, if $\hat{\varepsilon}(\omega)$ takes proper values.

In real metals, like Cu, Ag, and Au, the Drude electron excitations are hybridized with interband transitions, and hence the cluster plasmon polariton shifts from about $\hbar\omega \sim 10$ eV of the free Drude metal with the same s-electron density down to the visible region of the spectrum (see Figure 3).

Excitations by the magnetic partial waves also contribute to $E(\omega)$; they are missing in Equation 10. They do not induce resonance behavior similar to that of the electrical partial waves, but exhibit selective spectral features in the infrared (IR) (see Genzel and Kreibig [1980]), and, e.g., in the case of Ag-particles, they also reflect the onset of interband transitions (see Kreibig and Vollmer [1995]). It should be mentioned, again, that they are the response of the electronic system to the magnetic part of the incident wave (magnetic material reactions are zero in the optical region) and are interpreted to be due to eddy currents of the respective electrons.

The resulting extinction spectra $E(\omega)$, an example of which is given in Figure 6, are composed of a multitude of different excitations (Figure 6b). The extinction E and the scattering S, both angle resolved and angle integrated, can be measured separately by conventional optical methods. Recently (see Kreibig and Vollmer [1995]), it was shown that the absorption A can also be determined directly by using photothermal methods (the "mirage effect").

Figure 6a presents, as an application of Mie's theory, the optical extinction spectra for clusters of aluminum, depending on the cluster size. With increasing size, both scattering and higher polar mode excitations appear, rendering the spectra, which for the case of $2R \ll \lambda$ are governed by the dipolar plasmon absorption alone, more and more complex. Figure 6b exhibits these contributions separately for the 100-nm-diameter Al particles by decomposition of the total extinction E into the various plasmon polariton and eddy current modes. Table 2 compiles the composition of the total Mie extinction.

In principle, it is necessary in calculations to sum over an infinite number of different multipolar contributions, i.e., $L = 1,2,...,\infty$. This is a consequence of the mathematical method of expanding the fields into orthogonal functions. Physically, however, even for larger clusters (yet, with $R < \lambda$), only few multipoles really contribute to the loss spectra in measurable amounts, as shown in Figure 4.

As stated before, the Mie solution is exact in the sense of Maxwell's theory. It describes the linear response of one cluster, which commonly is measured in the far-field regime. It, however, also describes the near-field response (see, e.g., Kreibig and Vollmer [1995]), which is important for electromagnetic cluster–cluster interactions (as we will describe in Section 3.3), for SERS of surface-adsorbed species and will become probably more important in the near future in connection with scanning optical near-field spectroscopy (SNOM), etc.

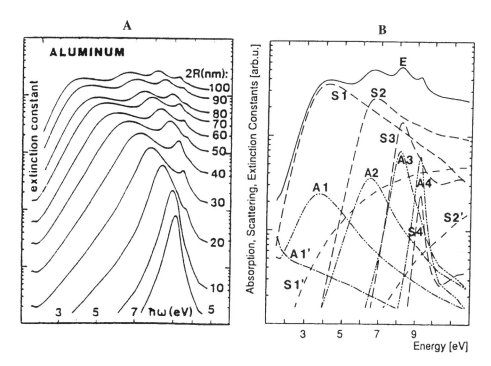

Figure 6 Calculated Mie spectra for aluminum nanoparticles. (a) Size dependence of the extinction. (b) Decomposition of the uppermost extinction spectrum of (a) ($2R$ = 100 nm) into the contributions of the multipolar modes L = 1 to 4. (A, S = absorption, scattering; primes denote "magnetic" partial wave contributions.)

Figure 7 clearly demonstrates fundamental differences between near- and far-field behavior, exemplified here for the scattering Poynting vector $\vec{S}$ (Dusemund [1991], Quinten, [1995]). The contributions of radial field components produce tangential energy flow from sources to sinks, while the far-field scattering is governed by the transverse electromagnetic field with the Poynting vector normal to the cluster surface. Accordingly, the angle-resolved scattering characteristics, also shown in Figure 7, strongly differ. The dashed curves in indicate that scattering energy flows back into the cluster in the near-field regime.

It is possible to reduce the widths and to develop sharp Mie resonances also for materials where the free clusters exhibit broad or even no spectral selectivities, by shifting them to other resonance frequencies. This can be done by embedding the clusters in matrix materials of suitable dielectric constant. By this means, the Mie resonances of, e.g., Cu or Au clusters can be enormously enhanced. In general, the spectral features of metal clusters are more drastically changed by this "dielectric" or "electromagnetic" effect of the cluster surrounding than was recognized previously. Figure 8 demonstrates this "immersion effect" for several, less common cluster materials. (These spectra are discussed in Kreibig and Vollmer [1995] in more detail.)

In all cases of Figure 8 the matrix materials were chosen to be nonabsorbing with a real ε_m. The extension of Mie's theory to absorbing embedding media is not straightforward because of

- the eigenabsorption of the matrix and
- the changed boundary conditions

The strongly simplifying approximation of inserting a complex-valued $\hat{\varepsilon}_m(\omega)$ in Equation 10 was published earlier (Kreibeg [1976]). More serious attempts were performed also, and an almost exact solution was published recently (see Quinten and Rostalski [1996] and references therein).

Shortly after Mie's paper appeared, Gans and Happel [1909] complemented the theory by adding the calculation of the corresponding refractive index of the cluster system which measures the real part of the cluster polarization:

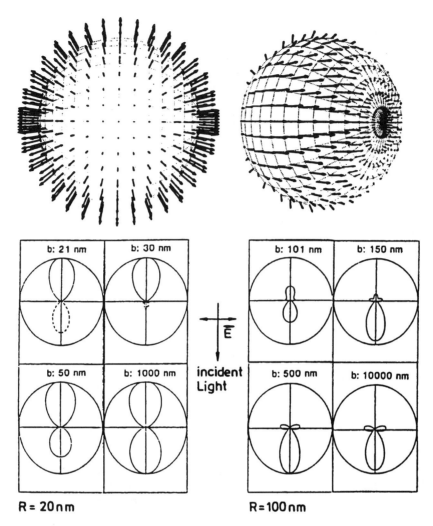

Figure 7 Near fields and far fields around a spherical metal cluster, from Mie's theory. Upper graphs: Poynting vectors; 2R = 20 nm. Left: far field; right: near field with tangential energy flow (Quinten, M., 1995.) Lower graphs: scattering characteristics of Au nanoparticles of 20 and 100 nm radius immersed in water. b: distance from the cluster center. The intensities are normalized. Dashed curves: Poynting vectors directed toward the cluster. (Dusemund, 1991).

$$n\left(\omega, R, \hat{\varepsilon}, \varepsilon_m, f\right) = \sqrt{\varepsilon_m} \left(1 + \frac{3}{4} \frac{3 \cdot f \cdot c^2}{2\pi\left(\sqrt{\varepsilon_m} \cdot R\right)^3 \cdot \omega^2} \cdot \mathrm{Im}\left\{a_1 - a_2 - b_1\right\}\right) \tag{11a}$$

$$L(\omega, \ldots) = \left(n - \sqrt{\varepsilon_m}\right) / \sqrt{\varepsilon_m} \tag{11b}$$

f is the volume concentration ("filling factor") of the particles in the system, a_1, a_2, and b_1 are the Mie coefficients of Equation 8. The authors only included two electric and one magnetic partial waves, and, hence, Equation 11 is restricted to $2R \leq 10^2$ nm.

Again, this quantity describes colloidal Mie systems, i.e., macroscopic systems of many (well-separated, noninteracting) clusters of low filling factor. It is not clear, how — in analogy to the replacement of the extinction constants by cross sections — this quantity might be related to the single cluster. In fact, Gans and Happel's theory is an example of effective medium theories, which will be treated in Section 3.2.

Table 2 Hierarchy of the contributions of absorption and scattering of different modes to the Mie extinction (The mode index is designated I here, instead of L)

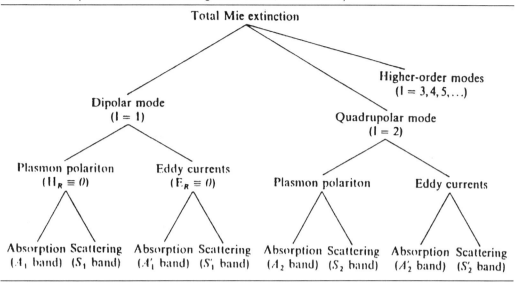

Figure 8 Mie extinction spectra of less common particle materials. $2R = 10$ nm. The clusters are embedded into matrix materials of different dielectric constant ε_m between 1 and 10 ("immersion spectroscopy"). Obviously, the occurence of peaks depends strongly on the matrix.

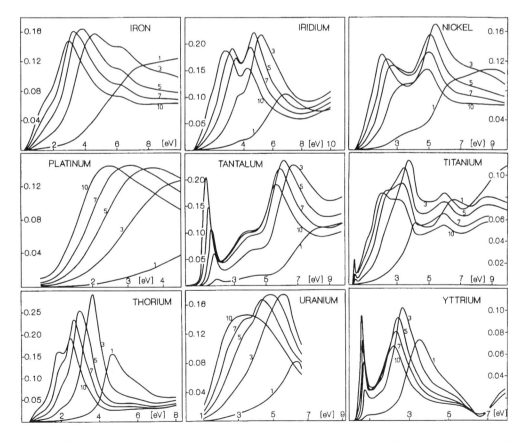

Because Mie and Gans and Happel focused on macroscopic colloidal suspensions, $\sqrt{\varepsilon_m}$ in Equation 11 refers to the pure embedding material (i.e., the solvent in the case of colloids). We will have to consider later that the influences of such surrounding media are not fully described this way. Also we remember that these clusters are assumed not to "feel" each other, this assumption is skipped when treating cluster matter in general in Sections 3.2 and 3.3.

2.2. Metal Clusters with Nonspherical Shapes

The importance of Mie's solution for spheres — a shape which will scarcely be met by larger realistic metal clusters — lies in the fact that no other cluster shape has been treated so successfully and comprehensively. Analytic solutions, beside Mie's, could only be obtained for ellipsoidal clusters. Ellipsoids are important as well because of their extremal case of high eccentricity, which approaches the cylindrical, almost infinite structures, i.e., quasi-one-dimensional wires. However, in most cases ellipsoidal cluster shapes were only treated in the quasistatic regime, and, hence, long wire structures can only be approximately described.

A general, not fully analytic, approach to clusters of arbitrary size and shape is the T-matrix method, the T-matrix relating, by linear transformation, the expansion coefficients of the incident and the scattered fields and allowing numerical solutions (Barber and Yeh [1975]; Ström [1975]).

As another kind of treatment, the discrete-dipole-approximation method was introduced successfully. The numerous literature on nanoparticles of nonspherical shapes has been repeatedly compiled. So, reference is given to Bohren and Huffman [1983] and Kreibig and Vollmer [1995].

2.2.1. Ellipsoidal Nanoparticles

In the frame of the quasistatic regime, this problem was solved by Gans [1912], who gave the complex electric polarizability $\hat{\alpha}_i$ along the ith principal axis o

$$\hat{\alpha}_i(\omega) = \varepsilon_0 \frac{\hat{\varepsilon}(\omega) - \varepsilon_m}{\varepsilon_m + \left[\hat{\varepsilon}(\omega) - \varepsilon_m\right] \cdot L_i} \cdot V_{cluster} = \varepsilon_0 \frac{1}{\varepsilon_m / \left(\hat{\varepsilon}(\omega) - \varepsilon_m\right) + L_i} \cdot V_{clusrter}; \quad V_{cluster} = \frac{4\pi}{3} a \cdot b \cdot c \quad (12)$$

L_i are the three geometric depolarization factors. Gans also developed the average polarizability of a three-dimensional many-cluster system with arbitrarily distributed orientations. The relevant physical difference from the case of spherical clusters ($L_i = 1/3$) is the occurrence of, in general, three plasma resonance peaks, instead of the single Mie resonance. For the special case of prolate spheroidal Drude particles, the resulting eigenmode frequencies are summarized in Figure 9. Here, the eccentricity is measured by $e = \sqrt{a^2 - b^2} / a$. For large $1/e$ the value of the Drude sphere $\omega_r = \omega_p / \sqrt{3}$ is reached for the dipolar plasmon, while, for smaller $1/e$ two modes are clearly separated into the higher one with E vector along the small axis and the lower one with E vector parallel to the long axis. Experimental results from Na particles on LiF were recently published (Götz [1993]).

2.2.2. Cubic Particles

Cubic shapes were also treated [Fuchs, 1975], but for nonmetallic clusters. The extension to metallic cubes appears to be straightforward. There are interesting edge and corner modes appearing in these clusters. Fuchs derived for the mode-averaged polarizability in the quasistatic approximation

$$\alpha = \varepsilon_0 \cdot V \sum_{i=1}^{N} \frac{C_i}{\varepsilon_m / \left(\varepsilon - \varepsilon_m\right) + L_i} \quad (13)$$

where C_i are the oscillator strengths and L_i the geometry factors due to depolarization, which are given numerically in the original paper.

2.3. Core-Shell Clusters (Spherical Heterosystems)

As an example of inhomogeneous clusters, the spherical nucleus-shell cluster was treated [Aden and Kerker, 1951], and, recently, Sinzig et al. [1993] extended to infinite numbers of "onion" shells of arbitrary materials.

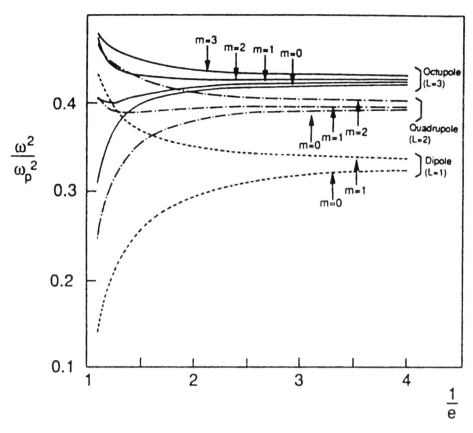

Figure 9 Ellipsoidal (prolate) Drude metal particles. Center frequencies of the multipolar modes with $L = 1,2,3$ as function of shape. ($1/e = 2,3,4$ correspond to axial ratios 0.866, 0.943, 0.968.) Calculations in the quasistatic approximation. (From Kreibig and Volmer [1995]).

This latter theory also allows us to handle the case of shell structures varying continuously along the radius direction, by assuming large numbers of only slightly varying shells. As will be shown in Section 4.2, the cluster surface or cluster–surroundings interface can, in fact, not be assumed to be a sharp two-dimensional separation plane, as suggested by the Maxwell boundary conditions. More realistically, the cluster surface has been modeled by a such three-dimensional shell structure.

Multishell clusters are, in principle, the spherical analogon of planar multilayer systems. Core and shell materials can be dielectrics, semiconductors, or metals; hence, a broad variety of unusual materials can be thus designed.

For the special case of one shell, with dielectric function $\varepsilon_s(\omega)$, of thickness d around a core with $\varepsilon(\omega)$ and radius R, embedded in a matrix with $\varepsilon_m(\omega)$, the quasistatic polarizability is [Aden and Kerker, 1951]

$$\alpha = \varepsilon_0 \frac{(\varepsilon_s - \varepsilon_m)(\varepsilon + 2\varepsilon_s) + \left(\dfrac{R}{R+d}\right)^3 (\varepsilon - \varepsilon_s)(\varepsilon_m + 2\varepsilon_s)}{(\varepsilon_s + 2\varepsilon_m)(\varepsilon + 2\varepsilon_s) + \left(\dfrac{R}{R+d}\right)^3 (\varepsilon - \varepsilon_s)(2\varepsilon_s - \varepsilon_m)} \frac{4\pi}{3}(R+d)^3 \qquad (14)$$

In the recently developed recurrence solution [Sinzig et al., [1993] for the general z-shell problem ($z = 0,1,2,3...h$), the Mie coefficients a_L and b_L of the homogeneous sphere (Equation 8) are replaced by the following expressions

$$a_L = -\frac{m_z \psi_L(m_z x_z)\left[\psi'_L(x_z) + T_L^h \chi'_L(x_z)\right] - \psi'_L(m_z x_z)\left[\psi_L(x_z) + T_L^h \chi_L(x_z)\right]}{m_z \xi_L(m_z x_z)\left[\psi'_L(x_z) + T_L^h \chi'_L(x_z)\right] - \xi'_L(m_z x_z)\left[\psi_L(x_z) + T_L^h \chi_L(x_z)\right]} \tag{15a}$$

$$b_L = -\frac{\psi_L(m_z x_z)\left[\psi'_L(x_z) + S_L^h \chi'_L(x_z)\right] - m_z \psi'_L(m_z x_z)\left[\psi_L(x_z) + S_L^h \chi_L(x_z)\right]}{\xi_L(m_z x_z)\left[\psi'_L(x_z) + S_L^h \chi'_L(x_z)\right] - m_z \xi'_L(m_z x_z)\left[\psi_L(x_z) + S_L^h \chi_L(x_z)\right]} \tag{15b}$$

where the functions T_L^z and S_L^z are given by

$$T_L^z = -\frac{m_z \psi_L(m_z x_z)\left[\psi'_L(x_z) + T_L^{z-1} \chi'_L(x_z)\right] - \psi'_L(m_z x_z)\left[\psi_L(x_z) + T_L^{z-1} \chi_L(x_z)\right]}{m_z \chi_L(m_z x_z)\left[\psi'_L(x_z) + T_L^{z-1} \chi'_L(x_z)\right] - \chi'_L(m_z x_z)\left[\psi_L(x_z) + T_L^{z-1} \chi_L(x_z)\right]} \tag{16a}$$

$$S_L^z = -\frac{\psi_L(m_z x_z)\left[\psi'_L(x_z) + S_L^{z-1} \chi'_L(x_z)\right] - m_L \psi'_L(m_z x_z)\left[\psi_L(x_z) + S_L^{z-1} \chi_L(x_z)\right]}{\chi_L(m_z x_z)\left[\psi'_L(x_z) + S_L^{z-1} \chi'_L(x_z)\right] - m_L \chi'_L(m_z x_z)\left[\psi_L(x_z) + S_L^{z-1} \chi_L(x_z)\right]} \tag{16b}$$

The recurrence starts with $z = 1$. Similar to Equation 8, L, m_z, and x_z denote the order of the Riccati–Bessel functions (ψ, χ, and ξ), the ratio of wave vectors k_{z+1}/k_z, and the size parameter $k_z R_z$, respectively, with R_z being the radius of the cluster with z shells. Because of the singularity at $r = 0$, χ_L is not a physical solution within the core and does not need to be considered for the boundary conditions of the first shell; T_L^1 and S_L^1 are, hence, reduced to the first terms in the brackets of Equation 16a and b. For $z = 0$ the general Mie solutions, Equation 8, follow as $T_L^z = 0$ and $S_L^z = 0$. Recently, experimental results were published which show that it is possible to distinguish optically between nucleus-shell clusters and homogeneous alloy clusters (see Kreibig and Vollmer [1995]). Figure 10 gives the example of an Na core ($2R = 2$ nm) surrounded by, alternatively, dielectric and Na shells of thickness 2 nm. From bottom to top the number of shells increases from zero to six.

2.4. Electrodynamic SERS

SERS was observed with various molecules adsorbed at the surface/interface of nanoparticles, and from the particles themselves. Some of the numerous investigations are compiled in Kreibig and Vollmer [1995]. Besides the "chemical" origin [Otto, 1991; Otto et al., 1992], the enhancement is also due to an electromagnetic effect; the enhancement of the inner field close to the Mie frequency causes the cluster to act as a nanoamplifier, both for the incident and the Raman-scattered fields [Kerker et al., 1980].

In the quasistatic approximation the process is described in two steps:

1. Excitation of the molecules at the frequency ω_i;
2. Emission of the molecules with Raman-shifted frequency ω_R.

Hence, the total enhancement factor amounts to

$$G = \left|\frac{E_{loc}(\omega_i)}{E_{inc}(\omega_i)} \frac{E_{loc}(\omega_R)}{E_{inc}(\omega_R)}\right|^2 = \text{const.} \left|\frac{\varepsilon(\omega_i)}{\varepsilon(\omega_i) + 2\varepsilon_m} \frac{\varepsilon(\omega_R)}{\varepsilon(\omega_R) + 2\varepsilon_m}\right|^2 \tag{17}$$

where E_{loc} denotes the local fields. As $\omega_i - \omega_R \ll \omega_i, \omega_R$, we substitute $\varepsilon(\omega_i) \approx \varepsilon(\omega_R)$ to find

$$G = \text{const.} \left|\frac{\varepsilon(\omega_i)}{\varepsilon(\omega_i) + 2\varepsilon_m}\right|^4 = \text{const.} \left|\frac{\varepsilon_1^2 + 2\varepsilon_1 \varepsilon_m + \varepsilon_2^2 + i2\varepsilon_2 \varepsilon_m}{(\varepsilon_1 + 2\varepsilon_m)^2 + \varepsilon_2^2}\right|^4 \tag{18}$$

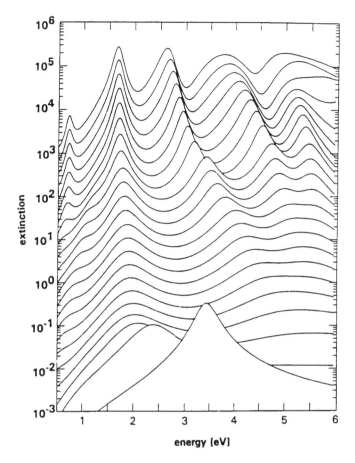

Figure 10 Extinction spectra of spherical heterostructures. Around a sodium core (R = 2 nm) there are six shells of alternating dielectric and sodium material. Total shell thickness: increasing from bottom to top between 0 and 10 nm. (From Sinzig, J. 1993.)

The resonance behavior of Equation 18 at the minimum of the denominator is reflected in the optical absorption cross section of such spherical metallic particles (without the Raman-active molecule) (see Equation 10).

3. MANY-CLUSTER SYSTEMS/CLUSTER MATTER

While single clusters are the preferred objects for fundamental research concerning the theoretical modeling of their electronic, atomic, and, hence, optical properties, common optical experiments require macroscopic samples with large numbers of clusters (typically >10^{10}) to obtain measurably large signals. Novel techniques, like optical emission in the STM (scanning tunneling microscopy) or SNOM (scanning near field optical microscopy), will allow optical experiments on single clusters of even small size and will reduce this problem.

Up to now, preparation of many-particle systems, i.e., of "cluster matter," appears inevitable for experimental research. This is facilitated by the fact that most clusters in nature, as well as in practical and technical applications, form such macroscopic cluster matter samples. Nevertheless, preparation and characterization of well-suited samples is the largest problem in cluster science. Usually, these systems are densely packed in one, two, or three spatial dimensions, and, hence, the particles cannot (as hypothetically done in Mie's theory) be treated as being independent from each other. Instead, mutual influences like intercluster electron tunneling, electrodynamic cluster–cluster coupling, or Ostwald ripening or coalescence may be important, causing special "ensemble" properties which differ more or less from the sum of "individual" properties.

This holds, in particular, for the optical properties because of the strong collective cluster excitations with their large electromagnetic near fields around each particle. Some examples of the electromagnetic cluster–cluster coupling effects resulting from these fields are shown in the experimental spectra of

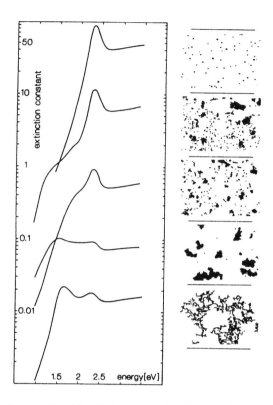

Figure 11 Optical extinction spectra of Au cluster matter of various topologies shown by the adjacent TEM micrographies.

Figure 11, where — from top to bottom — the packing density of otherwise identical samples is changed, as shown by the corresponding electron micrographs. The Mie resonances are concomitantly broadened and are split into several peaks, demonstrating dramatic influences of the particular geometric arrangements of the clusters.

Careful characterization and analysis of the topology of each cluster matter sample is thus prerequisite for a successful separation of ensemble and individual properties. Often this separation proves, however, to be impossible, and alternative kinds of samples then have to be chosen, if one wants to analyze their optical properties.

Conversely, if it is possible to arrange the clusters directedly in special topologies, samples with a broad field of different optical properties can be created, which may be important, e.g., for filters, solar absorbers, dyes, and paints.

Some arbitrarily chosen topologies of cluster matter (based here, as an example, on the spherical shape of the single building unit) were depicted schematically in Figure 1. Taking into consideration the broad chemical variability of the constituents — which has been done up to now, only to a small extent — the potential for future research is further extended.

In light of this variety it appears impossible to obtain a unified description of such inhomogeneous composite material, as has been done so successfully in the case of (on an atomic scale) homogeneous materials. Instead, different systems have to be treated differently.

A classification can be performed by the structural and compositional (i.e., chemical) correlations. The former include the broad range from fully statistical to regular ("supercrystal") arrangements. The latter distinguish between monocomponent systems and multicomponent systems, where the different components may be clusters, cluster shells, substrate films, homogeneous matrix materials, etc.

Examples of monocomponent material are the nanoceramics and, more generally, the nanocrystalline, nanophase, or nanostructured materials of compressed and/or sintered single clusters [Gleiter, 1990]. In the case of metallic units the intergrain boundary conductivity prevents or reduces typical cluster-induced optical excitations, as the collective cluster resonances.

Hence, the optical properties of monocomponent nanocrystalline metallic cluster matter rather resemble the ones of fine-grain polycrystalline metals; they will be discussed only briefly at the end of this chapter.

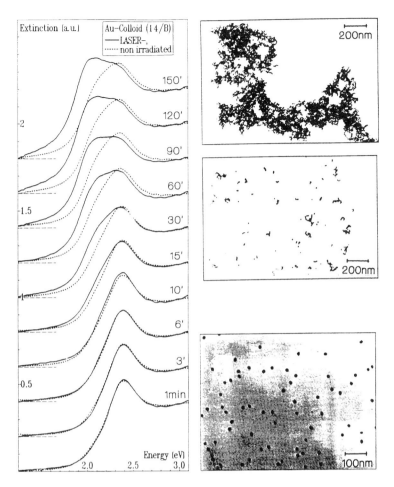

Figure 12 Light-induced aggregation in an aqueous gold colloid ($2R = 10$ nm), shown in the TEM (transmission electron micrograph) lower right. By Ar-514.5 nm laser irradiation, aggregation is induced ("light-induced van der Waals forces" or photochemical effects, see upper right figure), while the nonirradiated sample shows only weak effects (see middle right). The optical extinction spectra of the irradiated and the nonirradiated sample are shown at left. The double peak is due to electromagnetic coupling in the aggregates. (Eckstein, H. and Kreibig, 1993.)

Examples of multicomponent materials are the inclusion-matrix or porphyric materials with clusters of uniform or varying geometric size and shape and chemical compositions. The nature of the matrix includes vacuum, e.g., the well-separated free units in a cluster beam. In general, Mie-like collective optical excitations are well observed in this class of materials, if a nonconducting matrix allows the formation of electric cluster surface charging.

In most real examples of the multicomponent cluster matter, complex cluster–cluster correlations are observed: a purely statistical local arrangement is altered by the tendency of all clusters to form aggregates by their mutual van der Waals attraction forces. Hence, the building units (clusters) tend to form larger building blocks (cluster aggregates) if they are mobile (e.g., during the sample preparation) according to cluster–cluster attraction. The van der Waals attraction between metal clusters being mainly due to virtual plasmon excitations, similar dipolar forces are additively induced by irradiation with light which excites real plasmons. Probably, this effect was recently verified experimentally [Eckstein and Kreibig, 1993] and is shown in Figure 12.

In general, two kinds of independent correlations between the building units and the building blocks, respectively, describe the topology of the macroscopic cluster matter sample. If neighboring clusters in an aggregate are in direct electrical contact (e.g., by formation of grain boundaries), this aggregation is specified as "coalescence." By this process new building units with irregular, larger shapes are formed. If, in contrast, the metallic clusters remain separated, e.g., by the DLOV (Derjaguin, Landau, Overbeek,

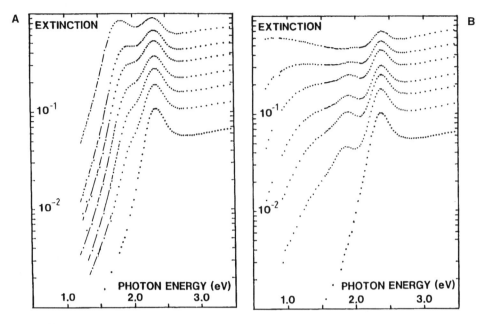

Figure 13 Extinction spectra of Au cluster matter ($2R$ = 17 nm). In an aqueous colloidal system coagulation and coalescence are induced, respectively, by chemistry. The amount of aggregation increases from bottom to top. (A) Coagulation causes two-peak structure by electromagnetic coupling; (B) coalescence causes additional broad features at low frequencies similar to the compact material. (Courtesy of M. Quinten.)

and Vervey [Sonntag and Strenge, 1988]) equilibrium of interaction forces or by insulating interlayers, and thus the electrons are strictly confined in each cluster, the aggregates are called "coagulation" aggregates in the following. Only in the latter case, surface charges can be built up by electromagnetic fields, and Mie-like resonances (though different from those of well seperated clusters) can be excited; then, the optical individuality of the clusters is kept.

The building blocks (the aggregates) themselves are arranged in particular topolog, which may give rise to various kinds of interactions. Hence, modeling of the sample topology requires several steps, in some cases a fractal sequence. The macroscopic optical properties then are described by proper averaging at each of these steps. Different models for this averaging have been developed. One special group among these models is established by the "effective medium models (Section 3.2)."

3.1. Some Experimental Examples

Before treating optical multiple-cluster effects in cluster matter in detail, the experimental examples in Figures 13 and 14 are presented with the purpose of demonstrating general effects. Figure 13 shows two series of extinction spectra of colloidal Au clusters, which were forced to increased coagulation aggregation (Figure 13a) and to coalescence (Figure 13b), respectively [Quinten and Kreibig, 1986].

Optical properties of gold clusters are also shown in Figure 14c. Their mean size is $R \cong 10$ nm, each separated by 1 to 2-nm-thick insulating phenylphosphine ligand layers which were produced from aqueous colloidal solutions of low f. (This stabilization method was developed by G. Schmid [see Schmid, 1994]). While the building units are not influenced by increased packing in these samples, other cluster matter production, like, e.g., Ag island films, produced by evaporation onto alkali–halide substrates, exhibit dramatic changes of the building units during increase of the packing density (Figure 14e).

Several optical properties of the sample of Figure 14c are compiled in Figure 14a, b, and d. While the well-separated Au clusters exhibit the typical gold-ruby color by the Mie resonance at about $\hbar\omega =$ 2.3 eV (Figure 14a, curve 1), this resonance is both shifted and broadened in the course toward dense packing and formation of coagulation aggregates (Figure 14a, curve 2). At the highest possible packing of f $\approx$ 0.3 (limited by the volume of the ligand shells) the samples are hardly distinguished by visual inspection from compact Au films: due to the "Oseen-effect" (see Kreibig and Vollmer [1995]) the trajectories of the scattered light are compressed into the regular reflected beam, the sample, hence,

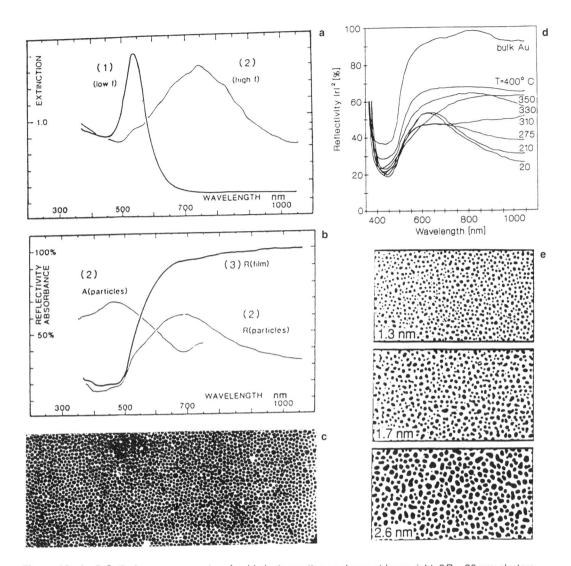

Figure 14 (a-d) Optical response spectra of gold cluster matter as shown at lower right. $2R = 20$ nm; clusters stabilized by triphenylphosphine shells. Sample (1): well-separated single clusters. Sample (2): densely packed. Sample (3): compact gold film. Spectra are shown of the extinction, the absorption (measured photothermally), and the reflection (which is compared with bulk gold). Spectra at lower left demonstrate increasing coalescence by heat treatment. The lower reflectivity compared with bulk stems from the remaining shell material. (e) Transmission electron micrographs of Au-island films, evaporated on substrate (Norrmann et al. 1978).

looking like a metallic mirror of Au Figure 14b shows the reflectivity spectrum, and, for comparison, the reflectivity of a compact film. In the frame of the Oseen-effect, the intensity of this regular beam is proportional to $\langle f \rangle^2$, with f the "global" filling factor, while the remaining diffuse scattering depends on $\langle f - f_e \rangle^2$, f_e being the "local" filling factors which may vary according to the sample topology and are, e.g., strongly different from f if distinct and separated aggregates are formed. However, the reflectivity of curve 2 in Figure 14b decreases towards low frequencies, again, in contrast to the compact material. As shown in Figure 14d, this reminiscence to the cluster behavior vanishes, if, by annealing, coalescence of the clusters of Figure 14c is induced.

Like the reflectivity, the optical energy absorption of the cluster matter films, measured by photo-thermal (mirage effect) spectroscopy, already indicates the characteristic spectral features in the intermediate region between the response by conduction electrons and interband excitations, which gives bulk gold its characteristic color (Figure 14b).

3.2. Effective Medium Models/The Bergman Theory

The effective medium approach is characterized by *a priori* statistical cluster distributions which are included in the calculation of a proper macroscopic ("effective") dielectric function of the whole sample by assuming mean electric fields. Thus, the inhomogeneous material is replaced by a fictitious homogeneous material with identical macroscopic optical properties.

All effective medium models are restricted to the quasistatic approximation $(2R << \lambda)$ (which is well applied to nanoparticles in the FIR [far infrared] but is less well suited for the VIS [visible spectral region]). The occurrence of elastic light scattering, which represents one of the fundamental differences between optically homogeneous and inhomogeneous materials, remains disregarded in this approximation. The response on the incident wave is reduced to instantaneous (i.e., retardation-free) interaction with the electric part of the fields only, which is assumed not to vary over the length scale of the particle size.

Various effective medium models were recently compiled [Kreibig and Vollmer, 1995]; the present discussion will, hence, be restricted to the model of Bergman [1978], which appears, concerning the sample topology, as the most general one. By the extensions of Felderhof [1984] multipolar interactions are included. The underlying idea is that the optical response of an arbitrary macroscopic inhomogeneous composite material (here assumed to consist of two component cluster matrix matter) is given by the superposition of the resonance-like response of a variety of clusters of continuously varying shapes but identical material properties. (A variation of the size is not considered in the frame of the quasistatic approximation.) The formal structure of the optical polarizabilities of, e.g., ellipsoids (Equation 12) and of cubes (Equation 13) being identical, these relations were generalized to interpret the L_i as a continuous variable β in the interval [0,1]. As demonstrated earlier [Granqvist and Hunderi, 1977], the influence of the electric cluster–cluster interaction fields can also be included by changing, formally, L_i of the clusters. Hence, contributions of the kind of Equation 12 are integrated with β being the variable in [0,1]: the resulting scalar effective medium dielectric function is then

$$\hat{\varepsilon}_{eff}(\omega) = \varepsilon_m \left(1 - f \int_0^1 g(\beta) / (\hat{t} - \beta) d\beta \right) \tag{19}$$

with ε_m the real dielectric function of the embedding material, f the volume concentration of the clusters, $\hat{t}(\omega) = \varepsilon_m / (\varepsilon_m - \hat{\varepsilon}(\omega))$, and $g(\beta)$ the distribution of oscillator strengths, called the *spectral function*. (In fact, only for $|\hat{t}| < 1$, i.e., Re $\hat{\varepsilon} < 1$, we observe resonances, as discussed for the single cluster in the previous sections.) Formally, the two essential properties of the sample, the dielectric and the geometric, topological ones, are separated in Equation 19, being included in the independent functions $\hat{t}$ and $g(\beta)$, respectively.

$g(\beta)$ obeys a conservation rule (the conservation of total oscillator strength) given by its zero moment

$$\int_0^1 g(\beta) d\beta = 1 \tag{20}$$

Also, for the first moment follows a simple relation (however, restricted to isotropic or cubic media)

$$\int_0^1 \beta \cdot g(\beta) d\beta = (1 - f) / 3 \tag{21}$$

Equation 19 is consistent with the postulate that the electrical field energy density of the fictitious effective medium, described by $\hat{\varepsilon}_{eff}$ should equal the one of the real, inhomogeneous matter (Theiss [1989]). (Due to the quasistatic approximation, the magnetic fields of the electromagnetic waves remain disregarded, here.)

The amazing and advantageous feature of this model is the general and open structure of the spectral function $g(\beta)$ which allows the concise description and comparison of different and complex sample

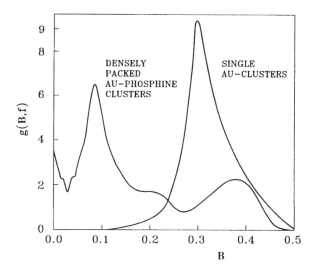

Figure 15 Bergman spectral functions of diluted gold clusters ($2R = 35$ nm) in a colloidal system of low filling factor, and of the aggregated sample of Figure 14c. They were evaluated from the extinction spectra. (Henkel, S., 1996.)

topologies. Thus, Equation 19 formally includes other effective medium models derived for particular topologies, which may be described by special $g(\beta)$.

From experimental evaluations it follows that $g(\beta)$ is a continuous function of β along the whole interval $0 \leq \beta \leq 1$ and exhibits more or less distinct maxima, characterizing special, frequently present structural elements. For example, a sample containing spherical clusters with low f is represented by a peak at $\beta = 1/3$, while percolation (i.e., structures of extreme eccentricities) shows up in contributions close to $\beta = 0$. Statistically oriented linear chains of clusters of proper material exhibit two such peaks (which reflect the splitting of coupled oscillators).

However, the implications of the structures in $g(\beta)$ on $\hat{\varepsilon}_{eff}$ are not straightforward, as is obvious from Equation 19. The integrand contributes much where the denominator, i.e., $|\hat{t} - \beta|$, is close to zero, while large values render the respective contributions small, and structural pecularities of $g(\beta)$ in these regions do not show up in $\hat{\varepsilon}_{eff}$. Since β varies in [0,1], spectral regions where $|\hat{t}| \gg 1$ are irrelevant.

Conversely, it is, in general, not easy to derive from the optical absorption (i.e., from $\hat{\varepsilon}_{eff}$), evaluated from measured spectra, the spectral density of an unknown sample topology.

As an example, Figure 15 gives the spectral densities of the sample of Figure 14c, which were recently evaluated numerically from the measured optical absorption spectra of the colloidal Au clusters, well separated and densely packed, respectively [Henkel, 1996].

3.3. Generalized MIE Theory (GMT)

An alternative to the quasistatic approximations of the effective medium models is to renounce analytic relations for the dielectric function and, instead, to compute numerically the realistic fields of the transmitted and scattered light for the individual structural building units, to sum them numerically, and, therefrom, to determine the extinction, scattering, and absorption of the whole ensemble.

This way has been pursued for cluster matter consisting of spherical embedded clusters which are allowed to form coagulation (but not coalescence) aggregates [Ausloos and Gerardy, 1980; Quinten and Kreibig, 1986]. Examples of this kind of cluster matter are depicted in Figure 11. The calculation is based on the single-cluster case of the Mie theory, which was generalized to include the electromagnetic near-field coupling among all clusters of the ensemble at their real positions. (The positions can be experimentally determined, e.g., by SXM or TEM.) These electromagnetic fields are self-consistently determined and added to the field of the incident wave for each cluster.

The degree of precision thus equals that of Mie's theory and includes arbitrary cluster materials and sizes, field retardation effects, higher-order multipolar excitations, near-field effects, size- and surface-depending (ω) of the cluster material, etc. The only drawback is still the restriction to spherical particle shapes. Similar to the theory of Mie, the GMT yields expressions for the true energy absorption, the (angle-integrated or angle-resolved) elastic scattering, and the total extinction.

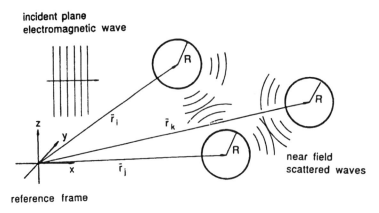

incident plane
electromagnetic wave

z

y

x

$\vec{r}_i$

$\vec{r}_k$

$\vec{r}_j$

R

R

R

near field
scattered waves

reference frame

Figure 16   GMT: Coordinates for calculation of electromagnetic interactions.

Figure 16 gives an overview of the strategy of the theory: in order to include the mutual scattering fields which cause the electromagnetic coupling between all single clusters in the aggregate, a fourth potential has to be added to the ones of Equation 3. It describes the scattered near fields of all clusters $j \neq i$ at the position of particle i which interact with this particle. The additional potential is created, formally, by transforming all individual scattered waves of the clusters j into one additional electromagnetic interaction field at each cluster i:

$$\Pi^{\text{int}} = \sum_{j \neq i}^{N_a} \Pi^{sca}(j) = \frac{1}{\left|\vec{k}\right|^2 r_i} \sum_{L=1}^{\infty} \sum_{m=-1}^{+1} \psi_L\left(\left|\vec{k}\right| r_i\right) Y_{L,m}(\theta_i, \phi_i) \times \sum_{j \neq i}^{N_a} \sum_{q=1}^{\infty} \sum_{p=-q}^{+q} A_{Lm}^{qp} b_{qp}(j) \qquad (22)$$

Here, $\vec{k}$ denotes the wave vector, Y spherical harmonics, ψ spherical Bessel functions of polar order L, b_{qp} the complex amplitude coefficients of the scattered waves, and A_{Lm}^{qp} the transformation matrix of the spherical coordinates of particle j into those of particle i. The explicit form of the latter is given in Ausloos and Gerardy [1980].

Now, all four potentials have to obey Maxwell's boundary conditions. Hence, a system of $N_a(2L_{\max} + 1)$ equations (N_a = number of clusters per aggregate; $L_{\max}$ = maximum number of multipolar modes taken into account) is obtained which allows us to calculate self-consistently the complex amplitude coefficients b_{Lm} of the wave scattered from particle i. One ends up with the scattering and the extinction constant of the cluster aggregate, the latter reading

$$E(\omega) = \frac{3c^2}{2\omega^2 \varepsilon_m R^3} \operatorname{Re}\left\{ \sum_{i \neq 1}^{N_a} \sum_{L=1}^{\infty} \sum_{m=-q}^{+q} (-1)^L b_{Lm}(i) \right\} \qquad (23)$$

by summing up over all N_a-particles, all multipoles, and the polarization states of the incident light.

The resulting fields may, alternatively, become larger or smaller than the incident field alone, and interference effects like the Oseen effect [Kreibig and Vollmer, 1995] and the "weak localization" in the back-scattering [Maret, 1992] follow directly.

The reactions of the optical extinction on several important topological parameters are for some simple model structures compiled in Figure 17, which gives numerical results for an arbitrarily selected cluster material. (In fact, the choice is not purely arbitrary: Ag clusters exhibit among all metals the sharpest resonance.) These parameters are

- the aggregate shape
- the aggregate size
- the single-cluster size
- the neighboring cluster distance

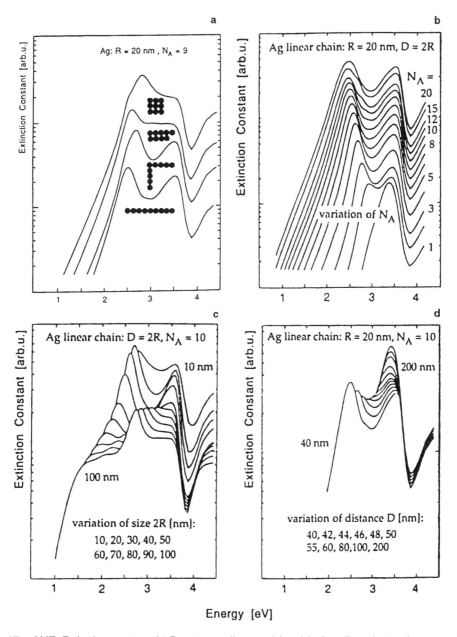

Figure 17 GMT: Extinction spectra of $2R$ = 40 nm silver particles. (a) nine silver clusters in aggregates of different topologies. (b) Linear silver chains of various lengths. (c) Linear chain of ten clusters of varied size. (d) Linear chain of ten clusters with varied cluster–cluster distance (for distances larger than five; say, diameters the coupling effects vanish). (Courtesy of M. Quinten.)

The graphs describe optical properties of distinct, small coagulation aggregates, the positions of which are assumed to be statistically distributed in three dimensions. They would be different for two-dimensional or one-dimensional distributions. Macroscopic samples are assumed to consist of many of these building blocks in well-separated arrangements.

Even at a constant number n of clusters per aggregate, the spectra differ strongly with the individual shape of the cluster aggregate (Figure 17a). In general, the coupling of n oscillators should give $n' \leq n$ individual eigenfrequencies. In fact, however, we have no eigenfrequencies of free modes, but (as discussed already in the frame of Mie's theory in Section 2) plasmon polariton modes, i.e., cluster modes coupled to external electromagnetic fields, are excited. Hence, the lowest-order, "in-phase" modes are

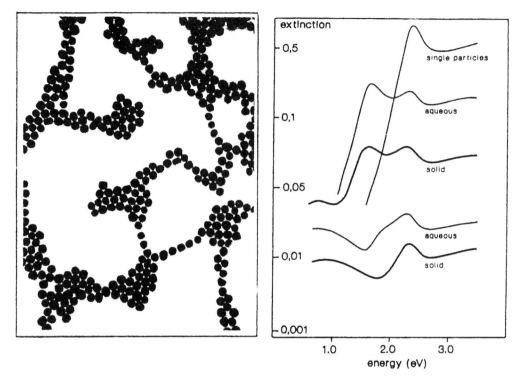

Figure 18 TEM micrograph of quasi-fractal gold cluster aggregates ($2R$ = 38 nm) and the according extinction of separated (coagulated as shown in the micrograph) and partly coalesced aggregates (from top to bottom). Shown are spectra with both aqueous and solid gelatin embedding of the clusters. The double peak due to coupling changes into the peaks of elongated percolation precursors in the two power curves (see Kreibig and Vollmer [1995]).

the important ones, primarily the dipole, since $R < \lambda$ and $d < \lambda$, d = size of the aggregate, were chosen for these examples. The two peaks of the linear aggregates in Figure 17a are thus to be attributed to the two polarization states of the electric field, $E\|$ chain axis (lower frequency) and $E\perp$ chain axis (higher frequency). For more-complex structures and larger single clusters such a clear ascription is not possible.

The effects of the other parameters are visualized for the special case of linear dense-packed chains in Figure 17b, c, and d.

Figure 17b demonstrates that the typical two-peak structure is increasingly split and shifted toward low frequencies when the length of the chain grows. There is a crossover of this behavior to constant splitting for long chains.

The extinction spectra of more complex two- or three-dimensional aggregates differ from the linear chain. In realistic samples with various aggregate structures most of the calculated spectral features are therefore smeared out in measured spectra by superimposition effects. Yet, the two principal peaks have repeatedly been detected (Figure 18).

Figure 17c demonstrates the cluster-size dependence for the example of the linear chain of $n = 10$. Obviously, the peak splitting, as well as widths, positions, and heights of the peaks, depend strongly on particle size. However, the splitting is reduced rapidly, because of the decrease of near-field strengths, when the next-neighbor distance D (D: center-to-center distance) is increased (Figure 17d). At $D \geq 6R$ the interaction effects have almost vanished.

It should be remembered that these aggregates are all assumed to be nonconductive due to the finite separation of the clusters, though the thickness of the inter layers. $D - 2R$ was, formally, extrapolated to zero in the figures to obtain the largest possible electromagnetic cluster–cluster interaction. Tunneling is excluded, as well (see, however, Kreibig [1996b]).

As mentioned before, the geometric position of each considered cluster is an explicit input parameter of the GMT; hence, each irregular arrangement of spherical clusters of arbitrary size can be simulated.

In principle, although not yet performed, cluster size distributions and mixed chemical compositions can be included, too. For several reasons, the GMT is superior to the effective medium models. At

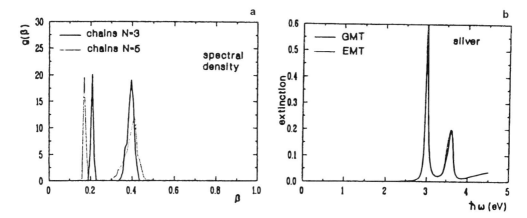

Figure 19 Comparison of Bergman theory and GMT for a system of coagulated linear silver cluster triplets and quintuplets. From absorption spectra, calculated by GMT, the spectral density $g(\beta)$ was evaluated (a) and inserted into the Bergman Equation 19. The thus-obtained absorption spectra (b) coincide quantitatively with the input spectra (K. Sturm, M. Quinten, W. Theiss, and U. Kreibig, unpublished. See Kreibig and Vollmer [1995]).

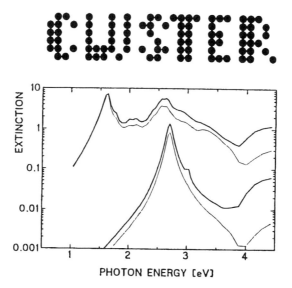

Figure 20 GMT: extinction and scattering spectra (bold and weak lines) of silver clusters ($2R = 40$ nm) of the above depicted particle aggregate. (Courtesy of M. Quinten.)

present, its most stringent limitations are due to the computer facilities; they are not principal ones. The maximum size of coagulation aggregates, simulated by us with GMT, amounts to some 10^2 particles where dipole and quadrupolar Mie modes were included. Larger aggregates are interesting to investigate, e.g., the quasi-fractal structures produced by DLA (diffusion limited aggregation). This upper limit appears to be still too small to assure, in every case, that the computable section of a macroscopic sample is large enough to be representative of the whole sample.

Figure 19 gives more insight into the relations between the Bergman theory and the GMT. For the case of very small cluster sizes, the optical absorption of arbitrarily oriented linear cluster chains was computed from the GMT. Then, the spectral function $g(\beta)$ was derived therefrom, and, applying Equation 19 of Bergman, the respective absorption spectrum was rebuilt. The consistency of this control shown in Figure 19b is almost quantitative.

To conclude this section, another numerical example of a more complex cluster aggregate is shown in Figure 20. The remarkably high contribution of scattering to the optical extinction spectrum should be noted.

4. SIZE AND SURFACE EFFECTS IN THE OPTICAL PROPERTIES OF METAL NANOPARTICLES

The optical excitations of free, embedded, and ligand-stabilized metal clusters, consisting of few atoms only, exhibit multipeak features typical of molecules, which are well pronounced at low cluster temperatures. Good correspondence was obtained with "first principles" molecular theories, if properly broad line widths were chosen. The extended literature on these optical properties is compiled in Bonacic-Koutecky et al. [1986], Haberland [1993], de Heer [1993], Kreibig and Vollmer [1995], and several other reviews.

In the frame of LDA and TDLDA (time-dependent local density approximation) applied to jellium, the optical polarizability of somewhat larger clusters (up to $N \sim 2 \times 10^2$) has also been successfully treated [Ekardt, 1986]. The assumption of one central potential for the whole cluster suffices to create discrete, potential-box-like electronic eigenstates, which, being similar to the states of atoms, are called "shell states." The experimental excitation of these shell states was successfully investigated with mass spectrometric methods, combined with directed optical fragmentation ("beam depletion spectroscopy"). In realistic metal cluster materials, optical excitation and decay channels are more complex, and a comprehensive description beyond the jellium model is difficult. We refer the reader to the above-cited reviews.

Direct optical absorption or transmission spectroscopy of free, mass-selected clusters of this kind has only been possible for very small clusters ($n \leq 10$), but not with large ones because of their low number density in producible cluster beams. The alternative way of embedding mass-selected clusters into matrices will be treated in the next section.

At high cluster temperatures, which have their origin in the cluster production method, molecular fine structures are smeared out [Haberland, 1993]. The long-lasting discussion about the transition from the "molecular" to "metallic" behavior of the metal clusters with increasing cluster size has been thus complicated.* This transition means that

- electron wave functions extend over the whole particle in analogy to the behavior of conduction electrons in bulklike metallic samples, and
- energy eigenstates are densely packed close around the Fermi level, so that, including their lifetime and dynamic broadening, a "quasi-continuous" conduction band is created and the electrons can, by external fields, change their energies continuously. Introduction of the mean level spacing ΔE and the level broadening δE, for levels close to the Fermi edge, limits the classic metallic state concerning excitation by external electromagnetic fields to

$$\Delta E < \delta E \tag{24}$$

This limit separates a lower size region of discrete electron energy levels from the region of quasi-continuous conduction band because of dynamic level overlap. It is a lower limit and does not mean that the metallic behavior above is bulklike, since excitation selection rules (e.g., wave vector conservation), transition matrix elements, and band structure develop their own size dependences. Above this limit, however solid state theories may be extrapolated to describe such metallic clusters. There are experimental hints (e.g., Tiggesbäumker et al. [1993]; Harbich et al. [1993]) that the transition between both size regions occurs at about 5×10^1 electrons per cluster and is "soft." A bulklike electronic band structure only exhibits somewhere in the size region of about 5×10^2 per cluster (e.g., Kreibig and Vollmer [1995]).

It is generally assumed that beyond the limit of Equation 24 the Mie description can be applied, i.e., the separate treatment of electrodynamics and material properties. Cluster size effects, nonlocal effects, inhomogeneity of the cluster material due to the surface/interface, etc. are then formally incorporated in the average dielectric function, $\hat{\varepsilon}(\omega)$ only, while the electrodynamic formalism based upon Maxwell's theory is assumed to hold irrespective of cluster size. (Problems with this assumption will be discussed in the following section.)

As is obvious, e.g., from Figures 3 and 6, the optical spectra of metal clusters strongly vary with cluster size. There are two distinct reasons.

---

* Here, "metal" means the kind of chemical substance, while "metallic" points to the physical properties of an extended metal, like the electrical conductivity.

Table 3 Determination of the dielectric function from optical experiments on bulk solids, films, or clusters

surface of bulk solid	thin solid film	spherical clusters
$R(\lambda, \varphi)$ $\quad$ $\delta(\lambda, \varphi)$	$R(\lambda, \varphi, d)$ $\quad$ $T(\lambda, \varphi, d)$	$\gamma(\lambda, R)$ $\quad$ $L_{disp}(\lambda, R)$
Kramers Kronig	Kramers Kronig	Kramers Kronig
Fresnel formulae	Fresnel formulae	Mie-Gans-Happel formulae
$\varepsilon_1(\lambda), \varepsilon_2(\lambda)$	$\varepsilon_1(\lambda, d), \varepsilon_2(\lambda, d)$	$\varepsilon_1(\lambda, R), \varepsilon_2(\lambda, R)$
R: reflectivity δ: phase difference φ: angle of incidence	T: transmittivity d: film thickness	γ: absorption / extinction constant R: cluster radius L_{disp}: relative dispersion

First, the electrodynamics develop strong size dependences, which are, in the frame of classic electrodynamics, exactly included in Mie's theory. They are called *extrinsic* or *electrodynamic* size effects. They are caused by phase and retardation effects which are unimportant when $R \ll \lambda$, but become increasingly effective for clusters of increasing size. They manifest themselves in

- size dependences of peak positions and widths of the Mie resonances due to increased radiation damping ("Mie scattering")
- the occurrence of higher electric multipole plasmon modes in addition to the fundamental dipole mode (see Figure 6), and
- the contributions of magnetic modes

No information, whatsoever, is available from Mie's theory about the second kind of size effects, the *intrinsic* or *material* size effects: deviations of the optical material properties, expressed by the average dielectric function $\hat{\varepsilon}(\omega, R)$.

Maxwell's theory being phenomenological, these properties have to be introduced separately and treated on the basis of modern solid state and surface physics. In close analogy to the Fresnel and Murmann formulae, the Mie–Gans–Happel formulae can be inversely applied to evaluate $\hat{\varepsilon}$ spectra from measured quantities like E and L. v. Fragstein et al. (1958, 1967) were the first to derive $\hat{\varepsilon}(\omega, R)$ directly from experimentally determined $E(\omega)$ and $L(\omega)$ of metal clusters. The alternative way of applying a Kramers–Kronig analysis (Table 3) has later been used [Kreibig, 1970].

In contrast to the former, these latter size effects and their numerous origins are not yet understood and described sufficiently. Conversely, determining them from experiments in more detail will enable deeper insight into the behavior of confined many-electron systems, into general surface and interface properties, etc. Again, the assumption of spherical symmetry of the clusters makes modeling easiest, and, so, investigations are mostly based on such cluster shapes. However, the results are not specific for this particular topography and can — at least qualitatively — be applied to cubic, rodlike, or platelet shapes, etc., as well.

An important fact is that *material* size effects and *electrodynamic* size effects can be experimentally separated by investigating particles of different sizes. At small sizes the material effects are large, but the Mie formulae become size independent ("quasistatic" approximation of Mie's theory, Equation 10); at large cluster sizes, the electrodynamic size effects are larger, while $\hat{\varepsilon}(\omega, R)$ has already converged toward $\hat{\varepsilon}$ of the bulk. As an example, the size region where the two effects exchange their relative importance in noble metal clusters is 15 to 20 nm diameter.

Modeling the frequency-dependent linear response of a metal on an exciting optical field around the visible, we obtain the contributions (frequently assumed to be additive) of the electrons in the conduction band, described usually by the collective Drude–Lorentz–Sommerfeld ansatz, and the single electron–hole excitations between different bands, the interband transitions, to be calculated from Fermi's "Golden Rule" and the band structure of the material. In this frame we obtain for the contributions to $\hat{\varepsilon}$ of Equation 1

$$\hat{\chi}^{Drude}(\omega) = \omega_p^2 / (\omega^2 + i\omega\gamma) = \omega_p^2 / (\omega^2 + \gamma^2) + i\omega_p^2 \cdot \gamma / (\omega(\omega^2 + \gamma^2)) \tag{25}$$

with $\omega_p^2 = ne^2/m_{eff} \cdot \varepsilon_0$ and [Bassani and Parravicini, 1975]

$$\hat{\chi}^{interband}(\omega) = \frac{8\hbar\pi e^2}{m_{eff}^2} \sum_{i,f} \int_{BZ} \frac{2d\vec{k}}{(2\pi)^3} \left| eM_{if}(\vec{k}) \right|^2$$

$$\left\{ \frac{1}{\left[E_f(\vec{k}) - E_i(\vec{k}) \right]\left[\left(E_f(\vec{k}) - E_i(\vec{k}) \right)^2 - \hbar^2\omega^2 \right]} + i\frac{\pi}{2\hbar^3\omega^2} \delta\left[E_f(\vec{k}) - E_i(\vec{k}) - \hbar\omega \right] \right\} \tag{26}$$

Even in nanoparticles of "good" metals like Na, K, Ag, or Al, the relaxation frequency γ is of important magnitude and restricts the number of oscillations of cluster plasmons to less than, say, 10^2 and sometimes even to less than 10^1. In principle, all included material characteristic quantities may change when going from the bulk to the nanosized specimen: density n, effective mass m_{eff}, relaxation frequency γ of the conduction electrons, matrix elements M_{if} (i.e., excitation probabilities), and states of the band structure E_i, E_f of the interband transitions. The Bloch states with well-defined linear $\vec{k}$-vectors, being the basis of conventional band structures, are less well suited than spherical waves, in the case of clusters. Hence, softening of $\vec{k}$-selection rules is expected which may be included in a size-dependent M_{if} while additional surface or interface states would have to be added to the band structure parameters E_i, E_f. Experimental results are rare, the few ones existing (for a summary, see Kreibig and Vollmer [1995]) indicate that the cluster size limits, below which size effects become important, are strongly different for these parameters. Roughly, we can estimate from experimental results on noble metal particles, i.e., metals with small γ and large $(-d\varepsilon_1/d\omega)$ that

- $\gamma \to \gamma(R)$ for $2R \leq 20$ nm ($N \lesssim 10^5$ electrons/cluster),
- n, $m_{eff} \to n(R)$, $m_{eff}(R)$ for $2R \leq 3$ nm ($N < 10^3$),
- $M_{if} \to M_{if}(R)$ for $N < 10^3$, and
- $E_{i,f} \to E_{i,f}(R)$ for $N < 10^3$ (as a particular case, $E_F \to E_F(R)$).

The latter relations are estimated from recent results on free Ag clusters and results on Au clusters embedded in glass matrices [Kreibig et al., 1996a]. Figure 21 shows recent evaluations of the interband contribution to $\hat{\varepsilon}(\omega,R)$ for free Ag clusters of 250 atoms in the mean and of similar Au clusters produced in photosensitive glass. They were obtained by Kramers–Kronig analysis of measured absorption spectra. The obvious size dependence may be caused by band structure break down, by changes of the matrix elements, or by the excitation of surface states. At 10^3 atoms per cluster, these effects have vanished.

Hence, the dominating size effect, by far, is $\gamma \to \gamma(R)$, and so — for good metals with low γ — we have mainly to deal with this. All the others are then expected to appear only in very small specimens. More precisely, the most important critical size is the mesoscopic limit, $R < \ell$, ℓ = mean free path of the conduction electrons.

More than 20 different theoretical models of $\hat{\varepsilon}(\omega,R)$ have been investigated in the course of time; some of them are listed in Kreibig and Vollmer [1995]. They may be divided into

- models based on the classic conductivity ("mean free path limitation," FPL), for which $\Delta E < \delta E$ of Equation 24, holds and
- quantum theoretical models ("quantum-size effects," QSE), which regard the splitting of the bulk conduction band into discrete energy eigenvalues ($\Delta E < \delta E$), transitions between which cause additional increase of resonance bandwidths. This is sometimes compared with Landau damping. Up to now the importance of these two effects for metal clusters has been under discussion.

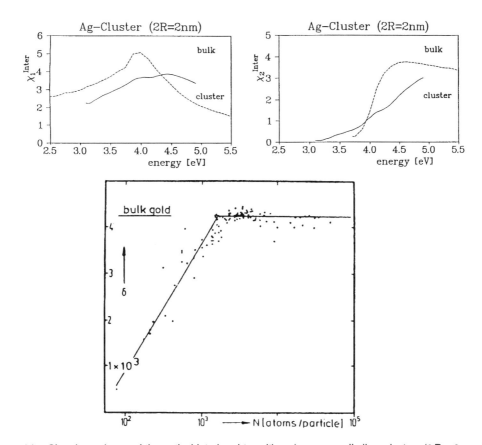

Figure 21 Size dependence of the optical interband transitions in very small silver clusters ($2R$ = 2 nm; free cluster beam in UHV; Hövel, H., in Kreibig et al. [1996] and in gold clusters of varied size which were embedded in a glass matrix (see Kreibig and Vollmer [1995]). The upper spectra show real and imaginary parts of the interband susceptibility of Equation 1. The lower graph shows the size dependence of the slope δ of the interband transition edge of $\chi_2^{interband}$. The data were evaluated by Kramers-Kronig analysis from measurements.

One can equally regard these electron confinement effects alternatively as due to the limited volume or as due to the potential step at the limiting surface of the cluster. These pictures appear as the two sides of the same can. Following the common use, they are called *size effects*.

There are, in addition, numerous models which treat the atomic and electronic structure of the surface in detail and also the interface between cluster and surroundings when the cluster is contaminated by adsorbates, deposited on substrates, or embedded in some condensed matter matrix. Such are additional *surface effects*. They will be treated separately in Section 4.2. It will be shown that their influences may be essentially larger than those of the respective size effects.

4.1. Size Effect Corrections of the Average Dielectric Function

In order to keep Mie's theory applicable, cluster size effects (and also effects of the inhomogeneous structure of the cluster material) are formally incorporated into the average $\hat{\varepsilon}(\omega,R)$ only.

Limiting postulates for the size-dependent $\hat{\varepsilon}(\omega,R)$ are:

- The changes are described as changes of the bulk $\hat{\varepsilon}(\omega)$ with

$$\lim_{R \to \infty} \hat{\varepsilon}(\omega, R) = \hat{\varepsilon}^{\text{bulk}}(\omega) \tag{27}$$

- Size effects of γ, and of M_{if} of Equations 25 and 26 and the discretization effect manifest themselves essentially in increased dissipation of field energy, i.e., increase of Im $\hat{\varepsilon}$. Re $\hat{\varepsilon}$ is then changed as well, in order to keep the Kramers–Kronig rules. Size dependences of n, m_{eff}, M_{if}, E_i, E_f, in addition, change Re $\hat{\varepsilon}$.

- Connection to the excitation spectra of molecular clusters and, finally, the atom is not the aim, since for the molecular state different atomic structures are to be expected. (In fact, an alternative criterion to distinguish between molecular and metallic cluster equilibrium states may be that, in the former, the cluster shape depends strongly on the number of atoms and may change drastically when this number is changed by ±1.) Instead, there is a lower size limit of applicability (estimated to be about 5×10^1 electrons/cluster), which coincides with the "transition" to molecular clusters.
- Cluster effects should, as well, occur in the atom dynamics contribution χ_l of Equation 1: at one side, special cluster phonons are expected, similar to the cluster plasmons; at the other side, substrate and matrix phonons passing the cluster interface will influence the dynamics of the cluster atoms and also the temperature dependence of the cluster electron excitations. However, these effects will be disregarded in the following discussion of the optical frequency region.

Both the conductivity theories and the quantum-size-effect theories concerning the cluster electrons yield an additive term $\Delta\varepsilon_2$:

$$\varepsilon_2(\omega, R) \approx \varepsilon_2^{bulk}(\omega) + \Delta\varepsilon_2(\omega, R) \tag{28}$$

which is proportional to R^{-1}.

In the frame of these approaches, this is derived either from the contribution of inelastic cluster surface collisions to the Mathiessen rule or from the average splitting of electronic levels assumed to be proportional to $N^{1/3}$ (N: number of electrons per cluster).

The conductivity concept is based on the trajectories of electron wave packets. In sufficiently small bodies, the probability of scattering by static or dynamic lattice defects in the volume of the clusters is small compared with scattering at their surface. Hence, the condition for a mesoscopic system, $\ell_\infty > 2R$, ℓ_∞ = bulk mean free path, is fulfilled, and at low temperatures the electron transport is ballistic. Even if in the whole volume filled by the clusters of a many-cluster system, the total density of static lattice defects is the same as in the bulk material (there are experimental hints that, because of surface tension and segregation effects, this density is smaller), there are clusters which are free of this kind of defects, while others still contain some of them, due to number density fluctuations. In static-defect-free particles the remaining electron scattering is at the surface. (Time averaging over the electron dynamics in that one cluster does thus not equal the corresponding number averaging over the many-cluster system.)

Of course, this freedom of bulklike scattering centers does not hold for the dynamic lattice defects (atomic vibrations) which can be created during each electronic excitation event, provided the selection rules are fulfilled.

The mean-free-path concept includes the additional surface scattering events (with variable angular electron-scattering characteristics), which, according to Mathiessen's rule contribute to the relaxation frequency γ of Equation 25

$$\gamma = 1/\tau_D = \sum_j (1/\tau_{D,i}) = (1/\tau_D)^{bulk} + A \cdot v_F / R \tag{29}$$

where v_F = Fermi velocity and A is a parameter.

Then, inserting the modified dielectric function into Mie's theory, Equation 10, we obtain

$$\Gamma(R, A) \approx \Gamma_{bulk} + \frac{A}{R} \cdot \left(2\omega_p^2 v_F / \omega^3\right)\left[\left(\partial\varepsilon_1 / \partial\omega\right)^2 + \left(\partial\varepsilon_2 / \partial\omega\right)^2\right]^{-1/2} \tag{30}$$

i.e., the cluster plasmon bandwidth contains a contribution which also is proportional to A/R — besides Γ_{bulk}, the one due to the bulklike $\hat{\varepsilon}$.

Collective excitations with limited lifetime of the electrons in a metal cluster should, in principle, be described by starting from the all-electron wave function. Even for small clusters of about 10^2 electrons this is a hopelessly complex task, and reduction to single-electron models for the lifetime limitation is

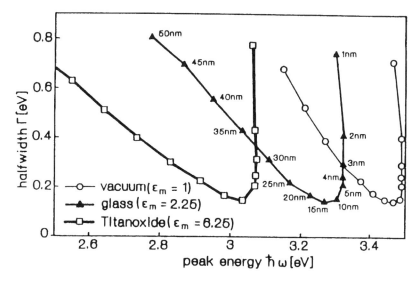

Figure 22 Calculated dipolar Mie-plasmon extinction of silver clusters (sizes given as parameters): half-width of the peak-plotted against peak position. Size effect of Equation 29 with $A = 1$ included. Different embedding media.

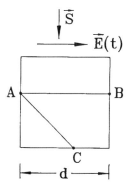

Figure 23 Electron wave packet trajectories in a cubic particle (see text).

inevitable. So, relaxation of the collective excitations is usually described by single-electron-scattering events, cascades of which occur during the plasmon lifetime.

All physical details about the extra amount of field energy dissipation in clusters are then enclosed in the quantity A. Several models, listed in Genzel and Kreibig [1985], yield values between 0 and 1. In one case a dependence on frequency ω was detected. Figure 22 shows, for the example of Ag clusters, embedded in various matrices, both the plasmon peak positions $\hbar\omega_s(R)$ and $\Gamma(R)$ when $A = 1$ is assumed. The small size region is dominated by the intransic size effect, while at large sizes the electrodynamic size effect is relevant.

Assuming the surface curvature radius to be large compared with the Fermi wavelength and, further, the surface to be fully smooth and rigid, we find the wave packets suffer elastic scattering, i.e., regular reflection. Easiest to discuss is the case of a cubic particle with its planes parallel or normal to the external electric field $\vec{E}$. We begin with the case of a constant field in x direction, which also is assumed constant in time t for the short time scales considered in the following: $\vec{E} = $ const. (x,t). (By the way, metallic nanoparticles are too small for any kind of skin effects.)

An electron is then accelerated along the trajectory AB of Figure 23 to the field-induced velocity

$$\left|\vec{v}_D(AB)\right| = \left[(2e/m)\cdot\left|\vec{E}\right|\cdot d\right]^{1/2} \tag{30a}$$

adding to the statistical electron velocity, which we assume, in application of the restrictive selection rules in the bulk, to amount to the Fermi velocity, v_F. During the elastic reflection, the momentum $2m|\vec{v}_F + \vec{v}_D|$ but no energy would be transferred to the surface, provided this surface is fully rigid. On the way back from B to A the electron is decelerated against the field, giving the additional kinetic energy $(m/2)v_D^2$ back to the electric field. After another reflection at A, the story is repeated. Hence, averaging over many such closed trajectories, no field energy is dissipated by these surface collisions. The analogue holds for other trajectories (e.g., AC in Figure 23) which may be open or closed (Balian–Bloch trajectories (see Kreibig and Vollmer [1995]). As a consequence, $A = 0$ in Equation 29.

Assuming, as the other extremum, complete transfer of the field-induced additional electron energy to the surface by fully inelastic scattering, at maximum the energy $(m/2)v_D^2$ is dissipated per collision while the amount of the statistical Fermi velocity is assumed to remain unchanged.

In general, electron momentum is also dissipated in such collisions, leading to varying directions of the trajectories after the collision described by a scattering characteristic. Two examples have been given previously: the mean free path due to surface collisions in a spherial particle of radius R amounts to $l = R$ if diffuse scattering is assumed and amounts to $l = 4/3R$ if a Lambert cosine law is considered for the scattering characteristic (see Kreibig and Vollmer [1995]).

In fact, the implicit assumption of an electric field that is constant over the cluster volume and in time does not meet the optical conditions, even for small bodies. While the spatial constancy is fulfilled for small particles, $R \ll \lambda$ (quasistatic case), the time variations of the fields are rapid in this spectral region, in relation to the electron motion and the plasmon lifetimes (which, for "good" metals amount to $\leq 10^2$ oscillations).

As an example, the path AB in Figure 23 takes about $\tau = d/v_F = 6 \times 10^{-15}$ s for an electron with v_F in a silver cluster of $d = 10$ nm. The oscillation time of light at Mie frequency amounts to $\tau^* = 1 \cdot 2 \times 10^{-15}$ s and, hence, the field changes its phase several times while the electrons travel from A to B. The spatial variations, in contrast, are negligible since the corresponding wavelength amounts to 400 nm.

Hence, the quasistatic approximation, both in space and time, usually applied for estimations of the free path effect only holds, e.g., for silver clusters as small as 2 nm. For larger particles the field-induced additional electron velocity is reduced to

$$v_d^* = \int_0^{\tau} \frac{e}{m} E_0 e^{i\omega t} dt \qquad \text{with } \tau = d/v_F \qquad (31)$$

The maximum field energy to be dissipated per surface collision thus is smaller at high frequencies, since, on the way AB, the contributions to v_d during each complete oscillation of the field cancel and only the last, incomplete oscillation contributes to v_d^*. This affects the value of A in Equation 29, which now depends on frequency, $A(\omega)$. While the mean free path and, hence, τ are usually independent of frequency, the transferred maximum amount of energy per surface collision is reduced by the factor $[(\tau^*/2)/\tau]^2$.

The energy transfer at some surface may take place via electronic excitations into bulk or surface states, atomic vibration excitations, or electromagnetic "bremsstrahlung" and "thermal" radiation. Concerning the atomic vibrations, the electron reaching B in Figure 23 penetrates the surface potential barrier and leaves — due to its momentum — the ionic cluster up to distances on the order of the spill-out length (see Figure 25), thus pulling — by Coulomb force — some surface ions off their equilibria and inducing surface and volume vibrations. This effect, corresponding to strong electron–phonon coupling, was earlier proposed by Fröhlich (private communication, 1971).

In the next section will be shown that energy, momentum, and phase relaxations depend sensitively on details of the surface or interface.

In particular, the phase relaxation is important for the lifetime of the Mie resonance excitation. Since the Mie resonance is a collective excitation of all conduction electrons, they gain coherently their $\vec{v}_d$ from the exciting field. This is the simplest picture of how the collective excitation is superimposed on the individual electron motion. It is open to question in which way and to what extent the dissipation of this excitation energy is to be described by cascades of single quasi-particle excitations during the plasmon lifetime.

In larger bodies where the concept of Bloch waves holds the whole Fermi sphere in k-space oscillates with ω, the amplitude being limited by the single electron relaxation processes around v_F. (An alternative model yielding identical frequency-dependent polarization of the cluster assumes, both in k-space and r-space, two in-phase surface currents with opposite circular polarization.)

At smaller sizes, i.e., discrete $\vec{k}$-values, the scattering processes are reduced as long as $\vec{k}$-conservation rules both for the electrons and the lattice are still stringent and electron transitions between the discrete levels occur. According to earlier experiments [Kreibig, 1974], this effect reduces the temperature-dependent part of the relaxation processes toward small cluster sizes.

This picture of the surface collision of wave packets (or its "classical" pendant, the "free path effect") was opposed by Kawabata and Kubo [1966], by stating that the electron confinement determines the quantized electron states rather than leading to surface collisions. In fact, the introduction of standing wave eigenstates in the potential box models include the "reflection" of traveling waves at the potential walls, and the detailed modeling of this potential wall based on the atoms at the cluster interface gives, in principle, way to introduce also inelastic processes.

The effect of splitting the conduction band into discrete energy levels was observed in size-limited semiconductors like quantum well structures or quantum wires and causes energy dissipation by electronic excitations between these levels. In metallic systems, however, the experimental search for such effects was not yet decisive. Previously, the identification of discrete electron density spectra by scanning tunneling spectroscopy of ligand-stabilized Pt^{309} clusters was announced [van Kempen, 1996], however, only as a tiny effect, compared with the predictions of theoretical models. Only in the "molecular" size region, below, say, 20 electrons per metal cluster, clear spectral fine structure was observed, as pointed out at the beginning of this section. The search for quantum size effects (QSE) by discrete level structure in metal nanoparticles with optical experiments is complicated by the fact that theoretical predictions of the QSE lead to an additional contribution to $\hat{\varepsilon}_2$ of Equation 28 which exhibits formally a similar $1/R$ size dependence as the classic free path limitation (FPL) model and only the predicted numerical values of A differ somewhat [Genzel and Kreibig, 1985].

Since, however, as will be shown in the next section, this quantity is very sensitive to a multitude of special surface effects, too, no clear distinction between FPL and QSE could be drawn from optical experiments on metal clusters up to now.

Hence, there is obviously an unsolved basic problem. Possibly the level-spacing effect does not show up in excitation spectroscopies of many-electron systems of metallic densities at all, because of the extremely strong electron correlation in metals compared with semiconductors. Thus, the resulting lifetime broadening of excited electronic states, δE, of Equation 24 may exceed the level separation ΔE, thus smearing out the discrete features in the optical spectra. Already the determination of discrete eigenstates from simple single quasi-particle confinement models may be inadequate for metallic electron densities. Potential box models should therefore be extended to include the strong electron correlation effects and to include realistic lifetime broadening of the levels involved in excitations. This problem awaits further theoretical investigation.

4.2. Surface/Interface Effects in Optical Properties of Clusters

As stated in the previous sections, the extraordinary optical properties of many nanosized metals, in the main, manifest in special collective conduction electron excitations as a result of surface charging.

The collective excitations are most clearly observed and most completely described if the samples are close to spherical shape (or oriented ellipsoids), so for the following focus will, again, be on spherical metal clusters whenever quantitative relations are considered. The general trends and features, of course, also hold for nanosized metals of other shapes.

As was seen in the previous section, the confinement of metal electron systems can either be treated via limited volume (QSE) or, alternatively, via collisions at the surface (FPL).

There are, in addition, strong electronic and optical cluster effects, which mostly also change with cluster size, that are caused by particular details of the surface (of free particles in vacuum) or of the interface (of particles in contact with some surrounding foreign material) and that are exclusively classified as *surface* effects. Among these are

- special electronic and atomic surface structures and
- chemical surface effects

Some of the latter are of enormous importance because of their technical use, such as the catalytic activities of Pt, Pd, Au, or other metal clusters [see, for example, Schmid, 1994]. Besides, a short look at Figure 1 indicates that in all enclosed macroscopic nanoparticle systems embedding materials are involved. Since clusters are, due to their high surface energy, thermodynamically unstable, these matrices

SURFACE/INTERFACE: "PLANE" GEOMETRY

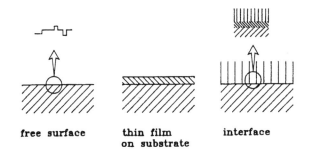

"SPHERICAL" GEOMETRY

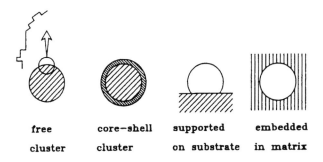

Figure 24 Surface and interface topologies of planar and spherical symmetry.

act as stabilizers, preventing the clusters from spontaneous coalescence processes. Many common cluster production methods are based on chemical surface reactions. They lead, e.g., to cluster growth in colloidal systems based upon aqueous or organic solutions. It is impossible to obtain uncontaminated clusters from such colloidal systems. As another example, silver grains in photographic systems are always in chemical interface contact, before fixation with surrounding metal halides and afterward with gelatin. It appears surprising that systematic physical (in contrast to chemical) research on metal cluster surfaces as alternative to the planar extended surface common in surface science is still at its beginning.

On the one hand, special and complex surface topologies, used in plane geometry in surface science, can be built up as well in cluster systems (Figure 24); on the other hand, there are effects unique in the clusters and unknown from planar samples, like surface curvature effects and charge conservation effects in metal clusters which are free (in vacuum) or embedded in nonconductive matrices. Advantageously, because of the extremely high specific surface of clusters, no sophisticated experiments with enhanced surface sensitivity are required, as is the case with planar surfaces. For example, the optical excitations of clusters act as sensitive sensors of surface and interface properties — this holds particularly for the Mie resonances in metallic clusters.

More-complex particles like the core-shell clusters (see Section 2.3) exhibit interesting special effects: in order to obtain uniform and low chemical potential throughout the whole particle, charge transfer may occur across the core-shell interface leading to strong local Coulomb forces. Charge transfer can also take place when chemical bonds are formed at the interface. Due to the limited numbers of charge carriers in the cluster nucleus and the shell, this may cause essential change of the electron densities and, hence, of both collective excitation frequencies and interband excitations, provided the Fermi level is involved in the latter.

As described in Section 2, the Mie theory is based (as well as its planar analogon, the Fresnel formulae) on the Maxwellian boundary conditions which postulate a sharp step of the dielectric function at the surface. This is strongly simplifying on the size scale of nanoparticles. In fact, there is a smooth transition of the electron density at the free surface from metallic to zero, which extends to the order of

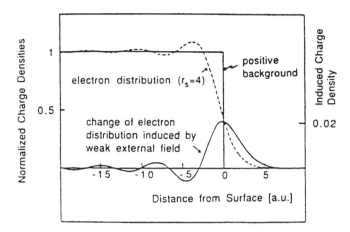

Figure 25 Normalized electron charge distribution $n(x)$ at the surface of a jellium cluster (parameters of Na), and its changes by a weak external electric field normal to the surface. Changes are localized inside and outside the ionic surface. (After Liebsch 1993.)

the Fermi wavelength. This "spill-out" of the electron density forms an electric double layer at the cluster surface as shown in Figure 25 since the ionic cluster forms a sharper, steplike surface [Eckerd, 1986]. Hence, the "electronic cluster" is larger than the "ionic" one. This spill-out should be sensitive to ions from material surrounding the cluster.

Figure 25 also shows the additional charge density induced by a spatially constant external electric field (like the one of the electromagnetic wave exciting a Mie plasmon in the quasistatic regime). It is concentrated at or even outside the geometric (ionic) surface and, hence, is extremely sensitive against details of the topology and chemistry at the surface. Hence, the optical excitation of Mie resonances probes as a highly sensitive tool the electronic structure of the particle surface/interface, provided the cluster material exhibits well-defined Mie resonances.

These resonances are changed by surface/interface effects compared with the predictions of Mie's theory (based upon the ideal spherical surface with steplike potential). That means the measured shape of the resonance and the spectral band position and width contain information about surface details. We will show that it is possible to model these effects theoretically by formally changing the average dielectric function of the cluster material (i.e., the total cluster polarizability including the interface) which then is inserted into the unchanged Mie formalism.

Hence, the investigation of surface properties is — as was the case for the size effects of the previous section — traced back to the modeling of a properly changed and adapted $\hat{\varepsilon}(\omega, R)$. The modeling of this $\hat{\varepsilon}$ for some special cluster system is assumed to have been successful if thus-computed Mie spectra quantitatively coincide with the measured ones. In the case of strong and selective (i.e., narrow) Mie bands, this condition is very strict. $\hat{\varepsilon}$ then describes a small volume of inhomogeneous matter.

Concerning the cluster surface in contact with surrounding foreign material (i.e., adsorbates, substrates, or embedding materials), two different types of changes of the optical properties and, in particular, the Mie resonances are then distinguished:

- The "dielectric effect" due to the dielectric function ε_m of the surrounding material deviating from the vacuum value $\varepsilon_m = 1$. This is due to the Maxwell boundary conditions causing changes of the Mie resonances by electrodynamics (already without any corrections of the cluster-material-$\hat{\varepsilon}$).
- The contributions of the surrounding material to the cluster-material-$\hat{\varepsilon}$. As a consequence of the latter it makes little sense to regard embedded clusters as isolated entities of their own. Instead, the cluster plus surrounding atoms close to the interface is the appropriate entity. As an example, the famous ligand-stabilized "Schmid clusters" Au_N, Ru_N, Rh_N, Pd_N, and Pt_N with the number of atoms $N = 13, 55, 147, 309, 561$, etc. [Schmid, 1994] cannot be treated only as metal clusters to explain their optical properties, but the ligands are integral constituents.

These two kinds of effects have recently been verified experimentally with Ag clusters [Hövel et al., 1993; Kreibig et al., 1996c], and we present several experimental results, before explaining optical cluster interface effects by a simple model.

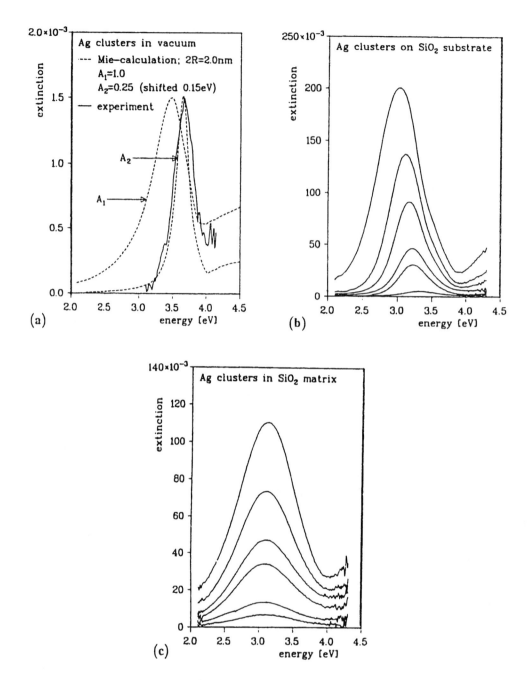

Figure 26 Absorption spectra of silver clusters ($2R = 2$ nm) produced in UHV as a free beam (a), subsequently deposited on SiO_2 substrate, (b) and after embedding in a co-evaporated SiO_2 matrix (c). In (a) Mie calculations are included; Figures (b), (c): spectra with increasing coverage from bottom to top. (From Hövel, 1993.)

4.3. Some Experimental Results

A free, cold Ag cluster beam was produced in our ultra-high vaccum apparatus (UHV) from a thermal source [Hövel et al., 1993]. These clusters were subsequently deposited on solid substrates and could then be contaminated by reactive gases. In addition, the clusters were embedded in solid matrices, co-evaporated during the cluster deposition on a substrate from one or two electron beam evaporators. Optical absorption spectra were recorded at each step. One series is represented in Figure 26. For the *same* Ag clusters, the Mie resonance shifted from 3.65 to 3.3 and 3.1 eV when the free clusters were deposited on a quartz glass substrate and embedded in an SiO_2 matrix. At same time, the width increased from 0.25 to 1.0 eV.

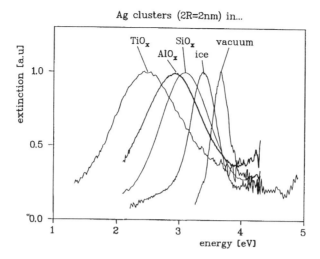

Figure 27 Optical extinction spectra of silver clusters as in Figure 26, embedded in various solid embedding matrices. (From Kreibig, et al. 1996c.)

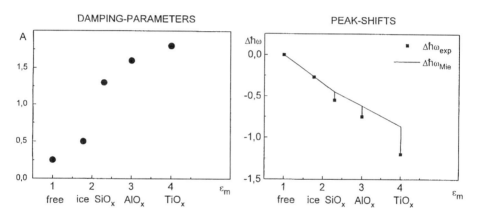

Figure 28 Damping parameters A (from Equation 29) and peak shifts evaluated from the spectra of Figure 27. The Mie theory predicts $A = 0$ and the "dielectric" peak shifts of the solid line.

The mean cluster diameters were 2.0 nm (~250 atoms, in each cluster) in one sequence of experiments and 2.6 nm (~500 atoms, each) in another one shown in the following. Standard deviations were in both cases ~20%. Using the absorption of the free clusters as a reference, the effects of the changed cluster surroundings could be quantitatively separated from cluster size and structural imperfection influences.

Figure 27 summarizes Mie absorption spectra of the silver clusters without and with different dielectric matrix materials [Kreibig et al., 1996c]. Both position $\hbar\omega_s$ and width Γ of the Mie resonance are strongly influenced by the matrix, and the effect is highly specific to its material. The free clusters exhibit the narrowest peak at the highest frequency. All manipulations at the surface caused red shifts and broadening. Both interface effects appear to correlate to the matrix dielectric constant ε_m as shown in Figure 27; the major part of the peak shift is due to the "dielectric effect", i.e., the change of Maxwell boundary conditions by ε_m. However, the observed shifts exceed this classic electrodynamic effect slightly (Figure 28).

In regard to the bandwidth parameter A, experimental data of which are compiled in Figure 28, it is mentioned that the classic Mie theory assumes $A = 0$, independent of the nature of the surrounding material.

Also, chemical interface reactions can be followed by the investigation of the resonance absorption. Two examples are shown in Figure 29. In both experiments the Ag clusters were deposited on quartz glass substrates before gaseous O_2 (Figure 29a) and sulfur (Figure 29b), respectively, were added. In both cases, ω_s and Γ are influenced. The formation of oxides appears to be limited to the cluster surface, while in the case of sulfur the transformation of the whole cluster into Ag_2S was observed.

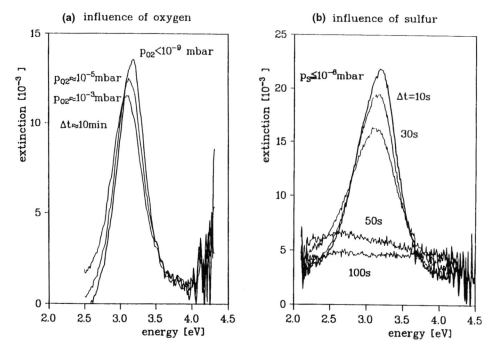

(a) influence of oxygen

(b) influence of sulfur

Figure 29 The deposited Ag clusters of Figure 26, exposed to oxygen (a) and sulfur (b). The optical plasmon resonance absorption spectra indicate a surface-limited reaction with the former, but complete chemical reaction with the latter. (After Nusch, see Kreibig et al., 1996d).

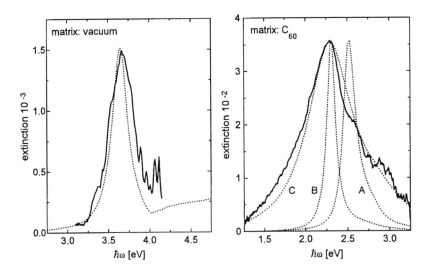

Figure 30 Absorption spectra of 2.6 nm silver clusters. Left: measured at free beam in vacuum (solid line) and calculated from Mie theory with adapted $\varepsilon(\omega, R)$ (dotted line). Right: measured spectra of the clusters embedded in solid C_{60} film and adaption by Mie theory (A: bulk-$\varepsilon(\omega)$; B: shifted by a 20% reduction of the conduction electron density; C: $A = 1$. (Gartz, in Kreibig et al., 1996d).

As another example, Figure 30 presents the absorption spectrum of Ag clusters (here containing about 500 atoms each) embedded at room temperature in a solid C_{60} film. This system was treated in detail and will be described elsewhere. Here the results are merely summarized:

• The shift of ω_s is quantitatively described by assuming that each of the C_{60} molecules adjacent to the cluster surface takes one electron from the Ag cluster into its lowest unoccupied electronic level. This is in full correspondence with recent charge-transfer investigations in C_{60}–Ag layer systems [Wertheim

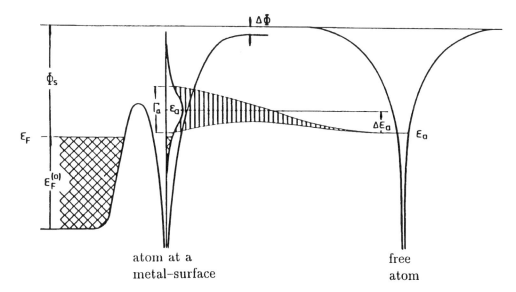

Figure 31 Electron energy scheme for chemisorption at a metal surface. Right: adatom level broadened and shifted at the surface. Left: conduction band of the metal. (After Hölzl, 1979.)

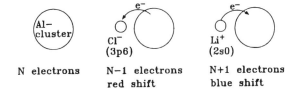

Figure 32 Schematics of static charge transfer by chemisorption at a metal cluster surface. (In parentheses: the closed-shell final ion states.) The shifts concern the respective Mie resonances.

and Buchanan, 1994; Hunt et al., 1995]. As a consequence the 5sp-electron density in the metal cluster is lowered by about 20%!

• The width Γ_{exp} of the Mie band is consistent with a value of $A = 1$. This means that the s-electron relaxation time is reduced by about a factor of 5 compared to the free clusters by *phase destruction* due to the dynamic charge-transfer processes described below. It is only mentioned here that the eigenabsorption of solid C_{60} in the spectral region of the Mie resonance rendered the evaluation of A more difficult, since the extension of Mie's theory to absorbing embedding matrix had to be used (see Section 2.1).

4.3.1. Chemical Interface Effects: Static Charge Transfer

Energy levels of adatoms change energetic position and are heavily broadened when the adatoms are attached to a metal surface. Usually, they are separated by a tunnel barrier from the metal cluster (Figure 31). Depending on the occupation and on the overlap of the adsorbate states with the Fermi level in the metal cluster, electrons may tunnel to or from the adsorbate, thus changing the electron density in the cluster. New cluster wave functions are created including the adsorbate level. Their lifetimes are widely spread. For example, the tendency toward closed atomic subshells is predicted to cause an Li and a Cl adsorbate on an Al cluster to inject an electron into or to eject an electron from the metal cluster, respectively (Figure 32). The resulting *static* (i.e., permanent) *charge transfer* causes strong double-layer Coulomb forces which are included in the energy minimization of the cluster. The removal of one electron from our Ag clusters creates a Coulomb energy of, roughly, 0.1 eV. In the example of Ag_N/C_{60} of Figure 30, a total charge transfer of about 10^2 electrons is estimated from the Mie peak shift.) Various such transfer processes are expected if clusters are embedded in liquid or solid materials with dense surface coverage (see e.g., Henglein et al. [1991; 1992]).

These systems of metal cluster matter, hence, are as an interesting consequence open to electron density manipulations, usually assumed to be a domain of semiconductors only. Charge transfer changes the optical properties of the clusters. First, the spectral Mie resonance position ω_s is shifted since it is proportional to $N^{1/2}$, N being the number of "conduction" electrons in the cluster

$$\omega_s \approx \left[N / \varepsilon_0 m_{eff} \right]^{1/2} \cdot \left[2\varepsilon_m + 1 + \chi_1^{inter} \right]^{-1/2} \tag{32}$$

with ε_m and χ^{inter} the dielectric constant of the matrix and the susceptibility contribution by interband transitions in the metal, respectively. Second, the Fermi level E_F is changed, and E_F influences as well the effective mass of the conduction electrons and the low-frequency interband transition edge, which is due, in the case of Ag, to $4d \rightarrow 5sp(E_F)$ transitions.

4.3.2. Dynamic Charge Transfer (Tunneling)

If, after equilibration, the cluster Fermi level cuts a partly unoccupied, broadened adatom level, cluster metal electrons with energies around E_F or higher ("hot electrons") may tunnel through the interface barrier into the adsorbate level. After some residence time there, the electron will return to the cluster. Such *dynamic charge transfer* induces fluctuations of the cluster electron density. It affects the optical properties, if taking place while a Mie plasmon is excited, since these tunneling electrons are rendered out of phase of the collective excitation (field-induced enhancement of tunneling is excluded in this model), and, hence, this dynamic charge transfer reduces the excitation lifetime of the collective resonance because of destruction of the phase coherence of the electron drift motion.

In order to describe this effect, which was called *chemical interface damping*, an additional contribution to the size-dependent conduction electron relaxation time of Equation 29 is introduced in the dielectric function of the cluster material:

$$\gamma = \gamma_{bulk} + v_{Fermi} \cdot A / R \tag{33}$$

A expresses now both the strength of the size effects described in the previous section and, by additional terms, the chemical interface damping, i.e., the effect of the tunneling electrons. For one fixed cluster size R — as in the described experiments — A characterizes directly the probability of the temporal electron transfer and the number of adsorbate atoms at the cluster surface and, hence, properties of the electronic adsorbate levels involved.

In a series of papers, Persson [1992] quantitatively modeled the dynamic charge transfer, i.e., the magnitude of $A(\omega)$ for a given energetic position and density of electronic adsorbate states. (Here, the term adsorbates includes, as well, liquid and solid matrix atoms.) His numerical results are listed in Table 4, together with all the experimental data available up to now. In particular, the free Ag clusters of our experiments described in Section 4.3 are characterized by $A = 0.25$. This coincides with the value of the Ne matrix. According to the Persson theory the Ne matrix should, due to the absence of adsorbate states, exhibit no chemical interface damping. Thus, the remaining experimental $A = 0.25$ is ascribed to the size effect contribution, only. Hence, the relative contributions of size effects and of the chemical damping effect can quantitatively be separated.

Since, for usual light intensities, the time intervals between individual plasmon excitations in one cluster exceed the plasmon lifetimes $\tau = \hbar/\Gamma_{exp}$ by far, we can model the single excitation event. The number Z of tunneling processes during τ is approximatively

$$Z \leq S \cdot \frac{4\pi}{5} \hbar \cdot v_F \cdot \rho \cdot R^2 / \Gamma_{exp} \tag{34}$$

with S the transfer probability, ρ the electron density, v_F the Fermi velocity, and Γ_{exp} the measured width of the plasma resonance band.

Since adsorbate states close to E_F are predominantly contributing in our model presented here, the electron transfer measured optically from metal clusters is very similar to the one observed in planar metallic structures covered by adsorbats, by measuring the static and AC electric conductivity [Wißmann, 1975; Schumacher, 1993].

In summary of this section, static charge transfer processes toward or from the cluster surroundings cause the Mie resonances to shift, while dynamic charge transfer, i.e., tunneling processes involving

Table 4 Chemical cluster interface effect: *A* parameters

	Cluster surrounding	Δ*A* (Persson theory)	*A* (experiment)	Ref.
Ag$_N$ in	Solid Ne	~0	0.25	Charlé et al. [1984]
	Solid Ar	0.1	0.3	Charlé et al. [1984]
	Solid O$_2$	0.3	0.5	Charlé et al. [1984]
	Solid CO	0.8	0.9	Charlé et al. [1984]
	Na–SiO$_2$ glass	1.0	1.0	Charlé et al. [1984]
Au$_N$ in	Na–SiO$_2$ glass		1.5	Kreibig et al. [1996c]
Ag$_N$ in	Ice		0.5	Kreibig et al. [1996c]
	SiO$_x$ (~SiO$_2$)		1.3	Kreibig et al. [1996c]
	AlO$_x$ (~Al$_2$O$_3$)		1.7	Kreibig et al. [1996c]
	TiO$_x$ (~TiO$_2$)		1.8	Kreibig et al. [1996c]
	Solid C$_{60}$		1.0	Kreibig et al. [1996c]
	ITO		0.7...1.2	Kreibig et al. [1996c]
	SiO$_2$ substrate		0.7	Kreibig et al. [1996c]
	+ O$_2$		0.5	Kreibig et al. [1996c]
	+S		1.2...1.7	Kreibig et al. [1996c]
	+Ether			Kreibig et al. [1996c]
	+HCl			Kreibig et al. [1996c]
Au$_{55}$ in	Triphenylphosphine		≥2	Kreibig et al. [1996c]

Note: Δ*A* = *A*(embedded cluster) — *A*(free cluster).

adsorbate states, reduces the plasmon lifetime. Numerical values of the *A* parameter describing this latter effect are listed in Table 4 and are, in part, compared with the theoretical values evaluated by Persson.

5. NONLINEAR OPTICAL PROPERTIES OF METALLIC NANOPARTICLES

As nonlinear optical effects are dealt with in other parts of the present book, only some peculiarities of metallic nanoparticles will be treated here. A compilation is given, e.g., in Kreibig and Vollmer [1995].

The drastic enhancement of the inner electromagnetic field by Mie-like resonances, which is unique for particles of several metals, strongly increases nonlinear effects, like second harmonic generation (SHG), or third-order effects, like optical phase conjugation, degenerate four-wave mixing, Kerr effect, saturable absorption, and nonlinear index of refraction.

As a consequence, nonlinear effects already occur at low external fields. In addition, the symmetry-selection rules for the second-order susceptibility $\chi^{(2)}$ are influenced by the symmetries of the clusters and of the multipolar excitation modes.

Results of a recent theoretical investigation by Bennemann and co-workers [Östling et al., 1993] are shown in Figure 33. The power of second harmonic radiation was calculated from classic electrodynamics for particles with Drude $\hat{\epsilon}(\omega)$. It is obvious that the nonlinear intensity follows the linear Mie resonance modes. At maximum, the enhancement compared with planar samples amounts to 10^2 to 10^3. It is predicted that circularly polarized incident light only produces second harmonic radiation via higher polar modes beyond the dipolar one. For the third harmonics generation even enhancement factors of 10^5 compared with planar sample geometry were predicted.

Experimentally, the search for SHG was successful with various kinds of clusters, like Ag, Cu, and Na, deposited on solid matrices. The SH was recorded during the growth of Na clusters and was ascribed to the contribution of the conduction electrons [Götz, 1993]. Strong enhancement was found for frequencies of the SH in the range of the cluster Mie resonance.

The experimental observation of third-order processes was repeatedly described, like nonlinear refraction and Kerr effects. Figure 34 compiles several experimental results. In Figure 34a the conjugate signal from Au clusters in aqueous colloid follows the (calculated) absorption [Hache et al., 1986].

The cluster $\chi^{(3)}$ evaluated from saturable absorption of Au clusters embedded in glass, shown in Figure 34b, amounts to $2.8 \cdot 10^{-8}$ and $4.2 \cdot 10^{-8}$ ESU for mean sizes of 2.2 and 3.5 nm, respectively. This small size dependence was explained by assuming that interband excitations and hot electrons exceed the conduction electron contributions [Hache et al., 1988].

The nonlinear refractive index of Au clusters produced by ion implantation into fused silica was used to measure the short time, <10 ps, of the nonlinear response (Figure 34c) [Magruder et al., 1993].

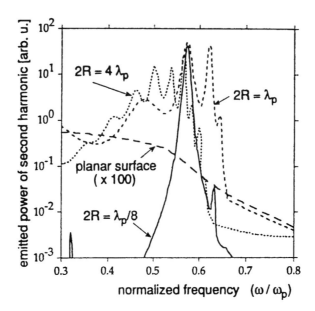

Figure 33 Radiated power of the second harmonic, generated in Drude metal clusters upon incident circular polarized light. The cluster size is varied compared with the plasma wavelength λ_p. For comparison the signal from the planar surface is added. (From Östling, D. et al., 1993.)

The nonlinear response of the inhomogeneous cluster matter as a whole was treated in the effective medium approach. For low concentrations of small spherical metal clusters the M. Garnett theory is well applicable. The resulting $\chi_{eff}^{(3)}$ reflects the local field enhancement of the linear $\hat{\varepsilon}_{eff}$ to the fourth power [Agarwal and Gupta, 1988]:

$$\chi_{eff}^{(3)} = \chi^{(3)} \cdot \frac{f}{|P|^2 \cdot P^2} \quad \text{with} \quad P = \frac{1 + f(x-1)}{x} \quad \text{and} \quad x = \frac{3\varepsilon_m}{\hat{\varepsilon} + 2\varepsilon_m} \tag{35}$$

This enhancement can be further increased by choosing ellipsoidal particles of proper eccentricity [Haus et al., 1989]. These systems of cluster matter are treated in another part of this book. Alternatively, the field enhancement might be optimized by proper "immersion" of spherical particles as shown in Section 2.1.

By combining optical linear and nonlinear materials in an effective medium, optical bistability can be created. In principle, this is possible with systems of nonlinear clusters in linear matrices and, inversely, linear clusters and nonlinear matrix [Bergman et al., 1993].

Recently, experiments were successful for the first kind of system. The nonlinearity of semiconductor clusters of CdS was increased by coating them with a silver shell and exciting at the plasma resonance frequency of the latter to increase the local field [Neuendorf et al., 1996]. Figure 35 shows the measured transmitted intensity exhibiting strong bistability.

6. TRANSITION TOWARD THE BULK/NANOSTRUCTURED MATTER

Effective medium models and the GMT, treated before, are both based upon optically polarizable metallic building units, being one, usually electrically isolated, component of cluster matrix matter. Both are subdued to restrictive assumptions, the former being limited by the quasistatic approximation to arbitrarily shaped but very small units, while the latter is applicable to arbitrary sizes but spherical shapes of the units. More general specimens of cluster matter are thus not included, and no theories appear to exist to describe them sufficiently. At dense packing, neighboring clusters, exclusively connected by coagulation, remain the building units, and, at the outermost case of almost touching units single electron tunneling between the clusters is possible (although excluded in the above aggregate theories).

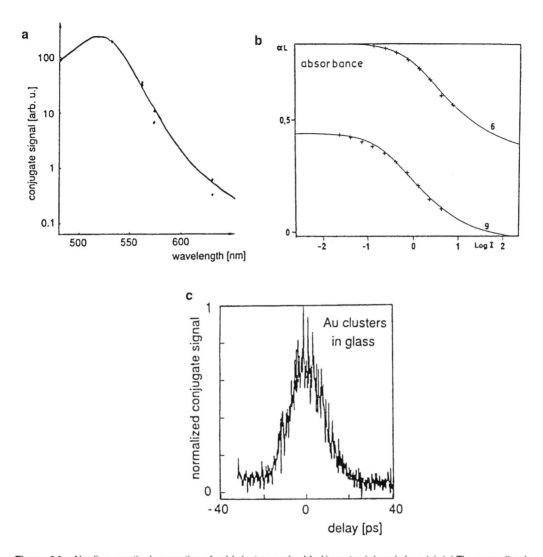

Figure 34 Nonlinear optical properties of gold clusters embedded in water (a) and glass (c). (a) The normalized conjugate signal (dots) follows the absorption spectrum (solid line). (From Hache, 1986.) (b) Saturated absorbance of 2.2-nm (upper curve) and 3.5-nm clusters versus intensity. (From Hache, 1988.) (c) Normalized conjugate signal as a function of pump-probe delay time. Response times at a scale of 10 ps were observed. (From Magruder, 1993.)

The individuality of the particles is lost, however, by coalescence, where conducting bridges are formed between touching neighbors. Coalescence, as an exothermic and, hence, spontaneously occurring transformation process, causes the formation of electrically conducting grain boundaries between the units and, thus, changes these building units into larger ones with, in general, irregular shapes and widely varying sizes which are no longer to be treated by any existing analytic electrodynamic theory.

In regard to the optical properties, the confinement of the conduction electrons within the clusters is then removed, and surface charges, as the typical feature of Mie-kind resonances, no longer occur where grain boundaries were built up. Figures 13 and 18 allow the comparison of the optical properties of otherwise equal Au colloids which were caused to coagulate and to coalesce, respectively.

If the filling factor f is increased in samples with statistically distributed metallic units, quasi-fractal percolation structures occur due to coalescence within a widely varying range of f. These structures develop interesting optical properties. Three-dimensional thick samples appear as "metal blacks," mainly as a result of multiple extinction of the penetrating light. While DC-conductivity experiments probing

188

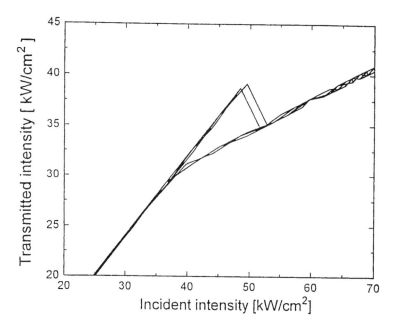

Figure 35 Nonlinear optics of silver-coated CdS nanoparticles; measured optical bistability of the intensity-dependent transmission in the spectral range of the shell resonance (λ-514 nm). (From Neuendorf, 1996.)

percolation require macroscopic conducting paths, the precursor states of percolation, i.e., the percolative correlation on a length scale of or below the light wavelength, can be investigated by optical means (see Kreibig and Vollmer [1995]).

When dense packing is approached in coagulation systems, interference effects of the electromagnetic waves transform the scattered light of the single units into the regularly reflected and transmitted geometric optical beams as in homogeneous compact metals (Oseen effect [Dusemund et al., 1991]). While, concerning the electromagnetic reradiation, these samples already resemble the bulk, they still differ strongly from the latter with regard to the frequency dependence of their optical properties. This was demonstrated in Figure 14. The samples still show peaklike spectral features which, although shifted and broadened, are a reminiscent of the Mie-like resonances of the single units. Once coalescence is actuated, these differences vanish and the optical spectra closely approach those of thin compact films (e.g., Figure 13b).

Recently, the concept of nanostructured (nanocrystalline, nanophase) matter unleashed an extended search for novel materials [Gleiter 1990]. This material is a monocomponent material of densely packed metal small clusters (compacted by external pressure) with excessive coalescence. Incoherent grain boundaries extending over several atomic layers contribute essentially to the total volume of such samples. Hence, nanostructured matter is as well classified as grain boundary–rich material. Promising mechanical and magnetic properties have been detected, while the optical properties still await detailed research. Spectacular behavior, in particular for technical purposes, is less expected in this field, the optical properties more closely resemble those of the polycrystalline bulk material than those of the clusters the nanostructured matter was made of. Hence, this material is beyond the scope of this chapter, which mainly focused on the unique and sometimes spectacular optical properties of metalic clusters and nanoparticles.

REFERENCES

Aden, A. and Kerker, M. 1951. *J. Appl. Phys.* 22: 1242.
Agarwal, G., Gupta, S. 1988. *Phys. Rev. A* 38: 5678.
Ausloos, M., Gerardy, J. 1980. *Phys. Rev. B* 22: 4950.
Barber, P., Yeh, C. 1975. *Appl. Opt.* 14:2864.
Bassani, F., Parravicini, G. 1977. *Electronic States and Optical Transitions in Solids.* Pergamon, London.
Bergman, D. 1978. *Physics Rep.* 43: 377.
Bergman, D., Levy, O., Stroud, D. 1994. *Phys. Rev. B* 49: 129.

Bohren, C. and Huffman, D. 1983. *Absorption and Scattering of Light by Small Particles,* Wiley, New York.

Bonacic-Koutecky, V., Fantucci, P., Koutecky, J. 1991. *Chem. Rev.* 91: 1035.

Born, M. 1972. *Optik.* 3rd ed. Springer, Berlin.

Charle, K., Frank, F., Schulze, W. 1984. *Ber. Bunsenges. Phys. Chem.* 88: 350.

Dusemund, B. 1991. Diploma work, University Saarbrücken.

Dusemund, B., Hoffmann, A., Salzmann, T., Kreibig, U., Schmid, G. 1993. *Z. Physik D* 20: 305.

Eckstein, H., Kreibig, U. 1993. *Z. Phys. D* 26: 239.

Ekardt, W. 1986. *Phys. Rev. B* 48: 526.

Felderhof, B. 1984. *Physica A* 126: 430.

Forstmann, F. 1967. *Z. Phys.* 203: 495.

v. Fragstein, C., Roemer, H. 1958. *Z. Phys.* 151: 54.

v. Fragstein, C, Schoenes, F. 1967. *Z. Phys.* 198: 477.

Fuchs, R. 1975. *Phys. Rev. B* 11: 1732.

Gans, R. 1912. *Ann. Phys.* 29: 881.

Gans, R. and Happel, H. 1909. *Ann. Phys.* 29: 277.

Genzel, L., Kreibig, U. 1980. *Z. Phys. B* 37: 93.

Genzel, L., Kreibig, U. 1985. *Surf. Sci.* 156: 678.

Gleiter, H. 1990. *Prog. Mater. Sci.* 33: 223.

Götz, T., Vollmer, M., Traeger, F. 1993. *Appl. Phys. A* 57: 101.

Granqvist, C., Hunderi, O. 1977. *Phys. Rev. B* 16: 3513.

Haberland, H. 1993, 1994. *Clusters of Atoms and Molecules* I, II, Springer Series in Chemical Physics 52, 56. Springer, Berlin.

Hache, F., Ricard, D., Flytzanis, C. 1986. *J. Opt. Soc. Am. B* 3: 1647.

Hache, F., Ricard D., Flytzanis, C., Kreibig, U. 1988. *Appl. Phys. A* 47: 347.

Harbich, W., Fedrigo, S., Buttet, J. 1993. *Z. Phys. D* 26: 138.

Haus, J., Kalyaniwalla, N., Inguva, R., Bloemer, M., Bowden, C. 1989. *Opt. Soc. Am.:* 797.

de Heer, W. 1993. *Rev. Mod. Phys.* 65: 611.

Henglein, A., Mulvaney, P., Linnert, T. 1991. *Faraday Discuss* 92: 31.

Henglein, A., Mulvaney, P., Linnert, T., Holzwarth, A. 1992. *J. Phys. Chem.* 96: 2411.

Henkel, S. 1996. Thesis: Die effektiven optischen Eigenschafte von Metall-Isolator Mischsystemen, RWTH, Aachen.

Hölzl, J., Schulte, F., Wagner, H. 1979. *Solid Surface Physics.* Springer Tracts in Modern Physics 85.

Hövel, H., Fritz, S., Hilger, A., Kreibig, U., Vollmer, M. 1993. *Phys. Rev. B* 48: 18148.

Hövel, H. 1996. In Kreibig, U., Hilger, A., Hövel, H., Quinten, M., *Large Cluster of Atoms and Molecules,* Martin, T. P., Ed., Kluwer, Dordrecht, in press.

v. d. Hulst, H. 1957. *Light Scattering by Small Particles. Wiley,* New York.

Hunt, M., Modesti, S., Rudolf, R., Palmer, R. 1995. *Phys. Rev. B* 51: 10039.

Kawabata, A., Kubo, R. 1966. *J. Phys. Soc. Jpn.* 21: 1765.

Kerker, M. 1969. *The Scattering of Light,* Academic, New York.

Kerker, M., Wang, D., Chew, H. 1980. *Appl. Opt.* 19: 4159.

Kreibig, U. 1970. *Z. Phys.* 234: 307.

Kreibig, U. 1974. *J. Phys. F: Met. Phys. F* 4: 999.

Kreibig, U. 1976. *Appl. Phys.* 10: 255.

Kreibig, U., Vollmer, M. 1995. *Optical Properties of Metal Clusters.* Springer Series in Materials Science 25. Springer, Berlin.

Kreibig, U., Hilger A., Hoevel, H., Quinten, M. Eds. 1996a. In *Large Cluster of Atoms and Molecules,* T. Martin, Kluwer, Dordrecht, in press.

Kreibig, U., Hilger, A., Schmid, G., Granqvist, C. G. 1996b. In *Science and Technology of Atomically Engineered Materials.* P. Jena, S. Khanna, B. Rhao, Eds. World Scientific, Singapore, in press.

Kreibig, U., Gartz, M., Hilger, A., Hoevel, H. 1996c. In *Fine Particle Science and Technology: From Micro- to Nano-particles.* E. Pelizzetti, Ed. Kluwer, Dordrecht, in press.

Kreibig, U., Gartz, M., Hilger, A., Hoevel, H. 1996d. In *Science and Technology of Atomically Engineered Materials.* P. Jena, S. Khanna, B. Rao, Eds. World Scientific, Singapore, in press.

Liebsch, A. 1993. *Phys. Rev. B* 48: 11317.

Magruder, R., Yang, L., Haglund, Jr., R., White, C., Dorsinville, R., Alfano, R. 1993. *Appl. Phys. Lett.* 62: 1730.

Maret, G. 1992. *Phys. Bl.* 48: 161.

Mie, G. 1908. *Ann. Phys.* 25: 377.

Neuendorf, R., Quinten, M., Kreibig, U. 1996. *J. Chem. Phys.* 104: 6348.

Norrmann, S., Anderson, T., Granquist, C. G., Hunderi, O. 1978. *Phys. Rev.* B18: 674.

Öestling, D., Stampfli, P., Bennemann, K. 1993. *Z. Phys. D* 28: 169.

Otto, A. 1991. *J. Raman Spectr.* 22: 743.

Otto, A., Mrozek, I., Grabhorn, H., Akeman, W. 1992. *J. Phys. Cond. Mater.* 4: 1143.

Persson, B. 1992. *Surf. Sci.* 281: 153.

Quinten, M., Kreibig, U. 1986. *Surf. Sci.* 172: 557.

Quinten, M. 1995. *Z. Phys. D* 35: 217.

Quinten, M., Rostalski, J. 1996. *Part. Part. Syst. Charact.* 13: 89.

Roemer, H., v. Fragstein, C. 1961. *Z. Phys.* 163: 27.

Schmid, G. 1994. *Clusters and Colloids — From Theory to Applications,* VCH, Weinheim.

Schumacher, D. 1993. *Surface Scattering Experiments with Conduction Electrons.* Springer Tracts in Modern Physics 128.

Selby, K., Kresin, V., Masui, J., Vollmer, M., de Heer, W., Scheidemann, A., Knight, W. 1991. *Phys. Rev. B* 43: 4565.

Sinzig, J., Radtke, U., Quinten, M., Kreibig, U. 1993. *Z. Phys. D* 26: 242.

Sonntag, H., Strenge, K. 1988. *Coagulation Kinetics and Structure Formation,* Plenum Press, New York.

Stratton, J. 1941. *Electromagnetic Theory.* McGraw Hill, New York.

Ström, S. 1975. *Am. J. Phys.* 43: 1060.

Theiss, W. 1989. *Optische Eigenschaften inhomogener Materialien.* Dissertation, RWTH Aachen.

Tiggesbäumker, J., Koeller, L., Meiwes-Broer, K. H., Liebsch, A. 1993. *Phys. Rev. A* 48: 1749.

Wertheim, G., Buchanan, D. 1994. *Phys. Rev. B* 50: 11070.

Wißmann, P. 1975. *The Electrical Resistivity of Pure and Gas Covered Metal Films.* Springer Tracts in Modern Physics 77, Springer, Berlin.

Chapter 8

Quantum-Dot Composites for Nonlinear Optical Applications

Richard F. Haglund, Jr.

Reviewed by R. W. Boyd and J. Butty

TABLE OF CONTENTS

ABSTRACT

Composite materials consisting of metallic or semiconducting, nanometer-diameter clusters embedded in dielectric hosts have unusual nonlinear optical properties of interest for both scientific and technological reasons. The nanocluster building blocks for the materials described in this review are large enough to have a physical structure virtually identical to that of the bulk material. However, their optical properties are influenced strongly either by classic or quantum effects which reflect the extreme spatial localization of the electron gas relative to its normal magnitude in bulk materials. These quantum-dot composites can be fabricated by a number of different chemical, physical, and microlithographic techniques, often in thin-film geometries suitable for the construction of waveguides or other optical or electro-optical structures. Measurements of nonlinear optical response, in both semiconductor and metal quantum-dot composites, indicate complementary but favorable characteristics in terms of switching speed, magnitude of nonlinearity, and switching energy. Thus, there appear to be significant possibilities for the deployment of these materials in photonic devices for optical switching and computing. However, due regard for system considerations suggests that the development of working technology will require more attention to functional as well as materials figures of merit. Working in or toward realistic device geometries may also have significant benefits for basic science studies of these novel materials.

1. INTRODUCTION

Suspensions of nanometer-size metallic particles in glassy hosts have been used literally for millennia for decorative and artistic purposes [Tait, 1991]. Michael Faraday was apparently the first to suggest that the beautiful colors of stained glasses were due to the electromagnetic properties of these nanoparticles and their unique optical characteristics [Faraday, 1857]. At the turn of the last century, Maxwell's equations were applied by Maxwell-Garnett [1904; 1906] in the electric-dipole approximation and to higher orders by Mie [1908] to produce a theoretical understanding of the scattering and absorption properties of these particles which has since been greatly refined and extended by many researchers. More recently, both metallic and semiconducting nanoparticles have found applications in passive optical elements such as polarizers and filter glasses, capitalizing on the ways in which the shape and the frequency response of the nanoparticles, respectively, can be controlled by suitable fabrication techniques. Increasingly, other host media, such as polymers and crystals, are also being used to enhance specific optical, mechanical and thermal properties of the composite material.

Although both classic and quantum-mechanical effects on the linear optical properties of metal nanocluster composites have been studied for many decades, the first experimental studies of the nonlinear optical properties of metal [Ricard et al., 1985] and semiconductor [Jain and Lind, 1983] nanoclusters are of relatively recent vintage. Extensive and intensive work on quantum-dot composites (QDCs) can be said to date from that time roughly a decade ago, coincident with the first serious attempts to fabricate well-characterized QDCs, [Flytzanis et al., 1991]. Driven by the interest in creating nonlinear optical elements for use in all-optical switching and computing applications, there has been an explosion of activity in the last decade not only in nanocluster composites prepared by classic chemical hydrosol techniques, but also using low-temperature processing methods (such as sol-gel); film deposition processes (e.g., metallorganic chemical vapor deposition); physical techniques, such as ion implantation and pulsed laser deposition; and microlithographic techniques borrowed from the microelectronics fabrication lines. Increasingly, these efforts have been directed at preparing materials suitable for deployment in waveguide devices, especially in thin-film geometries.

The QDCs which are the subject of this review are materials which comprise metal or semiconductor crystallites of nanometer dimensions embedded in a crystalline, amorphous, or polymeric host or matrix. The quantum dots are, for purposes of the review, defined somewhat arbitrarily to be nanoparticles, either crystalline or amorphous, with diameters greater than 1 nm. They can be fabricated by various chemical, physical, and microlithographic techniques, and generally have interatomic spacings close to or even identical to those of the bulk materials. However, the electronic properties of the nanocrystallites are dramatically different from those of the bulk because of the small size of the dot compared with the normal mean free path of the free electrons (in metals) or to the dimensions of electron–hole pairs (semiconductors); the limited number of conduction- or valence-band states; or both.

The scientific interest in these quantum-dot materials derives from the fact that a relatively small number of electrons in the dot form a strongly interacting many-particle system intermediate between the few-particle systems characteristic of isolated atoms and the infinite, translationally invariant electron systems of bulk matter. The application of nonlinear optics as a tool to study many-body effects in quantum-confined systems, for example, the interactions of a limited number of excited electrons with the background electron gas, is very much in its infancy [Perakis and Chemla, 1994]. QDCs hold great promise for technology because of the possibility of engineering their electronic [Kastner, 1993] and optical properties through materials design. In quantum-dot materials, the density of states is less than it is in the bulk because of the small number of electrons in the dot. However, nonradiative interactions, which in bulk matter relax optical excitations, are frustrated in quantum-confined systems, leading greater effective transition probabilities for a given optical frequency and thus directly to enhanced efficiencies and reduced power requirements in light sources, lasers, detectors, and switching elements.

In this review, the fundamental physical principles underlying the nonlinear optical response of these nanoclusters, as well as methods for measuring that response, are discussed first. The applications section describes some of the most important methods of fabricating, characterizing, and modifying the physical and optical properties of QDCs. Recent representative optical measurements of the properties of metallic and semiconducting dots are summarized, mostly those carried out using nonlinear spectroscopic techniques; however, since semiconductor quantum dots are treated in detail in a number of other chapters in this volume, the metal QDCs receive the greater emphasis, and results for semiconducting dots are adduced primarily to point out instructive contrasts. Finally, possible technological applications of these materials will be described along with the constraints which will govern the application of these materials in real optical communications or switching networks.

A number of recent review papers relevant to this subject are cited in the list of references. The changes in the field which warrant an updated look at the topic include: the successful application of sophisticated techniques to study the optical physics of quantum dots at nanometer dimensions and on picosecond and shorter time scales; the explosion of attempts to prepare both metal- and semiconductor-based QDCs by many different techniques; and the progress made in eliminating some of the difficult problems associated with the preparation of these materials (e.g., the use of sol-gel preparation techniques to defeat the photodarkening effect in certain semiconductor-dot-glass composites). Moreover, whereas 5 years ago the casual reader might have thought that only semiconductor quantum dots could have a realistic hope to gain entry to the world of photonic devices, it now appears that metal nanocluster-based materials may also have attractive, and in some respects complementary, properties to those of semiconductors. Thus, the fundamental questions about the evolution of material structure from atoms to condensed phases can be extended to both the metallic and semiconducting states.

The review necessarily leaves out a great deal of significant theoretical and experimental material, including metal island films, fractal clusters, and dielectric and superconducting clusters. It is also impossible, in a review emphasizing experimental developments, to give adequate attention to the theory of nonlinear optical effects in quantum-dot materials. It suffices to say that the exciting results achieved by recent experiments should increasingly attract theorists interested in applying the powerful techniques of many-body condensed-matter theory to nonlinear optical effects in quantum-confined systems.

2. FUNDAMENTALS: OPTICAL PHYSICS OF QUANTUM-DOT COMPOSITES

The nonlinear optical physics of metal or semiconductor QDCs is governed by both classic and quantum confinement effects. While classic mean-field theories are adequate to describe the optical response of relatively large nanoparticles in composite materials — say, particles larger than 20 to 25 nm diameter — the optical effects are then dominated either by classic or quantum-mechanical properties of the bulk materials. We shall confine our discussion to the smaller particles in which both classic and quantum effects play a role. For semiconductors, this typically means nanoclusters in the 2 to 30 nm diameter range, for which the nanoparticles are smaller than the bulk exciton radius. In the noble metals, particles larger than 12 to 14 nm may be considered to be bulklike and uninteresting for the purposes of this review. However, these smaller particles, to which the name *quantum dot* or *quantum box* has been given, have the *geometric* structure of the bulk material; it is only their electronic structure which is altered by the small size of the cluster.

As with bulk materials, the nonlinear optical effects on QDCs originate in changes induced in the medium by an applied optical field. For low incident intensities, only the linear reponse of the nanocluster composite is important; however, for continuous-wave or pulsed laser light in waveguides and for high-intensity pulsed laser irradiation in other geometries, higher-order effects play an important role. The ith component of the polarization **P** induced in a medium by the optical field can be expanded up to third order in a power series in the applied electric field $\mathbf{E} = \mathbf{E}_o e^{i\omega t}$ as follows.

$$P_i = \sum_j \chi^{(1)}_{ij} E_j + \sum_{j,k} \chi^{(2)}_{ijk} E_j E_k + \sum_{j,k,l} \chi^{(3)}_{ijkl} E_j E_k E_l \tag{1}$$

where the summation indices refer to Cartesian coordinates in the material and the polarization direction of the applied optical field. The susceptibilities $\chi^{(q)}$ are functions of the optical frequency ω and can, in principle, be calculated from the materials properties of the composite medium. The second-order susceptibility vanishes in any centrosymmetric medium, which we can safely assume to be the case in all the materials to be discussed in this review. Thus, it is assumed from this point on that the composite media have nonvanishing first- and third-order susceptibilities only.

The first-order susceptibility is related to the linear index of refraction n_o and the linear (Beer's law) absorption coefficient α through the following equations:

$$n_o = \Re e\left[1 + \chi^{(1)}\right] \qquad \alpha = \frac{\omega}{n_o c} \Im m\left[\chi^{(1)}\right] \tag{2}$$

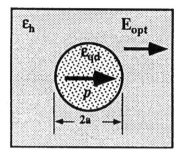

Figure 1 Schematic diagram of a nanoparticle of radius *a* embedded in a dielectric matrix, showing the direction of the induced dipole *p*.

For a material without a preferred axis of symmetry, such as spherical quantum dots embedded in an amorphous matrix, the third-order susceptibility of the composite has an analogous relationship to the nonlinear index of refraction and to the nonlinear absorption coefficient, as follows [Weber et al., 1978]:

$$n_2 = \frac{12\pi}{n_o}\Re e\left[\chi^{(3)}\right] \quad \beta = \frac{96\pi^2\omega}{n_o^2 c^2}\Im m\left[\chi^{(3)}\right] \tag{3}$$

The third-order susceptibility $\chi^{(3)}$ is, in general, a fourth-rank tensor with 81 components; however, because of the symmetry considerations applicable to most media, the actual number of components is usually substantially fewer. For example, in metal QDCs with cubic symmetry, there are only four independent components of $\chi^{(3)}$.

The classic or dielectric confinement effect on QDCs can be illustrated with the help of Figure 1. For visible and infrared wavelengths, the optical field can be viewed as having nearly constant magnitude over the entire dot. In a metal quantum dot, the conduction-band electrons oscillate against the ionic background charge, and the surface provides a resonant restoring force; the process has an interesting analog in the giant dipole resonance of nuclear physics [Broglia, 1994]. The motion of the electrons alters the field in the vicinity of the dot, as described by Maxwell-Garnett [1906] and Mie [1908] using what we would now call mean-field theories. The following treatment is close to that of Maxwell-Garnett in spirit.

Consider a composite consisting of small particles occupying a relative volume fraction $p \ll 1$ in a dielectric host. We denote the complex dielectric constant of the quantum dot by $\varepsilon_{qd}(\omega) = \varepsilon_1(\omega) + i\varepsilon_2(\omega)$ and the dielectric constant of the host medium by ε_h. The Lorentz local-field relationships derived in standard textbooks on electrodynamics show that the effective dielectric constant $\tilde{\varepsilon}$ of such a composite medium obeys the relationship

$$\frac{\tilde{\varepsilon}-\varepsilon_h}{\tilde{\varepsilon}+2\varepsilon_h} = p\cdot\frac{\varepsilon_{qd}-\varepsilon_h}{\varepsilon_{qd}+2\varepsilon_h} \tag{4}$$

For small volume fractions, this expression may be expanded to first order in the volume fraction, and recalling that the absorption coefficient is related to the average dielectric constant of the material by $\alpha = (\omega n/c)\cdot\Im m\left[\tilde{\varepsilon}-1\right]$, one finds that the absorption coefficient is given by

$$\alpha = 9p\cdot\frac{\omega\varepsilon_h^{3/2}}{c}\cdot\frac{\varepsilon_1}{\left(\varepsilon_1+2\varepsilon_h\right)^2+\varepsilon_2^2} \tag{5}$$

This expression has a resonance at the optical frequency ω_r, for which $\varepsilon_1(\omega_r) + 2\varepsilon_h(\omega_r) = 0$. This feature is known as the surface plasmon resonance, and it is responsible for the brilliant colors of Au and Ag nanocluster composites. An absorption spectrum for Au quantum dots embedded in fused silica, showing the growth of a resonant feature near 533 nm during ion implantation in a heated substrate, is shown in

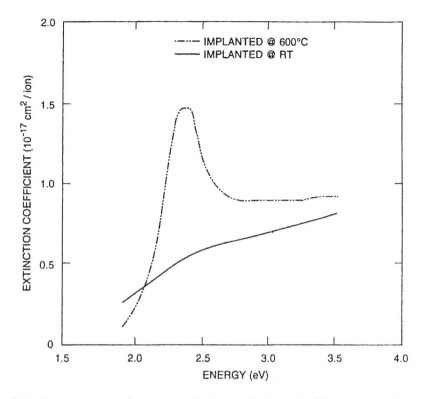

Figure 2 Extinction spectrum vs. photon energy in electronvolts for an Au:silica nanocomposite made by ion implantation at sample temperatures of 20 and 600°C. The surface plasmon resonance feature in the curve "Implanted @ 600°C" arises from the formation of Au quantum dots by nucleation in the hot substrate. (From Buchal, C. et al., *Annu. Rev. Mater. Sci.,* 24, 125, 1994. With permission.)

Figure 2. The resonant contribution to the absorption in metal QDCs diminishes linearly with the dot radius, so that size effects are difficult to distinguish in absorption spectra for nanoparticles with less than approximately 1 nm diameter.

While dielectric confinement is, in principle, exhibited in both metal and semiconductor QDCs, the enhancement due to the alteration of the local field is much larger for the former than for the latter, because the real part of the dielectric constant for the semiconductor nanoclusters is seldom large enough to cancel out the dielectric constant of the host medium in the denominator of Equation 5. As will presently become clear, this difference is somewhat compensated by the fact that the quantum effects on the third-order susceptibility of the quantum dot $\chi_m^{(3)}$, are in some respects larger (and less complex) in semiconductors than in metals.

The third-order susceptibility of the composite medium can be derived by considering the medium of Figure 1 and applying Maxwell's equations for the case in which the susceptibility is nonlinear, but the calculation is carried out to first order in the electric field [Stroud and Wood, 1989]. If we consider the cubic nonlinearity, for which the effective dielectric constant is $\varepsilon_{qd} = \varepsilon + \chi^{(3)} \cdot |\mathbf{E}|^2$, then the electric displacement is given by:

$$\mathbf{D} = \varepsilon\mathbf{E} + \chi^{(3)}|\mathbf{E}|^2\mathbf{E} \tag{6}$$

where the contraction over the components of the electric field $\mathbf{E}$ required to account for the tensor character of the dielectric constant is implied. For a charge free medium, the divergence of the displacement vanishes: $\nabla \cdot \mathbf{D} = 0$. Moreover, if the nanoparticle radius a satisfies the inequality $(\omega n_o a/c) \ll \delta$, where δ is the skin depth for the metal, and if $a \ll \delta$, then it is also approximately true that $\nabla \times \mathbf{E} = 0$. This is tantamount to assuming that the particle is so small that the field is essentially constant over the entire quantum dot. (This approximation is *not* valid in the far infrared, but for visible frequencies it is reasonably accurate for particles with diameters less than 50 nm.) Under these circumstances, it can be shown that the nonlinear optical susceptibility of the composite assumes a form perfectly analogous to the nonlinear conductivity of granular materials [Stroud and Hui, 1988]

$$\chi_{eff}^{(3)} = p \cdot \chi_{qd}^{(3)} \frac{\left\langle |\mathbf{E}|^2 \mathbf{E}^2 \right\rangle}{\mathbf{E}_o^4} \tag{7}$$

where the brackets denote a spatial and temporal average. For spherical metal nanoclusters, the field $\mathbf{E}$ inside the particle is related to the electric field far from the nanocluster by

$$\mathbf{E} = \mathbf{E}_o \frac{3\varepsilon_2}{\varepsilon_1 + 2\varepsilon_2} \tag{8}$$

By substituting Equation 8 in Equation 7, the effective third-order susceptibility of the composite becomes:

$$\chi_{eff}^{(3)} = p \cdot \chi_{qd}^{(3)} \cdot \left| \frac{3\varepsilon_h}{\varepsilon_1 + 2\varepsilon_h} \right|^2 \cdot \left(\frac{3\varepsilon_h}{\varepsilon_1 + 2\varepsilon_h} \right)^2 \equiv p \cdot \chi_{qd}^{(3)} \left| f_c \right|^2 \cdot \left(f_c \right)^2 \tag{9}$$

where the quantity f_c may be considered as a measure of the local field enhancement of the polarization. More rigorous, self-consistent treatments using a jellium model for the metal [Hache et al., 1988a] yield the same result for this special case. This local-field enhancement was observed in experiments by Magruder et al. [1993a] for Au nanocrystallites embedded in fused silica.

These results can be generalized to the case in which either the quantum dot or the host matrix may be linear or nonlinear, still within the framework of the Maxwell-Garnett approximation [Sipe and Boyd, 1992]. By building on this foundation, it is also possible to calculate the response of layered structures of nonlinear optical composites, including the nonlinear index of refraction [Boyd and Sipe, 1994]. This effect was verified in recent measurements of the nonlinear phase shift in a layered structure of TiO_2 and the nonlinear polymer poly(p-phenylene-benzobisthiazole). The nonlinear phase showed an enhancement of approximately 35% over that of the more nonlinear component predicted without the local-field corrections [Fisher et al., 1995].

Calculations of the optical response of metal QDCs embedded in nonlinear dielectrics indicate a substantial enhancement of the nonlinear susceptibility for gold nanocrystallites embedded in lithium niobate, for example, and for complex quantum dots, such as metal shells overlaid on dielectric cores [Neeves and Birnboim, 1989]. The largest calculated enhancements, of order 10^6, are for composite materials which would seem to be extremely difficult to realize in practice; however, more modest enhancements, in the range of 10^2, are predicted for less exotic materials. Enhanced nonlinear effects in these types of composite structures have been reported for Pt nanoislands in barium titanate [Haglund et al., 1995] and for complex nanoclusters with an Au shell derived from reduction of Au_2S colloids [Zhou et al., 1994].

The effects of quantum confinement can be approached using a common formalism for both metals and semiconductors, although the existence of a finite band gap energy for semiconductors and zero gap for metals produces quite distinctive optical effects in the two cases. For spherical quantum dots, the electron wave functions are the well-known functions for a "particle in a box," which are functions of momentum (n), spin and angular momentum quantum numbers (ℓ, m), and a band index b [Schmitt-Rink et al., 1987]:

$$\Psi_{bn\ell m}(\mathbf{r}) = Y_{\ell m}(\vartheta, \varphi) \frac{1}{R} \left(\frac{2}{r} \right)^{1/2} \frac{J_{\ell+1/2}(k_{n\ell} r)}{J_{\ell+3/2}(k_{n\ell} R)} \cdot U_b(r) \tag{10}$$

where the J_λ are Bessel functions of half-integer argument and the $Y_{\ell m}(\vartheta, \varphi)$ are spherical harmonics. The energy levels of a spherical particle with band gap energy E_{gap} are

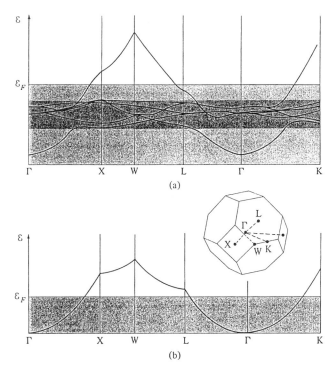

Figure 3 Calculated band structure for Cu (a) and calculated free-electron bands along the same directions in *k*-space for a face-centered cubic crystal (b). The specified directions in the Brillouin zone are shown in the inset. Note the free-electron-like character of the topmost band in the upper part of the figure. (From G.A. Burdick, *Phys. Rev.*, 129, 138–150, 1963. With permission.)

$$E_{bn\ell} = \pm \frac{E_{gap}}{2} + \frac{\hbar^2 k_{n\ell}^2}{2m_b} \tag{11}$$

Here the band index b denotes those quantities — such as the effective electron or hole mass — which depend on whether the charge carrier is in the valence (v) or conduction (c) band. For other dot shapes, the form of the wave function changes accordingly; the normalized wave functions of a cubical quantum dot, for example, are products of sine functions.

For metals, of course, the band gap vanishes, and the effective masses of electrons and holes are roughly equal to each other. For a nanocomposite with metals in a wide–band gap dielectric matrix such as fused silica or sapphire, the Fermi energy of the metal particles lies in the middle of the dielectric band gap; there is thus essentially no overlap between the wave functions of the metal electrons and the electronic states of the dielectric. Figure 3 shows the band structure of Cu, rather typical of the noble metals. The general shape of the metal bands is preserved in the quantum-dot band structure, but the density of states is substantially reduced; one can imagine discrete states distributed along the bulk bands, with a density of states roughly equal to the number of valence electrons divided by the Fermi energy. A more exact expression for the density of states is [Halperin, 1986]

$$D(E_F) = \delta^{-1} = \frac{V m_e k_F}{2\hbar^2 \pi^2} = \frac{2\pi a^3 (3\pi^2 n_e)^{1/3}}{2\hbar^2 \pi^2} \tag{12}$$

For Fermi energies of a few electron volts, the level spacing for a 1-nm-diameter quantum dot is of order tens of meV. The level spacing as a function of dot diameter for a number of metals is shown in Figure 4.

In metal nanoclusters, three kinds of electronic transitions are possible: intraband, interband, and hot-electron transitions. The *intraband* transitions schematically originate in the filled conduction-band states near the Fermi level and terminate in other conduction-band states which satisfy the selection rules for

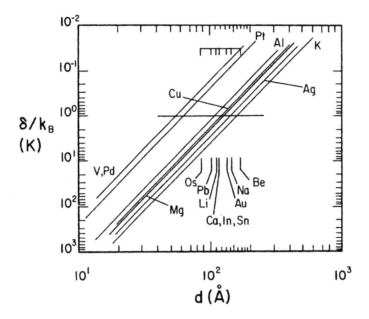

Figure 4   Level spacing as a function of nanocluster size for various metals. (From Halperin, W. P., *Rev. Mod. Phys.*, 58(3), 533, 1986. With permission.)

electric–dipole transitions. Because both the initial and final states are free-electron-like, these transitions show the strongest quantum-confinement effects, since the initial and final states both "feel" the effects of the boundary surface of the quantum dot. For an absorptive nonlinearity, a phenomenological treatment of $\chi^{(3)}$ for the intraband nonlinearity yields the result [Hache et al., 1988b]:

$$\chi^{(3)}_{\text{intra}} = -i\frac{64}{45\pi^2} T_1 T_2 \frac{1}{a^3} \frac{e^4}{m^2 \hbar^5 \omega^7} E_F^4 g_1(v)\left(1 - \frac{a}{a_o}\right) \tag{13}$$

where T_1 and T_2 are relaxation times (energy lifetime and dephasing time, respectively); $g_1 \, v$ is a function with a magnitude of order unity; and the constant a_o varies from 10 to 15 nm depending on the metal [Halperin, 1986; Flytzanis et al., 1991]. This contribution to the third-order nonlinearity diminishes with increasing particle size, vanishing identically in bulk materials.

Optical *interband* transitions are from the spatially localized d-like orbitals (states in shaded region of Figure 3) to the empty conduction-band states. These states are only weakly dependent on quantum-size effects because the initial state is already localized spatially at the parent ion. The electric dipole transitions between the d-band states and the quantum-confined conduction-band states produce a third-order susceptibility given in the same phenomenological model by

$$\chi^{(3)}_{\text{inter}} = -i\frac{2\pi A_4}{3} \tilde{T}_1 \tilde{T}_2 \frac{e^4}{m^4 \hbar^2 \omega^4} J(\omega) \cdot |P|^4 \tag{14}$$

where is A_4 an angular form factor, $\tilde{T}_1$ and $\tilde{T}_2$ are the energy lifetime and dephasing time for the interband transition, respectively; $J(\omega)$ is the joint density of states; and P_{ij} is a matrix element between initial and final states which is nearly constant given the flat character of the d-bands in which the interband transition originates. The energy and dephasing lifetimes are distinct from those for the intraband transitions, as signified by the tilde, since they presumably arise from different mechanisms. The transition probabilities are proportional to the joint density of initial and final states for a given photon excitation energy — that is, to the oscillator strengths of the allowed transitions. For gold nanoparticles in the 5 to 15 nm diameter range, the interband nonlinearity in the vicinity of the surface plasmon

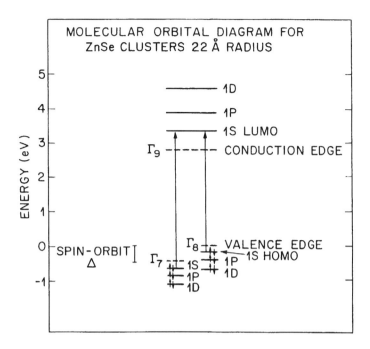

Figure 5 Electronic structure of a ZnSe compound-semiconductor quantum dot of radius 22 Å, showing the bulk band gap and the (blue-shifted) HOMO-LUMO gap which arises through quantum confinement. (From Bawendi, M. G. Steigerwald, M. L. and Brus, L. E., *Annu. Rev. Phys. Chem.*, 41, 477, 1990. With permission.)

resonance is approximately 50 to 100 times as the intraband contribution. Some size dependence for interband transitions may arise from the development of localized surface states above the "bulk" band edges in the smallest dots.

Finally, the *hot-electron* transitions are those in which an electron in the conduction band absorbs a photon and is heated, losing its energy ultimately by electron-phonon scattering or collisions with the walls. The hot-electron effect is sometimes called *Fermi smearing*, because it results in a broadening of the electron population distribution in the vicinity of the Fermi edge through electronic excitation.

$$\chi_{he}^{(3)} = i\,\frac{\omega \varepsilon_{D2} \tau_o}{24\pi^2 \gamma T}\,\frac{\partial \varepsilon_{2L}}{\partial T} \tag{15}$$

where γT is the electronic specific heat; ε_{D2} and ε_{2L} are, respectively, the imaginary parts of the free-electron gas (Drude) and interband contributions to the dielectric function ε; and τ_o is an electron cooling time. These hot-electron transitions make a positive imaginary (absorptive) contribution to the third-order susceptibility which is size independent, and can make a contribution to $\chi^{(3)}$ 10 to 10^3 times as large as the *intra*band transitions at a given wavelength. The excited vibronic states of the QDC resulting from hot-electron transitions may have relaxation times as long as hundreds of picoseconds.

The effects of quantum confinement on the optical response of semiconductor quantum dots are substantially altered by the finite band gap energy. The effective masses of electrons and holes are quite different, and because of the finite value of the static dielectric constant, the Coulomb interaction of electrons and holes is only partially screened, so that bound, excitonic (electron–hole) states are formed. The size of these excitons sets a natural scale length for quantum confinement [Efros and Efros, 1982; Brus, 1984], which can be divided into strong, intermediate, and weak regimes, depending on whether the ratio of nanoparticle size to exciton radius in the bulk material is much smaller than, comparable to, or much larger than unity. Because of the separation between valence and conduction bands (corresponding to bonding and antibonding molecular orbitals), the electronic structure of the semiconductor quantum dots resembles that of an "artificial atom" or "artificial molecule," more so than that of metal quantum dots.

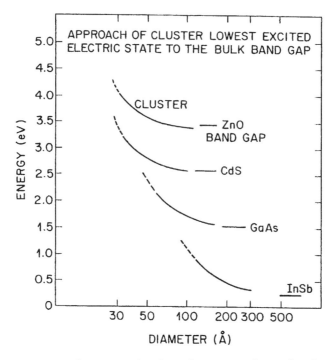

Figure 6 Effects of quantum confinement on the electronic structure of a semiconductor, showing the blue shifting of the lowest 1s–1s transition with decreasing quantum-dot size for various compound semiconductors as indicated. (From Bawendi, M. G., Steigerwald, M. L. and Brus, L. E., *Annu. Rev. Phys. Chem.*, 41, 477, 1990. With permission.)

The energy-level structure of semiconductor quantum dots, as calculated in a molecular orbital (MO) model, is sketched in Figure 5. The gap between the highest-occupied (HO) and lowest-unoccupied (LU) MOs corresponds to the energy gap of the bulk material, but is significantly blue shifted. The progressive shift in HOMO-LUMO gap with decreasing particle size is exhibited in Figure 6, calculated for a number of compound-semiconductor quantum dots of progressively decreasing size. Corrections to the simple MO picture accounting for nonparabolicity of the conduction band and for valence-band degeneracy can be made, giving a relatively exact account of the observed energy-level structure [Ekimov et al., 1993].

The third-order nonlinearities in semiconductor quantum dots arise from saturation, Coulomb interactions, and impurity effects, such as trapping of photogenerated charge carriers. The relative weights of these three mechanisms depend on the material type and also the strength of the confinement. In the case of *strong* confinement, for the *1S–1S* transition, saturation or band-filling effects are dominant and the leading term in the third-order susceptibility is well approximated by the expression [Flytzanis et al., 1991]:

$$\chi^{(3)}(\omega) = -i\frac{1}{V}\left|\frac{ep_{cv}}{m\omega}\right|^4 \frac{1}{\hbar^3} \frac{T_1 T_2^2}{1+i(\omega_{ab}-\omega)} \frac{1}{1+(\omega_{ab}-\omega)^2 T_2^2} \tag{16}$$

This equation can be derived from ordinary time-dependent perturbation theory and illustrates the two-level characteristic of the quantum-confinement regime in a semiconductor quantum dot.

Coulomb effects alter this straightforward expression in the case of *intermediate* confinement, because as the quantum dot becomes larger than the bulk exciton radius, one may have more than one exciton simultaneously in a single dot. In this case, the third-order susceptibility has a form similar to that of Equation 16, but is complicated by the addition of matrix elements and transition frequencies corresponding to two-carrier excitations [Banyai, et al., 1988]. In quantum dots which are large compared with the bulk exciton radius, one speaks of weakly confined electronic systems, and the third-order nonlinearity is essentially that of the bulk semiconductor material.

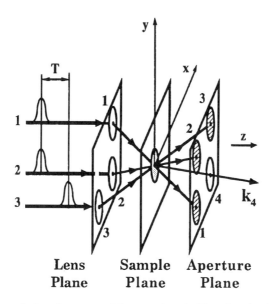

Figure 7 Apparatus schematic for a four-wave mixing experiment with codirectional beams (forward four-wave mixing). In the other typical geometry used in four-wave-mixing experiments, the pump beams are counterpropagating.

3. FUNDAMENTALS: EXPERIMENTAL METHODS FOR NONLINEAR OPTICAL INVESTIGATIONS

Two methods, and variations thereof, are in common use for measuring the nonlinear optical properties of metal quantum-dot materials. *Four-wave mixing* is the classic technique and is especially useful for the determination of the magnitude of the third-order susceptibility and the time dependence of the electronic modification to the nonlinear material induced by the incident laser light; in its most typical incarnation, the wave mixing is done at a single frequency, in which case it is referred to as *degenerate* four-wave mixing (DFWM). The second technique, the so-called *Z-scan* [Sheik-Bahae et al., 1990], is a particularly elegant and sensitive single-beam variation on earlier beam-profile measurement techniques [Weaire et al., 1979]. The Z-scan provides rapid information about the nonlinear refractive index and nonlinear absorption, in particular their signs; the Z-scan can also, in principle, deliver time-resolved information, but its physical interpretation is not necessarily as straightforward as the results of a four-wave mixing experiment.

In DFWM, illustrated in Figure 7, two pump beams and one probe beam are split off from a single laser beam; the pump beams typically contain 80 to 90% of the total pulse energy. The pump beams — which may be either co- or counterpropagating — are arranged to be coincident in time in the sample volume, while the probe beam arrival time in the sample volume is varied with respect to the pump beams by an optical delay line. The third-order susceptibility measured in such an experiment is $\chi^{(3)}(\omega,-\omega,\omega;\omega)$, where the negative sign indicates that the complex conjugate of the electric field is to be taken. The ith component of the nonlinear polarization associated with the experiment is then $P_i = \sum_{j,k,l} \chi^{(3)}_{ijkl} \, E_j(\omega) \, E_k^*(\omega) \, E_l(\omega)$, where the subscripts on the electric field components should be considered as including both the designation as pump or probe and the polarization of the beam. If one thinks of the beams j and k as the pump beams, this polarization has an appealing physical interpretation: it is the polarization induced by the probe beam in a material medium modified by the two pump beams; in that sense, the product of the third-order susceptibility with the two pump beams may be viewed as having generated an effective first-order susceptibility of the laser-modified composite material. The lifetime of this material modification is reflected in the DFWM time spectra shown in Figure 8.

DFWM measurements have been implemented in two geometries suitable for studies in thin nanocomposite layers. In one, the so-called phase-matched forward four-wave-mixing scheme shown in Figure 7 [Smirl et al., 1980], all of the beams pass through the sample in the forward direction; the probe beam is then advanced or retarded in time with repect to the pump beams. For a nanocomposite medium with cubic symmetry, the x-component of the third-order nonlinear polarization is

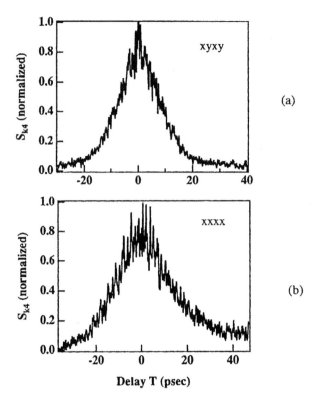

Figure 8 Time dependence of the phase-conjugate signals S_{K4} from a forward DFWM experiment on a Cu:silica QDC, contrasting the copolarized and cross-polarized cases. Note the thermal "tail" on the copolarized signal, indicating the decay of a thermal (or population) grating.

$$P_{4x}^{(3)} = \chi_{xxxx}^{(3)} E_{1x}^{*} E_{2x} E_{3x} + \chi_{xxyy}^{(3)} E_{1x}^{*} E_{2y} E_{3y} + \chi_{xyyx}^{(3)} E_{1y}^{*} E_{2y} E_{3x} + \chi_{xyxy}^{(3)} E_{1x}^{*} E_{2y} E_{3y} \tag{17}$$

where the order of the subscripts corresponds to the two pump beams 1 and 2, the probe beam 3, and the phase-conjugate, forward-diffracted beam. The first three terms can contain contributions resulting from the creation of a population or electron-density grating; therefore, these terms include the effects of thermal mechanisms resulting from differential absorption, while the fourth term involves only the electronic contribution. Hence, by judicious choice of input-beam polarization, it is possible to separate the thermal from the electronic contributions to $\chi^{(3)}$, as in the spectra of Figure 8. This is particularly important when four-wave-mixing measurements are made using high-repetition-rate laser systems.

The classic DFWM geometry with counterpropagating pump beams, used in experiments with filter glasses [e.g., Ricard et al., 1985] can also be made to work in extremely thin samples [e.g., Carter, 1987]. Nondegenerate three-wave mixing, one of the many variations of coherent Raman scattering, is an interesting technique which has been demonstrated to yield accurate results for the nonlinear refractive index of bulk glasses [Adair, et al., 1987]. In this scheme, the pump beams are adjusted to have slightly different frequencies, and the output intensity at the difference frequency, which is proportional to $\chi^{(3)}$, is measured. Thus far, however, this method has not been applied to studies of the nonlinear index of refraction in nanocluster composites. DFWM has also been demonstrated in a waveguide implementation, again using prism coupling to insert the pump and probe beams and extract the phase-conjugate signal [Gabel et al., 1987]. In this particular case, the waveguide was an ion-exchanged semiconductor-doped glass, and the phase-conjugate reflectivity showed a maximum near 50 MW · cm^{-2}, a value obtained rather easily in the confined geometry of the waveguide.

In the Z-scan technique, shown schematically in Figure 9, a single well-characterized Gaussian beam is directed through a fixed, long-focal-length lens to the sample, located near the focal plane. As the sample is scanned along the optical axis of the system through the focal plane, the intensity on the sample is varied. For a medium with a positive nonlinear index of refraction, the effective refractive

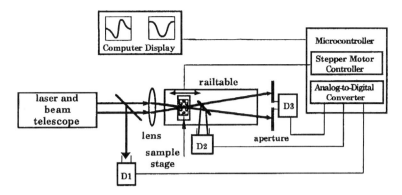

Figure 9 Apparatus schematic for a Z-scan experiment. The sample is translated through the focal plane of the long-focal-length lens. Intensity-dependent variations in the transmitted light are observed on detectors D_2 (for the whole beam) and D_3 (for light near the axis); the detector D_1 is used to correct the signals for fluctuations in the input laser intensity.

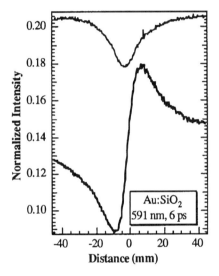

Figure 10 Near-field (top) and far-field (bottom) transmission spectra in the Z-scan geometry, for an Au:silica QDC, made by implanting $6 \cdot 10^{16}$ ions·cm^{-2} into a fused-silica substrate, then annealing the sample to approximately 1100°C. The dip in the near-field spectrum is indicative of positive nonlinear absorption, while the shape of the far-field spectrum indicates positive nonlinear refraction.

power of the combined sample and fixed lens is increased with decreasing distance to the focal plane. However, when the sample is to the left of the focal plane ($z < 0$), the far-field intensity on axis decreases as the beam covers a larger area, while to the right of the focal plane ($z < 0$), the far-field intensity increases as the beam is refocused to a smaller area. The combined effect for a detector sitting in the far field is to produce a Z-scan which appears like the one shown at the bottom of Figure 10. On the other hand, if one intercepts the entire beam, one is sensitive only to nonlinear absorption; for positive nonlinear absorption, the spectrum appears as shown at the top of Figure 10. Thus, a Z-scan measurement which measures both on-axis and whole-beam intensities can, in principle, make it possible to extract both the nonlinear index and nonlinear absorption from a single measurement. In addition, it is possible to use the Z-scan in a pump-probe configuration to extract time-resolved [Wang et al., 1994] and two-color [Sheik-Bahae et al., 1992] information about the sample response.

The far-field intensity distribution in the Z-scan geometry can be related in a completely rigorous way to the nonlinear index and nonlinear absorption through a Gaussian, partial-wave decomposition in the far field [Weaire et al., 1979, Sheik-Bahae et al., 1990]. However, a simpler approach based on self-focusing is accurate enough for many experiments, especially with thin nanocomposite films. For a thin nanocluster layer embedded in a transparent dielectric and for a high-repetition-rate laser, the relative far-field transmitted power is determined by both the optical response of the metal particle and the surrounding dielectric, and by the two-dimensional thermal relaxation processes through which the absorbed laser energy flows out of the illuminated region. It can be shown to be given by [Yang et al., 1996]

$$P_{det} = \left[\cfrac{1}{1 + \beta I_o L \cdot \cfrac{1 - \exp(-\alpha L)}{\alpha}} \right] \cdot \left[\frac{C_1 z}{\left(1 + \zeta^2\right)^2} + \frac{C_2 z}{\left(1 + \zeta^2\right)} - 1 \right]^{-2} \quad (18)$$

where

$$C_1 = \frac{2 L P_{inc}}{\pi n_o^2 r_o^4}, \quad C_2 = \frac{L P_{inc} K_r}{\pi n_o^2 r_o^4}, \quad \zeta \equiv \frac{z}{z_o}, \quad z_o = \frac{\pi r_o^2}{\lambda} \quad (19)$$

In this expression, P_{inc} is the total power on the sample; $I_o = P_{inc}/4\pi r_o^2$ is the intensity at the focal plane, with r_o the usual Gaussian beam radius; L is the sample thickness; K_T is the material-dependent, thermo-optic coefficient; and α and β are, respectively, the one- and two-photon absorption coefficients. The separation of the electronic and thermal contributions to nonlinear behavior is aided by the differing z-dependences of the two terms in Equation 18.

Among other widely used measurements for studying the nonlinearity of semiconductor quantum dots is the technique of spectral hole burning, based on nonlinear saturation effects. This is ideally suited to the study of semiconductors with their inhomogeneous size distributions, for spectral hole burning is based on the idea of selecting a single population from a distribution by using a narrow-band laser to saturate a single transition. In practice, this generally works with nanosecond (ns) pulsed lasers or even continuous wave (cw) lasers. The techniques are described in many practical texts on nonlinear spectroscopy. (See the Section entitled "For Further Information.") Photon echo techniques [Peyghambarian et al., 1989; Koch et al., 1993] and nondegenerate four-wave mixing [Woggon and Gaponenko, 1995] have been used to study quantum-confined semiconductors, but have not yet been employed to any significant degree on metal QDCs.

Electroabsorption spectroscopy, which is based on the optical Stark effect or quantum-confined Keldysh effect [e.g., Cotter et al., 1991; and Woggon et al., 1993], and magnetoabsorption pump-probe spectroscopy [e.g., Siegner et al., 1995] have played an important role in probing electronic structure and dynamics in semiconductor quantum-confined systems. It appears, however, that they have not yet been employed in studies of metal QDCs.

A less extensively used but extremely sensitive technique for nonlinear measurements, particularly in waveguide geometries and materials, is interferometry. The experimental configuration for a pulse-modulated interferometer using a cw mode-locked laser is illustrative of the technique and is shown in Figure 11, along with spectra of the power dependence of the interference fringes [Stegeman et al., 1988]. The mode-locked pulse train is modulated by a Pockels cell, split, and passed through a Mach–Zehnder interferometer, one arm of which contains the sample. The electronically induced phase shift was measured by setting the Pockels cell to transmit a low background signal from the cw mode-locked train (5%); a high voltage was applied to the Pockels cell once each millisecond to increase the transmission to 50%, thus producing one pulse each interval with ten times the intensity of the background. This technique takes advantage of the inherent stability of the cw mode-locked laser system and of the rapid sampling rate which is made possible by gated, phase-sensitive detection. To measure thermally induced effects in the waveguide, the Pockels cell was replaced by an acousto-optic modulator transmitting 10^3 pulses each millisecond. The stability of the interferometer is assured over this short time scale so that the data sets from each gate interval can be compared. While this technique has not been used for measurements on metal QDCs, it is likely to be useful when these are made in waveguide geometries.

4. FABRICATION AND CHARACTERIZATION OF QUANTUM-DOT COMPOSITES

The ideal properties of QDCs for nonlinear optical applications are stated easily: the dots should be homogeneous in composition, structure, and size; should be distributed in controlled lateral and vertical

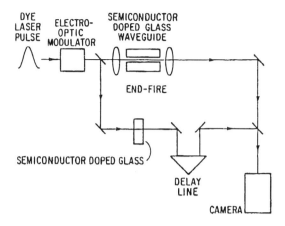

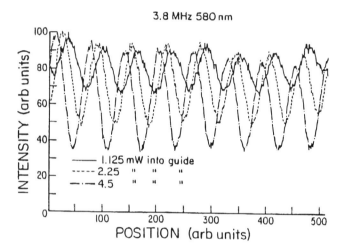

Figure 11 Schematic of a Mach–Zehnder interferometer (top) designed to measure the nonlinear fringe shift of waveguide materials (bottom). The beams are coupled into and out of the waveguide by the prisms cut so as to couple incident radiation directly into the waveguide. (From Stegeman, G. I. et al., *IEEE J. Lightwave Tech.*, 6(8), 953, 1988. With permission.)

patterns; and should be stable under long-term irradiation by laser light. The host matrix should be highly transparent, free of defects to the greatest extent possible, and should be crystalline or amorphous depending on the application. By far the most severe requirement is that one should be able to tailor the optical properties of the composite to suit any given scientific or technological interest — meaning that the fabrication technique should be in principle capable of producing arbitrary combinations of quantum dots and hosts.

These constraints have proved somewhat difficult to achieve in practice. While QDCs can be made by many different methods — including biochemical production by the action of yeasts in certain bacteria! [Dameron et al., 1989] — the discussion below is limited to chemical, physical, microlithographic, and self-organizing techniques. The discussion is further restricted to those techniques capable of producing a high-quality optical structure; thus, for example, those techniques by means of which metal or semiconductor dots are created in zeolite cages are not treated because, although the dots have uniform sizes and compositions, the scattering of light by the zeolite matrix is unacceptable for optical applications at present.

4.1. Fabrication of Quantum-Dot Composites
4.1.1. Chemical Techniques

Among the standard chemical techniques for preparation of QDCs are hydrosols; melt-glass techniques; sol-gel processes; chemical vapor deposition; and arrested precipitation of reverse micelles, followed by surface derivatization of the clusters to stabilize them against further interaction with the reactants in the micellar medium.

The earliest recipes for preparation of metal nanocluster hydrosols [e.g., Perenboom et al., 1981] begin with an acidic solution of the appropriate noble metal; reduce it with an agent such as sodium citrate to initiate the precipitation of particles; remove unwanted residual ions by dialysis; and add a protective compound, such as gelatin, to stabilize the hydrosol and prevent further aggregation. However, since the size and size distributions of the nanoclusters are determined almost entirely by equilibrium chemical kinetics, it is not possible to achieve much control over the optical properties of the nanoclusters. Indeed, in metals, these clusters are often larger than the size range within which quantum confinement effects are observed. Hence, these techniques are largely of historical interest.

Filter glasses containing metal, II-VI, and I-VII nanocrystallites can be made by a thermal diffusion process which gives relatively good control over the quality of the QDC [e.g., Borelli et al., 1987; Hache et al., 1988a]. In the case of semiconductor-based glasses, one begins with a silicate or borate glass heated to temperatures as high as 1500°C, together with the II-VI (e.g., Cd and Se) or I-VII (e.g., Cu and Cl) elements which form the semiconductor, along with other stabilizing materials required to yield acceptable glass properties. As the glass is heated, the individual elemental atoms diffuse either in atomic or ionic form to form a uniform distribution inside the melt, leaving a colorless material with nucleation sites containing up to 100 atoms. At these nucleation sites, the surface atoms are dominant. When the glass is heated again to more modest temperatures, of order 500 to 750°C, the clusters grow to an average size determined by the temperature and duration of the annealing or striking process.

The growth process has been modeled [Lifschitz and Slezov, 1959] under the assumption that there is a nucleation phase in the supersaturated solution, followed by a growth phase in which larger particles grow at the expense of smaller ones. This process produces particles of mean radius $\bar{a} = [(4/9)\sigma DC(T)\tau]^{1/3}$ where σ is the surface tension, D is the diffusion constant, $C(T)$ is a constant which depends exponentially on temperature, and τ is the duration of the annealing or striking process. The nanocrystallite radius has a distribution function given by

$$P(u) = \frac{3^4 e}{2^{5/3}} \frac{u^2 e^{-1/(1-2u/3)}}{(u+3)^{7/3}\left(\frac{3}{2}-u\right)^{11/3}} \tag{20}$$

where $u = a/\bar{a}$. The distribution function is asymmetric and evidently vanishes for $u \geq 1.5$. Whether or not the growth mechanisms are similar, for example, in the recently reported preparation of CdSe nanocrystals embedded in a ZnSe matrix by chemical vapor deposition [Danek et al., 1994] is also an interesting question.

Similar growth processes, such as Ostwald ripening, have been proposed for the growth of metal nanocrystallites in glasses synthesized both by conventional melt-glass techniques and by ion implantation [e.g., Magruder et al., 1994a]. However, there have been virtually no detailed comparisons of theory and experiment for metal QDCs, although it is presumed that they follow similar growth processes; in the case of metal QDCs made by ion implantation (vide infra), the nucleation and growth mechanisms are still more complex. Thus, the growth kinetics of metal quantum dots in composite materials must be viewed as an important open question to be settled by further research.

Recently, the synthesis of high-quality semiconductor [for example, Kao et al., 1994] and metal [De et al., 1995] QDCs by sol-gel techniques has been reported. In the sol-gel preparation of Ag:silica nanocluster composites in the form of thin, dip-coated films and monoliths, the sol was prepared with a 1:0.12:12:0.2:6:7 molar ratio mixture of the components $Si(OC_2H_5):AgNO_3:H_2O:HNO_3:C_3H_7OH:C_4H_9OH$. The final gel had an Ag::Si ratio of 0.12, and was quite lossy, with an attenuation coefficient in the vicinity of the surface plasmon resonance (410 nm) of some 100 cm^{-1}. A size distribution measurement made by high-resolution transmission electron microscopy (TEM) showed the great majority of the particles had diameters in the 1 to 4 nm range, while a very small number had larger diameters (see Figure 12). Temperature effects on the gelation process were studied by Fourier-transform infrared (FTIR) spectroscopy, and it was found that for anneals above 450°C, an asymmetric Si–O stretching mode associated with nonbridging Si–O groups disappeared, indicating a complete densification of the silica matrix. These materials can also be made with smaller concentrations of silver, which would be more nearly ideal for thin films or waveguide applications.

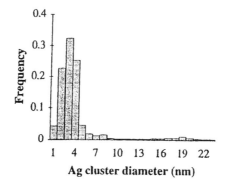

Figure 12 Size distribution of Ag nanoparticles embedded in a fused-silica matrix, obtained from direct observation and size determination from a TEM. The QDC is made by a sol-gel technique. (From De, G. et al., *J. Non-Cryst. Solids,* 1995. With permission.)

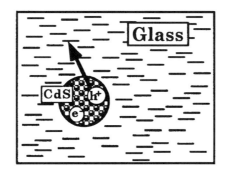

Figure 13 Schematic of a model for the interaction of localized traps in a matrix surrounding a semiconductor quantum dot with the dot electrons, proposed as a mechanism for photodarkening. (Adapted from Kang et al., *Appl. Phys. Lett.,* 64(12), 1487, 1983. With permission.)

A wide variety of other quantum dots — including Ag, Cu, Pt, Os, Co_3C, Fe_2P, Ni_2P, Ge, and an alloy of Pt and Sn — have been synthesized as xerogel composites [Burnam et al., 1995]. In this variation on the conventional sol-gel technique, low-valence complexes of transition metals or organometallic compounds of Ge containing ancillary ligands with silicate esters or (alkoxy)silicon functional groups are introduced into a conventional silica sol mixture. When the dopant molecules are introduced into a conventional silica sol-gel preparation, they are incorporated into the backbone of the xerogel by heterocondensation of Si–OH groups. Thermal annealing in a reducing or oxidizing atmosphere then leads to formation of nanocomposites in which the desired dopant aggregates form nanocrystalline particles ranging from 10 to less than 50 nm diameter. While published results show nanocomposites which are too small for optical applications, the most recent experiments have produced monoliths (~1 cm diameter) suitable for polishing and larger-scale optical applications.

Semiconductor quantum dots prepared in melt glasses have long been known to suffer from a problem known as *photodarkening,* in which the opacity of the QDC grows with exposure time to laser light. While the causes have been speculated upon for a number of years, only recently has it become possible to understand the mechanism and to alter the glass preparation technique to avoid it. It was found that whereas conventional glass-fusion-prepared samples underwent photodarkening, semiconductor QDCs made in a sol-gel process did not [Kang et al., 1994]. This was attributed to the much larger number of electron and hole traps available in the conventional glass mix which has a number of alkali or alkaline-earth metal oxides as glass modifiers to aid in lowering the glass temperature (long dashes in Figure 13). These charge-carrier-absorbing traps altered the matrix surrounding the dots so severely that the optical absorption characteristics were likewise drastically altered. When similar measurements were carried out with sol-gel glasses having much higher silica content and many fewer glass modifiers, the photo-darkening effect essentially disappeared.

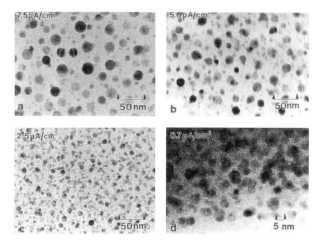

Figure 14 TEM micrograph of a series of four Cu:silica nanocomposites made by ion implantation, showing the variation in size and volumetric density of clusters as functions of ion current density during implantation. Each sample was made with the same total ion dose, nominally $6 \cdot 10^{16}$ ions·cm^{-2}. Note the change in scale on the micrograph for the lowest current density. Ion current density in μA·cm^{-2} is shown in the upper left corner of each photo.

While both sol-gel and melt-glass processes can produce semiconductor-nanocrystal-doped glasses of sufficient quality to serve as filters, control of the size distribution remains problematical because it is governed by equilibrium kinetics. Semiconductor clusters with much narrower size distributions have been prepared in solution by arrested precipitation in reverse-micellular processes [Murray et al., 1993]. Indeed, clusters of CdSe have been prepared with polymer end caps and then successfully grown into arrays of CdSe quantum dots in the polymer matrix [Murray et al., 1995]. While these macrocrystals composed of II-VI quantum dots are extremely sensitive to photochemical destruction by light above liquid helium temperatures, the possibility of QDCs in which the dots are arrayed in a regular spatial order is truly arresting. Thus far, no comparable degree of structural ordering has been achieved in metal QDCs.

4.1.2. Physical Techniques

Several groups in recent years have turned to physical methods in order to achieve a more arbitrary degree of control over the composition, size, and size distribution of both metal and semiconductor QDCs. In this category are included sputtering and other forms of physical vapor deposition, ion implantation, and pulsed-laser deposition. The most advanced of these techniques for optical applications is ion implantation.

Ion implantation is a simple way to create thin films of nanocomposite materials and has the additional advantage of being already compatible with standard microelectronics fabrication technology. It permits essentially arbitrary choices of host and nanocrystallite species, provided only that the free-energy considerations favor clustering rather than formation of compounds with the constituents of the host matrix. In spite of its various apparent drawbacks — such as residual damage from collisions which must be annealed away — ion implantation has proved to be surprisingly good for creating metal QDCs in thin-film configurations and has the additional advantage of being intrinsically compatible with microelectronics fabrication techniques.

For example, Magruder et al. [1994a] describe the implantation of Cu ions into fused silica, with size distributions controlled to a significant degree by adjusting the ion-beam current during implantation. As the TEM micrographs in Figure 14 and the size distributions in Figure 15 show, reducing the implantation current density reduces the mean diameter of the Cu nanoclusters but also narrows the size distribution. An analogous effect can be seen if one heats the sample during implantation: samples implanted with Cu at elevated temperatures (greater than about 300°C) show a rapid increase in the average nanocluster diameter and a broadening of the size distribution. These results appear to indicate that the nanocluster growth process during ion implantation is a diffusion-limited aggregation process which is controlled by the local temperature in the vicinity of the ion deposition sites. Magruder et al. [1993b, 1994b] have recently shown that it is possible to alter the fundamental surface plasmon resonance

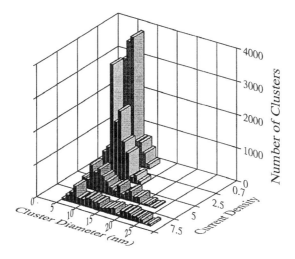

Figure 15 Size distribution of the Cu nanoclusters for the four samples shown in Figure 14 as a function of average cluster size and ion-beam current density. Note that the distribution narrows and steepens with *decreasing* ion-beam current.

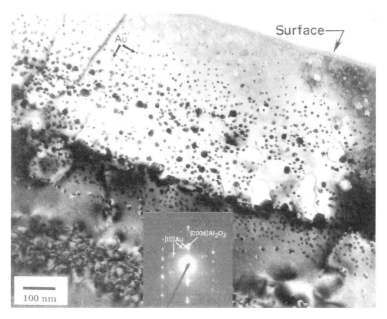

Figure 16 TEM micrograph of Au nanocrystallites in a sapphire substrate, along with electron diffraction pattern showing the ordered array of spots characteristic of oriented particles.

in metal QDCs by coimplanting two miscible metal species in the same layer. While Hosono has shown that the surface plasmon resonance of Cu nanoclusters can be shifted by co-implantation of F ions [Hosono et al., 1992a]. [Magruder et al., 1993]. In crystalline sapphire, recrystallization following Au-ion implantation not only induces nucleation and growth of the nanocrystallites, but also produces an elongation of Au quantum dots and aligns them with the local crystal axis. The ordered array of spots from the electron-diffraction analysis of the Al_2O_3 sample containing Au nanoclusters, shown in Figure 16, is experimental evidence of this thermal effect — which apparently occurs because the surface forces at the Au:sapphire interface stretch the Au nanocrystal as the surrounding sapphire recrystallizes.

An additional question related to nanocrystal synthesis by ion implantation is the role played by the chemical composition and local electronic structure of the substrate. While Cu aggregates spontaneously during ion implantation in fused silica, as manifested by the formation of a surface-plasmon resonance in the absorption curve around 565 nm, no such feature is found in ion-implanted borosilicate (Pyrex),

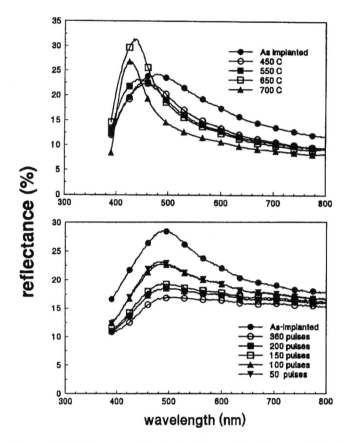

Figure 17 Absorption spectra for Ag nanoparticles in float glass subjected, respectively, to thermal annealing (upper graph) and annealing by KrF laser irradiation (lower graph). The peak shifts in the upper graph are associated with changes in the average size of the nanocrystallites; the lack of a shift in the laser-annealed sample suggests loss of nanocrystallite density without a change in average size. (From Wood, R. A. et al., *J. Appl. Phys.*, 74(9), 5754, 1993. With permission.)

aluminosilicate (Corning 0312), or soda-lime glasses [Mazzoldi et al., 1994; Bertoncello et al., 1995]. This suggests that there are chemical effects, or at least effects due to local electronic structure, which affect the nucleation process. Such effects probably also play a role in altering the optical response of nanocrystalline composites when an ion such as fluorine is implanted, since in some cases the fluorine may alter the optical response by incorporating itself into the glass host.

The size and size distribution of ion-implanted metal QDCs can be controlled to some degree by thermal annealing. Ordinary float glass has sufficient tin diffused into its top surface to constitute a good planar waveguide. Townsend and collaborators [Wood et al., 1993] have used this circumstance to make waveguides containing Ag and other metal nanoclusters by ion implantation, following which they attempted to control cluster size and size distribution by laser and by thermal annealing. Thermal annealing does influence the nanocluster size distribution, as shown by shifts in the surface plasmon resonance peak in the absorption spectrum (Figure 17). Annealing with a KrF laser (248 nm), on the other hand, appears to vaporize the Ag nanoclusters *in situ*, reducing the amplitude of the absorption peak without shifting it. It is not yet clear whether or not the laser annealing would have a different outcome if a laser of a different wavelength or pulse duration were used. Laser irradiation with 100-ps pulses at 532 nm, on the other hand, appears to enhance the nonlinear absorptive and refractive effects [Haglund et al., 1995].

Buchal et al. [1994] have recently described in detail the successful preparation of Au and Ag nanoclusters in sapphire by ion implantation and reviewed the implantation of a number of other metallic species — including tin and antimony — in various dielectrics. Amorphous semiconductor clusters have also been created in fused silica by ion implantation; quantum confinement leads to a strong red shift in the absorption edge and a large (albeit primarily thermal) third-order nonlinearity for ns laser pulses [Hosono et al., 1992].

The synthesis of nanoclusters of Group IV and compound semiconductors in dielectric hosts are of particular interest as light sources and because of the change from indirect to direct band gap character which evidently accompanies the changing electronic structure of the clusters as the cluster size decreases [Takagahara and Takeda, 1992]. The possibility of creating direct band gap materials which would be compatible with existing Si–Ge fabrication technologies, of course, is of enormous potential significance for electronics. Ge nanocrystals with dimensions on the order of 7 nm have been created by radio-frequency (rf)-magnetron cosputtering of Ge with SiO_2 in an Ar atmosphere at a pressure of 3 mTorr [Maeda et al., 1991]. The relative Ge content of the samples thus made was on the order of 43 atomic percent, as determined by inductively coupled plasma emission spectroscopy. Room-temperature photoluminescence spectra showed a strong enhancement in the region of 600 nm after annealing produced agglomeration of the Ge quantum dots. In samples which were not annealed, no defect luminescence was observed which could be traced to the formation of defects in the SiO_2 sample, while Raman spectra showed a strong vibrational mode in the annealed sample which corresponded to a known mode in Ge nanocrystallites with average diameters on the order of 6 to 8 nm. The photoluminescence and the extrapolated binding energies were found to be consistent with the Brus model for direct-gap materials.

Ion implantation has also been used to create nanoclusters of Si and Ge in fused silica matrices. For example, Ge nanocrystals have been synthesized by implanting Ge^+ ions at energies of 130, 140, and 150 keV at doses of 10^{16}, 4×10^{16}, and 10^{16} ions/cm^2, respectively, to create a virtually square profile of deposited ions [Shcheglov et al., 1995]. After annealing in vacuum at 600°C to precipitate the Ge in solid solution, a p^+ top layer was created afterward by doping the initial polycrystalline layer of Si with B at a temperature of 900°C. These materials are of particular interest as potential light sources for Si- and Ge-based optoelectronic devices, but it is clear that their deployment in nonlinear optical devices is possible in principle, based on the current results.

4.1.3. Physicochemical Techniques

Of course, both physical and chemical techniques have their respective advantages, and it is thus sensible to consider combining the two. In this category we include physical or chemical vapor deposition and their variants, as well as the recent combination of ion exchange and ion implantation.

Kay and his co-workers [Laurent and Kay, 1988] have created highly uniform distributions of gold dots with mean diameters in the range of 1 to 5 nm by intense ion bombardment of a gold target in the presence of a monomer which becomes the polymerized matrix for the gold clusters. The films are grown in a capacitively coupled coplanar diode to which an rf field at 13.56 MHz is applied. The gold substrate is the powered electrode, and the nanocomposite film is grown on the grounded electrode the temperature of which can also be controlled. Thin films of Au quantum dots have been successfully grown in an Ar-monomer atmosphere at a pressure of 20 mTorr; the monomer can be either C_3F_8 (perfluoropropane) or C_2F_4 (tetrafluoroethylene), but the latter seems to produce higher quality, more reproducible films. Of particular note in this case is that very high volume fractions of the Au can be produced, up to 0.45.

Gold nanocrystals dispersed in the much-studied nonlinear polymers poly(diacetylene), poly(diphenylbuadiyne), and poly-4-BCMU create composite media which can be applied by spin coating to form planar waveguides [Olsen and Kafafi, 1991]. Volume fractions of the Au clusters in the 5 to 15% range were observed, with cluster diameters of order 2 nm. Solvent-dependent aging effects on the surface-plasmon resonance amplitude and frequency were observed, as well as an enhancement of a factor of about 200 in nonlinear optical response for Au in poly-4-BCMU compared with the bare nonlinear polymer.

Another interesting technique for producing metal nanoclusters in a waveguide geometry relies on a combination of ion exchange and ion implantation [Garrido et al., 1995]. The method is based on the fact that soda-lime glass immersed in a bath of molten $AgNO_3$ or $CuNO_3$ salts held at the eutectic point (600 to 700°C, depending on the salt) will exchange Na^+ for Ag^+ or Cu^+, leading to the formation of a layer of Ag or Cu ions as deep as several microns in the glass. M-line spectroscopic analysis of an ion-exchanged sample showed that light could be guided in the ion-exchanged region. After ion exchange, He ions at 1.8 MeV energy (range 6.5 μm) at doses up to $1 \cdot 10^{16}$ cm^{-2} were implanted; the same procedure can be used with Ne ions in the implantation steps. Optical absorption spectra taken before

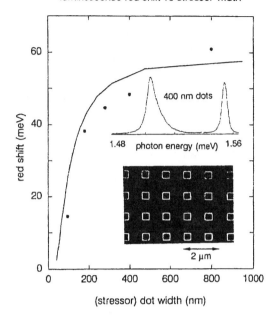

luminescence red shift vs stressor width

Figure 18 Luminescence shift as a function of size for quantum dots localized in GaAs/AlGaAs quantum wells. The localization is produced by the strain modulation from a layer of carbon dot stressors; the quantum well tends to contract near the edges of the stressor and expand under its interior region, confining the excitons by mechanical strain. The inset shows the stressor array generated by electron-beam lithography. (From Kash, K. et al., *J. Vac. Sci. Technol. B,* 10(4), 2030, 1991. With permission.)

and after implantation showed that the surface plasmon resonance for silver formed after the implantation, indicating that the energy deposited by the He ions was sufficient to initiate aggregation of the dispersed Ag or Cu ions.

4.1.4. Lithographic Techniques

The development of mesoscale heterostructure devices for electro-optics and photonics requires the capability for both vertical and lateral structuring of materials. Hence, there has been intense interest in adapting the microlithographic techniques for making extremely well-controlled quantum-dot materials.

One of the earliest attempts to create quantum dots in semiconductors was based on ion implantation [Cibert et al., 1986]. Masks were formed on top of a GaAs quantum well structure to allow patterned deposition of Ga^+ ions; rapid thermal annealing to 900°C resulted in enhanced interdiffusion of Ga and Al in the implanted region. After implantation, the masks were etched off by standard processing techniques, leaving behind an array of wires or dots in the quantum well. The structure was then analyzed using a high-resolution scanning electron microscope to produce spatially resolved cathodoluminescence. Multiple peaks in the cathodoluminescence spectrum confirmed the development of the quantum-dot energy-level structure. Typical box structures were of order 300 nm in cross section, but were somewhat nonuniform because of variations in mask sizes, interdiffusion, and ion-implantation-induced defects.

A significant variation on the standard lithographic techniques, generally identified with the work of Kash and her collaborators [1989], is the use of mechanical strain to create a zero-dimensional confinement in a quantum well. In the original scheme, a compressed stressor layer of InGaAsP was grown on an InP barrier sandwich enclosing an InGaAs quantum well. Arrays of wires and dots were then written by an electron beam onto a 500-nm-thick layer of polymethylmethacrylate, developed and etched by standard procedures. Figure 18 shows an array of carbon stressors arranged on a GaAs/AlGaAs quantum well, with the accompanying red shift of the luminescence peak as a function of stressor (quantum-dot) width [Kash et al., 1991]; the exciton confinement potential was 60 meV.

More recently, Ahopelto and his collaborators [1994] fabricated quantum dots without any lithographic processing whatever, using metallorganic vapor-phase epitaxy (MOVPE) instead of molecular-beam

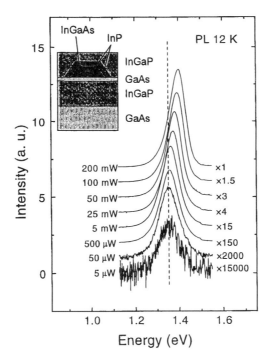

Figure 19   Schematic diagram of self-organized InP islands on a strained InGaAs quantum well, together with photoluminescence spectra for the islands showing the intensity-dependent wavelength shift induced as the Pauli exclusion principle prevents absorption into particular states (band-filling). (From Ahopelto, J. et al., *Appl. Phys. Lett.*, 65(13), 1662, 1994. With permission.)

epitaxy (MBE). This is accomplished by either growing a quantum well on top of self-organized InP islands [Ahopelto et al., 1994] or locally straining an InGaAs quantum well by growing self-organized InP islands on top of the well, similar to Kash's compressed stressor idea [Lipsanen et al., 1995]. In this procedure, the dots are produced *in situ* by three-dimensional growth on a two-dimensional wetting layer when the growth of the strained layer is interrupted at a critical thickness. This creates extremely well-defined quantum dots with narrow size distributions; as shown in Figure 19, the narrow peaks and wavelength shift confirm the existence of the excited states in a quantum dot.

An important recent development has been the discovery that dislocation-free semiconductor quantum dots in the form of strained, self-assembled islands of InAs or AlInAs are formed spontaneously during MBE on a GaAs substrate [Leonard et al., 1993, 1994]. At a critical thickness, the two-dimensional growth pattern characteristic of MBE shows a transition to island formation, producing islands of 150 to 250 Å diameter and 30 to 45 Å thick on top of a thin wetting layer. The spatial densities of the "quantum islands" grown in this way are as high as $4 \cdot 10^{10}$ cm^{-2} [Leon et al., 1995].

So far, lithographic techniques have been used only for relatively large metal clusters, by depositing silver on lithographically etched dielectric posts [Bloemer et al., 1988].

4.1.5. Laser Techniques

Tightly focused laser beams have also been used to create interdiffusion barriers defining a quantum dot in an existing quantum well. Brunner et al. [1992] describe the creation of localized interdiffusion of Ga and Al in a single GaAs/AlGaAs quantum well structure by focusing the beam of an Ar-ion laser (514.5 nm, 5.5 mW) to its diffraction limit (about 500 nm), relying on an external power stabilizer and an autofocus system to maintain a reproducible spot while drawing a dot structure on the quantum well using a computer-controlled x-y translation stage with 10-nm accuracy. This created square dot structures with dimensions ranging from 300 to 1000 nm, as well as a "0-nm" dot when the width of the annealing laser was just equal to half the side of the square traced out by the laser track. The band gap energy of the optical transition was shifted from 1.69 to 1.76 eV over this range; the blue shift only became evident, however, as the dot size shrank below 500 nm, the region of weak confinement.

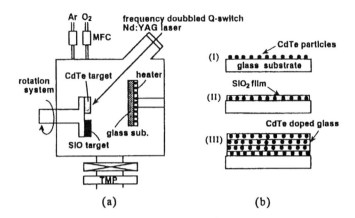

Figure 20 Schematic for the fabrication of multiple layers of CdTe quantum dots in a fused silica matrix by PLD. The atoms and ions in the ablation plume generated by the laser are deposited on the heated substrate. Film growth proceeds by depositing a layer of SiO$_2$, then a layer of CdTe, and then repeating the process; this produces a multilayer film in which the CdTe quantum dots are covered by the silica. (From Ohtsuka, S. et al., *Opt. Mater.*, 2, 209, 1993. With permission.)

Pulsed laser deposition (PLD) combines the advantages of epitaxial or layered vertical growth with the possibility for lateral structuring. Synthesis of layered structures comprising CdTe quantum dots in an SiO$_2$ matrix has been reported [Ohtsuka et al., 1992; 1993], and recently the same group has synthesized a one-dimensional photonic crystal containing a layer of quantum dots [Tsunetomo et al., 1995]. A schematic of the apparatus used to carry out this synthesis is shown in Figure 20; clearly, one can replace the quadrupled Nd:YAG laser in the diagram with other lasers, such as excimers, which are already widely used in the PLD synthesis of oxide films for nonlinear optical applications [Afonso et al., 1995]. Nanocrystals of Ge in an Si matrix have also been synthesized by PLD [Ngiam, et al., 1994]. Thus far, only one attempt to create metal nanoclusters by PLD has been reported [Haglund et al., 1995], although there is no reason in principle why this will not work well for metal-insulator composites.

In recent years, there has been growing interest in using high-flux, low-energy cluster beams to deposit metal and semiconductor nanocrystals directly on surfaces. This has the advantage in principle of separating the process of cluster growth — and with it questions of size distribution — from the process of synthesizing the composite. Using such a scheme, one can imagine synthesizing a composite by first depositing a layer of clusters of well-defined size onto a substrate; depending on the energy of the beam, the temperature of the surface, and the cluster diffusivity on the surface, the clusters might undergo some aggregation at this stage. The composite would then be formed by covering the clusters *in situ* by physical vapor deposition, sputtering, or PLD. High-flux sources which are already being employed in such schemes include the "smoke ion source," which combines thermal vaporization with inert gas condensation to form clusters [McHugh et al., 1989], and the laser-vaporization sources [Milani and de Heer, 1990], either of which can produce clusters with up to 10^5 atoms. The use of these very promising sources to produce cluster-assembled nanocrystalline composites, with particular attention to the synthesis of magnetic materials, has been reviewed recently [Melinon et al., 1995], but applications of the technique to produce nonlinear optical materials have not yet been reported.

4.2. Characterization of Quantum-Dot Composites

The characterization of the electronic, geometric, and physical properties of metal and semiconductor QDCs following fabrication is a major challenge, requiring many techniques and tools.

4.2.1. Physical Characterization Techniques

Ordinary and ultrahigh-resolution TEM have been standard tools for characterizing the size and size distribution of QDCs, but there are significant problems with the dielectric matrix background which scatters the electrons and makes it difficult to pick out the nanocluster in some cases. Moreover, the time required for sample preparation prior to microscopy insures that this technique cannot easily be employed for routine analysis. Nevertheless, Figure 21 shows the level of perfection which can be achieved by current techniques.

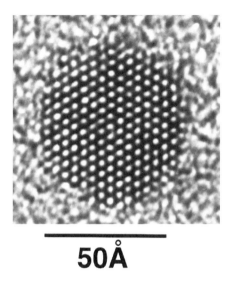

50Å

Figure 21 TEM micrograph of CdS quantum dots prepared by arrested precipitation in reverse micelles, then capped to prevent further nucleation and growth. (From Alivisatos, A. P., *MRS Bull.*, 20(8), 23, 1995. With permission.)

In the case of semiconductor quantum dots, the surface composition and electronic structure are of particular interest, since these are believed to have a strong influence on the relaxation processes which govern the speed of the nonlinear optical response and which also have an influence on photodarkening behavior. Nuclear magnetic resonance (NMR) has proved to be a particularly powerful tool for distinguishing bulk from interior surface sites. For example, Bawendi and co-workers [Becerra et al., 1994] used ^{31}P and ^{77}Se in a magic-angle-spinning experiment to identify the various tri-*n*-octylphosphine (TOP) oxide (O) or selenide (Se) species bound to various sites. Selective chemical processing established that the surface sites are capped 70% by TOP-O and 30% by TOP-Se species which form a close-packed, hydrophobic shell capping virtually all of the Cd sites but leaving the Se sites unpassivated. This may leave the Se sites uncoordinated and free to act as the shallow hole traps seen in photoluminescence experiments.

Studies of the pressure dependence of the crystal structure and other properties make it clear that details of the surface structure and composition in semiconductor quantum dots complicate the linear and nonlinear optical response in a variety of ways [Alivisatos, 1995]. Whether or not the optical properties of metal QDCs are similarly complicated by the details of structure and stoichiometry remains to be seen. In the case of CdSe quantum dots dispersed in a polymer, femtosecond photon-echo studies show that the excitonic response of the CdSe dots depends sensitively on size, temperature, and electron-phonon coupling [Mittleman et al., 1994]. While experiments of this type have not been carried out on metal QDCs, it is likely that both the electronic excitation and relaxation mechanisms will differ dramatically from the semiconductor case.

Optical *absorption spectroscopy* has been the workhorse measurement on metal nanocluster composites, since the location, amplitude, and width of the surface plasmon resonance are an excellent zeroth-order diagnostic of nanocluster species, size, and size distribution. However, the resonant plasmon response decreases in amplitude and broadens with increasing nanocluster size, making absorption spectroscopy a less useful tool for metal QDCs, precisely in the region where the nonlinear effects are presumed to be strongest [Kreibig and Genzel, 1985], as shown in Figure 22 for Cu nanocrystals synthesized by ion implantation. For ellipsoidal particles, absorption is measured as a function of polarization scales with the ellipticity of the particles; this variation on the standard absorption spectroscopy is potentially useful also for nanoparticles in uniaxial crystals such as sapphire. In principle, one should see quantum-size effects in far-infrared absorption of metal quantum dots; however, attempts to observe this effect have been frustrated by experimental problems, notably with interactions between nanoparticles which overwhelm the expected quantum effects of level density by orders of magnitude [Devaty and Sievers, 1985; Curtin and Ashcroft, 1988]. For semiconductor quantum dots, on the other hand, on absorption spectroscopy shows a much more dramatic size dependence, as seen from Figure 23.

Infrared absorption (IRA) and *reflectance* (IRR) *spectroscopy* are helpful for studying the binding arrangements of metal nanoclusters in transparent matrices. For example, IRRS has been used to study

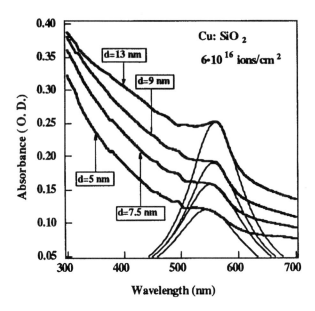

Figure 22 Effects of quantum-dot size on the absorption spectra of Cu QDCs of varying mean diameters fabricated by ion implantation. The smooth curves are fits to the surface plasmon resonance absorption of Equation 5 for each different diameter. Note the virtual disappearance of the plasmon resonance for the 5-nm-diameter nanocrystallites.

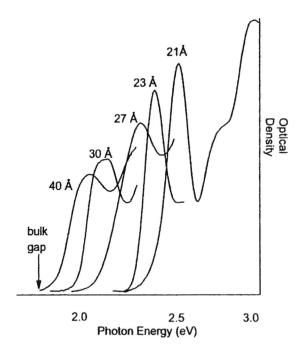

Figure 23 Effects of quantum-dot size on the absorption spectra of CdS dots prepared as in the sample of Figure 21, showing the blue shift of the absorption edge as quantum-dot size is reduced and the band gap energy increases. The progressive sharpening of the absorption features results from the development of discrete electronic states as quantum-dot size decreases. (From Alivisatos, A. P., *MRS Bull.,* 20(8), 23, 1995. With permission.)

the growth of Cu and Pb nanoclusters formed by ion implantation in fused silica [Haglund et al., 1992]. As Cu ion dose increased, the absorption peak corresponding to the Si–O stretch mode at 1111 cm^{-1} red-shifted, indicating a decrease in the Si–O force constant; the amplitude in this mode increased, however, showing that the number of bonds recovered as the nanocrystallites formed. This suggested that the Si–O network was being displaced by the ion implantation process, but was stretching around or accommodating the Cu clusters. The same stretch mode in the case of Pb nanoclusters in silica, however, showed a red shift accompanied by a decrease in amplitude, implying that the number of Si–O bonds decreased with dose, because of the formation of Pb–O bonds in the host matrix. Attempts to observe quantum-size effects in such metal nanocluster composites by far-infrared *absorption* spectroscopy have been unsuccessful, probably due to effects of inhomogeneous broadening at long wavelengths.

Raman spectroscopy is based on three-wave mixing and can reveal Raman-active, particle and optical modes of the irradiated solid. Raman spectroscopy can be used to show changes in local binding occasioned by the implantation and aggregation of the quantum dots. For example, Hosono et al. [1992] used Raman spectroscopy to show that fluorine ions, coimplanted in a region in which Cu quantum dots had been formed by ion implantation in fused silica, were forming Si–F–Si bonds in the host matrix rather than Cu–F bonds at the surface of the Cu nanoclusters. Raman techniques were also employed by Duval et al. [1986] to observe the size of mixed chromium–aluminum spinel microcrystallites in glass; the maximum in very low frequency Raman scattering bands was shown to be inversely proportional to particle size [Duval et al., 1986]. This technique has recently been employed to measure the size of Ag nanoclusters in an ion-exchanged glass and is found to give good agreement with TEM results for particles with diameters in the 1.5-nm range [Ferrari et al., 1995]. In these measurements, light was coupled into the waveguide by prisms, in the same geometry used for *m*-line spectroscopy of the guided-wave modes. The combination of Raman spectroscopy with high-pressure techniques has helped to isolate the effects of surface states and mechanical strain in the growth of semiconductor quantum dots [e.g., Silvestri and Schroeder, 1994; Zhao et al., 1994]; however, the technique has not yet been applied to the characterization of metal QDCs.

5. OPTICAL STUDIES OF QUANTUM-DOT COMPOSITES

5.1. Absorption Measurements

Absorption measurements have been widely used to study size effects through the surface plasmon resonance in metal QDCs, and from red or blue shifts in semiconductor QDCs. As observed by Vollmer and Kreibig [1992], there are some ambiguities in spectral shifts for metal quantum dots, depending both on size and mechanism. In the size region of interest in this chapter, near or below 10 nm diameter, red shifts in the plasmon resonance are caused by the spilling out of conduction-band electrons, while blue shifts are associated with quantum-size effects on the dielectric function of the quantum dot. Since the amplitude of the plasmon resonance decreases while the width increases as a function of size, interpreting these shifts is difficult at best, particularly in view of the effects of inhomogeneous line broadening. Absorption measurements, therefore, must be viewed as useful but hardly definitive measures of the optical properties of QDCs; in particular, they probably do not provide a reliable guide to nonlinear properties of these materials, as they sometimes do for bulk materials.

5.2. Luminescence Measurements

Luminescence measurements offer an extremely versatile suite of techniques for probing the details of optical response in semiconductor QDCs. Bawendi et al. [1992], for example, combined time-, temperature-, wavelength-, and polarization-resolved luminescence to understand the details of optical transitions in 3.2-nm-diameter CdSe dots prepared by an inverse-micellular technique and having less than 8% standard deviation in mean size. Measurements at 10 K showed two components in the luminescence, one deriving from a radiationless hole transition operating on a microsecond time scale, the other a fast component apparently derived from strong resonant mixing between an (interior) exciton-like state and an (exterior) surface state localized on Se lone-pair orbitals. There should, in principle, be no analog of luminescence measurements for metal QDCs; however, luminescent states may well exist in electronic defects in the composite materials.

5.3. Third-Order Susceptibility Measurements

There is a voluminous literature on measurements of the third-order susceptibility and of its real and imaginary parts, the nonlinear index of refraction and the nonlinear absorption or saturation, particularly for semiconductor QDCs. Since many of these measurements are described elsewhere in this volume, the following paragraphs concentrate on metal quantum-dot materials.

The earliest measurements of nonlinear optical response in a metal QDC were those by the Flytzanis group [e.g., Hache et al., 1988a]. They showed that in metal QDCs — ruby-gold filter glasses — the nonlinear optical response at 532 nm was primarily absorptive and dominated by the interband transition.

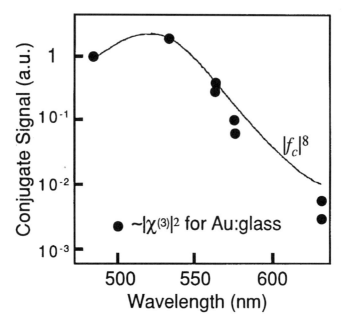

Figure 24 Phase-conjugate intensity measured by four-wave-mixing study in an Au QDC (ruby-gold filter glass) for frequencies near that of the surface-plasmon resonance at 533 nm, showing the enhancement as the eighth power of the field enhancement factor f_c. (From Flytzanis, C. et al., *Prog. Opt.*, 29, 321, 1991. With permission.)

Measurements of the phase-conjugate reflectivity as a function of wavelength at several points near the surface-plasmon resonance confirmed that the nonlinear component of the reflectivity did indeed exhibit the expected variation with $|f_c(\omega)|^8$. (See Equation 9 and Figure 24.) This fact by itself confirmed that, at this wavelength, interband transitions dominated the nonlinear response of the Au-doped glass. Moreover, it was found that the hot-electron contribution to $\chi_{qd}^{(3)}$ was dominant when the DFWM was carried out in the copolarized beam configuration, with a magnitude of approximately $\Im m[\chi_{he}^{(3)}] = -1.1 \cdot 10^{-7}$ ESU. When the experiment was repeated in the cross-polarized geometry, there was no sign of a size-dependent intraband transition over a range of Au particle sizes from 2.8 to 30 nm; only the interband contribution was measurable, having a magnitude of $\Im m[\chi_{inter}^{(3)}] = -1.7 \cdot 10^{-8}$ ESU.

Measurements of the phase-conjugate reflectivity by other groups [e.g., Bloemer et al., 1990] likewise showed no observable quantum-size effect in bare Au nanoclusters or QDCs at the wavelengths near the surface plasmon resonance. For Au quantum dots in glass, attempts to observe this effect with mode-locked, Q-switched lasers operating near 532 nm were almost certainly defeated by the fact that the intraband transition is so much weaker than the interband and hot-electron contributions to $\chi^{(3)}$. Indeed, using plausible values for the phenomenological theory whose results are quoted in Equations 14 through 16, Hache et al. [1988] calculated a value for the intraband contribution in Au filter glass of $\Im m[\chi_{intra}^{(3)}] = -10^{-10}$ ESU.

The utility of the Z-scan for rapid characterization of changes in material performance is illustrated by a measurement [Magruder et al., 1994b] which compares the nonlinear absorption and nonlinear saturation measurements in metal-alloy QDCs created by ion implantation of Cu and Ag ions. In this material preparation, a total of $1.2 \cdot 10^{17}$ ions $\cdot$ cm^{-2} were implanted into a pure fused silica matrix, and the samples were then annealed to induce nucleation of nanocrystallites. Whole-beam Z-scan measurements of three different samples with varying Ag::Cu ratios were taken (3Ag::9Cu, 6Ag::6Cu, and 9Ag::3Cu). In the Z-scan measurements, the 3Ag::9Cu sample showed nonlinear absorption, while the other two samples exhibited moderate nonlinear saturation, as shown in Figure 25. The nonlinear index was in the range of $(1.5 \pm 0.5) \cdot 10^{-9}$ cm$^2 \cdot$ W^{-1} for all three samples. However, increasing the ratio of Ag to Cu appeared to increase the nonlinear index slightly while at the same time there was a modest decrease in absorption at high intensities. As will become apparent in the last section of this chapter, this increases the performance figure of merit for the alloyed material significantly.

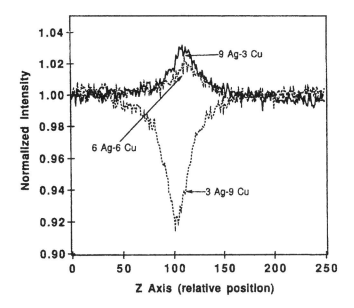

Figure 25   Nonlinear absorption and saturation in Cu:Ag QDCs for three different ratios of Ag to Cu concentration, measured by the Z-scan technique. Saturation is indicated by a rise in transmission near the focus of the Z-scan lens, which corresponds to the highest intensity. Nonlinear absorption, in contrast, is indicated by dip in transmission at the highest intensities. The total (Cu+Ag) ion dose in each case is $12 \cdot 10^{17}$ ions·cm^{-2}, with relative fractions x and y of Ag and Cu indicated by the label (xAg·yCu) for each transmission spectrum. (From Magruder, Osborne and Zuhr, 1994b; used by permission.)

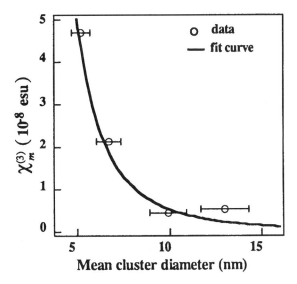

Figure 26   Magnitude of the third-order susceptibility $\chi^{(3)}$ as a function of quantum-dot size, for Cu nanocrystals in fused silica, using the same samples. The measurement of $\chi^{(3)}$ was made by forward DFWM in a cross-polarized geometry to eliminate thermal effects. The smooth curve is a normalized inverse-cube function.

5.4. Quantum-Size Effect

Measurements of quantum-size effects have been made in a number of different ways, including spectral hole burning, photon-echo and absorption measurements on semiconductor QDCs, and DFWM measurements of the size-dependent term in the third-order susceptibility for copper QDCs.

In metal nanocluster composites, the quantum-confined optical transition is the intraband transition which electrons from occupied conduction bands near the Fermi edge are excited to unoccupied states higher up in the conduction band. According to Equation 13, the third-order susceptibility for these transitions should increase with the inverse third power of the dot radius below a critical value which is of the order of 11 to 14 nm for the noble metals. In the case of Cu, the intraband and surface-plasmon resonances are close to 565 nm, so that one could hope to see the quantum-size effect with a picosecond laser operating at 532 nm. Using a set of four Cu:silica nanocomposites having mean diameters of 5, 7.5, 10, and 13 nm [Magruder et al., 1994a], the purely electronic component of $\chi^{(3)}$ was measured using a mode-locked, Q-switched frequency-doubled Nd:YAG laser (532 nm) with a pulse duration of 30 ps [Yang et al., 1994]. The third-order susceptibility was measured by forward, phase-matched DFWM, in which the *xyxy* component of $\chi^{(3)}$ was measured. (The subscripts on the electric field vectors and susceptibilities indicate, in this case, the polarization direction of the electric field.) For cross-polarized pumps, it is possible to avoid the formation of a thermal population grating in the sample, eliminating the hot-electron contribution and isolating the intraband, electronic transition. The results are shown in Figure 26 together with a fit using a cubic curve. Even though the standard deviations of the mean radius distributions are relatively large, it is clear that the data fit the inverse cube dependence on radius or diameter rather well. Recent reports of third-order nonlinearities of Cu and Ag nanoclusters embedded in melt glass have confirmed the size dependence expected for the absorptive nonlinearity near the surface plasmon resonance [Uchida et al., 1994].

5.5. Energy and Coherence Lifetimes

Among the important questions to be answered about the performance of quantum-dot materials are the lifetimes of the electronically excited states which lead to nonlinear effects. The energy lifetime, usually denoted T_1, is the time it takes the excited state to return to the ground state; this can be in the picosecond or even femtosecond range for nonresonant excitations, but on the nanosecond or longer time scale for resonant processes. Both radiative (R) and nonradiative (NR) processes contribute to this lifetime, according to the prescription $1/T_1 = 1/\tau_R + 1/\tau_{NR}$. The coherence, or dephasing, time, conventionally referred to as T_2, is the time it takes for the electron to lose phase memory of the excitation.

Not a great deal is known about these lifetimes. Heilweil and Hochstrasser [1985] measured the energy and dephasing lifetimes in Au nanoclusters in solution with mean diameters of $d = 20 \pm 1.5$ nm, and found, by measuring the dispersion of $\chi^{(3)}$, a plasmon relaxation time of approximately $\tau_T \approx 8 \cdot 10^{-15}$ s. Using several theoretical models for damping mechanisms in metal clusters, the radiative lifetime was calculated to lie in a range $\tau_R \approx 7 \cdot 10^{-14}$ to $8 \cdot 10^{-13}$ s. So far, there are no comparable four-wave-mixing measurements in metal nanocluster composites. The measurements of phase-conjugate signal as a function of pump-probe delay, from the quantum-size effect measurements by Yang et al. [1994], would seem to indicate a dephasing time less than 30 ps for electronic processes, and times on the order of 100 ps for the thermally dominated relaxation processes initiated by copolarized pump beams. Differential transmission measurements with 80 fs pump and 10 fs probe pulses on a Cu-nanoparticle composite by Bigot et al. [1995] show relaxation times of order 1 ps. Near the surface plasmon resonance these relaxation times increase, suggesting the possibility of long-lived electron-electron correlations in the 10-nm-diameter quantum dots.

The dephasing time in Cu nanocluster composites has been measured by Yang et al. [1996], using an indirect spectroscopic technique. If the metal quantum dot can be regarded as a two-level system, measurement of a resonant absorption process gives the width in frequency space of the excitation, which can be converted into a coherence lifetime if one knows the mechanism of deexcitation. For small nanoclusters, the dephasing time is dominated by electron collisions with the walls, and the dephasing time is related to the nonlinear susceptibility by

$$\beta \propto \Im m\left[\chi_{res}^{(3)}(\omega, -\omega, \omega; \omega)\right] \propto \frac{1}{\hbar\left(\omega_{fg} - 2\omega\right) - i\hbar\Gamma_{coh}} \tag{21}$$

Fitting the resonance in two-photon absorption in the neighborhood of the interband resonance using the energy difference ω_{fg} and the coherence width $\Gamma_{coh} = 1/T_2$ as adjustable parameters yields values for T_2 ranging from 80 fs for the large clusters ($d = 13$ nm) to 60 fs for smaller clusters ($d = 5$ nm); these times compare reasonably with the values of 120 fs measured for Γ_{coh} in thin Cu films. The measured

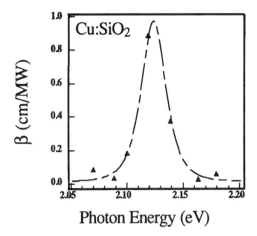

Figure 27 Measurement of the two-photon interband absorption resonance in a Cu:silica QDC using a tunable dye laser (pulse duration, 6 ps). The width of the resonance is proportional to the dephasing time, about 80 fs. The triangles are measured data; the dashed curve is a fit to the Lorentzian function of Equation 21.

data for the two-photon interband absorption in Cu quantum dots with an average diameter of 13 nm is shown in Figure 27, together with a curve which represents the fit to the data using Equation 21.

6. QUANTUM-DOT COMPOSITES IN PHOTONICS

Reduced-dimensional structures have a potentially significant role to play in photonic and electro-optic devices of the future, ranging from light sources to switches, modulators, and couplers. The advantages of devices based on QDC materials can already be guessed to some extent from the spectacular successes of quantum well materials in recent years: the capability for band gap engineering in these structures permits wavelength tuning, while their small size alters the electronic structure of the dots so as to guarantee greater pumping efficiency and lower overall thresholds for light emission and switching. For example, a recent report on the structural and optical properties of self-assembled InGaAs quantum dots indicates that the zero-dimensional characteristics of the quantum dot confer on it an enhanced photoluminescence efficiency compared with that of a standard quantum well [Leonard et al., 1994] — a development which may make possible an efficient quantum-dot laser. Indeed, optical gain in a sol-gel-synthesized CdS quantum-dot-loaded glass has recently been observed [Butty et al., 1995], albeit with a strong decrease in measured gain with rising temperature.

While it can be demonstrated theoretically that quantum-dot materials *should* improve on the desirable attributes of quantum wells, it will be necessary to greatly improve both the quality of the materials and our understanding of their performance characteristics if these advantages are to be realized. In this section, we consider some of the critical performance parameters which must be met and show how development of suitable materials could have a favorable impact on the behavior of QDCs in standard optoelectronic or photonic devices.

Two of the potential advantages of QDCs as photonic materials are substantial reductions in the speed and energy required to switch signals at the 100 GHz repetition frequencies expected in communications and computing systems of the 21st century. Figure 28 compares in graphical form the switching speeds and switching energies of a number of electronic and optical materials and devices [adapted from Saleh and Teich, 1991]. Within the broad range of parameters covered by "semiconductor microelectronics," current metal-oxide-semiconductor field-effect transistor (MOSFET) devices made in Si have low switching energies, but switching times in the nanosecond range, while GaAs FET devices also have low switching energy but, because of higher carrier mobilities and the ballistic mode of operation, also more-rapid switching times. Photonic devices based on multiple quantum well (MQW) structures — SEED and GaAs MQW devices and Fabry–Perot (FP) cavities based on ferroelectrics such as lithium niobate — have extremely low switching energies, but are relatively slow because they are based on charge-transfer or resonant relaxation processes. These speeds can be increased — for example, by using ion implantation to create traps in MQW structures — but this increases the complexity of the material fabrication process.

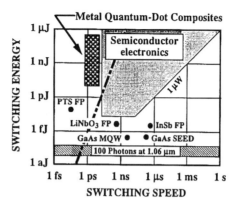

Figure 28 Plot of various photonic materials and devices showing their switching energies and switching speeds. The dashed line indicates the boundary between materials with acceptable and unacceptable thermal characteristics for high-frequency operation.

Metal QDCs fit into the shaded area to the right of current semiconductor electronics: they have quite rapid switching times, as low as picoseconds for metal QDCs, but present experiments show that they have relatively high switching energies, in the picojoule range. These can be lowered by taking advantage of some of the effects known for enhancing the third-order nonlinearity, but the expected enhanced sensitivities have yet to be demonstrated in working geometries. In addition, the problem of thermal loading must be faced: because transparent optical materials often have rather poor heat-transfer characteristics, the composite materials cannot operate at the same speed as the switching occurs in the quantum dot.

In addition to satisfying the criteria for rapid switching at high speeds, it will be necessary to develop practical device architectures based on various nonlinear optical effects. A number of different third-order nonlinear optical devices have been reviewed by Stegeman et al. [1988], starting from the premise that virtually every linear electro-optic coupling, modulation, or switching device can be reconfigured for all-optical operation by insertion of the appropriate nonlinear optical materials in waveguide geometries. Schematic representations of some of these devices and of their input/output characteristics are shown in Figure 29, including couplers, nonlinear reflectors, and interferometers. QDCs would appear to fit neatly into the fabrication of all of these device geometries, though the application would be particularly obvious in a Fabry–Perot resonator containing a nonlinear component, or in a Mach–Zehnder interferometer with a nonlinear material in one arm. Moreover, the third-order optical nonlinearity has intrinsic switching properties in addition to whatever additional characteristics are conferred on it through incorporation in a device geometry with optical feedback: optical bistability, for example, has been demonstrated experimentally and accounted for theoretically in QDCs [Haus et al., 1989].

One way of assessing the performance of a nonlinear optical material in a convenient but quantitative shorthand is by the use of figures of merit. Unfortunately, no single figure of merit seems to be sufficient to describe all of the desirable performance characteristics of nonlinear materials. Indeed, there are figures of merit which relate to fundamental properties, to the functional characteristics of specific nonlinear optical devices, and to general device characteristics, such as total signal throughput and thermal loading.

The first and most obvious figures of merit have to do with the magnitude of the third-order nonlinearity and its relationship to the linear optical properties of the material. In the case of purely dispersive or absorptive nonlinearities, Flytzanis et al. [1991] proposed the adoption of dispersive and absorptive figures of merit which can be used off- or on-resonance, respectively:

$$f_{disp} = \frac{\omega \cdot n_2}{n_o} \qquad f_{abs} = \frac{\beta}{\alpha \cdot \tau} \tag{22}$$

where the symbols have the meanings defined earlier. The relaxation time τ in the expression for f_{abs} is the energy relaxation time of the resonant or nonresonant process activated. Clearly, this will tend to be long in a resonant process and shorter for nonresonant processes. In these figures of merit, large values

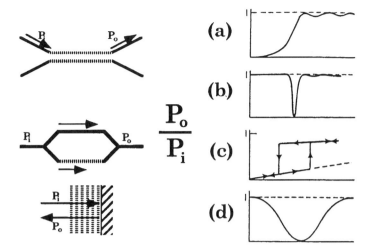

Figure 29 Schematic representation of third-order nonlinear optical devices which make use of or incorporate QDCs: (a) directional coupler (half-wave beat length); (b) directional coupler (full-wave beat length); (c) Bragg reflector; (d) Mach–Zehnder interferometer. The response of each device to an optical pulse is shown on the right of the device. Nonlinear material is shown by vertical/horizontal hatching. (Adapted from Stegeman and Stolen. 1989. With permission.)

indicate greater potential utility as optical switching materials: for dispersive nonlinearities, the figure of merit is larger for higher frequencies and larger nonlinear index, while for absorptive nonlinearities, the figure of merit is inversely proportional to linear absorption and to the relaxation time.

Another figure of merit arises through the limitations inherent in running devices at the intensities needed for nonlinear optical operation. For example, one can compute the figure-of-merit limits imposed by two-photon absorption (*TPA*) on any all-optical switching scheme [Mizrahi et al., 1989]. This figure of merit is defined by

Table 1 Comparison of nonlinear optical properties in various materials

Material system	n_2 (cm$^2 \cdot$W^{-1})	Δn_{sat}	α (cm^{-1})	β (cm$\cdot$GW^{-1})	τ (ps)	W	L_c (m)
GaAs — nonresonant	-10^{-4}	0.1	10^4	—	10^4	0.9	—
GaAs — resonant	-10^{-8}	$2 \cdot 10^{-3}$	30	40	10^4	10	—
Cd$_x$S$_{1-x}$-doped glass	-10^{-10}	$5 \cdot 10^{-5}$	3	10	10	0.3	—
SiO$_2$— nonresonant	10^{-16}	$<10^{-6}$	10^{-5}	$5 \cdot 10^{-2}$	10^{-2}	$>10^3$	—
PTS — resonant	$2 \cdot 10^{-11}$	0.1	10^5	—	2	—	—
Cu implanted in silica	$3 \cdot 10^{-10}$	$\sim 10^{-5}$	10^3–10^4	$<10^{-2}$	<5	0.1–2	$5 \cdot 10^{-6}$

$$f_{TPA} = \frac{\lambda \cdot n_2}{\beta} \qquad (23)$$

and expresses the potential for the nonlinear material to induce a phase shift without dropping below the critical intensity needed to keep an all-optical circuit operating. Indeed, proper accounting for the effects of two-photon absorption may make it possible or even necessary to favor one material with optimal two-photon absorption over another material with larger values of the third-order susceptibility [Aitchison et al., 1990]. For this reason, it is also necessary to consider the role of optically active defects which may mimic the behavior of two-photon absorbers [DeLong et al., 1990].

Other figures of merit relate to the functional characteristics of specific devices. Stegeman and Stolen [1989] have developed a figure of merit to compare various nonlinear materials for their suitability in the devices shown in Figure 29. For example, in a nonlinear directional coupler or Mach–Zehnder interferometer, complete switching requires that the condition $w = \Delta\beta_{sat}L/\lambda < 2$ be satisfied, where L is the interaction length, λ is the wavelength, and $\Delta\beta_{sat}$ is the change in the guided-wave index at saturation, which can often be reasonably assumed to equal the saturation in the material index of refraction, Δn_{sat}. However, high throughput also requires that $L\alpha < 1$, where α is the linear absorption coefficient. Combining these two conditions yields a dimensionless figure of merit for the directional coupler:

$$W = \frac{\Delta\beta_{sat} \cdot L}{\lambda} \times \frac{1}{L\alpha} = \frac{\Delta n_{sat}}{\lambda\alpha} \tag{24}$$

A comparison of semiconductor and metal QDCs in glass with a fused silica fiber shows that, by this figure of merit, optical fibers appear to be the "best" nonlinear material. However, metal QDCs and semiconductor QDCs in glasses appear to have comparable values of this particular figure of merit, as shown in Table 1.

A general performance criterion for QDCs leads to the introduction of another figure of merit, this one having to do with the thermal loading induced in nonlinear materials at very high pulse repetition frequencies. In all-optical switching schemes, one needs to have a phase change of π over some reasonable length (perhaps 1 m for fiber devices, less than 1 cm for waveguide integrated-optical devices). For a device operating on the nonlinear index, this phase change must satisfy $\phi = 2\pi n_2 I_o L/\lambda$. On the other hand, at high speeds, the thermal loading produced by absorption of a single pulse of duration at the intensity is $\Delta Q = \alpha L I_o a\tau$, where α is the absorption coefficient and $V = aL$ is the volume of the waveguide in which the pulse is absorbed. As pointed out originally by Friberg and Smith [1987], one can then construct from these parameters a simple figure of merit which defines the performance of a given material at high pulse-repetition frequencies where thermal loading becomes critical:

$$f_{thermal} = \frac{\text{Index change required for switching}}{\text{Thermal index change}} = \frac{n_2 C_p \rho}{\alpha\tau(dn/dT)} \tag{25}$$

where C_p is the heat capacity at constant pressure, ρ is the density, τ is the relaxation time, and dn/dT is the thermo-optic coefficient. High-quality nonlinear opticals must then satisfy the inequality $f_{thermal} \gg f_{device} \cdot \tau_{th}$, where τ_{th} is the thermal relaxation time for the material. Indeed, the calculation of this figure of merit produces the dashed line shown in Figure 28, to the right of which any operating device must lie if it is to meet the requirements for sufficiently low thermal loading.

The use of nonlinear optical measurements to assess the thermal figure of merit is illustrated in Figure 30, which shows the Z-scan for a Cu:silica QDC into which fluorine has been implanted subsequent to the formation of the Cu nanoclusters. The fluorine, incorporated into the glass host in Si–F–Si bonds, has the effect of reducing the thermal conductivity of the sample by approximately 40%. In taking Z-scans as a function of frequency, one finds that for laser pulse-repetition frequencies in excess of approximately 5 MHz, the thermal component of the nonlinear index shows a dramatic increase, while below that pulse-repetition frequency, the electronic component of the nonlinear index is dominant.

For device applications, semiconductor quantum dots have some significant advantages and disadvantages as material building blocks. Chief among the advantages at present is the fact that fabrication and characterization techniques and the general knowledge base are much more advanced in semiconductor QDCs, which are now benefiting from decades of development of electronic applications. In addition, at least the sol-gel technique is capable of producing spatially dense samples of quantum dots [Kao et al., 1994]. Among the disadvantages are the problems with uniform size distribution and the attendant inhomogeneous broadening; the relatively small spatial densities attainable for quantum dots in melt glasses and lithographic preparations; unwanted photochemical interactions, such as photodarkening, in some cases; and the slow response time resulting from the near-resonant transitions which one would like to exploit because of their high oscillator strength. This last problem can be overcome, however, by running slightly off resonance, as proposed by Cotter et al. [1992].

Metal quantum dots, in contrast, do not suffer from the problem of inhomogeneous broadening due to variations in size distribution to the extent of the semiconductor dots, both because they can be fabricated by non-equilibrium techniques and because inhomogeneous broadening does not affect their optical response in the same way as it does the spectra of the semiconductor quantum dots. However, fabrication techniques are less sophisticated for these materials compared with semiconductors; in particular, the questions of lateral and vertical structuring have hardly been addressed at all. Moreover, the growth process for the metal nanoclusters is poorly understood and characterized at this time, and

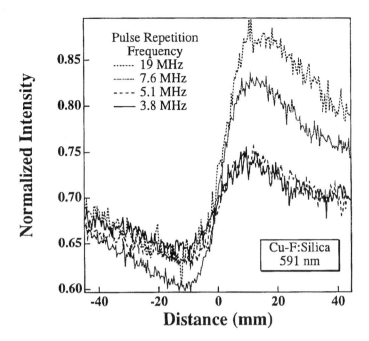

Figure 30 Z-scan spectra for a Cu:silica QDC, taken as a function of pulse repetition frequency for a 6-ps, mode-locked dye laser at 591 nm. The sample was implanted with F ions after the quantum dots were formed to degrade thermal conductivity. Increased thermal nonlinear refraction is signaled by the growing peak-to-valley asymmetry of the transmission function as pulse-repetition frequency increases.

such options as lithographic fabrication or self-organization are essentially unstudied. Finally, understanding of the physics of excitation and relaxation processes in these materials is at a relatively primitive stage.

Electro-optic and magneto-optic effects are also significantly less well studied in metal QDCs than in semiconductor materials. Most of the studies of magnetic QDCs, for example, have been directed at making composites which will exhibit giant magnetoresistive effects; because of the high density of metal in these materials, they are optically opaque and therefore unusable in electro-optical or magneto-optical applications.

7. SUMMARY AND CONCLUSIONS

In summary, it appears that QDCs have a significant future in photonic, electro-optic, and possibly in magneto-optic applications, but much work needs to be done on the materials to insure their compatibility with device requirements. Lateral and vertical structuring techniques are better developed for semiconductor quantum dots, but these have relatively slow response times and, under some circumstances, suffer from photodarkening and other maladies. On the other hand, the metal QDCs have not yet been prepared in ways which give confidence in control over the dimensions and spatial arrangement of the dots, even though it appears that most of these problems could be overcome in principle. In all cases, as has been observed by many researchers, the linear and nonlinear optical characteristics of the material are only part of the constraints; it is also necessary that the materials have excellent chemical, electrical, and mechanical properties, be readily processable, and exhibit long-term stability and immunity to long-term damage from the laser signals. The determination that these constraints are satisfied by QDCs awaits the results of implementation in waveguide and other device geometries.

While this improvement in our ability to control material configuration is absolutely necessary from the technological point of view, it is also probably a critical evolutionary step from the scientific perspective, for the ability to interact with nonlinear materials over long path lengths in waveguide geometries confers immense advantages in basic studies of materials properties. It is thus to be anticipated that continuing progress both in the synthesis of new materials and in their fabrication into device geometries will greatly enhance our understanding of three-dimensional electron confinement in both metals and semiconductors

— much as the fabrication of quantum wells touched off major advances in our understanding of semi-conductor properties and electron dynamics in one-dimensionally confined materials.

ACKNOWLEDGMENTS

It is a pleasure to acknowledge helpful conversations with Stephan Koch, Robert Magruder, Paolo Mazzoldi, Ilias Perakis, Roland Sauerbrey, and Frank Träger in the preparation of this review. The opportunity for learning more about semiconductor quantum dots and the common technological problems shared with metal QDCs was made available by the *Graduierten-Kolleg* of the Philipps-Universität, Marburg, through an invitation to present a series of lectures on "Mesoscopic Structures in Optoelectronics." Research on QDCs at Vanderbilt is supported by the Army Research Office under Grant DAAH-0493-G-0123.

REFERENCES

The literature on quantum-dot materials has grown nonlinearly, although not quite exponentially, in the last few years; the INSPEC data base lists some 15 abstract titles under "quantum dots" in 1990, over 180 in 1994, and nearly 130 in the first 6 months of 1995. Many of these refer primarily to electronic and magnetic materials and applications; however, the number of references to optical properties 4 quantum dots is growing proportionately. References cited here are restricted for the most part to the last 10 years; literature was surveyed through 1995.

Adair, R., Chase, L. L. and Payne, S. A., 1987. Nonlinear refractive-index measurements of glasses using three-wave frequency mixing, *J. Opt. Soc. Am. B* 4 (6), 875–881.

Alfonso, C. N., Gonzalo, J., Vega, F., Dieguez, E., Cheang Wong, J. C., Ortega, C., Siejka, J. and Amsel, G., 1995. Correlation between optical properties, composition and deposition parameters in pulsed laser deposited LiN60$_3$ films, *Appl. Phys. Lett.* 65(12), 1452–1454.

Ahopelto, J., Lipsanen, H., Sopanen, M., Koljonen, T. and Niemi, H. E.-M., 1994. Selective growth of InGaAs on nanoscale InP islands, *Appl. Phys. Lett.* 65 (13), 1662–1664.

Aitchison, J. S., Oliver, M. K., Kapon, E., Colas, E. and Smith, P. W. E., 1990. Role of two-photon absorption in ultrafast semiconductor optical switching devices, *Appl. Phys. Lett.* 56 (14), 1305–1307.

Alivisatos, A. P., 1995. Semiconductor Nanocrystals, *MRS Bull.* 20 (8), 23–32.

Banyai, L., Hu, Y. Z., Lindberg, M. and Koch, S. W., 1988. Third-order optical nonlinearities in semiconductor micro-structures, *Phys. Rev. B* 38 (12), 8142–8152.

Bawendi, M. G., Steigerwald, M. L. and Brus, L. E., 1990. The quantum mechanics of larger semiconductor clusters (Quantum Dots), *Annu. Rev. Mater. Sci.* 41, 477–496.

Bawendi, M. G., Carroll, P. J., Wilson, W. L. and Brus, L. E., 1992. Luminescence properties of CdSe quantum crystallites: resonance between interior and surface localized states, *J. Chem. Phys.* 96 (2), 946–954.

Becerra, L. R., Murray, C. B., Griffin, R. G. and Bawendi, M. G., 1994. Investigation of the surface morphology of capped CdSe nanocrystallites by ^{31}P nuclear magnetic resonance, *J. Chem. Phys.* 100 (4), 3297–3300.

Bertoncello, R., Trivillin, F., Cattaruzza, E., Mazzoldi, P., Arnold, G. W., Battaglin, G. and Catalano, M., 1995. *J. Appl. Phys.* 77, 1294–1301.

Bigot, J-V., Merle, J. C., Cregut, O. and Daunois, A., Electron dynamics in copper metallic nanoparticles probed with femtosecond optical pulses.

Bloemer, M. J., Ferrell, T. L., Buncick, M. C. and Warmack, R. J., 1988. Optical properties of submicrometer-size silver needles, *Phys. Rev. B* 37 (14), 8015–8021.

Bloemer, M. J., Haus, J. W. and Ashley, P. R., 1990. Degenerate four-wave mixing in colloidal gold as a function of particle size, *J. Opt. Soc. Am. B* 7 (5), 790–795.

Borelli, N. F., Hall, D. W., Holland, H. J. and Smith, D. W., 1987. Quantum confinement effects of semiconducting microcrystallites in glass, *J. Appl. Phys.* 61 (12), 5399–5407.

Borelli, N. F., Aitken, B. G. and Newhouse, M. A., 1995. Resonant and non-resonant effects in photonic glasses, *J. Non-Cryst. Solids* 185, 109–122.

Boyd, R. W. and Sipe, J. E., 1994. Nonlinear optical susceptibilities of layered composite materials, *J. Opt. Soc. Am. B* 11 (2), 297–303.

Broglia, R., 1994. What color is an atom?, *Contemp. Phys.* 35 (2), 95–104.

Brunner, K., Bockelmann, U., Abstreiter, G., Walther, M., Böhm, G., Tränkle, G. and Weimann, G., 1992. Photoluminescence from a single GaAs/AlGaAs quantum dot, *Phys. Rev. Lett.* 69 (22), 3216–3219.

Brus, L. E., 1984. Electron-electron and electron-hole interactions in small semiconductor crystallites: the size dependence of the lowest excited electronic state, *J. Chem. Phys.* 80 (9), 4403–4409.

Buchal, C., Withrow, S. P., White, C. W. and Poker, D. B., 1994. Ion implantation of optical materials, *Annu. Rev. Mater. Sci.* 24, 125–157.

Burnam, K. J., Carpenter, J. P., Lukehart, C. M., Milne, S. B., Stock, S. R., Jones, B. D., Glosser, R. and Wittig, J. E., 1995. Nanocomposites containing nanoclusters of Ag, Cu, Pt, Os, Co_3C, Fe_2P, Ni_2P, Ge, or Pt/Sn, *NanoStruct. Mater.* 5 (2), 155–169.

Butty, J., Hu, Y. Z., Peyghambarian, N., Kao, Y. H. and Mackenzie, J. D., 1995. Quasicontinuous gain in sol-gel derived CdS quantum dots, *Appl. Phys. Lett.* 67 (18), 2672–2674.

Carpenter, J. P., Lukehart, C. M., Stock, S. R. and Wittig, J. E., 1994. Formation of a nanocomposite containing particles of Co_3C from a single-source precursor bound to a silica xerogel host matrix, *Chem. Mater.* 7, 201–205.

Carter, G. M., 1987. Excited-state dynamics and temporally resolved nonresonant nonlinear-optical processes in polydiacetylenes, *J. Opt. Soc. Am. B* 4 (6), 1018–1024.

Cibert, J., Petroff, P. M., Dolan, G. J., Pearton, S. J., Gossard, A. C. and English, J. H., 1986. Optically detected carrier confinement to one and zero dimension in GaAs quantum wells, wires and boxes, *Appl. Phys. Lett.* 49 (18), 1275–1277.

Cotter, D., Girdlestone, H. P. and Moulding, K., 1991. Size-dependent electroabsorptive properties of semiconductor microcrystallites in glass, *Appl. Phys. Lett.* 58 (14), 1455–1457.

Cotter, D., Burt, M. G. and Manning, R. J., 1992. Below-band-gap third-order optical nonlinearity of nanometer-size semiconductor crystallites, *Phys. Rev. Lett.* 68 (8), 1200–1203.

Curtin, W. A. and Ashcroft, N. W., 1988. Theory of far-infrared absorption in small-metal-particle–insulator composites, *Phys. Rev. B* 31, 3287–3295.

Dameron, C. T., Reese, R. N., Mehra, R. K., Kortan, A. R., Carroll, P. J., Steigerwald, M. L., Brus, L. E. and Winge, D. R., 1989. Biosynthesis of cadmium sulphide quantum semiconductor crystallites, *Nature* 338 (13 April), 596–597.

Danek, M., Jensen, K. F., Murray, C. B. and Bawendi, M. G., 1994. Preparation of II-VI quantum-dot composites by electrospray organometallic chemical vapor deposition, *J. Cryst. Growth* 145 (1–4), 714–720.

De, G., Licciulli, A., Massaro, C., Tapfer, L., Catalano, M., Battaglin, G., Meneghini, C. and Mazzoldi, P., 1995. Silver nanocrystals in silica by sol-gel processing, *J. Non-Cryst. Solids* in press.

DeLong, K. W., Mizrahi, V., Stegeman, G. I., Saifi, M. A. and Andrejco, M. J., 1990. Role of color center induced absorption in all-optical switching, *Appl. Phys. Lett.* 56 (15), 1394–1396.

Devaty, R. P. and Sievers, A. J., 1985a. Possibility of observing quantum size effects in the electromagnetic absorption spectrum of small metal particles, *Phys. Rev. B* 31 (4), 1951–1954.

Duval, E., Boukenter, A. and Champagnon, B., 1986. Vibration eigenmodes and size of microcrystallites in glass: observation by very-low-frequency raman scattering, *Phys. Rev. Lett.* 56, 2052–2055.

Efros, A. L. and Efros, A. L., 1982. Interband absorption of light in a semiconducting sphere, *Sov. Phys. Semicond.* 16, 773.

Ekimov, A. I., Hache, F., Schanne-Klein, M. C., Ricard, D., Flytzanis, C., Kudryavtsev, I. A., Yazeva, T. V., Rodina, A. V. and Efros, Al. L., 1993. Absorption and intensity-dependent photoluminescence measurements on CdSe quantum dots: assignment of the first electronic transitions, *J. Opt. Soc. Am. B* 10 (1), 100–107.

Faraday, M., 1857. The Bakerian Lecture: experimental relations of gold (and other metals) to light, *Phil. Trans. R. Soc.* 147, 145–181.

Ferrari, M., Gonella, F., Montagna, M. and Tosello, C., 1995. Detection and size determination of Ag nanoclusters in ion-exchanged soda-lime glasses by waveguided Raman spectroscopy, submitted to *Phys. Rev. B*.

Fisher, G. L., Boyd, R. W., Gehr, R. J., Jenekhe, S. A., Osaheni, J. A., Sipe, J. E. and Weller-Brophy, L. A., 1995. Enhanced nonlinear optical response of a composite material, *Phys. Rev. Lett.* 74, 1871–1874.

Flytzanis, C., Hache, F., Klein, M. C., Ricard, D. and Roussignol, Ph., 1991. Nonlinear optics in composite materials, *Prog. Opt.* 29, 321–411.

Friberg, S. R. and Smith, P. W., 1987. Nonlinear optical glasses for ultrafast optical switches, *IEEE J. Quantum. Electron.* 23 (12), 2089–2094.

Fuchs, G., Melinon, P., Santos Aires, F., Treilleux, M., Cabaud, B. and Hoareau, A., 1991. Cluster-beam deposition of thin metallic antimony films: cluster-size and deposition-rate effects, *Phys. Rev. B* 44 (3), 9202–9208.

Fukumi, K., Chayahara, A., Kadono, K., Sakaguchi, T., Horino, Y., Miya, M., Fujii, K., Hayakawa, J. and Satou, M., 1994. Gold nanoparticles ion implanted in glass with enhanced nonlinear optical properties, *J. Appl. Phys.* 75 (6), 3075–3080.

Gabel, A., DeLong, K. W., Seaton, C. T. and Stegeman, G. I., 1987. Efficient degenerate four-wave mixing in an ion-exchanged semiconductor-doped glass waveguide, *Appl. Phys. Lett.* 51 (21), 1682–1684.

Garrido, F., Caccavale, F., Gonella, R. and Quaranta, A., 1995. Silver colloidal waveguides for nonlinear optics: a new methodology, *Pure Appl. Opt.,* in press.

Hache, F., Ricard, D., Flytzanis, C. and Kreibig, U., 1988a. The optical Kerr effect in small metal particles and metal colloids: the case of gold, *Appl. Phys. A* 47 (6), 347–357.

Hache, F., Ricard, D. and Girard, C., 1988b. Optical nonlinear response of small metal particles: a self-consistent calculation, *Phys. Rev. B* 38 (12), 7990–7996.

Haglund, R. F., Jr., Magruder, R. H., III, Morgan, S. H., Henderson, D. O., Weller, R. A., Yang, L. and Zuhr, R. A., 1992. Nonlinear index of refraction of Cu- and Pb-implanted fused silica, *Nucl. Instrum. Meth. Phys. Res.* B65, 405–411.

Haglund, R. F., Jr., Yang, L., Magruder, R. H., III, Wittig, J. E., Becker, K. and Zuhr, R. A., 1993. Picosecond nonlinear optical response of a Cu:silica nanocomposite, *Opt. Lett.* 11 (3), 373–375.

Haglund, R. F., Osborne, D. H., Magruder, R. H., White, C. W., Zuhr, R. A., Townsend, P. D., Hole, D. E. and Leuchtner, R. L., 1995. Fabrication and modification of metal quantum-dot composites by laser and ion beams, *Symp. Proc. Mater. Res. Soc,* in press.

Halperin, W. P., 1986. Quantum size effect in metal particles, *Rev. Mod. Phys.* 58 (3), 533–606.

Haus, J. W., Kalyaniwalla, N., Inguva, R. and Bowden, C. M., 1988. Optical bistability in small metallic particle composites, *J. Appl. Phys.* 65 (4), 1420–1423.

Haus, J. W., Kalyaniwalla, N., Ingura, R., Bloemer, M. and Bowden, C. M., 1989. Nonlinear optical properties of conductive spheroidal particle composites, *J. Opt. Soc. Am.* B 6 (4), 797–807.

Heilweil, E. J. and Hochstrasser, R. M., 1985. Nonlinear spectroscopy and picosecond transient grating study of colloidal gold, *J. Chem. Phys.* 82 (11), 4762–4770.

Hosono, H., Abe, Y., Lee, Y. L., Tokizaki, T. and Nakamura, A., 1992b. Large third-order optical nonlinearity of nanometer-sized amorphous semiconductors: phosphorus colloids formed in SiO_2 glass by ion implantation, *Appl. Phys. Lett.* 61 (23), 2747–2749.

Hosono, H., Abe, Y. and Matsunami, N., 1992a. Coalescence of nanosized copper colloid particles formed in Cu-implanted SiO_2 glass by implantation of fluorine ions: formation of violet copper colloids, *Appl. Phys. Lett.* 60(21), 2613–2615.

Jain, R. K. and Lind, R. C., 1983. Degenerate four-wave mixing in semiconductor-doped glasses, *J. Opt. Soc. Am.* 73 (5), 647–653.

Kang, K. E., Kepner, A. D., Gaponenko, S. V., Koch, S. W., Hu, Y. Z. and Peyghambarian, N., 1993. Confinement-enhanced biexciton binding energy in semiconductor quantum dots, *Phys. Rev. B* 48 (20), 15449–15452.

Kang, K., Kepner, A. D., Hu, Y. Z., Koch, S. W., Peyghambarian, N., Li, C.-Y., Takada, T., Kao, Y. and Mackenzie, J. D., 1994. Room temperature spectral hole burning and elimination of photodarkening in sol-gel derived CdS quantum dots, *Appl. Phys. Lett.* 64 (12), 1487–1489.

Kao, Y. H., Hayashi, K., Yu, L., Yamane, M. and Mackenzie, J. D., 1994. Sol-gel fabrication of semiconductor quantum dots, in *Sol-Gel Optics III,* ed. J. D. Mackenzie, *Proc. SPIE* 2288, 752.

Kash, K., Bhat, R., Mahoney, D. D., Lin, P. S. D., Scherer, A., Worlock, J. M., Van der Gaag, B. P., Koza, M. and Grabbe, P., 1989. Strain-induced confinement of carriers to quantum wires and dots within an InGaAs-InP quantum well, *Appl. Phys. Lett.* 55 (7), 681–683.

Kash, K., Mahoney, D. D., Van der Gaag, B. P., Gozdz, A. S., Harbison, J. P. and Florez, L. T., 1991. Observation of quantum dot levels produced by strain modulation of GaAs-AlGaAs quantum wells, *J. Vac. Sci. Technol.* B 10 (4), 2030–2033.

Kastner, M. A., 1993. Artificial atoms, *Phys. Today* (1), 24–31.

Koch, S. W., Hu, Y. Z. and Binder, R., 1993. Photon echo and exchange effects in quantum-confined semiconductors, *Physica B* 189, 176–188.

Kreibig, U. and Genzel, L., 1985. Optical absorption of small metallic particles, *Surf. Sci.* 156, 678–700.

Laurent, C. and Kay, E., 1988. Properties of metal clusters in polymerized hydrocarbon versus fluorocarbon matrices, *J. Appl. Phys.* 64 (1), 336–343.

Leon, R., Safard, S., Leonard, D., Merz, J. R. and Petroff, D. M., 1995. Visible luminescence from semiconductor quantum dots in large ensembles, *Appl. Phys. Letter.* 67 (4), 521–523.

Leonard, D., Fafard, S., Pond, K., Zhang, Y. H., Merz, J. L. and Petroff, P. M., 1994. Structural and optical properties of self-assembled InGaAs quantum dots, *J. Vac. Sci. Technol.* B 12 (4), 2516–2520.

Leonard, D., Krishnamurthy, M., Reaves, C. M., Denbaars, S. P. and Petroff, D. M., 1993. Direct formation of quantum-sized dots from uniform coherent islands of InGaAs on GaAs surfaces, *Appl. Phys. Lett.* 63 (23), 3203–3205.

Lifschitz, I. M. and Slezov, V. V., 1959. Kinetics of diffusive decomposition of supersaturated solid solutions. *Sov. Phys. JETP* 35 (8), 331–339.

Lipsanen, H., Sopanen, M. and Ahopelto, J., Luminescence from excited states in strain-induced $In_xGa_{1-x}As$ quantum dots, *Phys. Rev. B* 51 (19), 13868–13871.

Maeda, Y., Tsukamoto, N., Yazawa, Y., Kanemitsu, Y. and Masumoto, Y., 1991. Visible photoluminescence of Ge microcrystals embedded in SiO_2 glassy matrices, *Appl. Phys. Lett.* 59 (24), 3168–3170.

Magruder, R. H. III, Osborne, D. H., Jr. and Zuhr, R. A., 1994b. Nonlinear opitcal properties of nanometer-dimension Ag-Cu particles in silica formed by sequential ion implantation, *J. Non-Cryst. Solids* 176 (3), 299–303.

Magruder, R. H. III, Wittig, J. E. and Zuhr, R. A., 1993b. Wavelength tunability of the surface plasmon resonance of nanosize metal colloids in glass, *J. Non-Cryst. Solids* 163(2), 162–168.

Magruder, R. H., III, Yang, Li, Haglund, R. F., Jr., White, C. W., Yang, Lina, Dorsinville, R. and Alfano, R. R., 1993a. Optical properties of gold nanocluster composites formed by deep ion implantation in silica, *Appl. Phys. Lett.* 62 (15), 1730–1732.

Magruder, R. H., III, Haglund, R. F., Jr., Yang, L., Wittig, J. E. and Zuhr, R. A., 1994a. Physical and optical properties of Cu nanoclusters fabricated by ion implantation in fused silica, *J. Appl. Phys.* 76 (2), 708–715.

Maxwell-Garnett, J. C., 1904. Colours in metal glasses and in metallic films, *Philos. Trans. R. Soc.* A 205, 237–288.

Maxwell-Garnett, J. C., 1906. Colours in metal glasses, in metallic films, and in metallic solutions. — II, *Philos. Trans. R. Soc.* A 203, 385–420.

Mazzoldi, P., Caccavale, F., Cattaruzza, E., Tramontin, L., Boscolo-Boscoletto, A., Bertoncello, R., Trivillin, F., Arnold, G. W. and Battaglin, C., 1994. Peculiarities and application perspectives of metal-ion implants in glasses, *Nucl. Instrum. Meth. Phys. Res. B* 91, 505–516.

McHugh, K. M., Sarkas, H. W., Eaton, J. G., Westgate, C. R. and Bowen, K. H., 1989. The smoke ion source: a device for the generation of cluster ions via inert gas condensation, *Z. Phys. D* 12, 3–6.

Melinon, P., Paillard, V., Dupuis, V., Perez, A., Jensen, P., Hhoareau, A., Perez, J. P., Tuaillon, J., Broyer, M., Vialle, J. L., Pellarin, M., Baguenard, B. and Lerme, J., 1995. From free clusters to cluster-assembled materials, *Int. J. Mod. Phys. B* 9 (4/5), 339–397.

Mie, G., 1908. Beiträge zur Optik trüber Medien, speziell kolloidaler Metallösungen, *Ann. Phys.* 25 (3), 377–445.

Mittleman, D. M., Schoenlein, R. W., Shiang, J. J., Colvin, V. L., Alivisatos, A. P. and Shank, C. V., 1994. Quantum size dependence of femtosecond electronic dephasing and vibrational dynamics in CdSe nanocrystals, *Phys. Rev. B* 49 (20), 14435–14447.

Mizrahi, V., DeLong, K. W., Stegeman, G. I., Saifi, M. A. and Andrejco, M. J., 1989. Two-photon absorption as a limitation to all-optical switching, *Opt. Lett.* 14 (20), 1140–1142.

Murray, C. B., Kagan, C. R. and Bawendi, M. G., 1995. Self-organization of CdSe Nanocrystallites into three-dimensional quantum-dot supperlattices, *Science* 270 (21), 1335–1338.

Murray, C. B., Norris, D. J. and Bawendi, M. G., 1993. Synthesis and characterization of nearly monodisperse CdE (E = S, Se, Te) semiconductor nanocrystallites, *J. Am. Chem. Soc.* 115, 8706–8715.

Neeves, A. E. and Birnboim, M. H., 1989. Composite structures for the enhancement of nonlinear-optical susceptibility, *J. Opt. Soc. Am. B* 6 (4), 787–796.

Ngiam, S.-T., Jensen, K. F. and Kolenbrander, K. D., 1994. Synthesis of Ge nanocrystals embedded in a Si host matrix, *J. Appl. Phys.* 76, 8201–8203.

Ohtsuka, S., Koyama, T., Tsunetomo, K., Nagata, H. and Tanaka, S., 1992. Nonlinear optical property of CdTe microcrystallites doped glasses fabricated by laser evaporation method, *Appl. Phys. Lett.* 61 (25), 2953–2954.

Ohtsuka, S., Tsunetomo, K., Koyama, T. and Tanaka, S., 1993. Ultrafast nonlinear optical effect in CdTe-doped glasses fabricated by the laser evaporation method, *Opt. Mater.* 2, 209–215.

Olsen, A. W. and Kafafi, Z. H., 1991. Gold cluster laden polydiacetylenes: novel materials for nonlinear optics, *J. Am. Chem. Soc.* 113 (20), 7758–7760.

Perakis, I. and Chemla, D., 1994. ac stark effect of the Fermi edge singularity:observation of "excitonic polarons," *Phys. Rev. Lett.* 72 (20), 3202–3205.

Perenboom, J. A. A. J., Wyder, P. and Meier, F., 1981. Electronic properties of small metallic particles, *Phys. Rep.* 78 (2), 172–292.

Peyghambarian, N., Fluegel, B. D., Hulin, D., Migus, A., Joffre, M., Antonetti, A., Koch, S. W. and Lindberg, M., 1989. Femtosecond optical nonlinearities of CdSe quantum dots, *IEEE J. Quantum Electron.* 25, 2516.

Ricard, D., Roussignol, Ph. and Flytzanis, Chr., 1985. Surface-mediated enhancement of optical phase conjugation in metal colloids, *Opt. Lett.* 10 (10), 511–513.

Rochford, K. B., Yanoni, R., Stegeman, G. I., Krug, W., Miao, E. and Beranek, M. W., 1992. Pulse-modulated interferometer for measuring intensity-induced phase shifts, *IEEE J. Quantum Electron.* 28 (10), 2044–2050.

Saleh, B. E. A. and Teich, M. C., 1991. *Fundamentals of Photonics,* John Wiley and Sons, New York. Chapter 21.

Schmitt-Rink, S., Miller, D. A. B. and Chemla, D. S., 1987. Theory of the linear and nonlinear optical properties of semiconductor microcrystallites, *Phys. Rev. B* 35 (15), 8113–8125.

Shcheglov, K. V., Yang, C. M., Vahala, K. J. and Atwater, H. A., 1995. Electroluminescence and photoluminescence of Ge-implanted Si–SiO$_2$/Si structures, *Appl. Phys. Lett.* 66 (6), 745–747.

Sheik-Bahae, M., Said, A. A., Wei, T.-H., Hagan, D. J. and Stryland, E. W., 1990. Sensitive measurement of optical nonlinearities using a single beam, *IEEE J. Quantum Electron.* 26 (4), 760–769.

Sheik-Bahae, M., Wang, J., DeSalvo, R., Hagan, D. J. and Van Stryland, E. W., 1992. Measurement of nondegenerate nonlinearities using a two-color Z-scan, *Opt. Lett.* 17 (4), 258–260.

Siegner, U., Mycek, M.-A., Glutsch, S. and Cherula, D. S., 1995. Ultrafast coherent dynamics of Fano resonances in semiconductors, *Phys. Rev. Lett.,* 74(3), 470–473.

Silvestri, M. R. and Schroeder, J., 1994. Pressure- and laser-tuned Raman scattering in II-VI semiconductor nanocrystals: electron-phonon coupling, *Phys. Rev. B* 50 (20), 15108–15112.

Sipe, J. E. and Boyd, R. W., 1992. Nonlinear susceptibility of composite optical materials in the Maxwell-Garnett model, *Phys. Rev. B* 46 (3), 1614–1629.

Smirl, A. L., Boggess, T. F. and Hopf, F. A., 1980. Generation of a forward-traveling phase-conjugate wave in germanium, *Opt. Commun.* 34 (3), 463–468.

Stegeman, G. I., Wright, E. M., Finlayson, N., Zanoni, R. and Seaton, C. T., 1988. Third order nonlinear integrated optics, *IEEE J. Lightwave Tech.* 6 (8), 953–970.

Stegeman, G. I. and Stolen, R. H., 1989a. Waveguides and fibers for nonlinear optics, *J. Opt. Soc. Am. B* 6 (4), 652–662.

Stroud, D. and Hui, P. M., 1988. Nonlinear susceptibilities of granular matter, *Phys. Rev. B* 37 (15), 8719–8724.

Stroud, E. and Wood, V. E., 1989. Decoupling approximation for the nonlinear-optical response of composite media, *J. Opt. Soc. Am. B* 6 (4), 778–786.

Tait, H., Ed., 1991. *Glass, 500 Years,* New York, Harry N. Abrams.

Takagahara, T. and Takeda, K., 1992. Theory of the quantum confinement effect on excitons in quantum-dots of indirect-gap materials, *Phys. Rev. B* 46 (23), 15578–15581.

Tsunetomo, K. and Koyama, T., 1995. Optical nonlinearity of semiconducting microcrystallite-doped glass in a one-dimensional photonic crystal, OSA Annual Meeting, Portland, OR.

Uchida, K., Kaneko, S., Omi, S., Hata, C., Tanji, H., Asahara, Y., Ikushima, A. J., Tokizaki, T. and Nakamura, A., 1994. Optical nonlinearities of a high concentration of small metal particles dispersed in glass: copper and silver particles, *J. Opt. Soc. Am. B* 11 (7), 1236–1243.

Vollmer, M. and Kreibig, U., 1992. Collective excitations in large metal clusters, in *Nuclear Physics Concepts in the Study of Atomic Cluster Physics*, eds. R. Schmidt, H. O. Lutz, and R. Dreizler, Berlin, Springer-Verlag, 285–276.

Wang, J., Sheik-Bahae, M., Said, A. A., Hagan, D. J. and Van Stryland, E. W., 1994. Time-resolved Z-scan measurements of optical nonlinearities, *J. Opt. Soc. Am. B* 11 (6), 1009–1017.

Weaire, D., Wherrett, B. X., Miller, D. A. B. and Smith, S. D., 1979. Effect of low-power nonlinear refraction on laser-beam propagation in InSb, *Opt. Lett.* 4 (10), 331–333.

Weber, M. J., Milam, D. and Smith, W. L., 1978. Nonlinear refractive index of glasses and crystals, *Opt. Eng.* 17 (5), 463–469.

Wilson, W. L., Szajowski, P. F. and Brus, L. E., 1993. Quantum confinement in size-selected, surface-oxidized silicon nanocrystals, *Science* 262 (20), 1242–1244.

Woggon, U., Bogdanov, S. V., Wind, O., Schlaad, K.-H., Pier, H., Klingshirn, C., Chatziagorastou, P. and Fritz, H. P., 1993. Electro-optic properties of CdS embedded in a polymer, *Phys. Rev. B* 48 (16), 11979–11986.

Woggon, U. and Gaponenko, S. V., 1995. Excitons in quantum dots, *Phys. Stat. Sol. (b)* 189, 285–343.

Wood, R. A., Townsend, P. D., Skelland, N. D., Hole, D. E., Barton, J. and Afonso, C. N., 1993. Annealing of ion implanted silver colloids in glass, *J. Appl. Phys.* 74 (9), 5754–5756.

Yang, L., Becker, K., Smith, F. M., Magruder, R. H., III, Haglund, R. F., Jr., Yang, L., Dorsinville, R., Alfano, R. R. and Zuhr, R. A., 1994. Size dependence of the third-order susceptibility of copper nanoclusters investigated by four-wave mixing, *J. Opt. Soc. Am. B* 11 (3), 457–461.

Yang, L., Osborne, D. H., Haglund, R. F., Jr., Magruder, R. H., White, C. W., Zuhr, R. A. and Hosono, H., 1996. Probing interface properties of nanocomposites by third-order nonlinear optics, *Appl. Phys. A*, 62 (4), 403–415.

Zhao, X. S., Ge, Y. R., Schroeder, J. and Persans, P. D., 1994. Carrier-induced strain effect in Si and GaAs nanocrystals, *Appl. Phys. Lett.* 65 (16), 2033–2035.

Zhou, H. S., Honma, I., Komiyama, H. and Haus, J. W., 1994. Controlled synthesis and quantum-size effect in gold-coated nanoparticles, *Phys. Rev. B* 50 (16), 12052–12056.

FOR FURTHER INFORMATION

There are many fine texts on nonlinear optics and optical physics, including: Y. R. Shen, *The Principles of Nonlinear Optics* (New York, John Wiley & Sons, 1984); Robert W. Boyd, *Nonlinear Optics* (New York, Academic Press, 1992); and Amnon Yariv, *Quantum Electronics,* 4th ed. (New York, 1991). Nonlinear optical spectroscopy also is a mature field, with many fine textbooks and monographs, such as Marc D. Levenson and Satoru S. Kano, *Introduction to Nonlinear Laser Spectroscopy,* rev. ed. (New York, Academic Press, 1988); Shawl Mokamel, *Principles of Nonlinear Optical Spectroscopy* (New York, Oxford University Press, 1995); Wolfgang Demtröder, *Laser Spectroscopy: Basic Concepts and Instrumentation*, rev. ed. (New York, Springer-Verlag, 1995), Vol. 5 in the Chemical Physics series; and Roland Menzel, *Laser Spectroscopy; Techniques & Applications* (New York, Marcel Dekker, 1994).

The classic reference on the linear optical properties of metallic granular composites is the book by Bohren and Huffman, *Scattering and Absorption of Light by Small Particles* (New York, Academic Press, 1983). The most comprehensive treatment of semiconductor quantum-dot materials, emphasizing theory and recent optical experiments but also including a chapter on potential applications, is the book by L. Bányai and S. W. Koch, *Semiconductor Quantum* Dots (Singapore, World Scientific, 1993). The analogous compendium on metal quantum dots is found in the monograph by U. Kreibig and M. Vollmer, *Optical Properties of Metal Clusters* (Berlin, Springer-Verlag, 1995), Vol. 25 in the Springer Series in Materials Science.

A wealth of information about the use of ion implantation to create guided-wave structures for optical applications, including the use of ion implantation to create metal nanocluster composites, is in the book *Optical Effects of Ion Implantation*, by P. D. Townsend, P. J. Chandler, and L. Zhang (Cambridge, Cambridge University Press, 1994).

The NATO Advanced Study Institute program has sponsored a number of meetings on the topic of low-dimensional materials; one representative conference volume of interest is *Science and Engineering of One- and Zero-Dimensional Semiconductors*, S. P. Beaumont and C. M. Sotomayor Torres, Eds., ASI vol. 214 (New York, Plenum Press, 1990). A more recent conference volume with a number of general-interest papers on optical properties of semiconductor quantum dots is *Nanostrucures and Mesoscopic Systems*, edited by W. P. Kirk and M. A. Reed (New York, Academic Press, 1992).

Information about practical constraints on photonic systems, as well as a comprehensive introduction to nonlinear optics from the engineering point of view, may be found in E. B. A. Saleh and M. C. Teich's

Introduction to Photonics (New York, John Wiley, 1992). More-detailed information on optical communications systems is presented by M. N. Islam in *Ultrafast Fiber Switching Devices and Systems* (Cambridge, Cambridge University Press, 1992). A survey of the problems of quantum-confined materials in photonic devices may be found in the book *Nonlinear Photonics*, edited by H. M. Gibbs, G. Khitrova, and N. Peyghambarian (Heidelberg: Springer-Verlag, 1990), Vol. 30 in the Springer Series in Electronics and Photonics. A more recent work covering related themes is *Photonics in Switching*, John E. Midwinter, Ed. (Orlando, Academic Press, 1992), in two volumes.

The interdisciplinary character of the field of quantum-dot materials guarantees that apart from very general-interest publications, the scientific literature will be scattered in many different places. The journals *Journal of Vacuum Science and Technology* and *Superlattices and Microstructures* are frequent repositories of relevant articles from conference proceedings, dealing with the fabrication of QDCs; more-definitive archival articles tend to appear in *Optics Letters*, *Applied Physics Letters*, *Journal of Applied Physics,* and *Physical Review B*. Other journals which frequently handle nonlinear materials studies include *Optics Communications*, *Journal of the Optical Society of America* B, and *Applied Physics A*. Systems issues tend to be discussed most frequently in *IEEE Journal of Quantum Electronics* and *IEEE Journal of Lightwave Technology*. Increasingly, the *Journal of Non-Crystalline Solids* lists articles relating to QDCs because of the prominent role of glassy host materials.

Of particular note are two recent special issues of journals devoted to aspects of nonlinear optical phenomena in confined systems: a special issue of *Chemical Physics,* 210 (1–2), Eds. S. Mukamel and D. S. Chemla, "Confined excitations in molecular and semiconductor nanostructures," Oct. 1996; and a special number of *Journal of the Optical Cociety of America,* 13 (6), entitled "Radiation and dephasing processes in semiconductors."

Chapter 9

Nonlinear Optics of Microdroplets Illuminated by Picosecond Laser Pulses

Janice L. Cheung, Justin M. Hartings, and Richard K. Chang

Reviewed by R. L. Armstrong

TABLE OF CONTENTS

ABSTRACT

Liquid microdroplets have been shown to be convenient systems for the study of nonlinear optics in microcavities. The curved liquid–air surface of these droplets, with typical radius ≈50 μm, acts as a lens, concentrating the incoming radiation and increasing the internal intensity. The cavity modes of the spherical droplet trap the nonlinearly generated radiation and thus provide feedback and modify the gain for scattering processes. The increase in the internal intensity, the provision for feedback, and the modification of the gain associated with the quantum electrodynamic effect, result in lowered thresholds and several unique features for nonlinear optical processes. The cavity modes of a droplet have lifetimes on the order of 10 ns, and the asphericity of droplets produces a precession of the cavity modes on a time scale of ≈500 ps.

In this chapter we summarize recent studies of nonlinear optics on microdroplets that are illuminated by a train of 100-ps pulses from mode-locked Q-switched Nd:YAG lasers. The illumination of micron-sized liquid droplets with the higher intensity and lower fluence input pulses of picosecond, rather than nanosecond, duration has revealed several nonlinear optical effects not previously observed in micro-particles. Furthermore, illumination by a train of picosecond pulses has allowed for time-dependent measurements, both on the 100-ps as well as the 1–400-ns time scales. In particular, a train of mode-locked pulses allows one to successively probe the droplet over a period of >100 ns.

We will review experimental observations of second-harmonic generation, third-harmonic generation, and sum-frequency generation in single microdroplets. How phase-matching considerations in droplets led to a study of the spatial overlap of cavity modes will also be discussed. Stimulated low-frequency scattering, associated with the rocking/reorientational motion of CS_2 molecules, from droplets illuminated by mode-locked pulses will be presented. The radiation lifetime and precession period of cavity modes, measured from stimulated Raman and Brillouin scattered light excited by picosecond pulses, will be reviewed. Illuminating specific portions of the droplet with a focused beam and then collecting the scattered radiation from selective portions of the droplet to isolate and identify the various scattering processes will also be discussed. For example, backward circulating stimulated Brillouin scattering can be isolated from the forward circulating elastic scattering. Q spoiling or extra leakage at the Descartes ring will be presented as evidence of laser-induced cumulative shape perturbations to the droplet surface.

1. INTRODUCTION

Liquid microdroplets have been shown to be excellent systems for the study of nonlinear optics. The spherical shape of these microdroplets gives rise to a number of features which make microdroplets an attractive medium for the study of nonlinear optical processes. The curved liquid–air surface of these droplets, with typical radius ≈50 μm, acts as a lens, focuses the incoming radiation at a "hot spot" just within the shadow side of the droplet, and increases the incident intensity by a factor of 100. The cavity modes of the spherical droplet allow internal radiation to be retained within the droplet and provide feedback for nonlinear optical processes. This input intensity increase and feedback, together with the quantum electrodynamic (QED) enhancement of the stimulated transition rates, results in much lowered thresholds for nonlinear optical processes, such as stimulated Raman scattering (SRS) and stimulated Brillouin scattering (SBS).

Several different types of lasers have been used as the source of input radiation in microdroplet cavity experiments. These lasers are operated in the **mode-locked,*** the **Q-switched**, or the continuous-wave (cw) modes of operation. The three different modes of operation produce radiation with different time profiles and associated Fourier transform-limited bandwidths, which in turn affect the observation of a variety of nonlinear optical processes.

The recent use of mode-locked, picosecond pulses in illuminating these micron-sized liquid droplets has revealed some interesting nonlinear optical phenomena not previously observed with Q-switched nanosecond pulses or cw radiation. The time duration of the mode-locked pulses (≈100 ps) is shorter than the typical droplet-cavity lifetime (≈10 ns) and permits the direct measurement of the growth and decay of nonlinear optical signals generated within droplets. The intensity of mode-locked laser pulses, however, is oftentimes too low for exciting nonlinear optical processes. By combining the Q-switched and mode-locked modes of operation, the pulse intensity can be increased. A train of 30 to 50 high-intensity picosecond pulses can then be produced which have the same advantages as mode-locked pulses mentioned above. The 100-ps pulses in the train are separated by 13.2 ns, and allow the droplet to be successively probed with picosecond pulses over a period of >100 ns. The observations of several nonlinear optical processes developing on the 1-ns and the 10-ns timescale will be summarized in Section 5. The low fluence and high irradiance of these Q-switched, mode-locked pulses, the high repetition rate of the laser, and the droplet size stability afforded by the droplet generator, together permit the observation of weak nonlinear optical processes. These optical processes include anti-Stokes Raman scattering, second-harmonic generation, and third-order sum frequency generation (see Section 4 of this chapter).

---

* Terms in bold type are defined in Section 7.

In this chapter, we summarize recent experimental results from microdroplets which are illuminated by a train of 100-ps pulses from mode-locked Q-switched Nd:YAG lasers. Using these experiments as examples, we hope to convey some of the unique spectral and temporal properties of microdroplet cavities.

2. MORPHOLOGY-DEPENDENT RESONANCES OF THE DROPLET CAVITY

The elastic and inelastic scattered light from droplets is modified spectrally and temporally if the wavelength of the scattered light corresponds to one of the natural cavity resonances of the droplets. These cavity resonances depend only on the size, shape, and the index of refraction of the droplets, as well as the index of refraction of the surrounding medium. They are known as morphology-dependent resonances (MDRs), whispering gallery modes, or as structural resonances. These resonances can be envisioned as the consequence of the trapping of light through total internal reflection just within the liquid–air surface, such that after circumnavigating the droplet rim, the light comes back upon itself in phase. Because of diffraction at the curved interface, radiation can both leak into and out of the droplet. In the spectrum of the scattered light, these resonances appear as a series of peaks.

For a spherical droplet surrounded by air, the wavelength locations of these MDRs are functions of the size parameter, x, and the relative index of refraction m(λ). The size parameter is

$$x = \frac{2\pi a}{\lambda} \tag{1}$$

where a is the radius of the droplet and λ is the wavelength of the light in vacuum. The relative index of refraction, m(λ), is equal to the index of refraction of the liquid n(λ) when the index of refraction of air is taken to be 1, such that

$$m(\lambda) = \frac{n(\lambda)}{1} . \tag{2}$$

By expanding the electromagnetic fields in vector spherical harmonics and matching boundary conditions at the liquid–air interface of the sphere, properties of droplet MDRs, such as their wavelength locations, spectral widths, internal and external fields, can be calculated exactly (see Section 3). In the following section, some of the important results of these Lorenz–Mie calculations are summarized. For a comprehensive review, see Hill and Benner [1988].

2.1. Mode and Order Numbers

Each cavity resonance of the droplet is characterized by three indices related to the spatial distribution of the resonance mode. These indices are: an order number ℓ, a mode number n, and an azimuthal number m. These indices can be thought of as quantum numbers characteristic of the resonance. The order number ℓ indicates the number of maxima in the angle-averaged radial intensity distribution inside the sphere. The intensity distribution of the cavity resonances are confined in the radial, r, direction within $a/m(\lambda) \leq r \leq a$. For ethanol droplets, with $m(\lambda) = 1.36$, the MDR intensity is confined to $0.7a \leq r \leq a$. The number of intensity maxima around the equator of the sphere is equal to twice the mode number, i.e., $2n$. The n values of high-Q modes are confined within $x \leq n \leq m(\lambda)x$. For visible light and a typical droplet radius of a ≈50 μm, n is on the order of 500. The value of m ranges from $\pm n$, $\pm(n - 1)$, $\pm(n - 2)$, …, to 0, and describes the azimuthal angle dependence of the internal field. MDRs of the same n and ℓ, with different m numbers, are oriented with their normals at different polar angles, θ, relative to the axisymmetric axis (labeled z in Figure 1). For an m mode of a fixed n number, the intensity distribution is maximum along a great circle whose normal is inclined at $\theta \approx \cos^{-1}(m/n)$ to the z-axis. The intensity distribution is mainly concentrated in a thin "ribbon" along the great circle for the large n associated with the droplets of a ≈50 μm. $m \approx 0$ mode great circles circulate around the poles of the droplet, while $m \approx \pm n$ mode great circles circulate around the equator. For a perfect sphere, the $2n + 1$ azimuthal m modes for a particular n and ℓ are degenerate in frequency because the direction of the z-axis is arbitrary in this case.

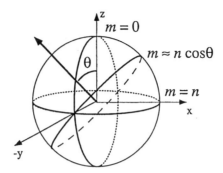

Figure 1 Schematic of a microdroplet, showing the spatial intensity distribution of three different m modes of the same n and ℓ.

For a droplet that is axisymmetrically distorted, however, the azimuthal modes are split [Chen et al., 1991] in frequency. The frequencies of these $2n + 1$ split m modes for a fixed n and ℓ have been determined through perturbation theory [Lai et al., 1990] to deviate from the original degenerate frequency, ω_o, according to

$$\omega(m) = \omega_o \left\{ 1 - \frac{e}{6} \left[1 - \frac{3m^2}{n(n+1)} \right] \right\}. \tag{3}$$

The distortion amplitude, $e = (r_p - r_e)/a$ is defined to be positive for prolate spheroids and negative for oblate spheroids. r_p and r_e are, respectively, the polar and equatorial radii. For an oblate spheroid, the $m = \pm n$ equatorial mode has the lowest resonance frequency and the polar $m = 0$ mode has the highest resonance frequency. For the prolate spheroid, the reverse is true. An m mode of a certain n can be envisioned as fitting n wavelengths around the perimeter of the spheroid along the great circle of the m mode. For an oblate spheroid, it then becomes physically intuitive to associate the smaller perimeter of the polar great circle $m = 0$ with a shorter wavelength and a higher frequency.

Experimentally, the frequency shift of MDR-related lasing and elastic scattering peaks has been observed for droplets undergoing quadrupolar shape oscillations [Tzeng et al., 1985; Arnold et al., 1990], and the frequency splittings of MDR-related SRS peaks have been measured for a deformed droplet [Chen et al., 1991]. The MDR wavelength variation (for a particular n mode) along the droplet rim of a slightly distorted droplet has been observed [Chen et al., 1993b] to follow Equation 3, with $m = n\cos\theta$.

From the m^2 dependence of $\omega(m)$ in Equation 3, it is clear that the degeneracy of the $\pm m$ modes are kept intact for an axisymmetrically distorted droplet. The $\pm m$ degeneracy is a consequence of the clockwise and counterclockwise modes being equivalent. Recently, the spectral splitting of the degenerate $\pm m$ modes [Weiss et al., 1995] due to the backscattering of the light on the m modes, has also been observed.

MDRs are also distinguished by their polarization. A mode that has no radial component of the electric field is known as a transverse electric (TE) mode, whereas a mode that has no radial component of the magnetic field is known as a transverse magnetic (TM) mode.

Light is coupled into an MDR when the input radiation coincides in space, in frequency, and in polarization with the MDR. Nonlinear optical processes in droplets are enhanced when the bandwidth of the generated radiation overlaps spectrally with an MDR. The next two sections describe the density and bandwidth of MDRs and discuss their spectral overlap with the bandwidths of the input laser and nonlinear optical processes such as SRS. For some nonlinear optical processes, phase matching is required for efficient generation. Section 2.4 explains the concept of phase matching in droplets and shows how phase matching is related to the spatial overlap of the generating and resultant MDRs.

2.2. Cavity Q Factor
The quality factor [Haus, 1984] of an MDR is defined as

$$Q = \frac{2\pi \cdot \text{stored energy}}{\text{energy loss per cycle}}. \tag{4}$$

A constant Q value leads to an exponential decay of the stored energy, W, according to

$$W = W_o \exp\left(\frac{-t}{\tau_{MDR}}\right) \tag{5}$$

with a cavity lifetime, τ_{MDR}, of

$$\tau_{MDR} = \frac{Q}{\omega_o} . \tag{6}$$

The MDRs with resonance frequency $\omega_o = 2\pi c/\lambda$, and cavity lifetime τ_{MDR}, have a Lorentzian line shape,

$$\frac{1}{(\omega - \omega_o)^2 + \left(\frac{1}{2\tau_{MDR}}\right)^2} . \tag{7}$$

The full width at half maximum (FWHM) of these resonances is,

$$\Gamma_{MDR} = 1/\tau_{MDR} = \omega_{MDR}/Q . \tag{8}$$

A high-Q MDR has a long measured lifetime (≈ 10 ns, corresponding to $Q \approx 5 \times 10^7$). Light leaks out of high-Q modes very slowly. Through the principle of reciprocity, light also couples slowly [Haus, 1984] into high-Q MDRs. Thus, for a laser pulse duration that is short relative to τ_{MDR}, the internal field does not have time to build up fully, even if the input-laser frequency is resonant with a high-Q MDR. For example, a 100-ps mode-locked laser pulse does not couple as efficiently to high Q ($\tau_{MDR} \approx 10$ ns, $Q \approx 10^7$) as to low Q ($\tau_{MDR} \approx 100$ ps, $Q \le 5 \times 10^5$) MDRs, which have shorter lifetimes.

An alternative way of describing the same principle is to look in the frequency domain. A short (Fourier transform-limited) laser pulse has a large bandwidth Γ_L, while a high-Q MDR has a narrow bandwidth $\Gamma_{MDR} \ll \Gamma_L$. The Fourier-limited bandwidth of 100 ps green pulses is $\Gamma_L = 0.3$ cm^{-1}, and the bandwidth of a $Q = 10^7$ MDR, is $\Gamma_{MDR} = 2 \times 10^{-3}$ cm^{-1}. Even if the center frequencies of the laser and the MDR are coincident, the spectral overlap is small, so only a small fraction of the laser light energy can be coupled into the MDR. This fraction is approximately Γ_{MDR}/Γ_L. This principle will be demonstrated in Section 3.1 through numerical calculations of the droplet internal energy upon illumination by a short pulse. If the Q were decreased, however, the amount of input coupling would increase. Similarly, the amount of input-coupled light would also be greater if another MDR with a lower Q factor were to overlap spectrally with the laser.

The actual Q of an MDR is determined by several factors: (1) radiative leakage losses (Q_{rad}), (2) internal absorption losses (Q_{abs}), (3) internal scattering losses (Q_{scatt}), (4) losses caused by some external perturbation that affects the leakage rate (Q_{pert}), and (5) losses associated with photon depletion by nonlinear optical processes within the droplet (Q_{depl}). The total Q of the MDR for these independent loss mechanisms is thus given by

$$\frac{1}{Q_{MDR}} = \frac{1}{Q_{rad}} + \frac{1}{Q_{abs}} + \frac{1}{Q_{scatt}} + \frac{1}{Q_{pert}} + \frac{1}{Q_{depl}} \tag{9}$$

The term Q_{rad} is the intrinsic Q factor that is associated with the MDR of a lossless spheroid. The intrinsic leakage rate is caused by the diffraction of the internal waves at the curved interface and depends strongly on the order number, ℓ, of the MDR. Q_{rad} and Q_{pert} affect both the rate that light couples into the MDR as well as the cavity lifetime. The terms, Q_{abs} and Q_{depl} only affect the lifetime of the light retained in the cavity mode of the droplet, while Q_{scatt} can cause both increased leakage from the droplet and coupling between spectrally overlapping MDRs.

Perturbations that can affect the leakage rate include thermally induced surface ripples, and laser-induced perturbations. Section 5 discusses the effect of various *laser-induced* perturbative mechanisms

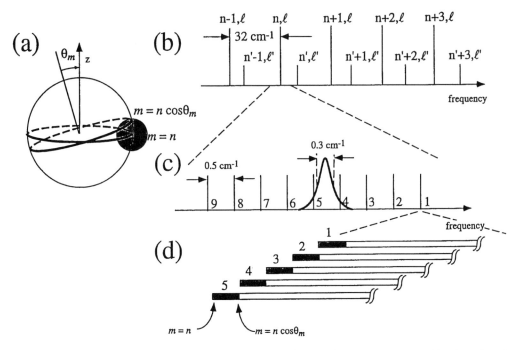

Figure 2 (a) Schematic of an axisymmetrically distorted oblate droplet with an ≈40 μm illuminated at the equator at one edge by a focused laser beam of beam waist 20 μm (shaded spot). Shown are two great circles of m-mode MDRs that spatially overlap with the focal spot. The input laser couples most effectively into m modes which have normals tilted to the z-axis $|\theta| \le \theta_m$. (b) Schematic of the mode spacing between high-Q MDRs, such as would be observed in a water SRS spectrum. Shown are modes of consecutive n for two different ℓ's. (c) Schematic showing the mode density of all MDRs with $Q \ge 10^3$. Shown also is the bandwidth (0.3 cm^{-1}) of the incident mode-locked laser. (d) Schematic showing the increase in mode density due to the splitting of the m degeneracy. The frequency spread of m modes that spatially overlap with the focal spot is indicated by the gray boxes.

on input coupling and cavity lifetimes. These mechanisms include **electrostrictive** or temperature-related surface bulging [Zhang and Chang, 1988; Lai et al., 1989] and index of refraction changes through the intensity-dependent index of refraction, n_2.

2.3. Density of Modes Relative to the Laser Bandwidth and the Nonlinear Optical Bandwidth

The separation between adjacent n modes of the same ℓ can be approximated by an analytical expression [Chylek, 1976]

$$\Delta x \approx \frac{\tan^{-1}\left(m(\lambda)^2 - 1\right)^{1/2}}{\left(m(\lambda)^2 - 1\right)^{1/2}} \tag{10}$$

For a droplet of radius, $a \approx 40$ μm, the mode spacing among modes of the same ℓ is about 35 cm^{-1}. The internal intensity distribution of the input radiation generally overlaps best with only one radial order MDR, i.e., with MDRs of a specific ℓ. Thus, within the 10-cm^{-1} gain bandwidth of SRS of ethanol, there can be at most two modes (of different polarizations, TE and TM) that support SRS. The SRS gain bandwidth of water, however, is broad (≈300 cm^{-1}). At least eight equally spaced MDRs are visible in the water SRS spectrum (see Figure 9) corresponding to TE and TM modes of the same ℓ's and consecutive n's. With higher input intensity, a second set of MDRs of another radial order, which has less overlap with the input intensity distribution, is able to provide sufficient gain for SRS, and appears in the water SRS spectrum. Figure 2b shows the mode spacing among high-Q MDRs, such as would be observed in a water SRS spectrum. Shown are the modes of consecutive n for two different ℓ values.

A numerical calculation, however, is required to find the density of *all* modes of different ℓ, and thus of different Q_{rad} factors. This density has been calculated numerically [Hill and Benner, 1988] for a perfect sphere (assuming degenerate m modes). For a 40-μm-radius sphere, the mode spacing is ≈ 0.5 cm^{-1} among all modes of different n and ℓ with $Q_{rad} \geq 10^3$. Thus, for a 100-ps pulse (0.3 cm^{-1} bandwidth), and an arbitrary-sized perfect sphere of a ≈ 40 μm, the laser may not spectrally overlap with a cavity resonance. The relation between the average mode spacing and the laser bandwidth is shown schematically in Figure 2c.

The flowing droplets studied in the experiments described in this chapter, however, are not perfect spheres. They are slightly oblate because of the inertial effects involved in their motion. Consequently, the mode density is increased by the lifted degeneracy of the m modes of fixed n and ℓ. For a ≈ 50 μm droplets, this mode-splitting ensures that a 100-ps pulse with a line width of 0.3 cm^{-1} will always overlap with some MDR. It is impossible in this case to avoid being on an input resonance.

Figure 2d shows the effect of this splitting on the mode density. An input laser focused to a small spot at the equator on one side of the droplet couples most efficiently into equatorial m modes with great circles that are tilted from the equator by an angle of $|\theta| \leq \theta_m$ (see Figure 2a). For a particular n, these equatorial m modes lie in the range of $n \cos\theta_m \leq |m| \leq n$ and have a frequency spread of $\Delta\omega = (-\omega_0|e|/2)$ $(\cos^2 \theta_m - 1)$ for large n. This frequency spread (gray box) is shown in Figure 2d, and compared with the frequency spread of *all* the m modes. The value of $e \approx 10^{-3}$ previously measured from the frequency splitting of SRS [Chen et al., 1991], a beam waist of 20 μm, and a droplet radius, a = 40 μm, are assumed. A comparison of the bandwidth (0.3 cm^{-1}) of 100-ps pulses shown in Figure 2c with the mode density (gray boxes) shows that there is always some MDR within the input beam line width.

2.4. Phase Matching

Many nonlinear optical processes involve the interaction of electromagnetic waves at two or more generating frequencies (for example, ω_1 and ω_2, with wavevectors $\mathbf{k}_1$ and $\mathbf{k}_2$) to produce waves at a resultant third frequency (ω_3, with wavevector $\mathbf{k}_3$). Efficient generation of radiation at the resultant frequency requires both energy conservation, $\omega_3 = \omega_1 + \omega_2$, and photon momentum conservation, $\mathbf{k}_3 = \mathbf{k}_1 + \mathbf{k}_2$. The photon momentum conservation condition is known as the phase-matching condition. Because of natural bulk dispersion in most materials, it is not always possible to achieve the phase-matched condition of $\Delta\mathbf{k} = \mathbf{k}_3 - (\mathbf{k}_1 + \mathbf{k}_2) = 0$.

For most of the well-studied nonlinear processes in droplets, such as lasing, SRS, and SBS, phase matching between the pump and the resultant waves is automatically satisfied. However, coherent anti-Stokes Raman scattering [Qian et al., 1985], stimulated anti-Stokes Raman scattering [Leach et al., 1992], and second- or third-order sum frequency generation [Acker et al., 1989; Leach et al., 1990; 1993] do require phase matching.

In droplets, the concept of phase matching was studied both experimentally [Leach et al., 1993] and theoretically [Hill et al., 1993] for the case of third-order sum frequency generation (TSFG). TSFG results from the nonlinear polarization $P^{NLS}_{(\omega_{TSFG})} = \chi^{(3)}E(\omega_1)E(\omega_2)E(\omega_3)$, where $\chi^{(3)}$ is the third-order non-linear susceptibility of the liquid, $E(\omega_i)$ are the electric field amplitudes of the generating waves, and $\omega_{TSFG} = \omega_1 + \omega_2 + \omega_3$. In the plane-wave formalism for TSFG, phase matching is achieved when the resultant frequency-shifted wave $E(\omega_{TSFG})$ copropagates at the same phase velocity as the $P^{NLS}_{(\omega_{TSFG})}$. Consequently, having the phases of $P^{NLS}_{(\omega_{TSFG})}$ and $E(\omega_{TSFG})$ matched is equivalent to having good spatial overlap between $P^{NLS}_{(\omega_{TSFG})}$ and $E(\omega_{TSFG})$ along the propagation direction. In droplets, the phase-matching condition is replaced by the spatial overlap condition of the input and output MDRs. The spatial overlap of the four MDRs (three input MDRs and one output MDR) must be integrated in r, $\sin\theta$ dθ, and dϕ. The dϕ integral forces the azimuthal mode numbers of the modes to satisfy $m_{TSFG} = m_1 + m_2 + m_3$, while there is no simple requirement for the n and ℓ values. Phase matching in droplets is thus synonymous with matching of the spatial distribution of the generating wave MDRs with the resultant wave MDR.

3. NUMERICAL CALCULATIONS OF THE INTERNAL INTENSITY OF A DIELECTRIC SPHERE ILLUMINATED BY SHORT LASER PULSES

The spherical geometry of droplets allows droplet internal intensity distributions to be expanded in vector spherical harmonics and simplifies the exact numerical calculations of optical intensities within these droplets.

Time-*independent* internal intensities of droplets have been studied in detail. Since it is the interplay between droplet internal intensity distributions and the cavity modes that make nonlinear optical pro-

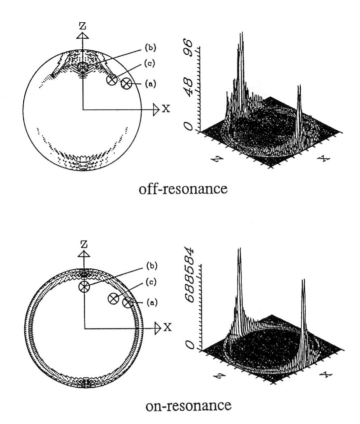

Figure 3 Points in the x–z plane where the internal intensities are computed. The coordinates of the points are (a) r = 0.87a, θ = 60°, φ = 0°; (b) r = 0.7a, θ = 0°, φ = 0°; (c) r = 0.707a, θ = 45°, φ = 0°. Points (a), (b), and (c) are superimposed upon contour plots of the internal intensities generated by resonant and off-resonant y-polarized incident plane waves. Surface plots of the internal intensities for resonant and off-resonant incidence are also shown. (From Chowdhury, D. Q., Hill, S. C., and Barber, P. W. 1992. *J. Opt. Soc. Am. A* 9(8):1364–1373. With permission.)

cesses in droplets unique, the understanding of nonlinear optics within droplets (Section 4) has benefited from these time-independent studies. Picosecond, mode-locked pulses, however, are shorter in duration than the cavity lifetimes of MDRs (see Section 2.2). Calculations of time-independent droplet internal intensities are therefore not suitable for modeling nonlinear optical processes in droplets illuminated by *picosecond* pulses. Thus, there has been a need for an understanding of the internal intensity distribution in the transient regime for modeling experiments with picosecond pulses.

3.1. Time-Dependent Droplet Internal Intensities for Plane Wave Illumination

Calculations of the time-*dependent* droplet internal intensities [Chowdhury et al., 1992a] were performed for plane wave illumination, which corresponds to the case where the beam waist of the focused laser beam is much larger than the droplet diameter. It was found that the primary factors determining the time dependence of the internal intensity are the pulse duration, the lifetime of the nearest high-Q MDR, the spectral detuning from that MDR, and the position inside the droplet where the intensities are calculated. In particular, these numerical results demonstrate the principle discussed in Section 2.2, that a short pulse couples more energy into a lower-Q MDR than into a high-Q MDR.

The time-dependent droplet internal intensities were calculated for incident radiation on resonance with a TE$_{94,2}$ MDR ($n = 94$, $1 = 2$) with Q = 8.8×10^6 and $\tau_{\text{MDR}} = 2.5$ ns. Figure 3 shows the three locations inside the sphere where the time dependence of the internal intensity for each location was calculated. These points are superimposed upon contour plots of the resonant and off-resonant internal energy distributions for the TE$_{94,2}$ MDR along with surface plots of the intensities for the two cases. Note that the resonant component of the internal intensity is confined near the rim of the droplet. Two maxima in the radial distribution are apparent, consistent with an MDR of mode order, $\ell = 2$. The

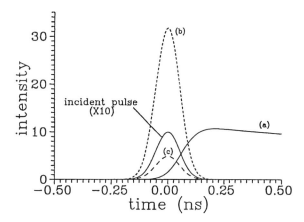

Figure 4 Transient internal intensities at the three points shown in Figure 3. The incident y-polarized plane wave has a Gaussian time dependence and propagates along the z axis. The scaled (×10) incident pulse is also shown. Curve (a) is at point (a), (b) is at point (b), and (c) is at point (c). The incident Gaussian pulse is on resonance with a pulse duration of 0.125 ns. (From Chowdhury, D. Q., Hill, S. C., and Barber, P. W. 1992. *J. Opt. Soc. Am. A* 9(8):1364–1373. With permission.)

nonresonant component is concentrated near the illuminated and shadow faces and is due to the lensing effect of the droplet surface. At location (a) the intensity is small unless the incident field is on an MDR of the sphere. Point (b) is on the front hot spot where the light intensity is concentrated by the focusing effect of the droplet surface. The contribution of the resonant field at point (b) is negligible. In a geometric optics picture, the intensity at point (b) is caused by the intersection of rays internally reflected once from the droplet shadow surface [Srivastava and Jarzembski, 1991; Chowdhury et al., 1992b]. Point (c) is at a location where the intensity is relatively small either on-resonance or off-resonance.

Figure 4 shows the transient intensity at the three different locations shown in Figure 3 for an incident pulse of duration $\tau_o = 0.125$ ns (full width at half-maxima). The incident pulse is on resonance with the $TE_{94,2}$ MDR, and the pulse duration, τ_o, is much shorter than the MDR lifetime of $\tau_{MDR} = 2.5$ ns. In other words, the incident frequency spectrum is much broader than the $TE_{94,2}$ MDR line width, although it is not broad enough to overlap significantly with other high-Q MDRs. Curve (a) in Figure 4 shows the transient buildup of the resonant intensity at point (a) of Figure 3. Because the incident pulse duration is much shorter than the cavity lifetime, the resonant field intensity does not have time to build up to a steady-state value. Thus, the peak intensity at point (a) is 3× weaker than the peak intensity at the front hot spot (b), which is shown in curve (b) in Figure 4. After the incident pulse intensity becomes negligible, the resonant intensity at point (a) continues to decay at a rate inversely proportional to the photon lifetime of the resonant mode. This decay rate is observable because the incident pulse is shorter than the cavity lifetime. After t = 0.15 ns (1.2 τ_o), essentially all the internal energy is in the resonant mode. In contrast, the transient intensity at point (b) follows the input pulse in the time scale shown on the plot. Curve (c) in Figure 4 shows the transient intensity at point (c) of Figure 3. The intensity at this point also follows the incident pulse.

Figure 5 shows the transient intensities at the same locations as Figure 4, but for which the incident pulse duration is 10× longer, i.e., $\tau_o = 1.25$ ns. The pulse duration of $\tau_o = 1.25$ ns is half the photon lifetime of the mode, $\tau_{MDR} = 2.5$ ns. Note the change of scale in the time axis between Figures 4 and 5. Again, the laser is on-resonance with the frequency of the $TE_{94,2}$ MDR. For the longer incident pulse duration, there is more time to couple into the MDR, so the resonant intensity approaches the steady-state value. The resonant intensity at (a) is now more than 3× *stronger* than the intensity at the front hot spot, (b). The resonant intensity also peaks later for a longer input pulse.

In both of the cases shown, the incident pulse duration is shorter than the resonant lifetime. Consequently, the exponential decay of the internal intensity caused by the resonance can be directly observed [Newton, 1966]. If the incident pulse rise and decay times were longer than the resonant lifetime, such as for the 10-ns pulse duration of Q-switched lasers, there would still be a delay between the peaks of the incident and internal intensities, but the slower switch-off rate of the incident pulse would make it difficult to observe the exponential decay associated with the resonance.

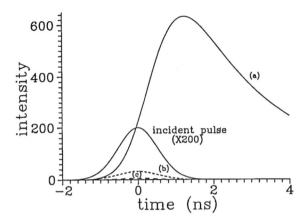

Figure 5 Transient internal intensities at the same locations as in Figure 4, except that the incident pulse duration is 1.25 ns. The scaled (×200) incident pulse is also shown. (From Chowdhury, D. Q., Hill, S. C., and Barber, P. W. 1992. *J. Opt. Soc. Am. A* 9(8):1364–1373. With permission.)

3.2. Gaussian Beam Illumination

The experimental observation by Zhang et al. [1988] that there is increased coupling into MDRs when the input radiation is tightly focused and incident at the edge of the droplet motivated numerical calculations of the droplet internal intensity for illumination by a focused beam with a Gaussian cross-sectional profile. For plane-wave illumination, only the $m = \pm1$ modes support input coupling. However, because of the wave vector spread of the incident Gaussian beam, the input radiation couples into many m modes. Therefore, calculations of the internal intensity for Gaussian beam illumination [Gouesbet et al., 1988; Barton et al., 1989; Lock and Hovenac, 1993] are more complicated than for plane-wave illumination.

When the frequency of the input radiation is coincident with that of a droplet resonance, numerical calculations show that there is greater intensity buildup [Barton et al., 1989; Khaled et al., 1993] on the MDRs when the Gaussian beam is focused at the edge rather than through the center of the droplet. The edge focus provides a better spatial overlap with the cavity mode volume that is situated just beneath the droplet surface. Numerical calculations [Barton et al., 1989] show that for edge illumination, even when the incident frequency is off-resonance, there is still a buildup of intensity in a band under the surface. When the input frequency is 500 MDR line widths away from the resonance, the best coupling position for a cw Gaussian beam was actually found to be beyond the surface of the droplet by 0.232 droplet radii [Khaled et al., 1993]. In the latter case, the wave vector matching between the MDR and the Gaussian beam is optimized.

The time-dependent internal intensity of a droplet illuminated by a Gaussian beam consisting of a single nanosecond pulse was also numerically calculated [Khaled et al., 1994]. The temporal effects were found to be similar to that of plane-wave incidence [Chowdhury et al., 1992a]. In addition, the coupling efficiency was found to be greatest when the beam is focused slightly beyond the droplet surface, consistent with the cw Gaussian beam results [Khaled et al., 1993].

The numerical calculations summarized in the last two sections are significant because of their ability to model real experimental conditions, such as using a focused beam and picosecond pulses. The results, however, are still preliminary, and possibilities of using these numerical techniques to verify experimental results, or to stimulate new experimental investigations, have yet to be explored.

3.3. Electrostrictive Effects Induced by Multiple Picosecond Pulses

In addition to providing the radiation for coupling into droplets and for pumping nonlinear optical signals within the droplet, picosecond input pulses can also generate a variety of physical perturbations on the droplet. These perturbations may spoil the Q factor of the droplet. The Q degradation in turn may be advantageous, allowing a larger fraction of the short input pulse energy to be coupled into MDRs. This section describes two theoretical studies of perturbations produced through electrostriction. Electrostriction refers to the tendency of liquids and gases to become more dense in regions of high electric field. Since the internal intensity distribution of the droplet is nonuniform, the electrostrictive effect may generate density inhomogeneities within the droplet. These density inhomogeneities may then lead to index of refraction changes that alter the Q factor of the droplet. In addition, electrostrictive forces may also induce small changes in the droplet shape where the droplet surface intensity is high, and so cause even further Q degradation.

The first theoretical work [Lai et al., 1993] was motivated by the experimental observation of nonlinear Mie scattering [Huston et al., 1990] from ethanol droplets. Nonlinear Mie scattering was proposed to be due to Q degradation involving the coupling of the optical field to the electrostrictively produced volume acoustic waves resonant with the droplet cavity (see Sections 5.3 and 5.5). These acoustic waves produce density fluctuations that lead to index of refraction fluctuations. The index of refraction fluctuations in turn can perturb the cavity modes and degrade the Q of MDRs.

This theoretical work [Lai et al., 1993] considered the interaction of a train of intense picosecond laser pulses with an ethanol microdroplet. The magnitude of the acoustic field electrostrictively produced by the input laser pulses was calculated and was found to be $d\rho/\rho \approx 2.3 \times 10^{-5}$. The effect of the acoustic field density fluctuations on the line width of the MDR was also calculated and found, however, to be very small, with $1/Q - 1/Q_o \approx 10^{-10}$, where Q_o is the unperturbed Q of the MDR.

The effect is small for two reasons. First, acoustic normal modes vanish at the surface (boundary condition), where the MDR mode is concentrated. Second, the Q degradation varies as the square of the index of refraction change, so a fractional change in the index of refraction results in an even smaller change in the Q factor.

A more recent Ph.D. dissertation [Ng, 1994] considered four possible mechanisms for the degradation of Q: (1) resonant volume acoustic waves; (2) acoustic waves generated by SBS in the mode volume of the droplet, which should have better overlap with the MDR mode volume; (3) electrostrictively produced surface bulging, similar to one of the mechanisms suggested for the extra leakage at the Descartes ring (Section 5.4); and (4) cavitation and coalescence of dissolved gas bubbles within the droplet. Surface bulging was found to be the only mechanism of the four to produce significant Q spoiling.

The two theoretical studies demonstrate the robust nature of these high-Q droplet cavity modes. The quadratic dependence of the Q degradation on the index of refraction change makes the Q factor very resistant to small changes in the index of refraction. In order to significantly alter the Q of the droplet modes, it is necessary to have a large index of refraction change together with a large spatial overlap with the cavity modes. These two requirements are satisfied for the surface bulge, making it a very efficient mechanism for degrading the overall cavity Q factor. A bulge amplitude $\approx 10^{-3}$ of the droplet radius was theoretically able to cause a change in the Q factor of $1/Q - 1/Q_o \approx 10^{-7}$, where Q_o is the Q value of the unperturbed sphere and assumed to be $Q_o \approx 10^8$.

4. EXPERIMENTAL OBSERVATIONS OF INELASTIC SCATTERING

The following several sections (Sections 4.2 through 4.6) describe nonlinear optical processes that have been observed in droplets illuminated by a train of picosecond pulses from a Q-switched mode-locked Nd:YAG laser. Many of these effects are very weak, and their observation is enabled by the high intensity and high repetition rate of the mode-locked laser pulses. In the time-integrated spectra shown, peaks corresponding to MDRs can be clearly resolved. Since MDR frequencies are very sensitive to the droplet size, the observation of MDR peaks in spectra that have been integrated over a period of minutes, demonstrates the extreme stability and reproducibility of the droplet generator. In fact, droplet size stability of 1 part in 10^4 has been reported [Leach et al., 1993] over a period of 90 min.

4.1. Experimental Geometry

The droplets are produced by a Berglund–Liu vibrating orifice generator. Liquid is forced through a small (≈ 35 μm dia.) orifice which is driven at frequency f_{osc} by a piezoelectric transducer. Driving the orifice within a certain frequency range breaks up the liquid column into a stream of monodispersed droplets of radius, $a \approx 40$ μm. Due to inertial effects, the final droplets are slightly oblate with a distortion amplitude of $|e| \approx 10^{-3}$.

The Q-switch repetition rate is synchronized to the driving frequency of the droplet generator such that the focused laser beam always illuminates the same region at the equator of the droplets. By illuminating only one edge of the droplet, the input-coupled light trapped on MDRs circulates the droplet counterclockwise (see Figure 6). Inelastic scattered light is generated in all directions, but only the inelastic scattered light generated in the near-forward or near-backward directions, near tangent to the droplet surface, will be supported by MDRs. Near-forward scattered light will have the same sense of circulation as the input-coupled light (counterclockwise in Figure 6), whereas the near-backward

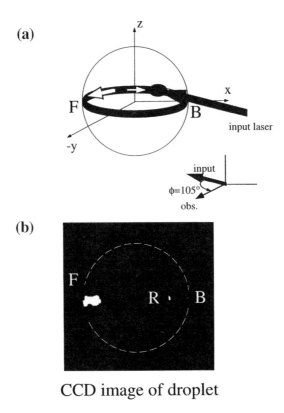

CCD image of droplet

Figure 6 (a) Schematic of an ethanol droplet illuminated at the equator by a focused input beam. The sense of circulation of the light on MDRs is indicated for forward circulating (F), and backward circulating (B). (b) CCD image of the green scattered light from the droplet. Viewing is from $\phi = 105°$. The dotted circle outlines the droplet rim.

scattered light will have the opposite sense of circulation (clockwise in Figure 6). For the rest of this chapter, radiation that has the same sense of circulation as the input-coupled light (counterclockwise) will be called *forward-circulating*, and labeled F, while light with the opposite sense of circulation (clockwise) will be called *backward-circulating*, and labeled B. The sense of circulation of radiation on an MDR can be identified by the location of light leakage from the droplet rim. On viewing the droplet from any angle, light with opposing circulation leaks from opposite rims of the droplet and appears as arcs in the magnified droplet image. These arcs can be separately detected by a photodiode, **streak camera**, or spectrograph to separately determine their temporal and spectral features.

4.2. Stimulated Raman Scattering

Stimulated Stokes Raman scattering occurs in droplets when the round-trip Raman gain exceeds the round-trip loss. The first-order **Stokes** SRS (with frequency $\omega_{1s} = \omega_L - \omega_{vib}$) can be sufficiently intense in droplets to act as a pump to generate a second-order SRS (with frequency $\omega_{2s} = \omega_L - 2\omega_{vib}$). The second-order SRS then pumps a third-order SRS in a cascade SRS process. Stokes SRS up to 14th-order (with frequency $\omega_{14s} = \omega_L - 14\omega_{vib}$) has been observed [Qian and Chang, 1986] from CCl_4 droplets illuminated with a single 10-ns green Q-switched laser pulse. While SRS is routinely observed from droplets illuminated by nanosecond pulses, it is also common in droplets illuminated by mode-locked pulses. SRS is generated on high-Q MDRs, has a long lifetime and a long interaction length, and acts as a pump for many of the weaker nonlinear optical processes described later in this chapter.

The SRS spectrum from CCl_4 droplets illuminated by a train of 100-ps green ($\lambda = 532$ nm) laser pulses, is shown in Figure 7 as a function of the Raman shift (in units of cm^{-1}). Several frames (intentionally offset from one another in the figure and shown with different intensity factors) are taken at different spectral regions within the broad-spectral range where the detector (**CCD**) is sensitive. The spectral positions of the jth-order SRS associated with the strongest vibrational Raman mode ($\nu_1 = 459$ cm^{-1}) are indicated by the series of black diamonds.

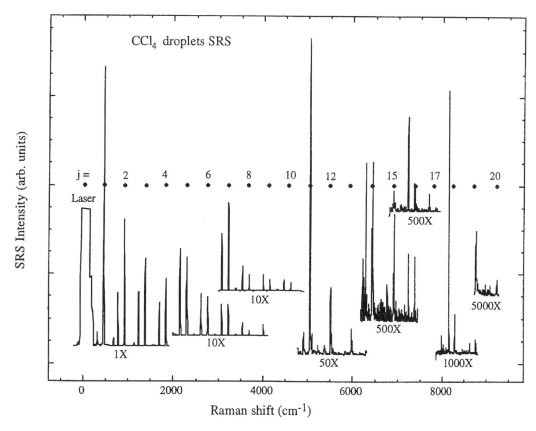

Figure 7   Detected SRS spectra from CCl_4 droplets plotted as functions of the Raman shift. The integer j (above the points) corresponds to the frequency $\omega_j = \omega_L - j\omega_{vib}$, where $\omega_{vib} = 459$ cm^{-1} is the frequency of the v_1 vibrational mode of CCl_4. Cascade multiorder Stokes SRS up to $j = 20$ is detected. Combinations and overtones of the v_2 and v_4 vibrational modes are also detected. The incident laser wavelength is $\lambda_L = 0.532$ μm, and the detector is a CCD. (From Leach, D. H., Chang, R. K., Acker, W. P., and Hill, S. C. 1993. *J. Opt. Soc. Am. B* 10(1):34–45. With permission.)

Several features of Figure 7 are noteworthy: (1) SRS peaks occur throughout a 9000-cm^{-1} frequency range resulting from a single input frequency (9000 cm^{-1} corresponds to wavelengths between 0.532 and 1 μm); (2) the intensity of the jth-order SRS from the v_1 mode decreases nearly monotonically as j increases; (3) there are many peaks corresponding to combinations of the v_1 vibrational mode with other weaker vibrational modes at $v_2 = 218$ cm^{-1} and $v_4 = 314$ cm^{-1}; (4) the highest-order SRS observed is $j = 20$; and (5) the SRS intensity at ω_{1s} is four or five orders of magnitude larger than the SRS intensity at ω_{20s}.

The difference in intensity between SRS peaks corresponding to adjacent j (for $j < 10$) is always less than a factor of two, demonstrating the high efficiency with which SRS radiation (on an MDR) pumps a subsequent-order SRS (also on an MDR). In an optical cell, the second-order Stokes SRS is at least 50 times weaker than the first-order Stokes SRS [Qian and Chang, 1986]. In fact, due to the long interaction length, first-order SRS at $j = 1$ is a more efficient pump for the second-order SRS of the v_4 mode than the incident laser is in pumping the first-order Stokes SRS of the v_4 mode.

4.3. Stimulated Anti-Stokes Raman Scattering

Stimulated anti-Stokes Raman scattering (SARS) is the first process to be described for which the SRS signal acts as a pump. The SARS signal in droplets is produced by the mixing of the pump and the first- and multiorder stimulated Raman signals. While nonlinear optical processes such as SRS and SBS are self-phase matched, SARS requires phase matching, with $\mathbf{k}_{SARS} = 2\mathbf{k}_L - \mathbf{k}_{SRS}$. In the bulk, SARS is a very weak process.

SARS and SRS are detected from water droplets illuminated by a train of 100-ps green pulses. Input radiation is focused at the edge such that the internal pump wave will propagate around the droplet rim,

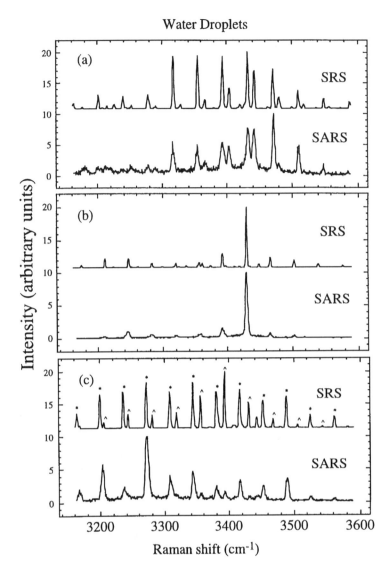

Water Droplets

Figure 8 (a) to (c) Simultaneously detected SARS and SRS spectra from water droplets for droplets with three slightly different radii all approximately 35 μm. In (c) the SRS spectrum contains two dominant sets of peaks (indicated by * and ^) that correspond to two different TE MDR mode orders ℓ. Note that in (c) the SARS spectrum does not mirror the SRS spectrum. (From Leach, D. H., Chang, R. K., and Acker, W. P. 1992. *Opt. Lett.* 17(6):387–389. With permission.)

as does the generated SRS, leading to better overlap. The SRS and SARS are simultaneously detected using two spectrographs.

Figure 8a to c display simultaneous SRS and SARS spectra from water droplets at three different values of the droplet radius. The large bandwidth of the water SRS Raman gain overlaps with several consecutive n modes of the same mode order, ℓ. SRS is achieved at these MDR frequencies and appears in the spectrum as a series of periodic peaks. The high resolution of these MDR peaks in the spectrum, which are integrated over a period of minutes, demonstrates the stability and reproducibility of the droplets produced by the Berglund–Liu droplet generator. While the many orders of SRS are a result of a cascade process (see Section 4.2), it is obvious that the same is not true for SARS because first-order SARS is weak. Vastly different SRS spectra can be obtained by changing the droplet radius. As the droplet size is changed between Figure 8a and b, the SARS spectrum mirrors the SRS spectrum. However, the coincidence of the SARS frequency with an output MDR affects the detected SARS intensity. For the same SRS intensity, the SARS intensity from droplets of a certain size can be more than two orders of magnitude larger than from different-sized droplets. This output resonance enhancement of the SARS signal on an MDR is further evident in Figure 8c. The two series of peaks in Figure 8c correspond to MDRs with two different ℓ values (distinguished by * and ^) and consecutive n values. The relative intensities of the two series of peaks in the SRS spectrum is not reflected in the relative intensities in the SARS spectrum. In fact, the SARS signal is much stronger for the ℓ^* series.

4.4. Stimulated Brillouin Scattering

SBS was observed [Zhang and Chang, 1989] from droplets illuminated by nanosecond pulses. The SBS signal was distinguished from the laser using a **Fabry–Perot interferometer**. SBS was not expected to be observed from droplets illuminated by mode-locked pulses because the 100-ps laser pulse duration is shorter than the lifetime of the Brillouin phonons (≈ 1 ns for 180° backward scattering, and > 100 ns for near-forward scattering). The phonon field does not have time to build up to a steady-state value during the 100 ps duration of the laser pulse, and the transient gain is therefore low. However, SBS has been observed from CS_2 and ethanol droplets illuminated by picosecond pulses when the focus of the laser is at the edge of the droplet. It is not feasible to use the Fabry–Perot interferometer to distinguish between the SBS and the laser since the line width of the laser is 0.3 cm^{-1} while the SBS shift is 0.2 cm^{-1}. Instead, the SBS light is distinguished using the directionality of SBS. By illuminating only one edge of the droplet with a focused beam, the input-coupled light trapped on MDRs circulates the droplet counterclockwise (see Figure 6). Brillouin scattered light is generated in all directions, but only the Brillouin scattered light generated in the near-forward or near-backward directions (both waves are tangent to the droplet surface) will be supported by MDRs and have sufficient overlap with the pump to build to a detectable signal. Near-forward SBS will have the same sense of circulation as the input-coupled light (counterclockwise in Figure 6), whereas the near-backward SBS will have the opposite sense of circulation (clockwise in Figure 6). Near-*backward* SBS is observed [Cheung et al., 1995] from ethanol droplets illuminated at the edge by a train of 100-ps pulses. The near-backward SBS signal is detected with an avalanche photodiode; the pulses are found to be stochastic and to have an input intensity threshold. Near-*forward* SBS is also conjectured [Wirth et al., 1992] from observations of stochastic scattering from CS_2 droplets. However, the near-forward SBS results were not conclusive, particularly because the near-forward SBS gain is much smaller than the near-backward SBS gain. Because near-forward SBS circulates in the same direction as the input-coupled light, directionality would not be useful in distinguishing near-forward SBS from the input. A method of isolating the near-forward SBS from droplets has not been found.

4.5. Stimulated Low-Frequency Scattering

This section describes another example of a nonlinear process for which the SRS signal acts as a pump. When CS_2 droplets are illuminated by multiple 100-ps pulses from a mode-locked, Q–switched Nd:YAG laser, extensive broadening (>500 cm^{-1}) is seen both on the Stokes side of the laser line as well as the first- and multiorder SRS lines [Cheung et al., 1993]. Figure 9 shows extensive broadening of the first-, second-, and third-order Stokes SRS from a pure CS_2 microdroplet (see Figure 9) illuminated by a *single-mode* pulse. The strong peaks at 656, 1312, and 1968 cm^{-1} are, respectively, the first-, second-, and third-order SRS lines of the ν_1 symmetric stretching mode of CS_2. For every input-laser shot, each order of the SRS lines is asymmetrically broadened, although the width of the broadening varies from shot to shot. The broadening extends more than 500 cm^{-1} toward the low-frequency (increasing Stokes shift) side of the first- and multiorder SRS lines. The observed spectra consist of sets of periodic peaks superimposed on a broad continuum. The periodic peaks are spaced by 38 cm^{-1} and correspond to MDRs with consecutive mode numbers (n-numbers) of the same radial mode order (ℓ-number). Similar asymmetric broadening is also seen from droplets of other liquids consisting of anisotropic molecules, such as benzyl alcohol, toluene, and chloroform.

This asymmetric broadening is attributed to a stimulated scattering process, in which the photon loses energy to the reorientational motion of the anisotropic liquid molecules and the photon frequency becomes shifted toward lower frequencies. The process is named stimulated low-frequency scattering (SLFS). It is generally accepted that there are at least three physical components of molecular reorientational motion that affect low-frequency shifts in the inelastically scattered light. These three components are diffusive, librational, and translational in nature [Hattori and Kobayashi, 1991]. Of the three components, diffusive motion occurs on the longest time scale and is responsible for the lower-frequency-shifted portion of the inelastically scattered light spectrum. The librational and translational components occur on a short time scale and, thus, are responsible for the higher-frequency portion of the inelastic

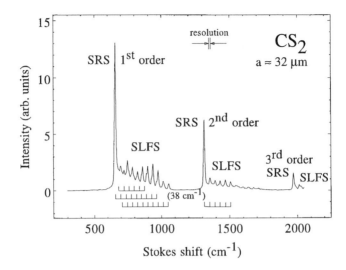

Figure 9 The stimulated inelastic spectrum from a CS_2 droplet with radius ≈ 32 μm, irradiated by a laser pulse at $\lambda_{in} = 532$ nm. The SLFS extends to ≈ 500 cm^{-1} on the Stokes side of each of the three orders of SRS associated with the v_1 mode of CS_2 (656 cm^{-1}). The periodicity of the MDR peaks (38 cm^{-1}) is indicated. The spectral resolution of the spectrograph and linear-array detector is shown. (From Cheung, J. L., Kwok, A. S., Juvan, K. A., Leach, D. H., and Chang, R. K. 1993. *Chem. Phys. Lett.* 213:309. With permission.)

spectrum (extends up to 100 cm^{-1} for CS_2). These three components of the reorientational motion are strongly coupled.

The observed broadening to more than 500 cm^{-1} is proposed to be caused by multiple orders of Stokes SLFS, i.e., a *cascade* SLFS process. In Figure 9, SLFS is pumped by the SRS into the two adjacent cavity modes. The MDR closer to the SRS in frequency will experience higher gain. The SLFS intensity in this MDR becomes so intense that it parametrically beats with the SRS light to pump SLFS in the next two cavity modes. Each higher-order SLFS is pumped by the lower-order SLFS at MDRs closer to the SRS in frequency.

The SLFS gain in bulk CS_2 has been measured to be an order of magnitude lower than the SRS gain. Thus, the strong SLFS signal observed illustrates the MDR enhancement of weak nonlinear optical processes. However, the SLFS spectra from droplets would not be useful for extracting the details of molecular motion, both because the SLFS gain profile is modified by the MDR mode structure and because of the cascade effect.

4.6. Third-Harmonic Generation and Sum-Frequency Generation

TSFG results from the interaction of three electromagnetic waves, at frequencies ω_1, ω_2, and ω_3, to produce a wave at the sum frequency, $\omega_{TSFG} = \omega_1 + \omega_2 + \omega_3$. The mixing of these three waves in a nonlinear medium generates a nonlinear polarization $P_{(\omega_{TSFG})}^{NLS} = \chi^{(3)}E(\omega_1)E(\omega_2)E(\omega_3)$, which acts as a source of radiation at ω_{TSFG}. $\chi^{(3)}$ is the third-order nonlinear susceptibility of the liquid, and $E(\omega_i)$ is the electric field amplitude at frequency ω_i. Energy conservation requires that $\omega_{TSFG} = \omega_1 + \omega_2 + \omega_3$, and photon momentum conservation requires that the resultant wave vector, $\mathbf{k}_{TSFG} = \mathbf{k}_1 + \mathbf{k}_2 + \mathbf{k}_3$, which is also known as the phase-matching condition. Third-harmonic generation (THG) at $\omega_{THG} = 3\omega_1$ results when the three input frequencies are degenerate, i.e., when $\omega_1 = \omega_2 = \omega_3$. Similar to SARS, the generating waves are at the frequency of the laser ω_L and/or the frequencies of the multiorder Stokes SRS. Discrete TSFG peaks with intensity $I_{TSFG}(\mathbf{p})$ occur at frequencies $\omega_{TSFG}(\mathbf{p}) = 3\omega_L - \mathbf{p}\omega_{vib}$, where ω_{vib} is the frequency of the vibrational mode with the largest Raman cross section and $\mathbf{p}$ is an integer. The highest $\mathbf{p}$ value is three times the highest-order Stokes SRS generated within the droplet. Although TSFG in droplets is relatively weak compared with SRS or lasing, it is still orders of magnitude more intense, per volume of liquid, than TSFG in bulk liquids. This relatively large intensity in droplets occurs because of the following features of the MDRs: (1) MDRs provide feedback so that both the generating and output TSFG waves grow more rapidly and are confined to the same regions of space for relatively long times; (2) certain MDRs of the generating waves and the MDRs of the TSFG wave overlap relatively well in space; (3) there is a relatively high spectral density of output MDRs.

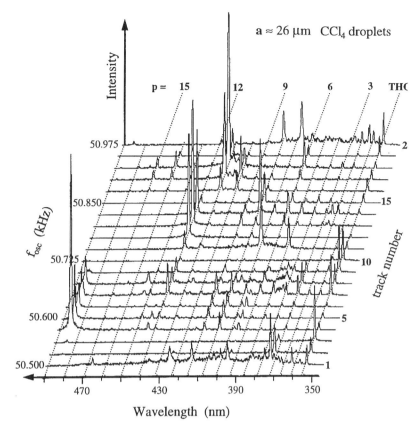

Figure 10 Detected TSFG spectra from CCl_4 droplets at 20 consecutive f_{osc} (shown on the vertical abscissa) starting at f_{osc} = 50.5 kHz and ending at f_{osc} = 50.975 kHz in 25-Hz increments. The horizontal abscissa indicates wavelength decreasing toward $\lambda_L/3$, and the series of dotted lines correspond to $\omega_{TSFG} = 3\omega_L - p\omega_{vib}$ where ω_{vib} = 459 cm^{-1} is the frequency of the ν_1 vibrational mode of CCl_4 and **p** is an integer. The incident-laser wavelength is λ_L = 1.064 μm, and the detector is a position-sensitive resistive-anode device referred to as a Mepsicron. (From Leach, D. H., Chang, R. K., Acker, W. P., and Hill, S. C. 1993. *J. Opt. Soc. Am. B* 10(1):34–45. With permission.)

Figure 10 displays the TSFG spectrum of **a** ≈ 26 μm CCl_4 droplets as a function of f_{osc}. CCl_4 droplets are chosen for two reasons. First, highly stable (1 part in 10^4) and reproducible CCl_4 droplets can be generated throughout a period of 90 min. Second, CCl_4 has a small ω_{vib} = 459 cm^{-1} and, thus, at each f_{osc} the optical multichannel analyzer used in this experiment can simultaneously detect a large number of TSFG peaks (up to **p** = 17, see the wavelength abscissa in Figure 10). In Figure 10, the droplet generator frequency is tuned in increments of Δf_{osc} = 25 Hz. Each change in f_{osc} corresponds to a change in droplet radius of approximately 1 part in 6000.

In Figure 10, the wavelength abscissa increases from $\lambda_L/3$. A series of dotted lines is drawn to guide the eye and connects points corresponding to $\omega_{TSFG} = 3\omega_L - p\omega_{vib}$ in each individual spectrum. The value of **p** increases from 0 to 17, and ω_{vib} = 459 cm^{-1}. As f_{osc} is tuned, different TSFG frequencies are seen to experience enhancement. The SRS always remains on one or more MDRs, because the SRS gain profile is broader than the average MDR spacing. The change of the TSFG spectra with f_{osc} is due to the tuning of the TSFG frequency on or near an output MDR.

5. EXPERIMENTAL OBSERVATIONS OF TEMPORAL EFFECTS ASSOCIATED WITH THE MICRODROPLET CAVITY Q FACTOR

5.1. Temporal Determination of the Photon Lifetime in a Microdroplet

The duration (100 ps) of mode-locked Nd:YAG laser pulses is considerably shorter than the typical cavity lifetime of MDRs (≈10 ns). These 100-ps pulses incident on a droplet can directly couple into

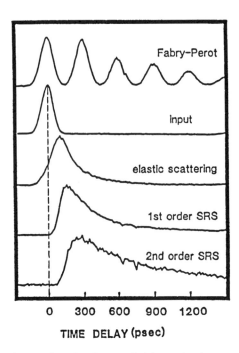

Figure 11 Time profiles of the following signals recorded by a streak camera: (1) transmission through a Fabry–Perot interferometer; (2) the input laser radiation channeled to the streak camera by an optical fiber; (3) backward-circulating SBS from edge B shown in Figure 6; (4) backward-circulating first-order SRS from edge B shown in Figure 6; (5) second-order SRS also from edge B. (From Zhang, J.-Z., Leach, D. H., and Chang, R. K. 1988. *Opt. Lett.* 13(4):270–272. With permission.)

MDRs (e.g., in the case of input resonance) or indirectly generate nonlinear optical radiation (such as SRS) on MDRs (e.g., in the case of output resonance). The measurement of the time profile of light leaking from these MDRs then gives a direct determination of the cavity lifetime. The lifetime of SRS from droplets is measured to be >5 ns [Zhang et al., 1988], giving a cavity Q > 2 × 10⁷. Previous estimates of the cavity Q were in the range of 10⁵ [Ashkin and Dziedzic, 1977; Tzeng et al., 1984; Snow et al., 1985; Thurn and Kiefer, 1985], and were deduced from the half-width of the MDR spectrum (limited by the spectrometer or tunable dye laser resolution) or by the energy transfer between two types of dye molecules [Folan et al., 1985]. A Q value of ≈10⁷ has also been deduced from later experiments involving absorption of dyes.

To increase the coupling of the incident radiation with an MDR, the incident beam is tightly focused to a spot diameter of ≈15 μm and irradiated along the droplet equator. The droplet is observed at 90° to the input laser direction. The collection optics imaged the backward-circulating radiation from the dimmer edge of the droplet onto the streak camera slit (from B in Figure 6).

Figure 11 shows the time profile of the backward-circulating green light recorded with a streak camera. The green light from B was identified originally as elastic scattering [Zhang and Chang, 1988] or input coupling, but should actually be attributed [Cheung et al., 1995] to near backward-scattered SBS because of the backward circulation of the green light with respect to the input laser. The SBS internal field continues to grow during the entire input pulse and reaches a maximum at the end of the input pulse. After the input pulse is shut off, the green backward circulating SBS radiation starts to decay with τ_{SBS}, which is measured to be 130 ps and corresponds to an MDR with Q ≈ 5 × 10⁵. The τ_{SBS} is less than the intrinsic τ_{MDR} because the SBS is depleted in pumping the SRS within the droplet.

The red SRS signal from ethanol droplets of this same size was isolated from the green signal by use of a red filter in the collection optics. The time profile of the first-order SRS from ethanol droplets (Raman shift = 2930 cm⁻¹) is shown also in Figure 11. The maximum intensity is reached slightly later than for the backward-circulating SBS, and the intensity decays with τ_{SRS} = 210 ps, corresponding to Q = 6 × 10⁵. Second-order SRS within the droplet is pumped by the internal intensity of the first-order SRS. The lifetime of the first-order SRS is therefore limited by depletion to second-order SRS, and the intrinsic MDR lifetime may be much longer. The time profile of the second-order SRS is seen

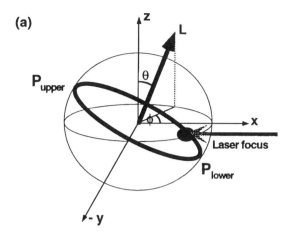

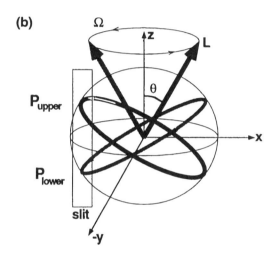

Figure 12 (a) Schematic of a spherical droplet illuminated by a focused laser beam (shaded spot) propagating along the –x axis. The internally generated SRS circulates around the rim of the droplet in a great circle that has a normal inclined at θ. (b) Schematic of an oblate droplet. The normal of the great ellipse precesses around the axisymmetric z axis at frequency Ω. The streak camera entrance slit simultaneously detects the SRS leaking from two spots P_{lower} and P_{upper} of the rim at one side of the droplet. (From Swindal, J. C., Leach, D. H., Chang, R. K., and Young, K. 1993. *Opt. Lett.* 18(3):191–193. With permission.)

in Figure 11 to reach a maximum intensity even later than the first-order SRS and to have a longer decay time of $\tau_{SRS} = 400$ ps, corresponding to $Q = 10^6$. As the input intensity is decreased, the effect of depletion caused by generating higher-order SRS is reduced. The rise and decay times of the SRS can become much longer. The longest SRS decay time was measured to be $\tau_{SRS} > 5$ ns, which translates to $Q > 2 \times 10^7$.

5.2. Precession and Mode Beating of Morphology Dependent Resonances in Nonspherical Droplets

By analogy with the angular momentum of an electron in a spherical potential, an MDR also has an "effective" angular momentum $\vec{L}$. As described in Section 2.1, the spatial distribution of an m–mode MDR is mostly confined to a great circle whose normal is inclined at $\theta = \cos^{-1}(m/n)$ with respect to the z axis (see Figure 12a). The angular momentum, $\vec{L}$, of the MDR is therefore defined to be along the normal of this great circle and has the following properties: $\vec{L}^2 = n(n+1)$; and $L_z = \vec{L} \cos\theta = m$. For a perfect sphere, the definition of the z axis is arbitrary and the great circles with normals inclined at $\theta > 0°$ are stationary. Each m-mode MDR has the same perimeter length and, thus, is spectrally $(2n + 1)$ degenerate.

For a slightly distorted droplet, the $(2n + 1)$ degeneracy of the m-modes of a sphere is split (see Section 2.1). Because of the distortion, the great circles of a sphere become great ellipses of a spheroid. Each great ellipse with its normal inclined at a different θ has a different perimeter length and, hence, a different resonant frequency. For the spheroid, the z component of the MDR angular momentum, L_z, is conserved, but neither L_x nor L_y is conserved. Thus, when the normal of the great ellipse is inclined at $\theta > 0°$, $\vec{L}$ will have a precession frequency Ω around the z axis. That is, the plane of the MDR great ellipse will precess.

This precession can also be understood in terms of a ray-optics viewpoint. The radiation that circulates once around the oblate droplet rim along a great ellipse does not end at the same point at which it began. The end point is shifted by a small angle dϕ (see Figure 12b) because of the droplet shape distortion. Therefore, if the normal of the plane of radiation is inclined at (θ,ϕ) at t = 0, then at t = π/Ω, after many trips around the droplet, the normal of the plane of radiation will be inclined at $(\theta,\phi + \pi)$. Light leaks tangentially from the great circles of the MDRs. When integrated over many precession cycles, the light leaking from the precessing MDRs appears as four arcs instead of two. These arcs are indicated by P_{upper} and P_{lower} in Figure 12b. A measurement of the oscillation period of the light leaking from P_{upper} and P_{lower} gives the precession period of the MDR.

The precession frequency, $\Omega = \omega(m + 1) - \omega(m) \approx d\omega/dm$, and is analogous to the group velocity $v_g = d\omega/dk$, where the linear momentum k is replaced by the MDR angular momentum m.

The precession frequency can be estimated [Swindal et al., 1993] in terms of the shape distortion $e = (r_p - r_e)/a$. Since the dependence of ω on m for a spheroid is given [Lai et al., 1990] by Equation 3 to be

$$\omega(m) = \omega_o \left\{ 1 - \frac{e}{6} \left[1 - \frac{3m^2}{n(n+1)} \right] \right\} \tag{11}$$

where ω_o is the frequency of the $(2n + 1)$ degenerate MDRs for a perfect sphere. The precession frequency is then

$$\Omega \approx \frac{d\omega}{dm} = \omega_o |e| \frac{m}{n(n+1)} \approx \frac{\omega_o |e|}{n} \cos\theta \approx \frac{\omega_o |e|}{m(\omega)x} \cos\theta \tag{12}$$

where $n(n + 1) \approx n^2$ for a sphere with a large size parameter x.

A measurement of the precession period is made for axisymmetrically distorted ethanol droplets [Swindal et al., 1993]. The second harmonic of a Q-switched mode-locked Nd:YAG laser ($\lambda = 0.532$ μm) is used to selectively excite SRS that is supported by MDRs with different m values (see Figure 12). The SRS radiation is then collected by a single lens at 90° to the z axis as well as 90° to the direction of laser propagation (along the −x axis). A red filter blocks the green elastically scattered incident-laser light and passes the red SRS. The lens forms a magnified image of the droplet at the vertical entrance slit of the streak camera, which simultaneously records the temporal evolution of the SRS from the two spots, P_{upper} and P_{lower}, on one side of the droplet.

Figure 13a shows the temporal profile of a single mode-locked laser pulse. Figure 13b shows the temporal profiles of the SRS pulses from P_{upper} and P_{lower} for three different inclination angles relative to the equator. When the droplet is illuminated 40° below the equator, the SRS is supported by a group of azimuthal modes with $m \approx \pm n \cos(40°)$. From the temporal oscillations of Figure 13b, it is apparent that the SRS intensities from the lower and upper spots (P_{lower} and P_{upper} of Figure 12b) are precessing at the same frequency of $\Omega = 21.6$ GHz. Furthermore, the oscillations of the two spots are noted to be 180° out of phase. As the input-laser beam is moved to illuminate 30° below the equator, the SRS is supported by a group of azimuthal modes with $m \approx \pm n \cos(30°)$ and the two red spots along one side of the droplet move toward the equator and one another. The SRS precession frequency from each red spot increases to 26 GHz. Again, the oscillations from the two red spots are 180° out of phase (see Figure 13c). When the equatorial rim is illuminated ($\theta \approx 0°$), the SRS is supported by a group of azimuthal modes with $m \approx \pm n$. Figure 13d shows no apparent oscillations. Portions of the two SRS spots near the equator cannot be spatially resolved by the streak camera. The two rapidly oscillating out-of-phase SRS signals are effectively summed, and the oscillations are not apparent. In Figure 13d, the exponential decay time, τ_{SRS}, of 1100 ps is due to the SRS cavity lifetime associated with leakage and other losses, such as depletion in generating second-order SRS.

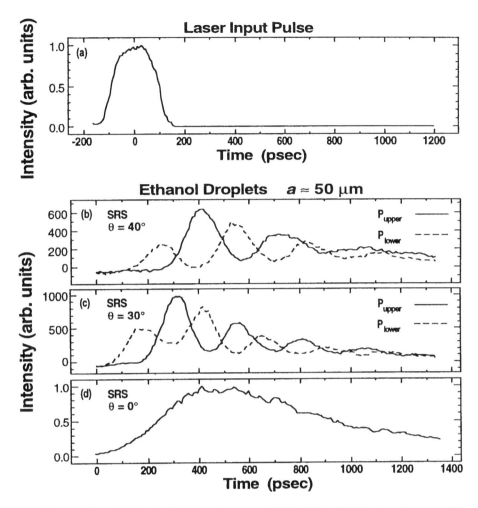

Figure 13 Time profiles of (a) the incident-laser beam, and (b) to (d) the SRS leakage from one side of the droplet. The incident-laser beam is centered (b) 40° below, (c) 30° below, and (d) on the equatorial plane. (From Swindal, J. C., Leach, D. H., Chang, R. K., and Young, K. 1993. *Opt. Lett.* 18(3):191–193. With permission.)

Precession frequencies measured with the input-laser beam inclined at different angles are compared with the theoretical precession frequency and yield a distortion amplitude |e| = 0.007.

5.3. Nonlinear Mie Scattering

Enhanced scattering was observed [Huston et al, 1990] in the latter half of a pulse train consisting of 100-ps pulses and was proposed to be due to a degradation of the cavity Q factor. The 100-ps pulses possess large Fourier-limited bandwidth and hence small spectral coupling with high-Q modes. With the reduction of the MDR Q factor, the MDR bandwidth is increased. The greater spectral overlap between the input spectrum and the MDR thus leads to a larger scattered signal. One of the proposed mechanisms for Q degradation involves the coupling of the optical field to the electrostrictively produced volume acoustic waves resonant with the droplet cavity. Acoustic waves produce density fluctuations that lead to index of refraction fluctuations. The index of refraction fluctuations can perturb the cavity modes and broaden the line width of high-Q cavity modes. These experimental observations motivated the theoretical study described in Section 3.3.

5.4. Laser-Induced Radiation Leakage from Dye-Doped Microdroplets

Ray optics and Lorenz–Mie calculations of droplet internal and surface intensity distributions have shown that some of the input rays of a parallel plane wave, after refraction by the spherical illuminated face, converge toward a ring on the droplet shadow face. This ring of high surface intensity is referred

Table 1 Possible mechanisms for light emergence from the descartes ring

Categories	Possible mechanisms	Characteristics
Index-of-refraction change	Intensity dependent index of refraction: $\Delta m(\omega, I_{inc}) = n_2 I_{inc}(t)$	Quasi-instantaneous
	Temperature dependent index of refraction: $\Delta m(\omega, \Delta T) = (\partial n/dT)\Delta T$	Cumulative
	Acoustic Brillouin phonons: short λ_p, near-backward	Quasi-instantaneous
	Acoustic Brillouin phonons: long λ_p, near-forward	Cumulative
	Plasma	Quasi-instantaneous
Surface bulge	Electrostriction	Cumulative
	Temperature lowering of surface tension	Cumulative
	Acoustic Brillouin phonons: short λ_p, near-backward	Quasi-instantaneous
	Acoustic Brillouin phonons: long λ_p, near-forward	Cumulative

to as the Descartes ring [Jarzembski and Srivastava, 1989; Srivastava and Jarzembski, 1991; Xie et al., 1991a; 1991b; Pinnick et al., 1992].

Extra leakage of SRS from the Descartes ring was reported when a water droplet, with a = 60 μm, was irradiated by a green laser beam with 10-ns pulse duration and λ_{input} = 0.532 μm. The droplet was viewed from the near-forward direction. The observation of laser-induced SRS leakage from the Descartes ring is significant because the extra emergence of SRS (other than tangentially from the rim of a perfect spherical droplet) has never been accounted for in the standard treatment of radiation leakage of nonlinear radiation on MDRs. Consequently, questions arise as to the difference, if any, between the characteristics of SRS (or any other internally generated nonlinear radiation) from the Descartes ring and from the usual droplet rim. In addition, questions arise as to the laser-induced perturbative mechanism or mechanisms that cause the internally generated radiation that is on MDRs to leak from the droplet. One form of surface perturbation is caused by the laser-induced electrostrictive force, already introduced in Section 3.

The initial experimental observations of SRS emerging from the Descartes ring and calculations of the surface intensity at the Descartes ring [Jarzembski and Srivastava, 1989] have motivated several other experiments with liquid droplets [Xie et al., 1991a; 1991b] and cylindrical jets [Pinnick et al., 1992]. These experiments involved illuminating the droplets and liquid columns with nanosecond pulses and showed that the separately measured SRS signals from a portion of the Descartes ring and from the other parts of the droplet rim have the same spectra, input-laser intensity thresholds, and delay times. The SRS intensity from the Descartes ring thus appears to be due to extra leakage from MDRs that is induced by the laser. This extra leakage was further shown to have a prompt component and a persistent component [Xie et al., 1991b].

The possible mechanisms for laser-induced radiation leakage from the Descartes ring involve either a localized index of refraction change or a surface bulge in a region of maximum surface intensity. These mechanisms can be further grouped according to their temporal responses, to account for the prompt and the persistent components [Xie et al., 1991b] in the laser-induced radiation leakage observed. The perturbations are divided into those with quasi-instantaneous or cumulative mechanisms. Quasi-instantaneous perturbations are essentially dependent only on the instantaneous irradiance of the laser radiation and are not dependent on the history of the previous pulses. Cumulative perturbations are dependent on the time integral of the input-laser pulse fluence, i.e., the pulse energy. Such perturbations can continue to develop after the laser pulse has ended. The various possible mechanisms are categorized in Table 1.

The use of a train of picosecond pulses (instead of a single nanosecond pulse as in previous experiments) can in principle allow the discrimination of the quasi-instantaneous mechanisms from the persistent, cumulative mechanisms. One single 100-ps pulse possesses sufficient irradiance to induce quasi-instantaneous perturbations, but has a sufficiently low fluence that eliminates cumulative effects such as heating. The combination of multiple pulses from a mode-locked train provides the ability to investigate cumulative perturbations. Instead of SRS, the leakage from the Descartes ring of lasing is investigated. The pulse irradiance required to generate lasing is lower than that for SRS. The associated dye absorption near the laser-wavelength also suppresses SBS and thus eliminates SBS as another possible perturbation. However, the addition of dye enhanced heating and temperature-related effects.

Extra leakage of lasing radiation at the Descartes ring is observed [Chen et al., 1993a] even when a dye-doped ethanol droplet is illuminated by a *single* 100-ps pulse. The single mode-locked pulse result shows the importance of quasi-instantaneous effects. However, even more leakage is observed with an input pulse train of five pulses, indicating that cumulative effects also play a role in the extra light leakage from the Descartes ring.

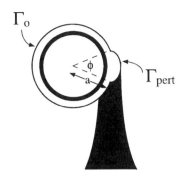

Figure 14 Schematic of a microdroplet with a laser-induced localized perturbation along one edge of the droplet perimeter (subtended by the angle ϕ). The perturbation causes an increased leakage rate of Γ_{pert} along the illuminated edge. The leakage rate along the remainder of the droplet perimeter is the unperturbed rate Γ_o. The incident radiation only experiences the perturbed leakage rate, Γ_{pert}, whereas the internal radiation experiences a leakage rate averaged over the droplet perimeter.

Despite the numerous experimental and theoretical studies on the extra light leakage from the Descartes ring, a single mechanism for the extra leakage could not be isolated. The effect appears to be due to several, possibly interconnected mechanisms. It is unclear whether or not further study would actually be able to single out one main mechanism for the extra leakage. The study of the Descartes ring, however, has increased the understanding of perturbative mechanisms that cause extra leakage from MDRs. This understanding provided a background for the study of Q spoiling which will be described in the next section.

5.5. Localized Q Spoiling of Droplet Cavity Modes

A localized laser-induced perturbation can be advantageous for increasing the coupling of short pulse radiation into cavity modes. As described in Section 2.4, light couples into a cavity mode at the same rate that it leaks out. Therefore, for short pulses on resonance with high-Q MDRs, the internal radiation does not have time to reach steady state within the duration of the input pulse. The fraction of input light coupled into high-Q MDR modes is small. A surface bulge or index of refraction change causes light to couple into a droplet mode at a faster rate. Hence, a larger fraction of the input pulse energy can be coupled into high-Q cavity modes.

The lifetime of the MDR radiation is determined by the spatially averaged leakage rate of radiation out of the droplet. For a leakage rate, Γ_{MDR}, that is spatially uniform along the whole perimeter of the droplet, the cavity lifetime is $\tau_{MDR} = 1/\Gamma_{MDR}$. In Figure 14 a localized perturbation induced by the laser causes an increased leakage rate along a short arc, ϕ, of the droplet perimeter. The internal radiation, in circumnavigating the droplet perimeter, sees both the unperturbed leakage rate (per unit path length), Γ'_o, as well as the perturbed leakage rate (per unit pathlength), Γ'_{pert}. The lifetime of the *internal* radiation is thus inversely proportional to the leakage rate averaged over the pathlength of the circulating light, such that

$$2\pi a \Gamma'_{avg} = (2\pi - \phi) a \Gamma'_0 + \phi a \Gamma'_{pert}$$

$$\tau_{MDR} = \frac{1}{2\pi a \Gamma'_{avg}} \tag{13}$$

For a focused input laser, however, input coupling only depends on the localized leakage rate along the *illuminated* edge of the droplet. Thus, the external radiation focused along the arc where there is increased leakage will couple into the droplet at the perturbed leakage rate, Γ'_{pert}, which is larger than Γ'_{avg}. In terms of cavity Q factors, the Q factor experienced by the input laser pulse is lower than the path-averaged Q factor experienced by the internal radiation on the MDR. The increased leakage/coupling due to the localized perturbation allows a larger fraction of the input pulse energy to be coupled *into* the droplet cavity mode to excite nonlinear optical processes. The excitation of nonlinear optical processes in droplets by short picosecond pulses is enhanced by this localized, laser-induced Q spoiling. The reduction in the

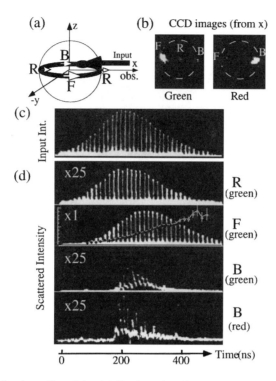

Figure 15 (a) Schematic of an ethanol droplet illuminated at the equator by a focused input beam. (b) CCD images of the green and red light scattered from the droplet. (c) Time profile of the 100-ps input-laser pulses, separated by 13.2 ns. (d) Time profiles of the green light from the spots R (reflected/refracted), F (forward circulating) and B (backward circulating), and the red light from B. The vertical scale of the oscilloscope traces have been magnified by the number shown to the left of the traces. The ratio of the pulse heights of F to the input is superimposed on the oscilloscope trace of the signal from F. The red light leaking from F is too weak to be detected by the photodiode.

overall Q of the droplet cavity mode is small, however, because the laser-induced perturbation is localized in the region of the illuminated spot. The long path-averaged lifetime of the droplet internal radiation still provides sufficient feedback for nonlinear optical processes such as SRS or SBS.

Increased input coupling of picosecond pulses into an ethanol droplet [Cheung et al., 1995] is observed over a period of ≈500 ns. The droplet is illuminated by a focused beam at the equator on the far side of the droplet as depicted in Figure 15a and viewed from $\phi = 180°$. In Figure 15a the input-coupled light is shown to circulate counterclockwise. To an observer in the x-direction, the input-coupled light leaks from the rim of the droplet at F (for forward-circulating). The input-coupled radiation can generate inelastic radiation (such as SRS and/or SBS) which is resonant with another MDR. Inelastically forward-scattered light on MDRs (F in Figure 15a and b) also circulates counterclockwise, whereas inelastically backward-scattered SBS or SRS circulates clockwise. To the same observer in the x-direction, the backward-scattered light will leak from the rim of the droplet at B (for backward-circulating).

When the droplet is viewed from the x-direction, both green and red (SRS) light are visible in the droplet image. Figure 15b shows separate green and red CCD images of the droplet. A red filter is used to distinguish between the green and red light. The arcs visible on the outermost edges of the droplet are from light leaking from MDRs, and are labeled as follows: F for forward circulating, and B for backward circulating. The forward-circulating light is predominately input-coupled light, while the backward-circulating light is inelastically scattered radiation such as SRS and SBS. The green spot in the center of the image in Figure 15b, labeled R, is reflected radiation from the tail of the Gaussian beam profile, i.e., reflected from the front and back surfaces of the droplet. The forward- and backward-circulating green light on MDRs shows markedly different time behavior, both from the input and from each other. This temporal difference is illustrated in Figure 15d.

At early times, the input-coupled light leaking from MDRs at F is weak. The input-coupled intensity from F then grows *smoothly* from pulse to pulse, reaching a maximum ≈150 ns *after* the maximum of the input Q–switched burst. The ratio of the input-coupled intensity to the input-laser intensity for each pulse in the train is superimposed on the oscilloscope trace in Figure 15d. The pulse-to-pulse increase of this ratio with time indicates that an increasingly larger fraction of the input pulse is coupled into MDRs. The increased input coupling is due to the laser-induced surface perturbation, which causes extra leakage into the droplet. The perturbation causing the extra leakage is persistent, i.e., it is retained during the 13.2-ns intervals between mode-locked pulses. Thus, the amount of extra leakage into the droplet modes is cumulative, with pulses in the front of the pulse train creating the perturbation such that pulses at the end of the train can be more efficiently coupled into the droplet modes. A similar mechanism involving increased input coupling expressed in terms of Q degradation was proposed to be the cause of nonlinear Mie scattering by Huston et al. [1990].

The SBS gain is much larger in the backward direction than in the forward direction, so the SBS light is predominately backward circulating and leaks from the droplet rim at B. The backward-circulating SBS light from B shows an *irregular* pulse-to-pulse growth in time above a smooth temporal continuum. The green irregular pulses are delayed in time with respect to the front of the input pulse train. Based on many shots (not shown), the pulse-to-pulse growth is stochastic. SRS, however, has equal gain in both the forward and backward directions. The red SRS pulses from B in the droplet image always start to grow after the green SBS pulses from B, and they also show a similar temporal continuum. Both the SRS and SBS light from B show a sharp input-intensity threshold. Note that in the red CCD image, the red light from B is stronger than the red light from F. This asymmetry is not observed when the droplet is viewed from $\phi = 105°$. Since the SRS process is isotropic, SRS light on oppositely circulating MDRs is expected to be of equal intensity. The brighter SRS light from B when viewed at $\phi = 180°$ is caused by a *localized* area of extra leakage at the point of incidence of the laser.

The temporal continuum in both the red and green signals from B is due to retained light on MDRs with photon retention times (τ_{MDR}) that are longer than the interpulse spacing of 13.2 ns. These long retention times are characteristic of MDRs with Q-factors of 10^7 to 10^8. The continuum occurs ≈ 200 ns after the start of the input pulse train, when the extra input coupling into MDRs is significant. The observation of these long τ_{MDR} values also implies a *localized* perturbation at the illuminated side of the droplet. As described before, the localized perturbation would cause increased input coupling of the laser light without decreasing the path-averaged cavity lifetime significantly. The increased input coupling leads to higher droplet internal intensity to excite SBS and SRS, while the long lifetime (transit time) of the trapped radiation leads to efficient pumping of SBS and SRS. As a result, SRS observed from the droplet is most intense when the input is focused at the edge of the droplet.

The perturbation causing the increased coupling is cumulative and persists during the 13.2-ns interpulse separation. Referring to Table 1, one can therefore see that this perturbation could be a temperature/heating effect, a surface bulge due to electrostriction, or the result of near-forward SBS phonons. The absorption of ethanol at the input-laser wavelength is very small, so temperature/heating effects are expected to be negligible. The effect of near-forward SBS phonons was found to be theoretically negligible [Ng, 1994]. Thus, the most likely perturbation for the cumulative extra input coupling observed would be that of the surface bulge. Quasi-instantaneous effects, however, should also be present, taking place within the 100 ps of the laser pulse duration.

6. CONCLUSION

Microdroplets form optical cavities with some of the highest optical Q factors presently achievable. The long lifetimes of radiation on the droplet cavity modes allow for a long interaction length for nonlinear optical processes. In this chapter we have reviewed several experiments that demonstrate the enhancement of weak nonlinear optical processes when either the generating and/or the resultant radiation spectrally overlaps with an MDR. The duration of picosecond pulses is shorter than the typical lifetime of the droplet cavity mode and provides a means of directly measuring the lifetime of the radiation within the cavity. The direct coupling of short pulses into a high-Q MDR with a long lifetime is very inefficient. We have shown how a localized perturbation can spoil the cavity Q factor experienced by the input radiation and increase the input coupling and yet retain a relatively high-Q factor for the *internal* radiation, such that nonlinear optical processes such as SRS and SBS are still enhanced.

The study of light interaction with microdroplets is a vast field, and space does not permit us to give an extensive review of all the exciting results that have been discovered in the past several years. Instead, we have concentrated exclusively on describing experiments and numerical calculations that involve illuminating the droplet with picosecond pulses. Using these experiments as examples, we hope to have conveyed some of the unique physical properties of these microdroplets.

7. DEFINING TERMS

CCD: A charge coupled device is a two-dimensional array of metal oxide semiconductor capacitors used for optical imaging.

Electrostrictive: Electrostriction is the phenomenon wherein some materials become compressed in the presence of an applied electric field. The electrostrictive force is independent of the polarization of the field, and is given by:

$$F = 1/2\alpha\nabla(E^2) \tag{14}$$

where α is the molecular polarizability.

Fabry–Perot interferometer: A multiple beam interferometer consisting of two reflective elements separated by a short known distance. This instrument is capable of very high resolution.

Mode locked: Mode locking is a means of producing short (picosecond) laser pulses. It involves modulating the loss or phase of an optical element in the laser cavity so as to cause different longitudinal modes of the cavity to oscillate together in phase. Mode locking occurs when the modulation period is equal to the cavity round-trip time.

Q-switched: Q-switching is a means of producing short (nanosecond) laser pulses of high intensity by modulating the quality of the laser cavity.

Stokes: A spectroscopic term describing a frequency shift toward the lower frequencies (longer wavelengths).

Streak camera: A high speed (picosecond resolution is possible) optical detector.

ACKNOWLEDGMENTS

We gratefully acknowledge the partial support of this research by the U.S. Air Force Office of Scientific Research (Grant No. F49620-94-1-0135) and by the U.S. Army Research Office (Grant No. DAAH 04-94-G-0031). We acknowledge the collaborations and helpful discussions with Prof. Kenneth Young and Dr. Steven C. Hill, both of whom have taught us and contributed much to our research. We are grateful to Prof. Robert L. Armstrong for reviewing this paper with such conscientiousness.

REFERENCES

Acker, W. P., Leach, D. H., and Chang, R. K. 1989. Third-order optical sum-frequency generation in micrometer-sized liquid droplets. *Opt. Lett.* 14(8):402–404.

Arnold, S., Spock, D. E., and Folan, L. M. 1990. Electric-field-modulated light scattering near a morphological resonance of a trapped aerosol particle. *Opt. Lett.* 15:1111.

Ashkin, A. and Dziedzic, J. M. 1977. Observation of resonances in the radiation pressure on dielectric spheres (and size measurement). *Phys. Rev. Lett.* 38:1351.

Barton, J. P., Alexander, D. R., and Schaub, S. A. 1989. Internal fields of a spherical particle illuminated by a tightly focused laser beam: focal point positioning effects at resonance. *J. Appl. Phys.* 65(8):2900–2906.

Chen, G., Chang, R. K., Hill, S. C., and Barber, P. W. 1991. Frequency splitting of degenerate spherical cavity modes: stimulated Raman scattering spectrum of deformed droplets. *Opt. Lett.* 16:1269–1271.

Chen, G., Chowdhury, D. Q., Chang, R. K., and Hsieh, W.-F. 1993a. Laser-induced radiation leakage from microdroplets. *J. Opt. Soc. Am. B* 10(4):620–632.

Chen, G., Mazumder, M. M., Chemla, Y. R., Serpengüzel, A., Chang, R. K., and Hill, S. C. 1993b. Wavelength variation of laser emission along the entire rim of slightly deformed microdroplets. *Opt. Lett.* 18(23):1193–1195.

Cheung, J. L., Kwok, A. S., Juvan, K. A., Leach, D. H., and Chang, R. K. 1993. Stimulated low-frequency emission from anisotropic molecules in microdroplets. *Chem. Phys. Lett.* 213:309.

Cheung, J. L., Hartings, J. M., and Chang, R. K. 1995. Different temporal behavior for the forward- and backward-circulating radiation within a microdroplet. *Opt. Lett.* 20:1089.

Chowdhury, D. Q., Hill, S. C., and Barber, P. W. 1992a. Time dependence of internal intensity of a dielectric sphere on and near resonance. *J. Opt. Soc. Am. A* 9(8):1364–1373.

Chowdhury, D. Q., Barber, P. W., and Hill, S. C. 1992b. Energy-density distribution inside large nonabsorbing spheres via Mie theory and geometrical optics. *Appl. Opt.* 31:3558–3563.

Chylek, P. 1976. Partial-wave resonances and the ripple structure in the Mie normalized extinction cross section. *J. Opt. Soc. Am.* 66:285.

Folan, L. M., Arnold, S., and Druger, S. D. 1985. Enhanced energy transfer within a microparticle. *Chem. Phys. Lett.* 118:322.

Gouesbet, G., Maheu, B., and Gréhan, G. 1988. Light scattering from a sphere arbitrarily located in a Gaussian beam, using a Bromwich formulation. *J. Opt. Soc. Am. A* 5:1427–1443.

Hattori, T. and Kobayashi, T. 1991. Ultrafast optical Kerr dynamics studied with incoherent light. *J. Chem. Phys.* 94:3332.

Haus, H. A. 1984. *Waves and Fields in Optoelectronics*. Prentice-Hall, Englewood Cliffs, NJ.

Hill, S. C. and Benner, R. E. 1988. Morphology-dependent resonances. In *Optical Effects Associated with Small Particles*, edited by Barber, P. W. and Chang, R. K., pp. 3–61. World Scientific, Singapore.

Hill, S. C., Leach, D. H., and Chang, R. K. 1993. Third-order sum-frequency generation in droplets: model with numerical results for third-harmonic generation. *J. Opt. Soc. Am. B* 10(1):16–33.

Huston, A. L., Lin, H.-B., Eversole, J. D., and Campillo, A. J. 1990. Nonlinear Mie scattering: electrostrictive coupling of light to droplet acoustic modes. *Opt. Lett.* 15:1176–1178.

Jarzembski, M. A. and Srivastava, V. 1989. Electromagnetic field enhancement in small liquid droplets using geometric optics. *Appl. Opt.* 28:4962.

Khaled, E. E. M., Hill, S. C., and Barber, P. W. 1993. Scattered and internal intensity of a sphere illuminated with a Gaussian beam. *IEEE Tran. Antennas Propagation* 41:295–303.

Khaled, E. E. M., Chowdhury, D. Q., Hill, S. C., and Barber, P. W. 1994. Internal and scattered time-dependent intensity of a dielectric sphere illuminated with a pulsed Gaussian beam. *J. Opt. Soc. Am. A* 11(7):2065–2071.

Lai, H. M., Leung, P. T., Poon, K. L., and Young, K. 1989. Electrostrictive distortion of a micrometer-sized droplet by a laser pulse. *J. Opt. Soc. Am. B* 6:2430.

Lai, H. M., Leung, P. T., Young, K., Barber, P. W., and Hill, S. C. 1990. Time-independent perturbation for leaking electromagnetic modes in open systems with application to resonances in microdroplets. *Phys. Rev. A* 41:5187–5198.

Lai, H. M., Leung, P. T., Ng, C. K., and Young, K. 1993. Nonlinear elastic scattering of light from a microdroplet: role of electrostrictively generated acoustic vibrations. *J. Opt. Soc. Am. B* 10(5):924–932.

Leach, D. H., Acker, W. P., and Chang, R. K. 1990. Effect of the phase velocity and spatial overlap of spherical resonances on sum-frequency generation in droplets. *Opt. Lett.* 15(16):894–896.

Leach, D. H., Chang, R. K., and Acker, W. P. 1992. Stimulated anti-Stokes Raman scattering in microdroplets. *Opt. Lett.* 17(6):387–389.

Leach, D. H., Chang, R. K., Acker, W. P., and Hill, S. C. 1993. Third-order sum-frequency generation in droplets: experimental results. *J. Opt. Soc. Am. B* 10(1):34–45.

Lock, J. A. and Hovenac, E. A. 1993. Diffraction of a Gaussian beam by a spherical particle. *Am. J. Phys.* 61:698–707.

Newton, R. G. 1966. *Scattering Theory of Waves and Particles*. Springer-Verlag, New York.

Ng, C. K. 1994. Ph.D. dissertation, Chinese University of Hong Kong, Shatin, Hong Kong.

Pinnick, R. G., Fernández, G. A., Xie, J.-G., Ruekgauer, T., Gu, J., and Armstrong, R. L. 1992. Stimulated Raman scattering and lasing in micrometer-sized cylindrical liquid jets: time and spectral dependence. *J. Opt. Soc. Am. B* 9:865.

Qian, S.-X., Snow, J. B., and Chang, R. K. 1985. Coherent Raman mixing and coherent anti-Stokes Raman scattering from individual micrometer-size droplets. *Opt. Lett.* 10:499.

Qian, S.-X. and Chang, R. K. 1986. Multiorder Stokes emission from micrometer-size droplets. *Phys. Rev. Lett.* 56:926.

Snow, J. B., Qian, S.-X., and Chang, R. K. 1985. Stimulated Raman scattering from individual droplets at wavelengths corresponding to morphology-dependent resonances. *Opt. Lett.* 10:37.

Srivastava, V. and Jarzembski, M. A. 1991. Laser-induced stimulated Raman scattering in the forward direction of a droplet: comparison of Mie theory with geometrical optics. *Opt. Lett.* 16:126–128.

Swindal, J. C., Leach, D. H., Chang, R. K., and Young, K. 1993. Precession of morphology-dependent resonances in nonspherical droplets. *Opt. Lett.* 18(3):191–193.

Thurn, R. and Kiefer, W. 1985. Structural resonances observed in the Raman spectra of optically levitated liquid droplets. *Appl. Opt.* 24:1515.

Tzeng, H.-M., Wall, K. F., Long, M. B., and Chang, R. K. 1984. Laser emission from individual droplets at wavelengths corresponding to morphology-dependent resonances. *Opt. Lett.* 9:499.

Tzeng, H.-M., Long, M. B., Chang, R. K., and Barber, P. W. 1985. Laser-induced shape distortions of flowing droplets deduced from morphology-dependent resonances in fluorescence spectra. *Opt. Lett.* 10:20.

Weiss, D. S., Sandoghdar, V., Hare, J., Lefevre-Seguin, V., Raimond, J.-M., and Haroche, S. 1995. Splitting of high-Q Mie modes induced by light backscattering in silica microspheres. *Opt. Lett.* 20:1835–1837.

Wirth, F. H., Juvan, K. A., Leach, D. H., Swindal, J. C., and Chang, R. K. 1992. Phonon-retention effects on stimulated Brillouin scattering from micrometer droplets illuminated with multiple-short laser pulses. *Opt. Lett.* 17:1334.

Xie, J.-G., Ruekgauer, T. E., Gu, J., Armstrong, R. L., and Pinnick, R. G. 1991. Observations of Descartes ring stimulated Raman scattering in micrometer-sized water droplets. *Opt. Lett.* 16:1310.

Xie, J.-G., Ruekgauer, T. E., Gu, J., Armstrong, R. L., Pinnick, R. G., and Pendleton, J. D. 1991b. Physical basis for Descartes ring scattering in laser-irradiated microdroplets. *Opt. Lett.* 16:1817.

Zhang, J.-Z. and Chang, R. K. 1988. Shape distortion of a single water droplet by laser-induced electrostriction. *Opt. Lett.* 13:916.

Zhang, J.-Z., Leach, D. H., and Chang, R. K. 1988. Photon lifetime within a droplet: temporal determination of elastic and stimulated Raman scattering. *Opt. Lett.* 13(4):270–272.

Zhang, J.-Z. and Chang, R. K. 1989. Generation and suppression of stimulated Brillouin scattering in single liquid droplets. *J. Opt. Soc. Am. B* 6(2):151–153.

FOR FURTHER INFORMATION

A comprehensive review of the properties of MDRs is presented in *Optical Effects Associated with Small Particles*, Chapter 1, by Steven C. Hill and Robert E. Benner (World Scientific, 1988). In addition, Chapters 2 to 4 in the same book describe various experimental studies involving optical radiation and microdroplets.

Several review papers have been written on the subject of microdroplets. An example of one such review paper is "Nonlinear Optics in Droplets" by Steven C. Hill and Richard. K. Chang, in *Studies in Classical and Quantum Nonlinear Optics*.

The *Principles of Nonlinear Optics* by Y. R. Shen and *Nonlinear Optics* by Robert W. Boyd are excellent teaching texts on the subject of nonlinear optics.

Chapter 10

Optical Properties of n-i-p-i and Hetero-n-i-p-i Doping Superlattices

Peter Kiesel and Gottfried H. Döhler

Reviewed by P. Wißmann

TABLE OF CONTENTS

1. INTRODUCTION

About 25 years ago a number of authors independently developed the idea of fabricating "artificial" semiconductors with novel electrical and optical properties by creating a superlattice [Esaki and Tsu, 1970; Kazarinov and Shmarsev, 1971; Stafeev, 1971; Ovsyannikov et al., 1971; Döhler, 1972]. Since then, the interest in such synthetic semiconductors has grown continuously, first of all because of the new possibility of testing concepts of quantum mechanics and novel phenomena in suitably designed model systems and because of their promising applications. Basically, two classes of superlattices have been considered. The "compositional superlattices" are composed of a periodic sequence of thin layers of different semiconductors. The "doping superlattices" are made of a uniform semiconductor material by growing alternating n- and p-doped layers, possibly with intrinsic layers in between. According to their doping profile they are also called n-i-p-i crystals [Döhler, 1972]. Both types of superlattices offer the fascinating possibility of "*tailoring*" the electrical and optical properties during growth by a suitable choice of their design parameters (material composition, layer thickness, number of periods, doping concentration, etc.). In contrast to their compositional counterpart, n-i-p-i doping superlattices offer an additional unique feature: the properties of a given sample are *tunable*. Band gap, subband structure, carrier concentration, and carrier lifetimes are not only properties which can be adjusted by an appropriate choice of the design parameters, but they are quantities which can be externally varied by light or an external bias within wide limits. Thus, doping superlattices form a class of semiconductors in which the electrical and optical properties are not fixed material parameters but tunable quantities. Neither bulk semiconductors nor compositional superlattices show a comparable strength of tunability.

Although these properties of n-i-p-i crystals are very appealing with respect to fundamental aspects, as well as for future device applications, and although they were predicted quite early [Döhler, 1972], their experimental verification came much later than the one for their compositional counterparts. This is quite surprising since the requirements for growing high-quality doping superlattices are less demanding than for the compositional ones.

The unusual properties of n-i-p-i crystals result from the different nature of the superlattice potential which is built up by the space charge of the ionized donors and acceptors in the n- and p-doped layers. This is in contrast to compositional superlattices, in which the superlattice potential originates from the different band gap values of the components. The periodic space charge potential causes a parallel modulation of the conduction and valence band edges resulting in an *indirect band gap in real space* (see Figure 1). As a consequence, electrons and holes are spatially separated, and, therefore, they exhibit recombination lifetimes which can be by orders of magnitude larger than those in unmodulated semiconductors. This lifetime enhancement depends strongly on the period and the amplitude of the space charge potential, which can be adjusted by the choice of the design parameters. Because of the long recombination lifetimes, even large excess carrier concentrations can be achieved by relatively weak optical or electrical excitation. The space charge of the excess electrons and holes partly screens the impurity space charge in the n- and p-layers, respectively. As a consequence, the superlattice potential and therefore also the indirect band gap in real space, the recombination lifetimes, the internal fields, and many other properties, which are correlated with these quantities, are tunable.

In contrast to optically excited n-i-p-i crystals whose dynamic behavior is governed by the internal electron–hole recombination lifetime, which can be as high as milliseconds, the relevant response time for structures provided with selective n- and p-type contacts is the externally adjustable RC-time constant. Thus, the unusual phenomena observed in doping superlattices can also be very fast (e.g., in the subnanosecond regime) and the widely spread prejudice — "n-i-p-i crystals are an excellent model system for the study of a broad range of physical phenomena but they are too slow for most of the device applications" — no longer holds for n-i-p-i crystals provided with selective contacts.

It was realized quite early that the possibilities of designing artificial semiconductors with tunable properties can still be drastically increased, if the concept of n-i-p-i structures is combined with that of compositional superlattices by fabricating "hetero-n-i-p-i superlattices" [Ruden and Döhler, 1983]. By inserting an undoped quantum well at the center of the doping layers, e.g., the "type-I hetero-n-i-p-i" or by choosing a lower band gap material for the intrinsic layers, the "type-II hetero-n-i-p-i" structures can be produced. One of the obvious advantages of hetero-n-i-p-i structures is a strong reduction of the effects of scattering and of potential fluctuations which are due to the randomly distributed ionized impurities in the doping layers. However, we will demonstrate that hetero-n-i-p-i structures also exhibit other interesting advantages.

Over the years several hundred papers on n-i-p-i structures have been published. In this review we will restrict ourselves to the discussion of luminescence and absorption in n-i-p-i and hetero-n-i-p-i systems. A number of interesting issues concerning optical properties of n-i-p-i structures, such as Raman scattering [Döhler et al., 1981; Zeller et al., 1982; Fasol et al., 1984], photoreflectance [Keil et al., 1991], and ambipolar diffusion [Gulden et al., 1991], had to be excluded.

This review is organized as follows. In Section 2 we will summarize the basic aspects about the electronic structure of n-i-p-i crystals and its tunability. Section 3 describes the preparation of doping superlattices by using a special growth technique, the epitaxial shadow mask molecular beam epitaxy (ESM-MBE). This technique allows us to provide n-i-p-i crystals with excellent ohmic contacts. This

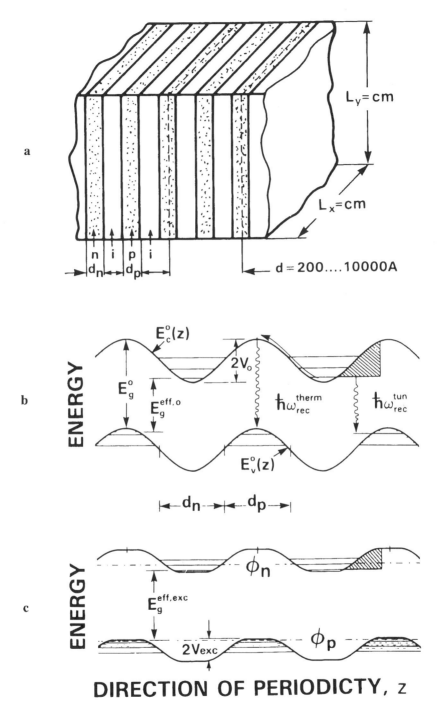

Figure 1 Schematic real space picture and band diagram of a n-i-p-i doping superlattice. The superlattice potential which modulates the conduction and valence band edges E_c and E_v is due to the space charge potential of the ionized donors and acceptors. Optical or electrical excitation leads to a partial screening of the superlattice potential by the space charge of the excess carriers. The shaded areas depict the barriers for the electron–hole recombination.

is essential for the application of n-i-p-i structures as tunable light sources or as electro-optical modulators. The tunable luminescence and the tunable absorption are dealt with in Sections 4 and 5, with regard to both basic aspects and device applications.

2. BASIC ELECTRONIC PROPERTIES OF n-i-p-i AND HETERO-n-i-p-i CRYSTALS

2.1. Tunable Subband Structure

The formation of two-dimensional subbands in doping superlattices was first briefly discussed by Esaki and Tsu [1970] and, in more detail, by Ovsiyannikov et al. [1971]. The first extensive theoretical investigation and, in particular, the discovery of the tunability of the electronic structure (i.e., the property that distinguishes doping superlattices from their compositional counterparts and from any other semiconductor) were due to Döhler [1972]. He carried out theoretical studies on a model structure consisting of a uniform semiconductor crystal which was only modulated by the introduction of n- and p-doped layers with δ-function-type distribution of dopants ("δ-doped n-i-p-i crystal"). The calculations of the tunable electronic structure which were presented by Döhler [1972] were not yet self-consistent. Self-consistent calculations for various doping profiles were reported later by many authors. (For calculations for zero temperature see, for example, Ruden and Döhler [1983b]; Döhler [1983a], and see Beyer et al. [1989]; Schrüfer et al. [1992] for finite temperatures.)

For the sake of clarity we outline in this section only the basics of the tunable electronic structure for the important specific case of a n-i-p-i structure consisting of uniformly n- and p-doped and undoped (i-) layers for $T = 0$ K. For more general treatments we refer to the literature [Döhler, 1983a; Ruden and Döhler, 1983b; Beyer et al., 1989; Schrüfer et al., 1992]. In the following we will also assume a homogeneously smeared-out space charge. We will refer to the effects due to the point-charge character of the impurity potentials and due to their random spatial distribution in connection with the luminescence of n-i-p-i crystals in Section 4.

In the upper section of Figure 1, a n-i-p-i crystal is shown schematically. The lower section of this figure depicts the real-space energy diagram for the ground state (part b) and an excited state (part c). The modulation of the conduction and valence band edges is due to the positive space charge of the donors and the negative space charge of the acceptors. For uniform doping within the n- and the p-layers and with no intrinsic layers, the strength of this modulation, $2V_0$, follows from Poisson's equation as

$$2V_0 = \left(2\pi e^2 / \kappa_0\right)\left[n_D\left(d_n / 2\right)^2 + n_A\left(d_p / 2\right)^2\right] \tag{1}$$

Here κ_0 stands for the static dielectric constant and e for the elementary charge. In Equation 1 it has been assumed that the product of doping concentration and layer thickness for the n-layers (donor concentration n_D and layer width d_n) is the same as for the p-layers (acceptor concentration n_A and layer width d_p), i.e.,

$$n_D d_n = n_A d_p \tag{2}$$

A term

$$\left(2\pi e^2 / \kappa_0\right) n_D d_n d_i \tag{3}$$

has to be added to the right-hand side of Equation 1 if undoped layers of thickness d_i exist between the n- and the p-layers.

The motion of the carriers in the superlattice direction is quantized by the space charge potential. Just as in the compositional superlattices, this leads to the formation of electronic subbands. Since we are dealing with parabolic potentials (instead of rectangular quantum wells), the energies of the low-index subband edges are harmonic oscillator levels for our simple example [Döhler, 1983b; Döhler and Ruden, 1984]

$$\varepsilon_{c,\mu} = \hbar\omega_c\left(\mu + 1/2\right) \quad \mu = 0,1,2,\ldots \tag{4}$$

$$\varepsilon_{i,\nu} = \hbar\omega_i\left(\nu + 1/2\right) \quad \nu = 0,1,2,\ldots; \quad i = \nu h, \nu l \tag{5}$$

for the conduction and valence subbands if measured from the conduction and valence band extrema at the center of the n- and the p-layers, respectively. The index i (= vl or vh) stands for the light and heavy mass valence bands (The mixing of heavy and light hole bands due to the potential $V(z)$ is neglected). The (harmonic oscillator) energies $\hbar\omega_c$ and $\hbar\omega_i$ are formally the same as the plasmon energies for electrons

$$\hbar\omega_c = \hbar\left(4\pi e^2 n_D / m_c \kappa_0\right)^{1/2} \tag{6}$$

and holes,

$$\hbar\omega_i = \hbar\left(4\pi e^2 n_A / m_i \kappa_0\right)^{1/2} \tag{7}$$

in uniformly doped semiconductors (m_c and m_i are the electron and light- and heavy-hole effective masses, respectively). The values obtained for GaAs ($m_c = 0.067m_o$; $m_{vl} = 0.074m_o$; $m_{vh} = 0.40m_o$; $\kappa_0 = 12.5$; $n_A = n_D = 10^{18}$ cm^{-3}) from Equations 6 and 7 are 40.4, 38.5, and 16.5 meV, for electrons, light holes, and heavy holes, respectively. Note that these energies depend on the doping level with the one-half power.

The value of the effective band gap $E_g^{eff,o}$, defined as the energy difference between the uppermost heavy-hole subband and the lowest conduction subband,

$$E_g^{eff,o} = E_g^o - 2V_o + \varepsilon_{c,o} + \varepsilon_{vh,o} \tag{8}$$

depends in most cases mainly on the value of V_o and, therefore, linearly on the doping concentrations, but quadratically on the thickness of the doping layers. $E_g^{eff,o}$ becomes zero if these quantities exceed certain values. A GaAs n-i-p-i crystal, for instance, becomes a "n-i-p-i semimetal" if the layer widths are larger than about $d_n = d_p = 65$ nm at a doping level of $n_D = n_A = 10^{18}$ cm^{-3}.

In the semimetal situation and also for the case where Equation 2 is not fulfilled, the calculation of the electronic structure has to be performed self-consistently [Döhler, 1978; Ruden and Döhler, 1983c]. This is also true for the excited n-i-p-i crystal, i.e., if there are electrons and holes populating subbands in the n- and p-layers, respectively.

By using the effective mass approximation, which is justified for energies close enough to the band edges, and the local density approximation of the local density functional formalism [Kohn and Sham, 1965], the following self-consistent Schrödinger-type equation is obtained for the case of the electron states in the lower subbands of a superlattice with a not too short superlattice period (for more general situations, see Ruden and Döhler [1983c]).

$$\left(-\left(\frac{\hbar^2}{2m_c}\right)\frac{d^2}{dz^2} + V_{sc}(z) - \varepsilon_{c,\mu}\right)\zeta_{c,\mu}(z) = 0 \tag{9}$$

The subband wave function is obtained by multiplying the envelope function $\zeta_{c,\mu}(z - md)$ for the m-th potential well with a Bloch factor consisting of the lattice periodic part $u_{c,k=0}(\vec{r})$ and the phase factor $\exp\{i\,\vec{k}_\parallel\,\vec{r}_\parallel + k_z md\}$ and subsequent summation over all potential wells.

The self-consistent potential,

$$V_{sc}(z) = V_o(z) + V_H(z) + V_{xc}(z) \tag{10}$$

is the sum of the bare potential of the impurity space charge, $V_o(z)$, as discussed before, and the Hartree, $V_H(z)$, and the exchange and correlation, $V_{xc}(z)$, contribution of the free electrons populating the subbands.

The Hartree contribution is given by the solution of Poisson's equation

$$d^2 V_H(z) / dz^2 = 4\pi e^2 n(z) / \kappa_o \tag{11}$$

Where $n(z)$ is the local electron density due to electrons populating subbands, $n(z)$ is given by

$$n(z) = \sum_\mu n_\mu^{(2)} \left| \zeta_{c,\mu}(z) \right|^2 \tag{12}$$

Here, the two-dimensional electron concentration in the μth subband, $n_\mu^{(2)}$, is determined by the requirement of a common quasi Fermi level, Φ_n, for all occupied electron subbands, which, for the case of zero temperature, reads

$$\varepsilon_{c,\mu} + \left(\hbar^2 / 2m_c \right) 2\pi n_\mu^{(2)} = \varepsilon_{c,o} + \left(\hbar^2 / 2m_c \right) 2\pi n_o^{(2)} \tag{13}$$

with

$$\sum_\mu n_\mu^{(2)} = n^{(2)} \tag{14}$$

For the exchange-correlation contribution to the self-consistent potential, only the exchange term was included in the calculations reported in Ruden and Döhler [1983c] since it turned out to be the dominant one because of the high local electron density in the n-layers. In the local density approximation, this yields

$$V_{xc} \cong -(3/\pi)^{1/3} e^2 \left(n(z) \right)^{1/3} / \kappa_o \tag{15}$$

In Ruden and Döhler [1983c], it has been shown that in the case of small effective electron masses and shallow impurity levels one can neglect the impurity bands and the spatial potential fluctuations due to the random distribution of the donor atoms in calculating the electronic structure of doping superlattices. In the same paper, it was also demonstrated that in most cases of III-V compound superlattices for a calculation of the hole subbands it has to be taken into account that at least at low temperatures the holes will usually populate an acceptor impurity band rather than valence subbands. This qualitative asymmetry is a consequence of both the larger effective mass of the heavy holes and the resulting larger acceptor binding energy. In this case, the self-consistent potential in the p-layers can be obtained by assuming a completely neutral central region of width

$$d_p^o = p^{(2)} / n_A \tag{16}$$

The position of the quasi Fermi level of the holes, Φ_p, can be approximated by the energy of the acceptor state in this region.

Figure 2 shows an example for the calculated electronic subband energies

$$E_{c,\mu} = E_c + \varepsilon_{c,\mu} \tag{17}$$

and the position of the electron quasi Fermi level, Φ_n, as a function of the two-dimensional electron concentration $n^{(2)}$ for a GaAs doping superlattice with $n_D = n_A = 10^{18}\,\mathrm{cm}^{-3}$ and $d_n = d_p = d/2 = 40$ nm. Also shown are the envelope wave functions $\zeta_{c,\mu}(z)$ for the lowest subbands for the ground state and for an excited state with $n^{(2)} = p^{(2)} = 1.4 \times 10^{12}\,\mathrm{cm}^{-2}$, corresponding to a state with electrons populating the lowest two subbands. The flattening of the self-consistent potential, $V_{sc}(z)$, and the widening of the wave functions, which is important for the tuning of recombination lifetimes to be discussed later on in Section 4, is easily seen.

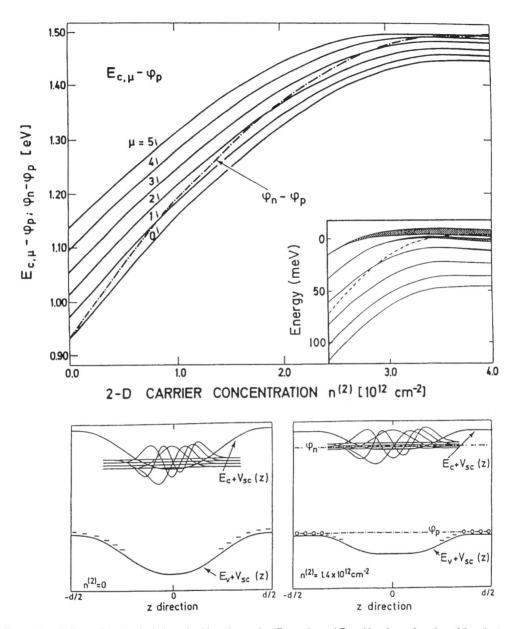

Figure 2 Self-consistent calculation of subband energies $E_{c,\mu}$ and quasi Fermi level as a function of the electron concentration per period $n^{(2)}$ for a GaAs doping superlattice ($n_A = n_D = 10^{18}$ cm⁻³, $d_n = d_p = 40$ nm, $d_i = 0$). The inset shows the situation for large carrier densities on an enlarged energy scale. The lower part of the figure shows the envelope functions $\zeta_{c,\mu}(z)$ for the lowest subbands for the ground state and for an excited state with $n^{(2)} = 1.4 \times 10^{12}$ cm⁻².

Instead of performing the self-consistent calculations, it is sufficient in many cases to have estimates for the relation between carrier density and effective band gap. For doping superlattices without intrinsic layers, for instance, such an estimation can be obtained by calculating the superlattice potential as if it were composed of flat and neutral central parts of width

$$d_n^o = n^{(2)}/n_D \tag{18}$$

and

$$d_p^o = p^{(2)}/n_A \tag{19}$$

with parabolic sections in between. The modulation of the band edges, which was given by Equation 2 for the compensated n-i-p-i crystal in the ground state, has now to be replaced by

$$2V_{exc} = \left(2\pi e^2 / \kappa_o\right)\left[n_D\left(d_n^+\right)^2 + n_A\left(d_p^-\right)^2 \right] \tag{20}$$

with

$$d_n^+ = \left(d_n - n^{(2)}/n_D\right)/2 \tag{21}$$

$$d_p^- = \left(d_p - p^{(2)}/n_A\right)/2 \tag{22}$$

and

$$n_D d_n^+ = n_A d_p^- \tag{23}$$

where Equation 23 expresses the requirement of macroscopic charge neutrality of the n-i-p-i structure. The important relation between the effective band gap in the excited state, $E_g^{eff,exc}$, and the two-dimensional carrier densities $n^{(2)}$ and $p^{(2)}$ is, in fact, rather well described by replacing V_o in Equation 8 with V_{exc} from Equation 20, as is apparent from a comparison with the nearly parabolic band gap vs. carrier density relation found in numerical self-consistent calculations (see Figure 2).

In the introduction it was mentioned that a periodic modulation of composition in addition to the periodic n- and p-doping adds a new dimension to the flexibility of tailoring the properties of n-i-p-i structures. Figures 3 and 4 depict schematically a type-I and a type-II hetero-n-i-p-i structure the properties of which are qualitatively similar to those of the "homo-n-i-p-i crystals." Details in which they differ from their counterparts will be discussed in Sections 4 and 5.

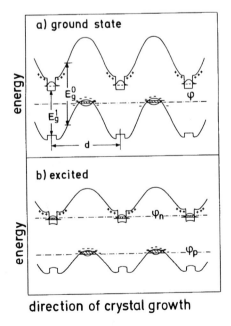

Figure 3   Schematic real space picture of the band diagram of a type-I hetero-n-i-p-i structure for the ground state (a) and an excited state (b).

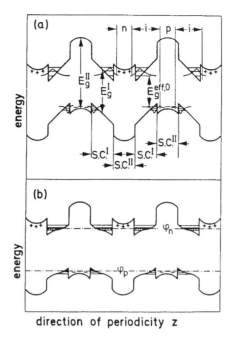

direction of periodicity z

Figure 4 Schematic real space picture of the band diagram of a type-II hetero-n-i-p-i structure for the ground state (a) and an excited state (b).

2.2. Tunable Lifetimes

In the introduction it was pointed out that the tunability of the electronic structure is a consequence of the extremely long excess carrier lifetimes. The long lifetimes, of course, result from the spatial separation of electrons and holes. In spite of the spatial separation, the lifetimes are not infinitely long as the carriers can recombine either by tunneling through or, at higher temperatures, by thermal excitation across the potential barrier (see Figure 1).

A useful analytical expression for estimating the lifetime enhancement for radiative and nonradiative tunneling recombination between electrons and holes as a function of design parameters and degree of excitation, i.e., carrier concentration and effective band gap, $E_g^{eff,exc}$, has been given in Döhler [1983b]

$$\tau_{tun}^{nipi} / \tau^{bulk} \cong \exp\left[\left(E_g^o - E_g^{eff,exc}\right) / \hbar\left(\omega_c + \omega_i\right)\right] \qquad (24)$$

with ω_c and ω_i following from Equations 6 and 7. A similar expression applies for the lifetime enhancement factor for vertical recombination of thermally excited carriers at finite temperatures:

$$\tau_{therm}^{nipi} / \tau^{bulk} \cong \exp\left[\left(E_g^o - E_g^{eff,exc}\right) / kT\right] \qquad (25)$$

A comparison of Equations 24 and 25 reveals that tunneling or thermal recombination processes are expected to predominate at a given temperature, T, depending on whether

$$kT < \text{or} > kT_o \equiv \hbar\left(\omega_c + \omega_i\right)/4 \qquad (26)$$

but, interestingly, independent of the value of $E_g^o - E_g^{eff,exc}(n)$. For a given host material, the condition (Equation 26) depends only on the doping level. The question of whether the heavy- or the light-hole energies, ω_{vh} or ω_{vl}, must be used in Equation 26 cannot be answered in general. From a more thorough consideration of the recombination channels, we can actually conclude that in many typical n-i-p-i configurations the tunneling recombination probability at room temperature will be determined by the uppermost light-hole subbands, although their population probability will be lower than for the corresponding heavy-hole subbands. This is due to the fact that the larger overlap of the light-hole subband wave functions with the electron subbands overcompensates for the lower population factor. Taking this into account, one finds that for GaAs, for example, tunneling will still predominate at room temperature, if $n_D = n_A = 3 \times 10^{18}$ cm^{-3} or larger. This is important with respect to the question of whether the room temperature luminescence is tunable or not, as will be seen in Section 4.

2.3. Optical and Electrical Modulation of n-i-p-i Structures

Because of the long electron hole recombination lifetimes it is possible to maintain large deviations from thermal equilibrium by rather weak optical excitation or electron and hole injection through selective n- and p-type contacts as shown schematically in Figure 5a and b, respectively.

For a given quasi Fermi level difference $\Phi_n - \Phi_p$ the recombination rate per period and areal cross section is determined by

$$\dot{n}_{rec}^{(2)} = \dot{p}_{rec}^{(2)} = n^{(2)} / \tau^{nipi} \tag{27}$$

To achieve a certain excited state of a n-i-p-i structure, this recombination rate has to be balanced by the optical generation rate per area and period in the former case or by the divergence of the injection current in the latter one. The details will be discussed in Sections 4 and 5. Here we would like to mention three fundamental differences of these two ways of modulation.

The excited state is uniquely characterized by $\Phi_n - \Phi_p$. In the case of optical excitation, this quantity depends in a complicated manner on the light intensity, the absorption coefficient, and the lifetime, which again changes on an exponential scale with $\Phi_n - \Phi_p$. In the case of modulation by an external voltage U_{pn} applied between selective n- and p-contacts, the steady state value of $\Phi_n - \Phi_p$ is determined rather exactly by eU_{pn} as long as the recombination current is sufficiently low so that a drop of the quasi Fermi levels within the layers due to their finite series resistances can be neglected.

As to the dynamic behavior, very short optical response times are possible for excitation processes if the optical power or the pulse energy is high. Because of the long recombination lifetimes, however, the recovery times for returning into the ground state are typically extremely long. The situation is fundamentally different if the sample is provided with selective contacts. The contacts do not only allow fast injection of nonequilibrium carriers, but fast extraction as well. Particularly fast response and recovery times can be achieved in structures with small lateral dimensions, resulting in very short RC time constants due to both low interlayer capacitance and low series resistance of the layers.

In contrast to the optical excitation, which always results in the generation of excess carriers, a depletion of the layers below their equilibrium densities can be achieved if a reverse bias U_{pn} is applied to a n-i-p-i semimetal (resulting in a negative effective band gap!). As a consequence, the total achievable changes of the carrier densities and the internal electric fields are much larger than those accessible by optical excitation. This will turn out to be particularly favorable in connection with the tunable absorption and optical nonlinearities. For the latter ones another advantage arises from the fact that the leakage currents between the layers are typically by orders of magnitude lower under reverse bias compared with forward bias. This property can be used to achieve extremely large optical nonlinear changes at extremely low optical power.

Finally, we will mention briefly a third possibility of modulating the electronic structure of n-i-p-i crystals by the application of sandwich contacts as depicted in Figure 5c. The effect of the external field, F, is, roughly speaking, a splitting of the effective band gap into $E_g^{eff} + eFd/2$ and $E_g^{eff} - eFd/2$. In this regime the modulation is extremely fast, as the relevant capacitance is that of the whole n-i-p-i structure of thickness Nd considered as a dielectric (with the same dielectric constant κ_0 as the bulk host material).

3. GROWTH OF n-i-p-i STRUCTURES WITH SELECTIVE CONTACTS

MBE is the most versatile and one of the most frequently used growth techniques to fabricate semiconductor heterostructures and superlattices. As already mentioned, the electrical and optical properties of

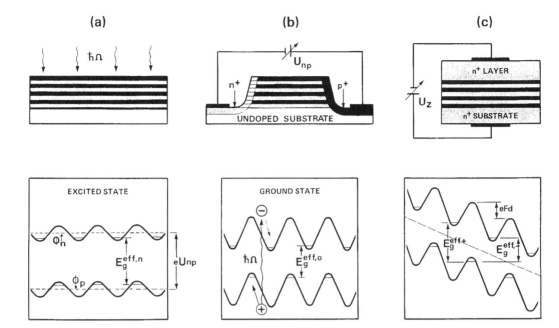

Figure 5 Different excitation processes for modulation of the electronic structure of n-i-p-i crystals. (a) By light; (b) by carrier injection or extraction via selective contacts; (c) by applying an electric field perpendicular to the layers.

n-i-p-i structures can be varied over a wide range by tuning the carrier density or the internal electric fields by applying an external voltage U_{pn} between selective n- and p-contacts. High-quality selective contacts, which exhibit ohmic behavior to one type of doping layer but are blocking with respect to the layers of opposite doping, are very important for many device applications, as well as for fundamental investigations.

First attempts to fabricate selective contacts consisted in alloying metal contacts from the top layer through to all the layers or sidewise onto mesa edges. These procedures have resulted in high contact resistances and in unacceptably high leakage currents under reverse bias, particularly for structures with high doping levels. In 1986 Döhler, Hasnain, and Miller [1986a] showed that a patterned silicon shadow mask can be used very successfully to laterally structure the doping profile in n-i-p-i systems. By using this technique, the first remarkable results on electroluminescence [Hasnain et al., 1986], electroabsorption, and electro-modulation were demonstrated [Chang-Hasnain et al., 1987; Döhler, 1987]. Despite the success, there was still a serious problem connected with this approach. The lateral device size is of the same order as the mask thickness. Since the silicon mask cannot be made smaller than some $100\,\mu m$ for reasons of mechanical stability, structures in the μm-range cannot be fabricated using this technique.

These limitations can be overcome by using an epitaxially grown shadow mask which is only several μm thick. Figure 6 shows the process flow for this ESM-MBE technique [Gulden et al., 1993; Malzer et al., 1996]. In a first growth step, the shadow mask, consisting of a few μm thick AlGaAs and a thin GaAs top layer, is grown. The wafer is then patterned using standard photolithography. In order to form the mask profile shown in Figure 6b, the AlGaAs layer is removed by a selective etch. Due to the highly selective etch, the original smooth substrate surface can be exposed on the bottom of the mask windows. After a standard cleaning procedure, the patterned wafer is reloaded into the MBE system. The fact that the different source materials in the MBE chamber are emitted under different angles can be used for a suitable alignment of the wafer with respect to the angle of incidence of the dopant beams. During the growth, the orientation of the stripe-shaped windows in the mask was chosen such that a regular n-i-p-i structure grows in the center of the window area. Toward the edges of the window, the structure changes into n-i-n-i and p-i-p-i structures as indicated in Figure 6c. Since the wafer has to be aligned in a proper

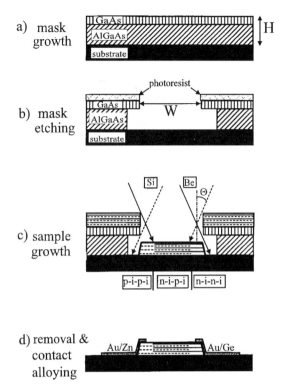

a) mask growth

b) mask etching

photoresist

c) sample growth

d) removal & contact alloying

Figure 6 Process flow for the ESM-MBE growth technique. (a) The wafer with the ESM is (b) photolithograph-ically patterned and selectively etched. (c) For the growth process the fact is exploited that at a fixed position of the substrate the molecular beams from the effusion cells are incident with different angles relative to the mask windows. Therefore, the lateral doping profile can be structured *in situ* during the growth creating exclusive n- or p-doped regions. (d) Highly selective ohmic contacts can be alloyed with standard methods after removing the shadow mask. (From Malzer, S. et al., 1996. With permission.)

position with respect to the doping cells, it cannot be rotated during growth. This leads to variations of the layer thickness and doping densities across the wafer. Further disadvantages are regions of bad growth and the abrupt variation of the composition of ternary or quaternary materials because of the shadowing effect. However, a continuous rotation of the substrate can be retained during the growth of undoped layers while the substrate has to be aligned for δ-doping of the layer. With a simultaneous control of shutter motion and substrate rotation, homogeneous doping layers can be achieved by the "flash doping technique," which means that the respective doping shutter is opened only for a short time when the substrate is in the optimum position. After the growth process the GaAs/AlGaAs mask is removed using the complementary photolithography and etching processes as during the mask prep-aration. The metallic contacts to the n-i-n-i and p-i-p-i regions can then be applied without any problems by standard methods, because the blocking pn-junctions to the respectively other doped layers are built in during growth. Thus, all the n-type and all p-type layers are contacted in parallel. Numerous n-i-p-i samples have been grown using this new technique. A similar yield and crystal quality is achieved compared with standard growth techniques. In particular, n-i-p-i crystals with excellent selective contacts are obtained. For instance, 15-period GaAs n-i-p-i structures with $N_a = 4 \times 10^{18}$ cm^{-3}, $N_d = 1 \times 10^{18}$ cm^{-3}, $d_n = d_p = 60$ nm and $d_i = 70$ nm were grown. Although the doping levels and the internal fields are very high (almost 10^6 V/cm at $U_{pn} = -6$ V), the reverse leakage currents are in the nA-range, whereas forward currents of mA are obtained [Kiesel et al., 1993; Malzer et al., 1996].

4. TUNABLE LUMINESCENCE

The luminescence in n-i-p-i doping superlattices provides a particularly nice demonstration of the tunable indirect band gap in real space. The peak position of the luminescence spectra at low temperatures is directly related to the tunable effective band gap E_g^{eff} and to the difference of the quasi Fermi levels

$\Phi_n - \Phi_p$. The intensity reflects the changing overlap between the occupied electron and hole subband wave functions. Finally, the disorder effects can be studied in great detail because of their strong influence on the shape of the luminescence spectra. It turns out that effects of individual subbands can be observed only in hetero-n-i-p-i superlattices, as the strong potential fluctuations in homogeneous n-i-p-i structures lead to the formation of broad low-energy tails.

4.1. Theory
4.1.1. Idealized Case
The calculation of luminescence spectra is straightforward for the idealized case [Ploog and Döhler, 1983; Beyer et al., 1989; Metzner et al., 1995]. The impurity space charge is assumed to be smeared out in the uniformly doped layer, or — in the case of a δ-doped n-i-p-i structure — in the doping planes ("jellium model," see Section 2.1). First, the electron and hole subband energies $\varepsilon_{c,\mu}(\vec{k}_\parallel)$ and $\varepsilon_{i,\nu}(\vec{k}_\parallel)$ ($i = \nu h$ or νl labels the heavy- or light-hole subbands), the corresponding envelope wave functions $\zeta_{c,\mu}(z)$ and $\zeta_{i,\nu}(z)$ and Fermi–Dirac occupation factors $f(\varepsilon_{c,\mu}(\vec{k}_\parallel) - \Phi_n)$ and $f(\Phi_p - \varepsilon_{i,\nu}(\vec{k}_\parallel))$ have to be calculated self-consistently for a given value of the quasi Fermi level splitting $\Phi_n - \Phi_p$ and temperature T. Subsequently, the optical recombination probabilities for electrons in the (c,μ)-th subband with holes in (i,ν)-th hole subband have to be calculated

$$w_{c\mu,i\nu}\left(\vec{k}_\parallel\right) \propto \left|\left\langle \zeta_{i\nu} \middle| \zeta_{c,\mu} \right\rangle\right|^2 \cdot \left|p_{c,i}\right|^2 \cdot \delta\left(E_{c,\mu}\left(\vec{k}_\parallel\right) - E_{i\nu}\left(\vec{k}_\parallel\right) - \hbar\omega\right) \tag{28}$$

($p_{c,i}$ is the bulk interband dipole matrix element; heavy–light-hole mixing is neglected).

Finally, in order to obtain the spectral intensity at the photon frequency ω, the contributions of all interband transitions at a given photon energy have to be summed up with the appropriate statistical weight and multiplied by $\hbar\omega$

$$I(\omega) = \sum_{c\mu,i\nu} \sum_{\vec{k}_\parallel} \hbar\omega \cdot w_{c\mu,i\nu}\left(\vec{k}_\parallel\right) \cdot f\left(E_{c,\mu}\left(\vec{k}_\parallel\right) - \Phi_n\right) \cdot f\left(\Phi_p - E_{i\nu}\left(\vec{k}_\parallel\right)\right) \tag{29}$$

It should be noted that — in contrast to the case of multiple quantum well structures of the GaAs/AlGaAs-type — there are no quasi selection rules for the transition matrix elements. As the electron and hole subband envelope functions $\zeta_{c,\mu}(z)$ and $\zeta_{i,\nu}(z)$ are shifted by half a superlattice spacing $d/2$ with respect to each other, the values of the optical recombination probabilities are dominated by the overlap of the tails of the envelope wave functions in the forbidden energy region (see Figure 3). These contributions may be small, but never zero. It is important to realize that the probability for nonradiative recombination processes involving deep impurity centers is also strongly reduced by factors of the same order of magnitude.

For higher subband indices the recombination probabilities may increase drastically because of a much larger overlap. This has two important consequences. At low temperatures the contributions of the higher-index subbands to the luminescence spectra will be much more prominent compared with those of low-index subbands if more than one electron and/or hole subband is occupied. Later on, we will see that disorder leads to low-energy tails for each of the (c,μ)–(i,ν) contributions. Therefore, it is not surprising that the contributions of the lower subbands may disappear completely in the luminescence tails related to these higher subbands.

The other interesting consequence concerns the competition between the tunneling recombination of carriers occupying low-index subbands with $f_{c,\mu}$ and $f_{i,\nu} \approx 1$ but exhibiting small overlap $\left|\left\langle \zeta_{i,\nu}(z) \middle| \zeta_{c,\mu}(z) \right\rangle\right|$ << 1 with carriers which are thermally activated into higher subbands with $f_{c,\mu}$ and $f_{i,\nu}$ << 1, but exhibiting much larger overlap. It is obvious that as a consequence of this competition the overall decay of the luminescence intensity $I(\omega)$ on the high-energy side of the spectrum will be less pronounced than the well-known "Boltzmann tail" in the bulk band-to-band luminescence. In addition, the peak of the luminescence spectrum may exhibit a more or less pronounced blue shift compared with $\Phi_n - \Phi_p$, depending on the temperature and the design parameters of the n-i-p-i structure. At sufficiently high temperatures and/or long superlattice periods, the luminescence peak will be close to the band gap of the host material, as "quasi-vertical" transitions of carriers in high-index subbands will dominate the luminescence. For n-i-p-i structures consisting of uniformly doped n- and p-layers (with $d_i = 0$), a simple analytical estimate [Döhler, 1983b] allows us to determine whether the luminescence is or is not tunable at given doping densities and a given temperature. For GaAs, e.g., the doping densities have to exceed values of $3 \times 10^{18}\,cm^{-3}$ for the occurrence of tunable room temperature luminescence, which agrees well with experimental observations (see Sections 4.2 and 4.3).

4.1.2. Realistic Case with Random Impurity Distribution

A rigorous theoretical calculation of the luminescence spectra becomes impracticable if the donor and acceptor impurity states and, in particular, their random site distribution are to be taken into account. Therefore, suitable simplifications are required. It is advantageous to distinguish between the cases of extremely low doping densities, where electrons and holes occupy impurity bands, and high doping densities, where the effects of impurity bands can be neglected compared with the subband effects ("metal limit").

It should be noted that this classification scheme defines quite different critical doping densities as "low" and "high" for donors and acceptors because of their different effective masses. The light effective mass of electrons results in low donor-binding energies ($E_D \approx 5$ meV in GaAs, e.g.), whereas, according to Equations 6, 7, and 13, confinement energies $E_{c,\mu}$ and Fermi energies ($\hbar^2/2m_c)2\pi n^{(2)}$) exceeding E_D are obtained at rather low donor ($n_D > 2 \times 10^{16}\,cm^{-3}$ for GaAs) and electron densities ($n^{(2)} > 1.3 \times 10^{11}\,cm^{-2}$ for GaAs). In contrast, due to the significantly higher effective heavy-hole mass, the acceptor binding energies ($E_A \approx 28$ meV for a Be acceptor in GaAs, e.g.) are much higher and the confinement energies $E_{vh,\nu}$ and Fermi energies ($\hbar^2/2m_{vh})2\pi p^{(2)}$) considerably smaller, exceeding E_A only at much higher acceptor ($n_A > 1.5 \times 10^{18}\,cm^{-3}$ for GaAs) and hole densities ($p^{(2)} > 4 \times 10^{12}\,cm^{-2}$ for GaAs). Therefore, many typical n-i-p-i systems with a comparable density of donors and acceptors per doping layer have to be considered at the same time as "strongly" n-doped and "weakly" p-doped.

4.1.2.1. Impurity Band Limit

This case is of particular interest for δ-doped n-i-p-i structures [Döhler, 1988; Metzner et al., 1992a; 1992b]. The situation resembles an ordered version of the well-known donor acceptor pair luminescence. For typical cases the thickness d_i of the intrinsic layers largely exceeds the effective Bohr radii of the donors and acceptors, but also exceeds the average donor–donor and/or acceptor–acceptor distances within the δ-doped layers. Therefore, the Coulomb term in the expression for the emitted photon energy

$$\hbar\omega = E_g^o - E_D - E_A + e^2 / \kappa_o r_{DA} \qquad (30)$$

exhibits only very small statistical fluctuations for the nearest-neighbor donor–acceptor pairs, as $(e^2/\kappa_o r_{DA}) \approx (e^2/\kappa_o d_i)$. Also, the pair recombination probability

$$W(r_{DA}) = W_o \exp\left(-r_{DA} / a_D^*\right) \cong W_o \exp\left(-2d_i / a_D^*\right) \qquad (31)$$

does not exhibit the huge variations observed in the pair luminescence in partly compensated bulk semiconductors (a_D^* is of the order of the effective donor Bohr radius).

If the (two-dimensional) donor or the acceptor densities are high enough to cause the formation of an impurity band due to the quantum mechanical impurity–impurity interaction, the luminescence spectrum basically reflects the density of states (DOS) of this impurity band. Specifically, at times which are long compared with exciton and band-to-band recombination lifetimes (which are typically of the order of nanoseconds) but short compared with $1/W(d_i)$ (which is typically in the higher microsecond range!), the luminescence spectrum after full excitation corresponds to the DOS of the neutral impurity band without Coulomb broadening. So far this is, to our best knowledge, the only way to obtain direct experimental information on the shape of the impurity DOS distributions.

4.1.2.2. High Doping Limit

If the doping density exceeds the critical values defined in the introduction to Section 4.1.2, the description of the electronic structure in terms of subbands is appropriate. It has, however, to be modified in order to take into account the spatially random distribution of the dopands. The potential fluctuations caused by the charged impurity atoms by which the actual space charge potential $V_{sc}(\vec{r})$ differs from the averaged self-consistent potential $V_{sc}(z)$ defined by Equation 10 are particularly large if they are not or are only weakly screened by free carriers.

In order to facilitate the discussion conceptually, we consider only the case of δ-doped structures in the following. A generalization to the case of doping layers of finite thickness is relatively straightforward. With increasing carrier density the free carriers will first populate states centered at local minima of the fluctuating potential. The eigenfunctions corresponding to these "tail states" will be localized in the (x,y)-plane, in addition to their confinement in z-direction, as described by the subband envelope functions $\zeta_{c,\mu}(z)$ or $\zeta_{i,\nu}(z)$. With increasing free carrier density the potential fluctuations will become more and more screened and $\vec{k}_\parallel$ will be a (reasonably) good quantum number for carrier densities above the Mott–Anderson insulator metal transition.

In a recent paper Metzner et al. [1995] have developed a model for the calculation of the luminescence spectra which takes into account the most important effects of the random distribution of the impurity atoms in the doping layers.

Starting from the self-consistent electronic structure of the jellium model, they calculated the screened potential of a single impurity charge fixed in the doping plane. As the screening length is rather insensitive to the carrier density, replacing the actual local carrier density by its (macroscopic) average value represents an acceptable approximation (at least for carrier densities significantly exceeding the classic percolation limit).

As a next step the probability distribution of the potential fluctuations is calculated. For this purpose the probability of finding the 1st, 2nd … $(N-1)$-th, N-th nearest neighbor within the distances $r_1 < r_2 < \dots r_{N-1} < r_N$ is evaluated by a Monte Carlo simulation. The probability of the corresponding potential fluctuation is obtained by summing up all the screened potentials of a given configuration and subtracting the corresponding jellium contribution.

In order to obtain the fluctuations of the local subband energies $E_{c,\mu}(x,y)$ and $E_{i,\nu}(x,y)$, envelope wave functions $\zeta_{c,\mu}(z)$, $z_{i,\nu}(z)$ and carrier densities $n^{(2)}$, $p^{(2)}$, the assumption is made that the random potential changes slowly in the (x,y)-plane compared with the spatial variation of $V_{sc}(z)$. In this case solving the same one-dimensional Schrödinger equation as before, but for the local potential $V_{sc}(x,y;z)$, and maintaining at the same time uniform quasi Fermi levels Φ_n and Φ_p in the respective layers will represent an adequate approach (which probably slightly overestimates the effect of the potential fluctuations).

As an example the probability distribution $P(V(z))$ for the conduction band edge in the doping plane is shown together with those for the two first electron subbands $P(\varepsilon_0)$ and $P(\varepsilon_1)$ in a δ-doped GaAs n-i-p-i structure in Figure 7. As expected, the subband energies, particularly for higher subbands, fluctuate much less than the conduction band edge. Also the distributions turn out to be strongly asymmetric (in contrast to the results of less-sophisticated approaches usually applied to similar problems).

An important "detail" which needs to be mentioned is the influence of the subband fluctuations on the envelope wave functions. In regions of higher than average density of impurity atoms ("clusters") the energies of the subband edges are not only lowered, but the wave functions are more strongly confined in z-direction than for the jellium case. Similarly, they are less confined in regions of lower than average density of impurity atoms ("voids") [Metzner et al., 1995].

Finally, the luminescence spectra are calculated by summing up all the possible spectra from all occupied subbands with the corresponding statistical weight. For both uniformly or δ-doped (homogeneous) n-i-p-i structures, it turns out that the contribution of each combination of conduction and valence subbands has a quite similar shape, in particular with regard to the low-energy tail. The results indicate that these tails are so broad that the tail-state contributions of higher subbands are typically of the same order as the peak contributions of lower-index subbands at the same photon energy. Therefore, according to our theoretical calculations, the chances for observing a clear-cut signature of the subband structure in the luminescence spectra of homo-n-i-p-i structures are not favorable.

The situation is quite different for the luminescence in type-II hetero-n-i-p-i structures (see Figure 4). In these systems the potential fluctuations in the electron and hole subbands are strongly reduced because of the spatial separation between the free carriers and their parent impurity atoms. Calculations for such structures using an appropriately modified version of our model are in progress.

4.2. Experimental Results

Tunable luminescence in n-i-p-i structures can be induced either by optical excitation (photoluminescence, PL) or by (lateral) electron and hole injection via selective n- and p-type contacts (electrolumi-

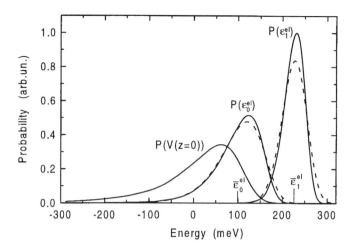

Figure 7 Probability distributions of the local conduction band edge and the electron subband energies $E_{c,0}(x,y)$ and $E_{c,1}(x,y)$. The dashed lines correspond to a first order perturbational calculation. (From Metzner, C. et al., *Phys. Rev. B* 51(8), 5106, 1995. With permission.)

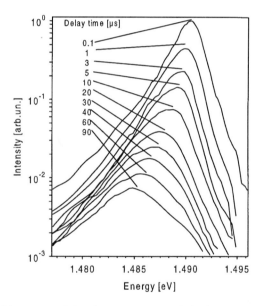

Figure 8 Time-resolved PL spectra of the donor–acceptor pair luminescence. The investigated $\delta n\text{-}i_1\text{-}\delta p\text{-}i_2$ structure consists of 10 periods ($n_D = 10^{11}$ cm^{-2}, $n_A = 10^{10}$ cm^{-2}, $d_{i1} = 50$ nm, $d_{i2} = 100$ nm). (From Schönhut, J. et al., *Solid State Electron.* 1996. With permission.)

nescence, EL). Apart from its interest for applications, the latter method has the advantage that a comparison between theory and experiment provides more information, as the difference of the quasi Fermi levels is directly related to the applied voltage through the expression $\Phi_n - \Phi_p = eU_{pn}$. The effects of optical and electrical modulation can also be combined. Results for all three cases will be presented in the following.

4.2.1. Tunable Photoluminescence

4.2.1.1. Weakly Doped δ-n-i-p-i Structures ("Impurity Band Limit")

Although the idea of n-i-p-i superlattices originated from a theoretical consideration of the donor–acceptor pair luminescence in such systems, it is only very recently that the first experimental studies have been performed [Schönhut et al., 1996]. The design parameters of the GaAs δ-n-i-p-i structure were

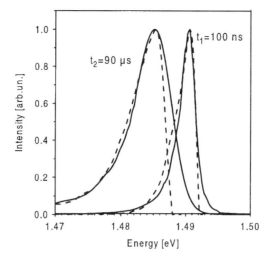

Figure 9 Comparison between experimental (solid line) and theoretical (dashed line) donor–acceptor pair luminescence spectra at t_1 = 100 ns and t_2 = 90 μs for the sample described in Figure 8. The shape of the spectra directly reflects the DOS of the donor impurity band.

$n_D = 10^{11}$ cm^{-2}, $n_A = 10^{10}$ cm^{-2} and $d_i = 50$ nm. The donor–acceptor pair luminescence was investigated in time-resolved PL experiments (see Figure 8). About 100 ns after excitation the excitonic and the band-to-band recombination had disappeared. For decay times up to 1 μs a narrow spectrum of only ≈2.5 meV was observed. With increasing decay time the spectra broadened as a result of the Coulomb contributions of the increasing number of ionized donors and acceptors ("classical broadening") and shifted to slightly lower energies as expected from numerical simulations [Metzner et al., 1992a; 1992b]. The shape of the experimental and the calculated spectra agreed very well over the whole range of decay times 0.1 μs $< \tau_{decay} <$ 100 μs if the quantum mechanical broadening was taken into account using a simple model. Two examples are given in Figure 9.

4.2.1.2. Highly Doped n-i-p-i Structures

PL studies on a large number of n-i-p-i structures have been reported, mostly with GaAs as host material [Döhler et al., 1981; Jung et al., 1983; Rehm et al., 1983; Schubert et al., 1988; Döhler et al., 1986b; Köhler et al., 1986a; Renn et al., 1993], but also with AlGaAs, InGaAs, and InGaAsP lattice matched to InP [Yamauchi et al., 1984; Carey et al., 1985], InAs [Phillips et al., 1993], and PbTe [Oswald et al., 1990]. Most of the investigated structures consisted of uniformly doped n- and p-layers without intrinsic layers in between, but starting with the work by Schubert et al. [1988] a number of PL studies on δ-n-i-p-i structures were also performed.

In the major number of investigations the steady state luminescence spectra and intensities were measured as a function of the excitation density, and a more or less pronounced tunability was observed. As an example, results obtained from a highly doped structure for both 90 K and room temperature are shown in Figure 10 [Renn et al., 1993]. By varying the excitation density from 0.03 to 1000 W/cm^2 by a factor of about 3×10^4 the peak position can be shifted between 0.7 and >1.4 eV by more than a factor of two.

A lot of interesting information can be deduced from intensity-dependent steady state PL studies. From the steady state condition that generation and recombination rate have to be equal, the total recombination lifetimes can be determined as a function of the tunable effective band gap E_g^{eff}. At low temperatures the luminescence peak energy agrees rather exactly with the effective band gap (or more precisely with the quasi Fermi level difference $\Phi_n - \Phi_p$). As the free carrier densities $n^{(2)}$ and $p^{(2)}$, the effective band gap, and $\Phi_n - \Phi_p$ are directly related according to Equations 18 through 23, the total excess carrier lifetime can be determined from

$$\tau_{tot} = \Delta n^{(2)} / \left(\dot{n}_{gen}^{(2)} \right) \tag{32}$$

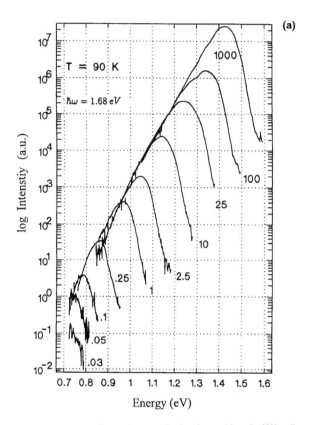

Figure 10 PL spectra of a n-i-p-i structure for various excitation intensities (in W/cm²) taken at 90 K (a) and at room temperature (b). Depending on the excitation level the peak position shifts from about 0.7 eV to >1.4 eV (for 90 K). The investigated n-i-p-i structure consists of 10 periods (n_D = 4 × 10¹⁸ cm⁻³, n_A = 1 × 10¹⁹ cm⁻³, d_n = 25 nm, d_p = 35 nm, and d_i = 0). (From Renn, M. et al. *Phys. Rev. B* 48, 11220, 1993. With permission.)

where the generation rate per n-i-p-i period is given by the photon flux density inside the semiconductor $I(\omega)/\hbar\omega$ multiplied by the product of the absorption coefficient $\alpha(\omega)$ and the superlattice period d

$$\dot{n}^{(2)}_{gen} = \left(I(\omega) / \hbar\omega \right) \cdot \alpha(\omega) \cdot d \tag{33}$$

The lifetimes deduced from the spectra shown in Figure 10 vary between 1 ns < τ_{tot} < 3 µs.

If all the recombination processes were radiative, i.e., if the internal quantum efficiency η_{int} was unity, these values could be compared directly with the radiative lifetimes following from theoretical calculations of the luminescence according to Section 4.1.1. From the measured dependence of the PL intensity as a function of excitation density, however, it follows, that η_{int} decreases with decreasing excitation density. From the fact that the luminescence intensity, which is proportional to $1/\tau_{rad}$, decreases by about a factor of 10^8 in Figure 10a, e.g., we deduce that η_{int} decreases by about a factor of 3 × 10³ if the excitation density is decreased by 4.5 orders of magnitude. The value of τ_{tot} at the highest excitation level, however, agrees quite well with the calculated one, which indicates that the radiative processes dominate over the nonradiative ones at high excitation level.

As outlined in Section 4.1.1, the high-energy tail of the n-i-p-i luminescence spectra at increased temperatures is expected to be wider than the Boltzmann-tail observed in bulk semiconductors because of strongly enhanced overlaps of the higher subband envelope wave functions. This is clearly demonstrated by results displayed in Figure 10b. If, at even higher temperatures, the recombination of the small number of thermally activated carriers occupying higher subbands will win in the competition with the large number of carriers near the quasi Fermi level, the luminescence peak will shift to higher photon energies. This prediction was confirmed by temperature-dependent PL experiments in structures of

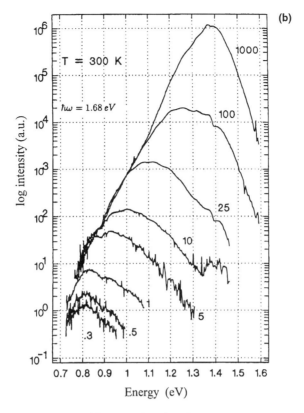

Figure 10 (continued)

varying doping level. In particular, it was found that the doping levels, in fact, have to exceed values of about 3×10^{18} cm^{-3} in order to observe tunable room temperature luminescence [Döhler et al., 1986b; Köhler et al., 1986a; 1986b].

Time-resolved PL measurements were also performed on highly doped n-i-p-i structures [Rehm et al., 1983]. They provide a very convincing illustration of the tunable band gap and the excitation-dependent total and radiative lifetimes τ_{tot} and τ_{rad}. Whereas the carrier density decreases with decreasing band gap more or less on a linear scale, the lifetimes are increasing on a logarithmic scale. This leads to a strongly subexponential time dependence of the decay of the luminescence intensity. A careful analysis of time-resolved PL measurements [Rehm et al., 1983] reconfirms the information which can be extracted from intensity-dependent steady state PL studies.

The luminescence spectra are very broad, as expected from our discussion in Section 4.1.2. They are also quite structureless, which confirms the theoretical expectation that observation of a signature of the subband structure is unlikely. At this point it should be mentioned, however, that there has been one report on a pronounced structure in the luminescence (and the absorption) spectra of highly doped δ-n-i-p-i structures, which was attributed to subband effects by the authors [Schubert et al., 1987].

4.2.1.3. Type-II Hetero-n-i-p-i Structures

So far only a few optical investigations on type-II hetero-n-i-p-i structures have been performed. Type-II hetero-n-i-p-i structures exhibit the ideal design for the observation of subband effects in the tunable luminescence. First of all, the spatial separation between the donor atoms and the electron states and between the acceptor atoms and the hole states leads to strongly decreased fluctuations of the subband edges. Second, the subband splitting is much larger than in purely space charge–induced quantum wells, due to the triangular confinement potential. In fact, the tunable subband structure could be seen very clearly in various steady state time-resolved PL experiments performed on an AlGaAs/GaAs type-II hetero-n-i-p-i crystal [Street et al., 1986; 1987]. As an example, we show in Figure 11 results of the temperature dependence of the PL spectra. With increasing temperature the contributions due to the thermal population of higher subbands are, in fact, becoming the dominant ones.

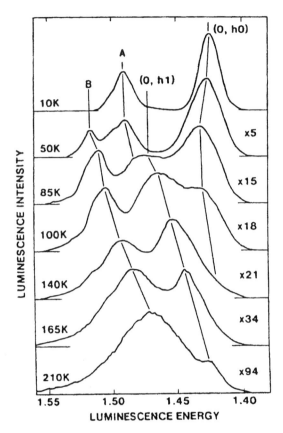

Figure 11 Temperature dependence of the PL spectra of an $Al_{0.3}Ga_{0.7}As$/GaAs type-II hetero-n-i-p-i structure ($d_i^{GaAs} = 15$ nm).

4.2.2. Tunable Electroluminescence

EL investigations of n-i-p-i structures are appealing with regard to various aspects. A voltage U_{pn} applied between selective contacts ideally results in a quasi Fermi level splitting $\Phi_n - \Phi_p = eU_{pn}$. In Section 4.1 it was shown that the luminescence peak energy is closely related to the quasi Fermi level splitting if the doping levels are sufficiently high, the superlattice period sufficiently small, and the temperature low enough. This means that under these conditions the luminescence spectra are voltage tunable according to the simple relation

$$\hbar\omega^{peak} \approx eU_{pn} \tag{34}$$

The quantity which determines the degree of excitation for the theoretical calculation of luminescence spectra is the quasi Fermi level difference. Whereas $\Phi_n - \Phi_p$ is an adjustable parameter for a comparison between theory and experiment in the case of PL experiments, for EL investigations it is a known quantity. Therefore, a comparison between theory and experiment represents a much more stringent test of the theory in the latter case.

Finally, the tunable luminescence is more appealing for applications. This is not only the case because of the direct transformation of electrical into optical energy; it is also true because of the much faster modulation dynamics, which does not rely on the (possibly very long) internal recombination lifetimes, but on the RC time constant, which, according to Section 5.3, can be made smaller than nanoseconds if the width of the n-i-p-i structure becomes small enough.

The first investigations of the tunable EL in n-i-p-i crystals were performed on structures provided with very poor selective Sn/Zn– (n-type) and Zn– (p-type) contacts obtained by alloying [Künzel et al., 1982]. A spectrally very broad tunable EL was observed near the contacts at liquid helium temperature at about a factor of ten higher voltages than expected from Equation 34.

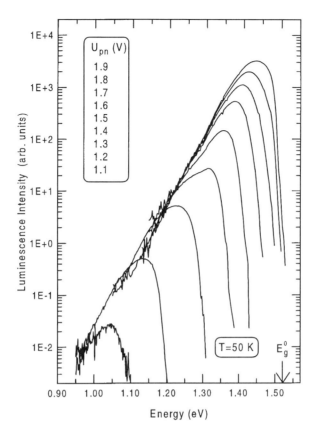

Figure 12 Experimental EL spectra for a δ-n-i-p-i crystal at 50 K for various applied voltages (design parameter: 20 periods $n_D = 6 \times 10^{12}\,cm^{-2}$, $n_A = 8 \times 10^{12}\,cm^{-2}$, $d_i = 14$ nm). (From Metzner, C. et al., *Phys. Rev. B* 51(8), 5106, 1995. With permission.)

Only after introducing the shadow mask technique described in Section 3 did it become possible to observe tunable and efficient EL at low voltages and at elevated temperatures [Hasnain et al., 1986]. Further improvement was achieved by using ESM-MBE as the width of the n-i-p-i region could be reduced down to a few μm, resulting in low series resistances of the doping layers and narrow recombination areas. In Figure 12 the results of EL measurements on a δ-n-i-p-i structure taken at 50 K are shown [Schrüfer et al., 1994; Metzner et al., 1995]. Over a wide voltage range the peak position follows the relation 34. Only at high excitation levels and correspondingly high injection currents the effects of the finite layer conductances, resulting in $\Phi_n - \Phi_p < eU_{pn}$, become apparent.

For these δ-n-i-p-i structures, finally, a detailed comparison between theory and experiment could be performed [Schrüfer et al., 1994; Metzner et al., 1995]. In Figure 13 an experimental spectrum obtained for $U_{pn} = 1.2$ V is compared with the theoretical results for $\Phi_n - \Phi_p = 1.17$ eV following from models of different level of sophistication. Spectrum J corresponds to the simplest model, the jellium model of Section 4.1.1, in which potential fluctuations are completely neglected. Obviously, it provides a very poor description of the reality. Spectrum B fully regards the potential fluctuations as discussed in Section 4.1.2.2, but neglects the effect of the potential fluctuations on the wave functions. Only spectrum A, which also includes the latter one, provides a really excellent agreement between theory and experiment. It should be noted that within a range of two orders of magnitude of intensity the deviations between theory and experiment remain below a factor of two, although no fitting parameter enters into the theory.

4.2.3. Electrically Modulated Photoluminescence

Very recently first PL investigations were also performed on samples in which the electronic structure was modulated by external bias simultaneously. A wide variety of interesting combinations of optical excitation and electrical modulation is possible. For instance, it was shown that a very fast modulation

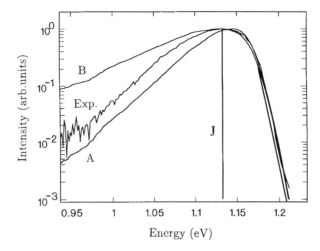

Figure 13 Luminescence spectra calculated with different theoretical models for an excitation level Φ_{np} = 1.17 eV, compared with an experimental spectrum at U_{pn} = 1.2 V (from Figure 12).

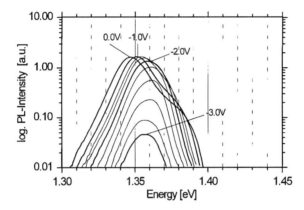

Figure 14 Electrically modulated PL spectra of a δ-doped hetero-n-i-p-i structure. The band profile of this structure is depicted in Figure 16. (From Schultz, J. et al., *Solid State Electron.* 1996. With permission.)

of the intensity and a shift of the spectral position of the PL peak can be achieved by an electric field in growth direction using the modulation scheme depicted in Figure 5c [Böhner, 1995].

In Figure 14, we show results of electrically modulated photoluminescence studies in a type-I hetero-n-i-p-i structure with undoped strained $In_{0.15}Ga_{0.85}As$ quantum wells at the center of the n-region (the electronic structure of the same sample will be discussed in more detail in connection with Figure 16). By variation of the voltage U_{pn} applied between the selective n- and p-contacts the electron density in the quantum wells, $n^{(2)}$ could be tuned between 0 and more than 4×10^{12} cm^{-2}. Using weak optical excitation, a low density of holes, occupying only states with $\vec{k}_\parallel \approx 0$, was induced. In this way the effects of band gap renormalization as a function of carrier density and the question how well $\vec{k}_\parallel$ conservation holds in the presence of disorder which is due to the remote donors in the neighboring δ-doping layers could be studied in a single structure [Schultz et al., 1996].

5. TUNABLE ABSORPTION AND OPTICAL NONLINEARITIES

Large changes of the absorption coefficient can be induced in n-i-p-i crystals either optically or electrically. In the former case, large steady state excess carrier densities in the doping layers can be created at low light intensities because of the long recombination lifetimes of the photogenerated carriers. These carriers reduce the absorption above the band gap directly due to the "bandfilling" effect. In addition, an indirect effect on the absorption is observed because the space charge of the photoinduced carriers lowers the internal fields. This results in just as strong absorption changes at photon energies near the band gap. Whereas light-induced changes of the absorption coefficient require high light intensities in bulk semiconductors and also in compositional superlattice structures, a strong light intensity–dependent absorption coefficient ("nonlinear absorption") — significant even at light intensities which are lower by orders of magnitude — is easily achieved in n-i-p-i structures.

In n-i-p-i crystals with selective contacts, the changes of the free-carrier density can be induced electrically by applying an external voltage U_{pn}, which yields the same absorption changes. In fact, the electrically induced absorption changes can be made much larger than the light-induced ones, by appropriate design of n-i-p-i semimetals (exhibiting free-carrier depletion under reverse bias). As the dynamics of electrical modulation does not depend on the very long internal recombination lifetimes, but on an adjustable RC-time constant, the modulation speed can be high in this case (nanosecond and subnanosecond range).

In Section 5.1, the absorption mechanisms in n-i-p-i crystals will be discussed. Section 5.2 refers to the nonlinear absorption in n-i-p-i structures without selective contacts. Electrically induced absorption changes in n-i-p-i crystals with selective contacts, including the dynamic behavior and applications as electro-optical modulators, are treated in Section 5.3. Finally, in Section 5.4 we present design considerations which indicate that an additional strong increase of the optical non-linearities at further reduced optical power is achievable in n-i-p-i structures with selective contacts under reverse bias.

5.1. Absorption Mechanisms

Changes of the absorption coefficient in semiconductors can be due either to field effects or to carrier-induced bleaching of absorption processes like excitonic or band-to-band transitions. The spatial sepa-ration of photogenerated electron–hole pairs in n-i-p-i systems always results in contributions of both of these phenomena: bleaching of the absorption in the n- and p-doped layers due to carrier-induced phase space filling and field effects in the intrinsic regions due to the partial screening of the superlattice potential by the excess carriers. These effects will be discussed in the following subsections. The construc-tive superposition of the two effects, which can be realized with specially designed hetero-n-i-p-i structures, will be the subject of Section 5.1.3.

5.1.1. Field-Induced Absorption Changes

In principle, various field effects can be used to modulate the absorption coefficient of semiconductors. Most of the concepts for electroabsorptive modulators are based on the absorption changes induced by an electric field perpendicular to the layers of a multiple quantum well (MQW) structure using the quantum confined Stark effect (QCSE) [Miller et al., 1985; Efron and Livescu, 1995]. Other approaches are based on the field-induced Wannier–Stark localization (WSL) [Mendez et al., 1988; Bar-Joseph et al., 1989] in short-period superlattices or on the electroabsorption in semiconductor bulk materials (Franz–Keldysh effect, FKE) [Franz, 1958; Keldysh, 1958]. Although the absorption changes obtained by the FKE are slightly lower, this effect is nevertheless very attractive for device applications. In contrast to the QCSE and the WSL, the field-induced absorption changes due to the FKE extend over a consid-erably broader spectral range both below and above the band gap energy. This implies a much better stability against temperature variations. In addition, the high-speed operation should be more favorable because in bulk material no problems concerning the trapping of photogenerated carriers in quantum wells are observed. Another interesting detail, especially for waveguide applications, is the considerably weaker polarization dependence of the absorption changes [Knüpfer et al., 1993]. The absorption changes due to the FKE have often been underestimated in the literature because they are due — according to the original single-particle theory [Franz, 1958; Keldysh, 1958] — to the field-induced broadening of the smoothly increasing square root absorption edge of the bulk semiconductor. However, the actual field-free absorption edge in direct semiconductors rises sharply because of excitonic effects, even at room temperature. As a consequence, the experimentally observed absorption changes as shown in Figure 15a are twice as high as expected from the simple Franz–Keldysh theory. By incorporating excitonic effects into the theory [Linder et al., 1990; 1993] excellent agreement between experiment and theory is obtained (Figure 15).

Since the FKE requires rather substantial electric field changes (several 100 kV/cm) and because of the slightly lower absorption changes, this phenomenon is less suitable for applications in normal incidence p-i-n modulators. It is, however, perfectly suited for the use in n-i-p-i structures in which high internal fields and long interaction lengths can be obtained simultaneously.

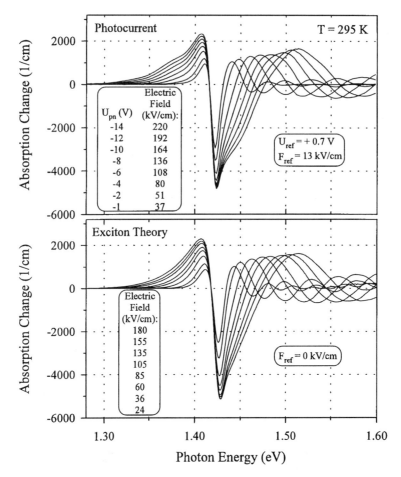

Figure 15 Field-induced absorption changes in GaAs due to the FKE. (a) Experimental values obtained by simultaneous evaluation of transmission and photocurrent experiments; (b) theoretical values according to Linder et al. [1990].

5.1.2. Absorption Changes Due to Phase Space Filling

The phase space filling in bulk semiconductors and MQW structures has been studied in detail (for a review see Schmitt-Rink et al. [1989] and references therein). It can be interpreted as the occupation of states by excitons in a real space picture and/or by free carriers in a momentum space picture (bandfilling), respectively. These effects result in the bleaching of the absorption near the band gap. In particular, the excitonic features in MQWs and modulation-doped MQWs (MDMQW) have been investigated in order to understand the bleaching phenomenon [Cingolani and Ploog, 1991].

In MQWs large optical intensities ($\approx$ MW/cm^{-2}) are required to bleach the band–band absorption significantly because of the short e-h lifetimes ($\approx$ ns) at relevant carrier densities of $n^{(2)} = p^{(2)} \approx 10^{12}$ cm^{-2}. Quenching of the exciton peak occurs at slightly lower densities ($n^{(2)} \approx 2.5 \times 10^{11}$ cm^{-2}) [Schmitt-Rink et al., 1989]. In n-doped MDMQW structures one can observe a blue shift of the absorption edge in the ground state, because the lowest conduction band states are occupied up to the Fermi level Φ_n (at $T = 0$ K). At finite temperatures, one has to take into account the broadening of the absorption edge. If band gap renormalization and screening effects are neglected, the absorption $\alpha(\omega)$ can be expressed as a sum of the contributions $\alpha_{iv,c\mu}$ of the transitions between the hole subbands iv ($i = vl, vh$) and the electron subband μ:

$$\alpha_{iv,c\mu}(\omega) = \alpha^o_{iv,c\mu}(\omega) \cdot \left(1 - f_c\left(\hbar\omega, T, \Phi_n\right)\right)$$

$$= \alpha^o_{iv,c\mu}(\omega) \cdot \left[1 + \exp\left\{\Phi_n - \left(m_{ic} / m_c\right)\left(\hbar\omega - E_{iv,c\mu}\right) / kT\right\}\right]^{-1} \quad (35)$$

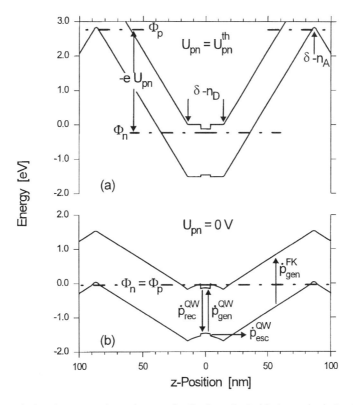

Figure 16 Schematic band structure in real space for the investigated hetero-n-i-p-i structure. This sample consists of 25 periods of 8-nm-wide strained $In_{0.15}Ga_{0.85}As$ quantum wells with 10-nm GaAs spacer embedded in a δ-doped n-i-p-i structure. The δ-n-doping was 4×10^{12} cm⁻² and the δ-p-doping 4×10^{12} cm⁻². The doping layers were separated by 70-nm intrinsic GaAs layers. (a) Under reverse bias with depleted quantum wells. (b) Ground state (no applied voltage) with free carriers in the quantum wells and p-layers. In this structure the occupation of the quantum wells can varied by the applied voltage from $n^{(2)} = 0$ to > 4×10^{12} cm⁻². (From Schultz, J. et al., *Solid State Electron.* 1996. With permission.)

where the superscript o relates to the unoccupied quantum wells. The $E_{iv,c\mu}$ are the corresponding effective band gaps of the QW material, m_c is the effective electron mass parallel to the layers, m_{ic} is the reduced e-vh and e-vl mass, and Φ_n the Fermi level of the electrons. A light-induced bleaching of the absorption in MDMQW structures requires a similarly high intensity as in the MQWs, for the same reasons as mentioned above.

If QWs are embedded into the n-layers of a n-i-p-i structure (for instance in a type-I hetero-δ-n-i-p-i as shown in Figure 16) two advantages are combined. At first, excitation at energies $\hbar\omega > E_g^{QW}$ leads to a strong absorption ($\alpha \approx 1.2 \times 10^4$ cm⁻¹), resulting in a high generation rate of carriers in the quantum wells. Second, the optical power required to maintain a certain particularly high ($n^{(2)} = 10^{12}$ cm⁻²) electron concentration is very low because the photogenerated holes can escape thermally (at elevated temperature) into the p-layers within some picoseconds at sufficiently high temperatures. This spatial separation of carriers leads to the long lifetimes which are typical for n-i-p-i structures. With reasonable values of $\tau_{rec} = 10$ ms and $I_{exc} = 100$ µW/cm² one obtains for the steady state $n^{(2)} \approx 5 \times 10^{12}$ cm⁻². The time scale of the absorption bleaching clearly depends on the generation rate of carriers and is about a few microseconds in the above example. By increasing the numbers of incoming photons per time, the phase space filling becomes faster. To attain a carrier concentration of ($\approx 10^{12}$ cm⁻²) within 1 ns, the required intensity is about 200 W/cm² (corresponding to 2 mW on a 50×50 µm² element).

In order to demonstrate the large bandfilling induced absorption changes, a δ-doped hetero-n-i-p-i structure with selective contacts was grown by ESM-MBE. Figure 16 shows the band diagram in real space for the ground state and under reverse bias. The design parameters of the investigated n-i-p-i structure comprising 8-nm-thick strained $In_{0.16}Ga_{0.84}As$ quantum wells are given in the figure caption of Figure 16. At a reverse bias of $U_{pn} = -3$ V the quantum wells are completely depleted. In the ground state ($U_{pn} = 0$ V) the quantum wells contain a carrier density of $n^{(2)} \approx 4.3 \times 10^{12}$ cm⁻². In n-i-p-i crystals with selective contacts the carrier density can be determined by transport and capacitance measurements independently from the optical experiments.

Excitonic absorption is expected if the quantum wells are depleted by applying a reverse bias larger than the threshold voltage ($U_{pn}^{th} = -3$ V). With decreasing reverse bias, electrons are injected into the quantum wells and the excitonic absorption will be bleached increasingly. By further decreasing the reverse bias, a blue shift of the absorption edge is expected. Figure 17a shows the absorption of the InGaAs quantum wells as deduced from transmission experiments for various applied voltages. The transmission curves have been corrected for the Franz–Keldysh absorption in the intrinsic GaAs layers. The resulting absorption spectra show the expected steplike characteristic. Both the bleaching of the excitonic absorption as well as the blue shift of the absorption edge are clearly observable. Figure 17b shows the absorption changes due to bandfilling for various voltage swings. We find a maximum value of the absorption changes of about 11,000 cm^{-1}. The experimental data are in good agreement with results of simple calculations based on Equation 35. The absorption changes presented above are obtained from transmission measurements at 2 K. The values obtained for room temperature measurements are only slightly lower ($\Delta\alpha_{max} \approx 8000$ cm^{-1}), but no excitonic peak at the absorption edge is observable, which leads to the conclusion that the quality of the sample could be improved further. Larson and Maserjian [1991] and Jonsson et al. [1994] have performed purely optical experiments on similar structures. They were able to observe the bleaching of the exciton very clearly. Therefore, we expect that even larger absorption changes due to bandfilling can be observed in the future in structures with selective contacts, especially at room temperature.

5.1.3. Constructive Superposition of Phase Space Filling and Field Effects in Hetero-n-i-p-i Structures

The carrier-induced absorption bleaching in the n- and p-layers and the field effects in the intrinsic region contribute simultaneously to the absorption changes in n-i-p-i structures. Both effects reveal a similar dependence on the (optical or electrical) excitation. However, in homogeneous n-i-p-i crystals the two effects yield maximum absorption changes at different photon energies. Furthermore, slightly above the band gap, where the FKE and the bandfilling effect are large, these two effects even are of opposite sign, which results in a destructive superposition.

Recently, we have shown [Kneissl et al., 1994a; Gulden et al., 1994] that in hetero-n-i-p-i structure both effects can be superimposed constructively if the band gap energy of the intrinsic layers is shifted to higher energies, such that the spectral position of the maximum changes in the field-induced absorption tail roughly coincides with the band gap energy of the doping layers. An especially large enhancement can be obtained by replacing the homogeneously doped layers by quantum wells in combination with δ-doping layers, as described in Section 5.1.2. Thus, one can make use of the stronger quasi 2-dim bleaching effects.

5.2. Optical Nonlinearities

As predicted by Ruden and Döhler [1985] large changes of the absorption coefficient have been observed in n-i-p-i structures by optical excitation at very low light intensities [Simpson et al., 1986; Kost et al., 1989; Ando et al., 1989; Law et al., 1989; Larson and Maserjian, 1991; Yoffe et al., 1993]. Since the recombination lifetimes in n-i-p-i crystals are very long, an extremely low optical excitation is sufficient to generate a large excess carrier concentration responsible for field- and carrier-induced absorption changes. The field contribution is caused by the decrease of the internal electric fields due to the screening of the space charge–induced superlattice potential by the photogenerated carriers.

Whereas in the article by Simpson et al. [1986] the FKE in the high-field region of a homo-n-i-p-i structure was used to modulate the absorption, other authors [Ando et al., 1989; Law et al., 1989; Larson and Maserjian, 1991] have replaced the intrinsic regions of the doping superlattice by MQW layers in order to take additional advantage of the larger absorption changes associated with the QCSE compared with the FKE. Promising results for optical modulation are obtained at low intensities. Unfortunately, the photoinduced changes depend only logarithmically on the light intensity. This fact, which will be discussed in more detail in the following paragraph, is due to the exponential dependence of the recombination lifetimes on the band modulation, which decreases under optical excitation.

As we know, the recombination lifetimes of photogenerated carriers in n-i-p-i structures can be extremely long, due to the spatial separation of electrons and holes by the built-in electric fields. In the structures of interest for the observation of nonlinear optical properties, the doping level is typically much lower than in those designed for the observation of the tunable luminescence. Therefore, at room

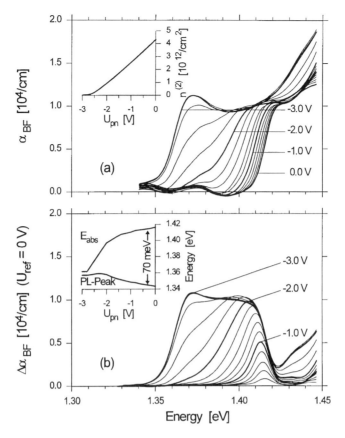

Figure 17 Dependence of the QW absorption on the electron sheet density at 2 K. (a) Absorption spectra from transmission experiments for various applied voltages. The insert shows the sheet electron density $n^{(2)}$ vs. U_{pn} which is determined by capacitance measurements. (b) Absorption changes due to bandfilling. The spectra refer to the ground state absorption at $U_{pn} = 0$. The inset shows the blue shift of the absorption edge related to lowest subband transition energy $(e_o–vh_o)$ taken from the electrically modulated PL spectra (see Figure 14).

temperature the tunneling recombination is usually negligible in comparison with the thermally activated electron–hole recombination, according to Section 2.3. Whereas Equations 24 and 25 refer to radiative recombination, the (total) recombination lifetime, which also includes nonradiative recombination via deep traps, can be expressed in the form

$$\tau_{rec}^{nipi} = \tau_{rec}^{bulk} \cdot \exp\left(\frac{\Delta}{nkT}\right) \qquad (36)$$

In Equation 36 $\Delta \approx E_g^o - E_g^{eff,exc} \approx V_{exc}$ (see Equations 22 and 25) is the band edge modulation. τ_{rec}^{bulk} is the recombination lifetime for uniform bulk material of similar doping levels, and n stands for the well-known pn-junction "quality factor" (typically between 1 and 2), which takes account of enhanced recombination due to deep impurities. In Figure 18 n-i-p-i structure with about zero effective band gap is shown in the ground state (a) and an excited state (b). In the ground state the barrier height has the value Δ_0, the exponent in the lifetime enhancement factor in Equation 36 is of the order of the band gap of the host material divided by nkT. The experimentally observed room-temperature lifetimes in GaAs n-i-p-i structures which are of the order of seconds, agree with Equation 36 if reasonable values for $\tau_{rec}^{bulk} = 1$ ns and $n \approx 2$ are used. Under optical excitation the photogenerated electrons and holes are accumulating in the n- and p-layers, respectively. The increasing carrier density results in the well-known increase of the effective band gap of the n-i-p-i crystal and increasing splitting between the electron and hole quasi Fermi levels (see Figure 18b).

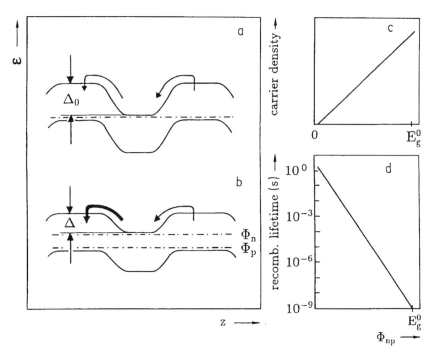

Figure 18 Schematic real space energy diagram of a n-i-p-i structure (a) in the ground state and (b) in an excited state. The arrows indicate the generation and recombination currents. The corresponding carrier density and typical recombination lifetimes as a function of Φ_{np} are shown in (c) and (d), respectively. (From Malzer, S. et al., *Phys. Status Solidi B* 173, 459, 1992. With permission.)

$$\Phi_{np} = \Phi_n - \Phi_p \tag{37}$$

The relation between the two-dimensional electron density $n^{(2)}$ and the quasi Fermi level splitting is shown schematically in Figure 18c. With increasing effective band gap the recombination barrier height decreases approximately according to

$$\Delta = \Delta_0 - \Phi_{np} \tag{38}$$

This results in an exponential increase of the recombination rate, as indicated by a bold arrow in Figure 18b. The recombination lifetime decreases exponentially according to Equation 36. This Φ_{np}-dependence is shown schematically in Figure 18d with typical experimentally found lifetimes. The steady state carrier density n at an optical generation rate of $\dot{n}^{opt}$ is given by

$$n = \dot{n}^{opt} \cdot \tau_{rec}^{nipi}\left(\Phi_{np}\right) \tag{39}$$

Because of the exponential changes of the lifetimes, the photoinduced carrier density and the internal fields depend only logarithmically on the optical power. This implies, of course, a dramatic reduction of the high responsivity observed at low optical excitation. In fact, this has been observed in the studies of the nonlinear optical transmission changes of n-i-p-i and hetero-n-i-p-i structures mentioned before [Simpson et al., 1986; Kost et al., 1989; Ando et al., 1989; Law et al., 1989; Larson and Maserjian, 1991]. At low optical power the observed nonlinearities were by many orders of magnitude larger than in other semiconductor structures. At higher optical power, however, the photoinduced optical changes were still much higher than in uniform semiconductors or MQW structures, but they increased only logarithmically, as expected from the arguments just presented. These undesired effects can be overcome

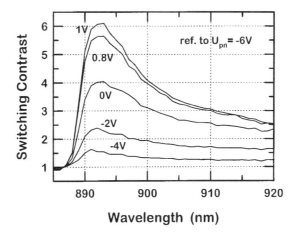

Figure 19 Reflection spectra normalized to the reflection at $U_{pn} = -6$ V for various voltage swings. The maximum switching contrast of 6:1 corresponds to an absorption change larger than 2200 cm^{-1} averaged over the whole sample thickness. The investigated homogeneous GaAs n-i-p-i structure consists of 15 periods. The 60-nm-thick n- and p-layers ($n_D = 2 \times 10^{18}$ cm^{-3} ; $n_A = 4 \times 10^{18}$ cm^{-3}) are separated by 70-nm-thick intrinsic layers.

in an elegant way by using n-i-p-i crystals with selective n- and p-type contacts. This possibility will be discussed in Section 5.4.

5.3. Electro-Optical Modulator Structures Based on n-i-p-i Crystals

For the application of n-i-p-i crystals as electro-optical modulators, tuning of the absorption via an external bias is necessary. We consider a n-i-p-i semimetal, i.e., a structure the doping level of which is large enough to exhibit free electrons and holes in the ground state. In such a n-i-p-i structure the internal fields can be varied within an enhanced range because of the possibility of applying reverse and forward bias. In n-i-p-i structures only relatively small external voltages are necessary to achieve high-field modulation depths. For a n-i-p-i structure with N periods the voltage swing required for a given change of the internal field is reduced by a factor $2N$ compared with a corresponding pin structure. An important consequence of the low potential drop is a reduction of the electric energy dissipated per optically generated carrier. Each photogenerated electron–hole pair contributes an amount eU_{pn}, which is in pin modulators usually significantly larger than the photon energy itself. Therefore, the electrical energy dissipation can be reduced by the same factor as the applied voltage. Moreover, the avalanche multiplication factor is kept small because of the low values of maximum energy gain of the accelerated electrons and holes. Thus, internal fields can be achieved that are far beyond the breakdown values of a linear pin junction, for which the intrinsic layer thickness has to be about $2N$ times larger. In the n-i-p-i modulator structure discussed in Section 5.3.1, e.g., internal electric fields in the intrinsic region of almost 10^6 V/cm are achieved without any remarkable increase of the reverse pn current.

5.3.1. Switching Behavior of n-i-p-i Modulator Structures

In order to demonstrate the suitability of n-i-p-i crystals as optical modulators, a number of homo- and hetero-n-i-p-i structures with selective contacts were grown. Figure 19 shows results of a GaAs homo-n-i-p-i structure which was designed to yield large changes of the internal field (for the design parameters see the figure caption). At a reverse bias of $U_{pn} = -6$ V the n-layers are depleted and a maximum internal field of 9.5×10^5 V/cm is achieved. A forward bias of 1 V reduces the internal field to 4×10^4 V/cm. Figure 19 displays reflection changes obtained for various voltage swings normalized to the reflection at -6 V. The incident light beam is being reflected on a mirror evaporated on the back side of the sample. Therefore, the reflection curves are basically double path transmission spectra. The on/off ratio measured with a voltage swing from $+1$ to -6 V is larger than 6:1. This corresponds to an absorption change, mainly due to the FKE, of more than 2200 cm^{-1} averaged over the whole sample thickness. As the Franz–Keldysh absorption is an off-resonance phenomenon, the obtained absorption changes are spectrally very broad. This is in contrast to the excitonic Stark shift in MQW structures, for which the spectral width is much narrower. For our modulator the contrast ratio remains larger than 3 within a wavelength range of 22 nm. This is an important advantage for device applications, as wavelength detuning by a thermally induced band gap shift due to optical and electrical energy dissipation will influence the performance of Franz–Keldysh modulators only at very high power levels. In consequence, a device requiring at least 50% of the maximum switching contrast will operate safely up to temperature variations of ±50 K.

The light modulation is mainly caused by the field-induced absorption changes in the intrinsic region. Therefore, the contrast ratio can be further increased by using δ-doped n-i-p-i structures. A replacement of the uniform i-layers by MQW structures results in an additional improvement because of the higher absorption changes associated with the QCSE. As a consequence, the applied voltage U_{pn} can be smaller or, for a given voltage swing, the thickness of the i-layers can be increased, resulting in a reduced device capacitance and an increased switching time. The spectral width of the absorption changes, however, reduces to a few nanometers, making the device more sensitive to temperature variations.

5.3.2. Dynamic Switching Behavior

In optically excited n-i-p-i structures, the carrier dynamics is governed by the internal recombination lifetimes. Due to the spatial separation of electrons and holes, these lifetimes can be as high as milliseconds or even larger. Under excitation the carrier lifetimes decrease exponentially as shown in Section 5.2. In n-i-p-i structures with selective n- and p-type contacts, however, this is not the case. Here the time constants are RC-times given by the resistance of the doped layers and the capacitance of the interdigitated n- and p-layers. Although the areal capacitances of n-i-p-i crystals are relatively large compared with pin structures, very short RC-times (e.g., less than a nanosecond) can be achieved for sufficiently small devices, since the RC-time constants basically scale quadratically with the device width [Kiesel et al., 1993a; Pfeiffer et al., 1996]. The RC-times of n-i-p-i structures are independent of the number of periods, as will be shown below. Therefore, high-speed operation in the multigigahertz range should be possible even for modulators with high switching contrast, if a suitable design is chosen.

The total recombination lifetime in a n-i-p-i crystal with selective contacts is given by the internal lifetime and an externally adjustable RC-time.

$$\frac{1}{\tau_{rec}} = \frac{1}{\tau_{rec}^{int}} + \frac{1}{\tau_{RC}} \tag{40}$$

The internal lifetime is rather long, according to Equation 36, as long as a reverse or only moderate forward bias is applied. Therefore, the total lifetime is governed by the RC-time. The high-frequency behavior is most easily understood in terms of equivalent circuits. Figure 20b shows such a circuit with the n-i-p-i structure represented as a capacitively loaded transmission line. In the following we consider a simplified circuit given in Figure 20c which turns out to give a surprisingly accurate description. It leads to the following expression for the switching time:

$$\tau_{RC} = RC = \left(R_{nipi} + R_{contact} + R_{source} \right) \cdot C_{nipi} \tag{41}$$

The physical significance and value of the various parameters entering this equation are as follows:

$R_{nipi} = (\rho_n + \rho_p) \, W/L$ is the combined resistance of the n- and p-layers. It is proportional to the device aspect ratio W/L and to the sum of the sheet resistivities of both layers, $\rho_n + \rho_p$. These quantities can be measured for a given n-i-p-i structure on devices that are provided with two n- and two p-type contacts.

$R_{contact} = \rho_c/L$ is the total contact resistance, including resistances occurring within the n-i-n-i and p-i-p-i region, and is inversely proportional to the device length L. The contact resistance per unit length can be extracted from the pn-characteristics under heavy forward bias.

R_{source} = the internal resistance of the driving voltage source, usually a constant: $R_{source} = 50 \, \Omega$

$C_{nipi} = LWC_{nipi}^{(2)}$ denotes the junction capacitance. It is proportional to the device area. The capacitance per unit area $C_{nipi}^{(2)}$ for a given n-i-p-i crystal can be measured directly on devices of various size; its slight dependence on the applied voltage can be ignored in most cases.

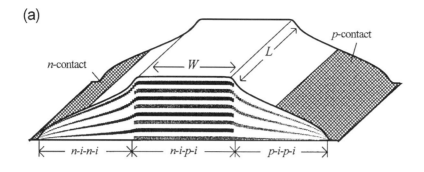

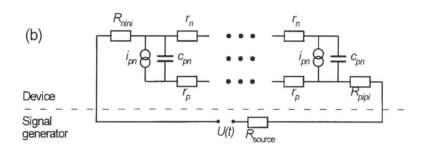

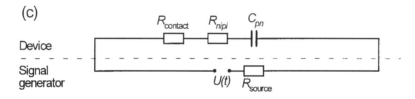

Figure 20 (a) Schematic cross section of a n-i-p-i modulator structure. W and L indicate the device dimensions as used in the text. (b) Electrical equivalent circuit of a n-i-p-i crystal. (c) Simplified equivalent circuit that has been used in the calculations.

It is clear that a substantial error is introduced by replacing the transmission by discrete elements R_{nipi} and C_{nipi}. In fact, the current has to go through a part of the n- and p-layers only, and neither has the n-i-p-i area to be recharged completely in order to achieve a transmission change at a spot typically near the center of the structure. As it turns out, these effects can be sufficiently accounted for by inserting $W/2$ instead of W into all equations given above. Combining them finally yields an expression for the time constant:

$$\tau_{RC} = \frac{W^2}{4}\left(\rho_n + \rho_p\right)C_{nipi}^{(2)} + \frac{W}{2}\,\rho_c C_{nipi}^{(2)} + \frac{LW}{2}R_{source}C_{nipi}^{(2)} \tag{42}$$

In Figure 21 experimentally obtained values for the switching time are compared with the theory above (dashed line). The excellent agreement with experiment over nearly two orders of magnitude is obvious, and it should be pointed out that it was achieved without any fit parameters. The parameters entering Equation 42 were measured for the investigated n-i-p-i modulator structure as indicated above. It should be mentioned that the obtained values ($\rho_n = 30\ \Omega$, $\rho_p = 320\ \Omega$, and $C_{nipi}^{(2)} = 75\ \text{fF}/\mu\text{m}^2$) are in good agreement with the values expected theoretically from the design parameters given in the figure caption

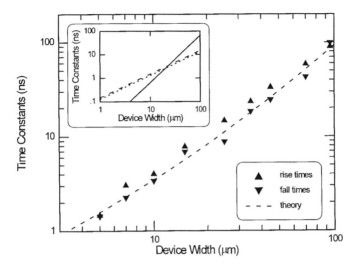

Figure 21 Time constants of a n-i-p-i modulator structure as a function of the n-i-p-i width W (length $L = 70$ mm). The dashed curve is given by Equation 42. The insert shows the contribution of the three terms in Equation 42 separately: solid line $\propto W^2$, dashed line $\propto W$, dotted line $\propto LW$. The investigated n-i-p-i structure consists of 25 periods. A single period comprises two δ-doped n-layers of $n_D = 4 \times 10^{12}\,\text{cm}^{-2}$ each, with an undoped region in between. This region contains a 8-nm-thick strained $\text{In}_{0.16}\text{Ga}_{0.84}\text{As}$ quantum well embedded between two 10-nm-thick i-GaAs spacer layers. The n-region is followed by a 70-nm-thick i-GaAs region and a δ-doped p-layer ($n_A = 1.2 \times 10^{13}\,\text{cm}^{-2}$). (From Pfeiffer, U. et al., *Appl. Phys. Lett.* 1996. With permission.)

of Figure 21. The switching times have been determined experimentally by AC transmission measurements on a series of n-i-p-i crystals grown by the ESM-MBE with various widths of the n-i-p-i region. The investigated structure shows a switching contrast of 2.2 within a voltage swing as low as 3.7 V. For the smallest device (W = 5 μm) time constants as low as 1.5 ns have been observed. The insert of Figure 21 separately shows the contributions of the three terms in Equation 42. For wide devices, the switching speed is limited by the resistances within the n-i-p-i area (solid line), their contributions being proportional to W^2. For narrow structures, the limiting factor is due to the contact resistance (dashed line), which adds a term $\propto W$ to the total RC-time. The dotted line arising from the 50-Ω driving impedance is proportional to the device area LW and lies much lower for shorter devices (in this case $L = 70$ μm) and, therefore, does not represent a real problem.

The switching times for the investigated n-i-p-i structure are very promising, although they were achieved with nonoptimized samples. The terms of Equation 42 clearly show at which points improvements could be made and what kind of results can be expected. The most obvious thing to do is to reduce the width of the n-i-p-i region. However, devices smaller than 5 μm are difficult to manufacture and unsuitable for most applications. Clearly, the most important quantity to work at is the contact resistance $R_{contact}$. The switching time for the 5-μm device is increased by about a factor of 10 due to the high contact resistance. With improvements on the growth technique (e.g., higher p-doping or homogeneously doped layers) the contact resistance, especially the resistance of the p-i-p-i region, should be reduced drastically, which leads to devices with switching times in the multigigahertz range.

5.4. Optical Nonlinearities in n-i-p-i Structures with Selective Contacts

The strength of the induced nonlinear absorption and refractive index changes at a given photon energy depend strongly on the design parameter of the n-i-p-i structure. Although it is not possible to establish universal criteria for an optimum design, the following requirements may apply quite generally:

I. The induced changes of the carrier density should be as large as possible in order to obtain large contributions due to the bleaching phenomena;

II. The field changes, which are directly correlated with the changes of the (two-dimensional) carrier density per superlattice period $\Delta n^{(2)}$, should also be as large as possible;

III. The superposition of the bleaching and the field effect should be constructive at a chosen photon energy;

IV. The dependence of the different phenomena on the carrier density should be designed such as to produce the largest possible nonlinear changes at the largest achievable excitation-induced carrier density and field;

V. The induced carriers should be used as efficiently as possible in order to minimize the optical switching energies. This requirement is fulfilled if the electron–hole pairs created by the absorption of photons have long recombination lifetimes;

VI. The quenching of the nonlinear changes should be possible within short times and with no or at least low additional energy needed.

All the requirements listed above can be achieved most effectively by using n-i-p-i structures with selective contacts. By electrical excitation the carrier density as well as the internal fields can be varied within a larger range compared with the optical case because of the possibility of applying forward and reverse bias voltages. The requirements V and VI seem to be incompatible with each other. However, in n-i-p-i crystals they can be met simultaneously. This results from the fact that the recombination lifetimes can be controlled externally if the n-i-p-i structure is provided with selective n- and p-type contacts. An external resistor across these contacts opens up an additional recombination path with a carrier lifetime given by the RC-time constant of the circuit. Thus, speed and sensitivity of the optical nonlinearities can be adjusted externally. As shown in Section 5.3.2, very low time constants can also be achieved. Therefore, n-i-p-i crystals with selective contacts exhibit large optical nonlinearities which can be adjusted to a given application. Furthermore, they also represent a unique system for fundamental studies of bandfilling phenomena, as, e.g., the Fermi level splitting Φ_{np}, and the carrier density as well can be controlled precisely by the applied voltage U_{pn}. A particular advantage of such systems is the possibility of determining the carrier density as a function of the applied voltage from transport and capacitance measurements, independently of the optical experiments.

In the following, the excitation and recombination processes in n-i-p-i crystals with selective contacts will be discussed in more detail. For this purpose we consider a n-i-p-i semimetal (see Figure 22). By applying a reverse bias to the selective contacts the n-layers will be depleted at an external potential eU_{pn}^{th}, the value of which depends on the design parameters of the sample. Under forward bias the long recombination lifetimes degrade again exponentially with increasing $\Phi_{pn} = eU_{pn}$ for the reasons given in Section 5.2. The recombination current density (which can be measured directly) is directly related to the excitation dependent lifetimes.

$$j = e \cdot \frac{n^{(2)}\left(\Phi_{np}\right)}{\tau_{rec}^{nipi}\left(\Phi_{np}\right)} \tag{43}$$

Under reverse bias, however, the recombination barrier height increases. Therefore, the recombination rate decreases exponentially. The generation processes, which balance the recombination processes in the ground state ($\Phi_{np} = 0$), are now dominating (see Figure 22a). The rate of the generation processes increases only slowly with increasing reverse bias (as in normal p-n junctions) as long as the internal fields remain significantly below the breakdown fields. As a result, the system returns into its ground state with a time constant which is similar to the recombination lifetime at zero bias (if the bias U_{pn} is switched off). In Figure 22c the absolute value of the current density and the corresponding (approximate) lifetimes are shown schematically on a logarithmic scale as a function of the quasi Fermi level splitting Φ_{np}. In Figure 22b the corresponding two-dimensional electron density is given. The dashed lines in the graphs (b) and (c) indicate the corresponding dependence for the n-i-p-i structure with depleted n-layers in the ground state which is usually used for optical excitation (see Section 5.2). It becomes obvious from the comparison of these two graphs that considerably larger changes of the carrier densities and of the internal fields are possible in the n-i-p-i semimetal and that the very long and nearly constant internal lifetimes apply over a very wide range.

We will now discuss the steady state kinetics of the photoinduced carrier density and field changes. Under illumination with light of the photon energy $\hbar\omega$ and intensity I_ω the photocurrent density is given by

$$j = e\eta \cdot \frac{I_\omega}{\hbar\omega} \tag{44}$$

where η is the quantum efficiency ($\eta = (1 - R)(1 - \exp(-\alpha d))$), which depends, among others, on the thickness of the superlattice d, and in particular on the absorption coefficient $\alpha(\omega)$ and its nonlinear changes. This photocurrent will recombine through the voltage source if the external bias is connected

294

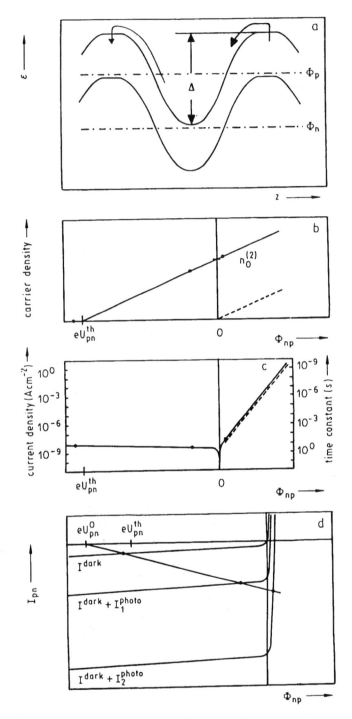

Figure 22 (a) Energy diagram of a n-i-p-i semimetal under reverse bias. In (b) the carrier density as a function of $\Phi_{np} = eU_{pn}$ is shown. The dashed lines are the corresponding graphs for a n-i-p-i structure with zero effective band gap. The pn-characteristics also represent the recombination lifetime as indicated by the time scale on the right hand side of (c). In (d) the current–voltage curve for the dark case and for two different illumination intensities is depicted. The actual voltage drop on the n-i-p-i structure is given by the intersection points of the pn-curves with the load line of the resistor R^{ext}. (From Malzer, S. et al., *Phys. Status Solidi B* 173, 459, 1992. With permission.)

directly to the n- and p-contacts. If, however, an external resistor R^{ext} is included into the circuit, the voltage drop U_{pn} on the n-i-p-i structure becomes

$$U_{pn} = U_{pn}^o + \left(I^{photo} + I^{dark}\right) \cdot R^{ext} \tag{45}$$

From Figure 22d we see that the value of U_{pn} depends strongly on the light intensity I_ω and on R^{ext}. If $U_{pn}^o < U_{pn}^{th}$, as depicted in Figure 22d, the voltage drop on the n-i-p-i structure in the dark state may exceed U_{pn}^{th}. With increasing I_ω the voltage drop decreases nearly linearly over the range of negative values of U_{pn}. In this range the electron and hole densities also increase nearly linearly, once U_{pn} becomes larger than U_{pn}^{th}. In this range the photocurrent recombines externally. Only if $U_{pn} > 0$, the internal recombination current increases exponentially and finally dominates, as discussed in Section 5.2. In this regime the light-induced carrier density increases only logarithmically with the light intensity. It is clear from Equation 45 that higher values of R^{ext} result in a lower light intensity required to reach $U_{pn} = 0$ and the corresponding carrier density n_o. In the limiting case of a very high resistance we find that the photocurrent has only to be of the order of the reverse bias leakage current I^{dark} in order to obtain carrier densities of the same order or even larger than those achievable in samples without contacts even under extremely high light intensities. Assuming a reasonable value of 1 µA/cm² for the leakage current density, for instance, one may obtain $n_o^{(2)} = (1 \text{ to } 10) \times 10^{12} \text{ cm}^{-2}$ as the steady state carrier density with a light intensity of only 1 µW/cm². Normal semiconductors and MQW structures require 9 to 12 orders of magnitude higher light intensities for the creation of comparable carrier densities.

Finally, we would like to sketch how the well-known dilemma between high nonlinearity and slow recovery can be overcome in n-i-p-i structures. For this purpose we discuss a system consisting of two identical n-i-p-i structures in series with a constant voltage U_{pn}^o as shown in Figure 23. In the ground state ① the n-layers in the two structures are depleted. If at the time t_1 a light pulse of the intensity I_ω is sent to the right device for at least a time interval τ,

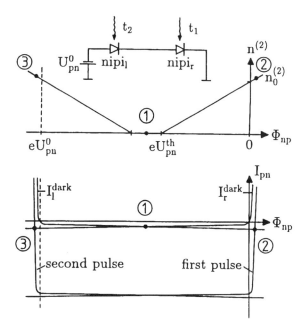

Figure 23 Carrier densities and steady state current–voltage characterictics of a system of two identical n-i-p-i structures in series with a constant voltage. The Φ_{np} axis corresponds to the right n-i-p-i device. In the dark case the system is in state ①. With an optical light pulse at t_1 on the right device, the system transfers to state ②. A second pulse at the time t_2 on the left device can move the system into state ③. (From Malzer, S., *Phys. Status Solidi B* 173, 459, 1992. With permission.)

$$\tau = N \cdot \frac{n_o^{(2)}}{\eta \cdot I_\omega / \hbar\omega} \tag{46}$$

which is necessary to create the carrier density $N \cdot n^{(2)}$, the system will come to state ②. It will remain in this state with a very long lifetime determined by

$$\tau_{rec}^{nipi} \approx \frac{eNn_o^{(2)}}{j^{dark}} \tag{47}$$

which may be in the range of seconds. By another pulse of twice the energy sent to the left device at the time t_2, the right device can be switched into the fully depleted state ③. Thus, the switching occurs with the minimum possible switching energy per area, while the switched state can be maintained over a period which is larger by many orders of magnitude.

6. OUTLOOK

In this review we have tried to present a balanced selection of interesting results on the basic and on the application-relevant tunable optical properties of n-i-p-i and hetero-n-i-p-i doping superlattices. We hope that we have been able to demonstrate that n-i-p-i structures represent a fascinating model system for the study of a wide variety of phenomena but are also ideally suited for optoelectronic device applications. Of course, we had to exclude the discussion of many aspects which are interesting with regard to basic physics as well as concerning device applications.

For instance, we did not mention a number of possibilities to improve on the performance of n-i-p-i modulators, such as incorporating the active material into an asymmetric Fabry–Perot cavity. Although closely related to both luminescence and absorption, we omitted the topic of stimulated light emission. This may become particularly interesting in the future in connection with microcavities. Population inversion is achieved very easily in n-i-p-i structures, due to the long recombination lifetimes. Thus, the conditions for low threshold lasing or superradiant emission normal to the surface at low emission into other modes appear favorable.

In our discussions of the changes of the absorption coefficient we have excluded completely the changes of the real part of the dielectric function. In fact, the nonlinear changes of the refractive index, although in the 10^{-2} range, turn out to be large in comparison with those of bulk or compositional superstructure semiconductors. The relatively large refractive index changes, which are mostly due to the very strong bandfilling effects, are particularly interesting, as they can coexist with very low absorption coefficients [Ruden and Döhler, 1985; Yoffe et al., 1993].

We have also excluded a discussion of our n-i-p-i-based "smart pixel" concept enabling optoelectrical and opto-optical switching. Extremely low switching energies and high values for the optical gain have been achieved [Kiesel et al., 1993b; Kneissl et al., 1994b]. As these devices can be integrated into two-dimensional arrays they are very appealing for many applications that require massively parallel signal and data processing. By using a suitable design, bistable switching can also be observed, which can be employed in optical memories.

Finally, we would like to mention some recent luminescence experiments which were started with the goal of demonstrating large tuning effects on the envelope wave functions and the interband transition dynamics induced by in-plane magnetic fields [Forkel et al., 1995]. These investigations represent just one example of the wide variety of possibilities that can be used to gain get new insights regarding the microscopic features of electron and hole states in semiconductors using investigations in n-i-p-i structures.

REFERENCES

Ando, H., Iwamura, H., Oohashi, H., and Kanbe, H. 1989. Non-linear absorption in n-i-p-i MQW structures. *IEEE J. Quantum Electron.* 25: 2135.

Bar-Joseph, I., Goossen, K. W., Kuo, J. M., Kopf, R. F., Miller, D. A. B., and Chemla, D. S. 1989. Room-temperature electroabsorption and switching in a GaAs/AlGaAs superlattice. *Appl. Phys. Lett.* 55: 340.

Beyer, H. J., Metzner, C., Heitzer, J., and Döhler, G. H. 1989. Temperature dependence of the tunable luminescence, absorption and gain spectra of NIPI doping superlattices — theory and comparison with experiment. *Superlattices Microstruct.* 6: 351–356.

Böhner, H. 1995. *Diploma thesis.* unpublished.

Carey, K. W., Döhler, G. H., Turner, J., and Vilms, J. 1985. Growth and characterisation of n-i-p-i doping superlattices in GaInAs grown by organometallic vapor phase epitaxy. *Inst. Phys. Conf. Ser.* 79: 385–390.

Chang-Hasnain, C. J., Hasnain, G., Johnson, N. M., Döhler, G. H., Miller, J. N., Whinnery, J. R., and Dienes, A. 1987. Tunable electroabsorption in gallium arsenide doping superlattices. *Appl. Phys. Lett.* 50: 915.

Cingolani, R. and Ploog, K. 1991. Frequency and density dependent radiative recombination processes in III-V semiconductor quantum wells and superlattices. *Adv. Phys.* 40: 535–623.

Döhler, G. H. 1972. Electron states in crystals with n-i-p-i superstructure. *Phys. Status Solidi B* 52: 79; Electrical and optical properties of crystals with n-i-p-i superstructure. *Phys. Status Solidi B* 52: 533.

Döhler, G. H. 1978. Ultrathin doping layers as a model for 2D systems. *Surf. Sci.* 73: 97.

Döhler, G. H., Künzel, H., Olego, D., Ploog, K., Ruden, P., Stolz, H. J., and Abstreiter, G. 1981. Observation of tunable band gap and two-dimensional subbands in a novel GaAs ("nipi") superlattice. *Phys. Rev. Lett.* 47: 864–867.

Döhler, G. H. 1983a. n-i-p-i doping superlattices–semiconductors with tunable electronic properties. *Jpn. J. Appl. Phys.* 22 (Suppl. 1): 29–35.

Döhler, G. H. 1983b. n-i-p-i doping superlattices — metastable semiconductors with tunable properties. *J. Vac. Sci. Technol. B* 1: 278–284.

Döhler, G. H. and Ruden, P. 1984. Theory of absorption in doping superlattices. *Phys. Rev. B* 30: 5932.

Döhler, G. H. 1986. The physics and applications of n-i-p-i doping superlattices. *CRC Rev. Solid State Mater. Sci.* 13: 97–141.

Döhler, G. H., Hasnain, G., and Miller, J. N. 1986a. In situ grown-in selective contacts to n-i-p-i doping superlattice crystals using MBE growth through a shadow mask. *Appl. Phys. Lett.* 49: 704.

Döhler, G. H., Fasol, G., Low, T. S., and Miller, J. N. 1986b. Observation of tunable room temperature photoluminescence in GaAs doping superlattices. *Solid State Commun.* 57: 563–566.

Döhler, G. H. 1987. Optoelectronic device applications of doping superlattices. *SPIE Proc.* 861: 21.

Döhler, G. H. 1988. Properties of impurity states in n-i-p-i superlattice structures. C. Y. Fong, I. P. Batra, and S. Ciraci, Eds., *NATO ASI Ser.* 183: 159.

Efron, U. and Livescu, G. 1995. Multiple quantum well spatial light modulators. In *Spatial Light Modulator Technology,* U. Efron, Ed., pp. 217–286. Marcel Dekker, New York.

Esaki, L. and Tsu, R. 1970. Superlattice and negative differential conductivity in semiconductors. *IBM J. Res. Dev.* 14: 61.

Fasol, G., Ruden, P., and Ploog, K. 1984. Raman-scattering from elementary excitations in GaAs with "n-i-p-i" doping superlattices. *J. Phys. C* 17: 1395.

Forkel, M., Heitzer, J., Ehrlich, J., Penner, U., Mackh, G., Ossau, W., Campmann, K. L., Gossard, A. C., and Döhler, G. H. 1995. Magnetic tuning of the spatially indirect interband transitions in parabolic and rectangular quantum wells. In *High Magnetic Fields in the Physics of Semiconductors,* D. Heiman, Ed., pp. 354–357. World Scientific, Singapore.

Franz, W. 1958. Einfluß eines elektrischen Feldes auf die optische Absorptionskante. *Z. Naturforsch.* 13a: 484.

Gulden, K. H., Lin, H., Kiesel, P., Riel, P., Ebeling, K. J., and Döhler, G. H. 1991. Giant ambipolar diffusion in n-i-p-i structures. *Phys. Rev. Lett.* 66: 373–376.

Gulden, K. H., Wu, X., Smith, J. S., Kiesel, P., Höfler, A., Kneissl, M., Riel, P., and Döhler, G. H. 1993. Novel shadow mask molecular beam epitaxial regrowth technique for selective doping. *Appl. Phys. Lett.* 62: 3180.

Gulden, K. H., Kneissl, M., Kiesel, P., Malzer, S., and Döhler, G. H. 1994. Enhanced absorption modulation in hetero n-i-p-i structures by constructive superposition of field effects and phase space filling. *Appl. Phys. Lett.* 64: 457–459.

Hasnain, G., Döhler, G. H., Whinnery, J. R., Miller, J. N., and Dienes, A. 1986. Highly tunable and efficient room-temperature electro-luminescence from GaAs doping superlattices. *Appl. Phys. Lett.* 49: 1357–1359.

Jonsson, B., Larrson, A. G., Sjölund, O., Wang, S., Andersson, T. G., and Maserjian, J. 1994. Carrier recombination in a periodically δ-doped multiple quantum well structure. *IEEE J. QE.* 30: 63–74.

Jung, H., Künzel, H., Döhler, G. H., and Ploog, K. 1983. Photoluminescence in GaAs doping superlattices. *J. Appl. Phys.* 54: 6965–6973.

Kazarinov, R. F. and Shmarsev, Yu, V. 1971. Optical phenomena due to the carriers in a semiconductor with a superlattice. *Sov. Phys. Semicond.* 5: 710.

Keil, U. D., Linder, N., Schmidt, K., and Döhler, G. H. 1991. Temperature dependence of type II hetero-n-i-p-i photo-reflectance spectra. *Phys. Rev. B* 44: 13504.

Keldysh, L. V. 1958. The effect of a strong electric field on the optical properties of insulating crystals. *Sov. Phys. JETP* 34: 788.

Kiesel, P., Gulden, K. H., Höfler, A., Kneissl, M., Knüpfer, B., Linder, N., Riel, P., Wu, X., Smith, J. S., and Döhler, G. H. 1993a. High speed and high contrast electro-optical modulator based on n-i-p-i doping superlattices. *Superlattices and Microstruct.* 13: 21–24.

Kiesel, P., Gulden, K. H., Höfler, A., Kneissl, M., Knüpfer, B., Riel, P., Wu, X., Smith, J. S., and Döhler, G. H. 1993b. Bistable opto-optical switches with high gain based on n-i-p-i doping superlattices. *Appl. Phys. Lett.* 62: 3288–3290.

Kneissl, M., Gulden, K. H., Kiesel, P., Luczak, A., Malzer, S., and Döhler, G. H. 1994a. Constructive superposition of field- and carrier-induced absorption changes in hetero-n-i-p-i structures. *Solid-State Electron.* 37: 1251–1253.

Kneissl, M., Kiesel P., Riel, P., Reingruber, K., Gulden, K. H., Dankowski, S. U., Greger, E., Höfler, A., Knüpfer, B., Wu, X., Smith, J. S., and Döhler, G. H. 1994b. Demonstration of low switching energies using new n-i-p-i based smart pixels. *SPIE Proc.* 2139: 115–129.

Knüpfer, B., Kiesel, P., Kneissl, M., Dankowski, S., Linder, L., Weimann, G., and Döhler, G. H. 1993. Polarization-insensitive high-contrast GaAs/AlGaAs waveguide modulator based on the Franz-Keldysh effect. *IEEE Photon. Technol. Lett.* 5: 1386–1388.

Köhler, K., Döhler, G. H., Miller, J. N., and Ploog, K. 1986a. Temperature-dependent luminescence of GaAs doping superlattices. *Solid State Commun.* 58: 769–773.

Köhler, K., Döhler, G. H., Miller, J. N., and Ploog, K. 1986b. Temperature dependence of tunable luminescence of GaAs doping superlattices. *Superlattices Microstruct.* 2: 339–343.

Kohn, W. and Sham, L. 1965. Self-consistent equations including exchange and correlation effects. *Phys. Rev.* 140: A1133.

Kost, A., Garmire, E., Danner, A., and Dapkus, P. D. 1989. Large optical non-linearities in GaAs/AlGaAs hetero n-i-p-i structures. *Appl. Phys. Lett.* 54: 301.

Künzel, H., Döhler, G. H., Ruden, P., and Ploog, K. 1982. Tunable electroluminescence from GaAs doping superlattices. *Appl. Phys. Lett.* 41: 852–854.

Larson, A. and Maserjian J. 1991. Optically induced excitonic electroabsorption in a periodically δ-doped InGaAs/GaAs multiple quantum well structure. *Appl. Phys. Lett.* 59: 1946.

Law, K. K., Maserjian, J., Simes, R. J., Coldren, L. A., Gossard, A. C., and Merz, J. L. 1989. Optically controlled reflection modulator using GaAs/AlGaAs n-i-p-i MQW structures. *Opt. Lett.* 14: 230.

Linder, N., El-Banna, W., Keil, U. D., Schmidt, K., and Döhler, G. H. 1990. Electro- and photo-modulation spectroscopy of long period doping superlattices. *Proc. SPIE* 1286: 359.

Linder, N., Gabler, T., Gulden, K. H., Kiesel, P., Kneissl, M., Riel, P., Wu, X., Walker, J., Smith, J. S., and Döhler, G. H. 1993. High contrast electrooptic n-i-p-i doping superlattice modulator. *Appl. Phys. Lett.* 62: 1916–1918.

Malzer, S., Linder, N., Gulden, K. H., Höfler, A., Kiesel, P., Kneissl, M., Wu, X., Smith, J. S., and Döhler, G. H. 1992. Optical nonlinearities in n-i-p-i and hetero n-i-p-i structures. *Phys. Status Solidi, B* 173: 459–472.

Malzer, S., Kneissl, M., Kiesel, P., Gulden, K. H., Wu, X. X., Smith, J. S., and Döhler, G. H. 1996. Properties and applications of the epitaxial shadow mask (ESM) MBE-technique. in press, *JVST B*.

Mendez, E. E., Agullo-Rueda, F., and Hong, J. M. 1988. Stark localization in GaAs/AlGaAs superlattices under electric fields. *Phys. Rev. Lett.* 23: 2426.

Metzner, C., Beyer, H. J., and Döhler, G. H. 1992a. Theory of donor acceptor pair luminescence in delta-doped n-i-p-i superlattices. *Phys. Rev. B* 46: 4128–4138.

Metzner, C., Schmidt, T., Müller, S. G., and Döhler, G. H. 1992b. Classical theory of impurity bands in δ-doped n-i-p-i superlattices. *Phys. Rev. B* 47: 10633–10647.

Metzner, C., Schrüfer, K., Wieser, U., Luber, M., Kneissl, M., and Döhler, G. H. 1995. Disorder effects on luminescnece in delta-doping n-i-p-i superlattices. *Phys. Rev. B* 51(8): 5106–5115.

Miller, D. A. B., Chemla, D. S., Damen, T. C., Gossard, A. C., Weigmann, W., Wood, T. H., and Burrus, C. A. 1985. Electric field dependence of optical absorption near the band gap of quantum well structures. *Phys. Rev. B* 32: 1043.

Oswald, J., Tranta, B., Pippan, M., and Bauer, G. 1990. PbTe–doping superlattices: properties of pnp and p-i-n-i-p structures. *J. Opt. Quantum Electron.* 22: 2433–2459.

Ovsyannikov, M. L., Romanov, Yu, A., Shabanov, V. N., and Loginova, R. G. 1971. Periodic semiconductor structure. *Sov. Phys. Semicond.* 4: 1919.

Pfeiffer, U., Kneissl, M., Knüpfer, B., Müller, N., Kiesel, P., Smith, J. S., and Döhler, G. H. 1996. Dynamical switching behavior of n-i-p-i modulator structures. *Appl. Phys. Lett.* 68: 1838–1840.

Phillips, C. C., Johnson, E. A., Thomas, R. H., and Vaghjiani, H. L. 1993. Interband and intersubband transitions in indium arsenide doping superlattices studied by absorption, nonlinear absorption and photoconductivity spectroscopies. *Semicond. Sci. Technol.* 8(1): 373–379.

Ploog, K. and Döhler, G. H. 1983. Compositional and doping superlattices in III-V semi-conductors. *Adv. Phys.* 32: 285.

Rehm, W., Ruden, P., Döhler, G. H., and Ploog, K. 1983. Study of time-resolved luminescence in GaAs doping superlattices. *Phys. Rev. B* 28: 5937–5942.

Renn, M., Metzner, C., and Döhler, G. H. 1993. Effect of random impurity distribution on the luminescence of n-i-p-i doping superlattices. *Phys. Rev. B* 48: 11220–11227.

Ruden, P. P. and Döhler, G. H. 1983a. Semiconductors with hetero n-i-p-i superlattices. *Surf. Sci.* 132: 540.

Ruden, P. and Döhler, G. H. 1983b. Electronic excitations in semiconductors with doping superlattices. *Phys. Rev. B* 27: 3547–3553.

Ruden, P. and Döhler, G. H. 1983c. Electronic structure of semiconductors with doping superlattices. *Phys. Rev. B* 27: 3538.

Ruden, P. P. and Döhler, G. H. 1985. Low-power non-linear optical phenomena in doping superlattices. In *Proc. of Conf. on the Physics of Semiconductors*, J. D. Chadi and W. H. Harrison, Eds., p. 535. Springer, New York.

Schmitt-Rink, S., Chemla, D. S., and Miller, D. A. B. 1989. Linear and nonlinear optical properties of semiconductor quantum wells. *Adv. Phys.* 38: 89.

Schönhut, J., Metzner, C., Müller, S., Schmidt, T., Förster, A., Lüth, H., and Döhler, G. H. 1996. Optical investigation of impurity bands in δ-doped n-layers. *Solid State Electron.* 40: 701–705.

Schrüfer, K., Eckl, S., Metzner, C., Beyer, H. J., and Döhler, G. H. 1992. Thomas-Fermi theory of n-i-p-i doping superlattices. *J. Appl. Phys.* 72(10): 4992–4994.

Schrüfer, K., Metzner, C., Wieser, U., Kneissl, M., and Döhler, G. H. 1994. Quantum effects of potential fluctuations in GaAs δ-doping superlattices. *Superlattices Microstruct.* 15: 413–420.

Schubert, E. F., Cunningham, J. E., Tsang, W. T., and Timp, G. L. 1987. Selectively δ-doped $Al_xGa_{1-x}As$/GaAs heterostructures with high two-dimensional electron-gas concentrations $n_{2DEG} \geq 1.5 \times 10^{12}$ cm^{-2} for field-effect transistors. *Appl. Phys. Lett.* 51(15): 1170–1172.

Schubert, E. F., Harris, T. D., and Cunningham, J. E. 1988. Minimization of dopant-induced random potential fluctuations in sawtooth doping superlattices. *Appl. Phys. Lett.* 53(22): 2208–2210.

Schultz, J., Malzer, S., Kneissl, M., Pfeiffer, U., Kiesel, P., Smith, J. S., and Döhler, G. H. 1996. Many body effects and charge carrier kinetics studied by electro-optical experiments in type-I hetero n-i-p-i structures with selective contacts. *Solid State Electron.* 40: 683–686.

Simpson, T. B., Pennise, C. A., Gordon, C. A., Anthony, J. E., and AuCoin, T. R. 1986. Optically induced absorption modulation in GaAs doping superlattices. *Appl. Phys. Lett.* 49: 590.

Stafeev, V. I. 1971. Many-layer structures consisting of a large number of pn-junctions. *Sov. Phys. Semicond.* 5: 359.

Street, R. A., Döhler, G. H., Miller, J. N., and Ruden, P. P. 1986. Luminescence of n-i-p-i heterostructures. *Phys. Rev. B* 33: 7043–7046.

Street, R. A., Döhler, G. H., Miller, J. N., Burnham, R. D., and Ruden, P. P. 1987. Luminescence transitions from excited subband states in n-i-p-i heterosturctures. *Proceedings of the 18th International Conference on the Physics of Semiconductors,* O. Engström, Ed., (World Scientific, Singapore: 215).

Yamauchi, Y., Uwai, K., and Mikami, O. 1984. Photoluminescence of InP doping superlattice grown by vapor phase epitaxy. *Jpn. J. Appl. Phys.* 23: 785–787.

Yoffe, G. W., Brübach, J., Karouta, F., and Wolter, J. H. 1993. Single-wavelength all-optical phase modulation in GaAs/AlAs hetero n-i-p-i waveguide: towards an optical computer. *Appl. Phys. Lett.* 63: 2318.

Zeller, Ch., Vinter, B., Abstreiter, G., and Ploog, K. 1982. Quasi-two-dimensional photoexcited carriers in GaAs doping superlattices. *Phys. Rev. B* 26: 2124.

Chapter 11

Stimulated Emission from Amplifying Random Media

Michael Kempe, Gedalyah A. Berger, and Azriel Z. Genack

Reviewed by Wolfgang Rudolph

TABLE OF CONTENTS

1. INTRODUCTION

In this chapter, we follow the progression of proposals, models, experiments, calculations, and potential applications that have stimulated interest and shed light on the photophysics of amplifying **random systems**.* Perhaps the most prominent new feature of such systems is that spontaneously emitted radiation which follows long paths in a scattering medium may be sufficiently amplified so that stimulated emission dominates the optical evolution in the medium and causes it to lase. Whether or not the threshold for lasing action may be lowered enough for the development of useful new display, imaging, and sensor technologies remains an outstanding issue. In this introduction we consider some key questions that are suggested by the conjunction of disorder and gain.

In standard resonators, lasing at low excitation levels is achieved by positive feedback [e.g., Saleh and Teich, 1991]. The field coupled back into the gain medium stimulates emission which reinforces the propagating field. This can lead to lasing in a single mode. The excitation threshold for coherent emission is reached when the gain in a round-trip exceeds the loss. Because disorder of any kind scatters light out of a lasing mode, its presence raises the threshold for lasers operating in a single mode or in a small number of modes. Thus, disorder and lasing at a low threshold appear to be incompatible.

In random amplifying media, however, the possibility of lasing even without mirrors becomes manifest as a result of multiple scattering which impedes the escape of photons from the medium. Photon paths within the gain region may become long enough that spontaneously emitted photons can typically stimulate the emission of more than a single photon before they escape the sample. At this point the **lasing threshold** is exceeded.

One of the first questions that must be addressed is whether or not coherence of the emitted light is a key feature of lasing in random amplifying media. A consideration of the nature of wave interference within a random medium suggests that, although interference is the source of local fluctuations in intensity [Golubentsev, 1984; Shapiro, 1986; Maret and Wolf, 1987; Genack, 1987], it does not play a prominent role in determining the average intensity within the medium. Within the scattering medium, the electromagnetic field varies on a scale of the wavelength, and the correlation length of the field is approximately $\lambda/2\pi = 1/k$. The complex trajectory of a particular partial wave can be described as a narrow thread of diameter $1/k$ which meanders through the medium following a specific scattering sequence. Interference

---

* Terms in bold type are defined in Section 8.

of the partial waves associated with these trajectories is the source of the rapidly fluctuating field in the medium. But only when these partial waves or Feynman paths return upon themselves does this interference influence average transport within the medium. In the diffusive regime, the probability that a Feynman path will return to a particular phase coherence volume, of dimensions $(1/k)^3$, within a three-dimensional sample, is of order $(1/k\ell)^2$, where ℓ is the transport mean free path in which the direction of propagation is randomized. But since $k\ell >> 1$ in the diffusive regime, the probability of return is small, $(1/k\ell)^2 << 1$ [Sheng, 1990]. It appears, therefore, that wave interference can be neglected in calculating optical transport in the medium. We may then simply treat the **random walk** of photons without reference to interference of the electromagnetic field. Such behavior is well described by a diffusion equation for photons.

The discussion above suggests that wave propagation in a random ensemble of samples can be described by the diffusion equation even when the incident light is coherent. One may ask whether or not the presence of gain alters this situation. In the absence of gain, probability theory and diffusion theory lead to the same distribution of paths of the light emerging from a random medium. The presence of gain merely leads to an enhancement of the intensity associated with longer paths. The same result is obtained by solving the diffusion equation with a gain term which enters on the same footing as the usual loss term but has a different sign [Letokhov, 1968]. The favoring of some paths over others in an amplifying medium still results in a broad peak in the modified path length distribution for photons. Thus, amplification does not lead to spatial coherence. These considerations are with regard to the case of a coherent incident beam and would not be vitiated when one considers the case in which the source itself arises from spontaneously emitted photons.

A number of authors have argued recently, however, that the occurrence of lasing in disordered systems is so singular a phenomenon, that it cannot be explained by photon diffusion theory, but can only be understood by treating the role of a subset of selected paths in the medium. The occurrence of lasing, even in the limit in which the sample is nearly transparent, $L \gtrsim \ell$, is taken as evidence that a model of randomly scattered photons is not sufficient to describe this phenomenon. The collapse of the emission line width and temporal profile for such samples is seen to be inconsistent with arguments based on the multiplication of randomly moving photons. One way of assessing the validity if an explanation based on the random walk of photons would be to see whether such a model is able to describe the broad range of phenomena that have been reported.

In this chapter, we model the multiply scattering regime, $L >> \ell$, in which photon migration can be described by photon diffusion [Sheng, 1990]. This is the limit in which the most dramatic effects are manifested and, consequently, the limit in which the most important applications are likely to emerge. The interpretation of measurements in the nearly transparent regime [Lawandy et al., 1994] is not dealt with in detail.

Even if feedback, in the strict sense of wave interference, is not of consequence in strongly scattering amplifying systems, the photon is nonetheless obliged to remain within the gain region for an extended period as a result of multiple scattering. This situation has been termed *nonresonant feedback* [Ambartsumyan et al., 1970] and is associated with the condition that the spacing between adjacent optical modes that exist in the gain region becomes smaller than the line width of the mode. The mode spacing is associated with the interaction volume probed by the emitted photons, while the line width of the mode is determined by the rate of disappearance of the light from this region due to leakage out of the region and absorption. The volume explored by a photon emitted at a depth d before it reaches the surface and leaves the sample is typically $\approx d^3$. The spacing between modes associated with this volume is the inverse of the spectral density of states in this volume, $\Delta v \approx (8k^2 d^3/\pi v)^{-1}$, where v is the phase velocity in the medium. On the other hand, the dwell time of a diffusing particle in a region of scale d is d^2/D, where the photon diffusion coefficient is given by $D = \frac{1}{3} v\ell$. But, since the width of the mode δv is the inverse of this dwell time, the line width is $\delta v \approx D/d^2$ [Thouless, 1977]. The number of modes within the line width of a single mode is thus $\delta v/\Delta v \approx k\ell kd >> 1$, and distinct modes cannot be discerned in the medium.

In the limit of strong scattering, in which $k\ell \lesssim 1$, wave interference is crucial and the light reaches the Anderson **localization** threshold [Anderson, 1958; John, 1984; Anderson, 1985; Sheng, 1990]. Constructive interference then enhances the return of the wave to any point within the medium, and light no longer spreads by photon diffusion. This enhances the dwell time of photons emitted at a given depth in the medium. Consequently, modes are discrete beyond the mobility edge, in contrast to the overlap of many modes in the diffusive regime.

In the next section, we give a historical overview of the study of stimulated emission in amplifying random systems. This is followed by a more detailed description of the fundamentals of emission from

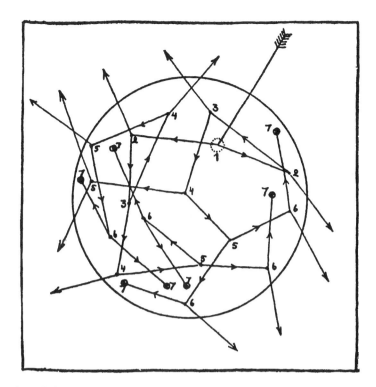

Figure 1 A nuclear chain reaction started in a spherical piece of fissionable material by a stray neutron. Although many neutrons are lost by crossing the surface, the number of neutrons in consecutive generations is increased, leading to an explosion. (From Gamow, G., *One Two Three ... Infinity,* Viking Press, New York, 1947. With permission.)

disordered media. In the subsequent section, we focus on measurements carried out on colloidal dye suspensions. These systems are ideal for exploring the mechanism for lasing in random amplifying media because key parameters can be independently varied. We then present a Monte Carlo simulation of the temporal, spectral, and spatial behavior of the emission and inversion in amplifying random media. These simulations incorporate the essential nonlinear aspects of the problem and allow us to understand a wide range of experimental observations and to address questions related to potential applications.

2. HISTORICAL OVERVIEW

The possibility of the conjunction of lasing and disorder was first considered by Letokhov [1968]. He explored the analogy of lasing in a scattering medium to sustainable nuclear reactions in a homogeneous nuclear reactor. In these systems the cross section for nuclear fission is uniform in space and time and the critical condition for chain reaction [Weinberg and Wigner, 1958] is analogous to the laser threshold condition. Letokhov considered the balance between neutron or photon multiplication and escape from the fissionable material or the region of inversion to be the threshold for thermonuclear reaction or lasing. An illustration from George Gamow's book *One Two Three ... Infinity* [Gamow, 1947] describing this situation is shown in Figure 1. The analogy to nuclear reactors leads to a model of an optical system with uniform gain and random scattering. Like the treatment of neutron diffusion in nuclear reactors, the photon density in the corresponding optical problem can be solved using a diffusion equation which includes a term, which is formally identical to the usual absorption term, but which incorporates a change of sign to account for the gain or negative absorption in the medium [Letokhov, 1968].

Letokhov considered a sphere with uniform gain as depicted in Figure 1. Unlike the nuclear analog, uniform gain is not intrinsic to the material. Important elements of the nonlinearity of the problem, such as the dependence of the degree of inversion on the emitted intensity, for example, are not included in this treatment.

A key application of lasing in random media, considered in early literature on this subject, was to optical frequency standards [Letokhov, 1968; Ambartsumyan et al., 1970]. Unlike lasers in cavities without scattering, the emission frequency from such a laser would be unaffected by resonator mode pulling, spatial and spectral hole burning, and resonator instabilities. Attention was given to the limiting line width of the resonator resulting from motion of the scattering particles. Several decades later this problem was reversed and the line width of light emerging from a multiply scattering medium without gain was used to investigate the dynamics of particle motion [Golubentsev, 1984; Maret and Wolf, 1987; Stephen, 1988; Pine et al., 1990].

Ambartsumyan et al. [1970] further investigated the behavior of multimode lasers. These authors considered multimode lasing which arises as a result of slight changes in a highly symmetric laser geometry, such as in a quasi-concentric resonator. They also considered cases in which a continuum of modes is formed as a result of disorder, as would occur when rough mirrors are used in a resonator, and finally they considered a mirrorless spherical geometry of random scatterers. They treated the case in which inversion is produced by excitation from outside a sphere and took the diffusion of the pump energy within the sphere into account. To follow the dynamics, they assumed the emission line was already narrowed in frequency. The problem was further simplified by ignoring the bleaching of the pump transition and thereby taking the degree of inversion to be proportional to the pump intensity. With this approach, the coupled optical diffusion and material excitation equations were solved to obtain the dynamics of the approach to stationary behavior near threshold.

These early investigators speculated that a possible realization of gain in a random system might be maser action in nonequilibrium interstellar clouds. This is a case of amplification in a turbulent medium which creates striking effects which have been explored in recent years. Though these discoveries did not bear out the initial speculation that maser action is a result of multiple scattering in interstellar space, the interplay of amplification and scattering plays a key role. Regions of small astronomical size and high microwave intensity with brightness temperatures of an equivalent black body exceeding 10^{15} K have been observed [Elitzur, 1995]. This can only result from substantial amplification in a system with inversion. Such inversion can occur near energy sources such as giant molecular clouds that house star-forming regions, as well as red giants. Amplification increases with the number of molecules that a photon encounters in its passage, but the densities of the maser region must be low enough that collision rates do not undo the inversion. Given the long lifetimes of low-lying excited states, the degree of inversion is maintained for densities that are 10^{11} times lower than atmospheric. Numerous observations have been made of clusters of masers which span 10^{12} km and with single maser spots which are 1000 times smaller. The line widths of the emission were found to be as narrow as one part in 10^6 of the center frequency of the line. The emissions from different spots in a cluster of masers are generally at different frequencies for a given molecular line. The frequencies at which maser emission is observed must arise from variations of the Doppler shift of the emitted radiation. Emission from a particular portion of the sky corresponds to velocity spreads of approximately 1000 m/s. The various frequencies observed within a cluster reflect variations in the peak of the distribution of the velocity component in the terrestrial direction. This phenomenon makes possible astronomical observation of molecular motion at great distances with unparalleled resolution. However, since a well-defined velocity component is selected for maser action, this phenomenon could not involve multiple scattering. Presumably, the gain length is substantially shorter than the scattering length, in this case. The length scale from which narrow-band stimulated emission is observed depends upon numerous aspects of the pumping source, the spatial variation in density of the emitting molecule or radical species, the velocity distribution, and the rate of nonresonant scattering from transitions not involved in the maser action. In this intriguing case of mirrorless amplification, scattering does not enhance stimulated emission, as was originally anticipated, but rather limits the amplification of spontaneously emitted photons.

In early discussions of lasing in random media, as well as in the subsequent literature, earthbound excitation is generally produced at the interface of the random gain medium. This is easiest to accomplish, but the excitation near the interface of the sample facilitates the escape of photons from the gain medium and, consequently, raises the lasing threshold. The ease with which stimulated emission is observed depends upon the means of excitation; the scattering lengths of the incident and emitted light; the density and optical cross sections of the active atoms, molecules, and ions; and the excitation geometry. Ideally, one would like to deposit the energy as deeply as possible into the medium and have the transport mean free path of the emitted photons as short as possible in order to extend the dwell time of photons in the gain region. One means of accomplishing this would be to exploit the dramatic changes in transport properties with frequency that occur near the mobility edge of the Anderson transition. We quote from a patent memo [Genack, 1986], which considered ways to minimize the lasing threshold

It is proposed here to pump a scattering sample containing fluorescing molecules at a frequency above the upper mobility edge but to have the fluorescence frequency within the window of localization. This will result in fluorescence being trapped in the sample and result in efficient stimulated emission into the localized photon states. The finite size of the sample will determine the coupling out. Since the photon wave function decays exponentially, the coupling out can be very weak.

Photon localization can be induced by strong disorder [John, 1984; van Albada and Lagendijk, 1985; Wolf and Maret, 1985; Garcia and Genack, 1991] or by a break in the translational symmetry of an otherwise periodic structure [John, 1987], which would possess a photonic band gap in the absence of disorder [Yablonovitch, 1987; John, 1987; Yablonovitch et al., 1991; Soukoulis, 1993]. Such disorder introduces localized states into the band gap. If the break in the periodicity occurs at a specific site in the middle of the structure, the rate of photon escape from the sample can be exponentially long while the transmission coefficient at the frequency of the localized state in the middle of the sample can approach unity.

Another way of achieving deep penetration into the medium, while maintaining strong scattering at the emission wavelength, is to exploit the frequency-dependent scattering properties of dielectric spheres. Scattering and emission from single dielectric spheres is reviewed in Chapter 9 by Cheung, Hartings, and Chang in this volume. Molecules which ordinarily have broad emission spectra exhibit a narrow peaked emission structure when they are in the vicinity of a dielectric sphere. This is a consequence of resonances of the emitted radiation with the sphere. From an analysis of the emission spectra, it is possible in some cases to determine the orientation of molecules on the surface [Folan and Arnold, 1988]. Sharp Mie resonances of polystyrene spheres have also been exploited to select a portion of a collection of spheres which interact strongly with a given excitation frequency. This has been utilized to perform photochemical spectral hole burning at room temperatures in collections of polystyrene spheres with surface layers of photochemically active molecules [Arnold et al., 1991]. Molecules that reside on the surfaces of spheres and are strongly excited by the incident radiation are photochemically transformed following excitation by the laser. This leads to a reduction of absorption and subsequent emission at the frequencies at which these molecules were excited. Arnold et al. [1991] observed spectral holes in the excitation spectrum of the molecular fluorescence with widths characteristic of the widths of the resonances of individual spheres. Such enhanced intensity at the site of some fraction of the spheres could lead to a lowering of the laser threshold in this inhomogeneous medium.

Emission from molecules which are dissolved in colloidal suspensions have also been studied at low excitation levels. Lawandy and co-workers [Martorell and Lawandy, 1990] have found that in colloidal polystyrene crystals, the spontaneous emission time from dye molecules in solution is lengthened. This has been ascribed to a collective propagation effect which leads to a lowering of the density of states into which the molecules might emit as a result of the presence of colloidal particles. For certain structures, a complete photonic band gap may exist. However, in samples such as polystyrene colloids, in which a photonic band gap does not exist, a partial suppression of the density of states might exist which would cause an inhibition of spontaneous emission. Related observations of inhibited spontaneous emission in systems which exhibit lasing at higher levels of excitation have not been reported.

Sharp resonances are not the exclusive province of highly symmetric structures; they may exist in randomly shaped microcrystals as well. Fork and Taylor [1979] have made a series of observations of anomalously fast and efficient optical emission in microcrystals containing Eu^{2+}, which are between one and several microns across. The emission exhibits a quantum efficiency of greater than 0.1 with an emission lifetime below 80 ps. Shifts in the emission spectrum and the radiative rates vary from crystal to crystal. If lasing action is discounted, this would correspond to an oscillator strength which is seven orders of magnitude greater than expected. A nonlinear dependence of emission intensity upon pump intensity occurs at incident power levels which correspond to the presence of only a few excitation quanta in the crystal. Emission may not be ascribable to laser action since the transition appears to be to the ground state of the ion. The nature of emission in these systems remains an open question.

The first observation of stimulated emission from random media was made in neodymium in sodium lanthanum molybdate powders composed of grains that were mostly within the range 1 to 10 μm at 77 K [Markushev et al., 1986]. Markushev et al. found that above a threshold in incident laser power, the duration of the luminescence from the $^4F_{3/2}$–$^4I_{11/2}$ transition in Nd^{3+} in powders of $Na_5La(MoO_4)_4$ was reduced by nearly four orders of magnitude. As this threshold was passed, the intensity of this line

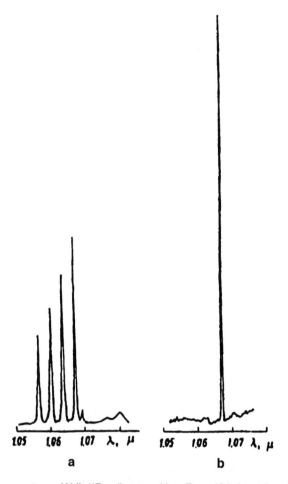

Figure 2 Luminescence spectrum of Nd^{3+} ($^4F_{3/2}$–$^4I_{11/2}$ transition, T = 77 K) below (a) and above (b) the threshold. (From Markushev, V. M. et al., *Sov. J. Quantum Electron.,* 16(2), 281, 1986. With permission.)

increased and the line narrowed to the instrumental width. The transformation of the spectrum is shown in Figure 2. Similar effects were later observed in measurements on europium pentaphosphate powders [Gouedard et al., 1993]. Gouedard et al. showed that the emission from these powders gave rise to a smooth intensity distribution. The large intensity fluctuations observed in reflection from static nonamplifying media were largely suppressed. They conclude that amplified emission from these powders is incoherent.

In measurement from powders of lasing crystals both the amplification and the scattering in the medium are associated with the grains of powder. In contrast to these studies, Lawandy et al. [1994] studied emission from colloidal solutions containing Rhodamine 640 (R640) perchlorate dye in methanol and rutile titania powder. The scattering is provided by the titania particles, but the gain is associated with dissolved dye molecules. As a result, the absorption of the dye and the scattering strength of the colloid can be varied independently by changing the dye or particle concentrations. Because this is an ideal situation for investigating the mechanism for emission in random media, it has stimulated numerous studies of these systems. Lawandy et al. [1994] found a narrowing of emission and a shortening of the emitted pulse above a certain threshold. A comparison of the emission spectrum in the neat dye solution and in the colloidal solutions is shown in Figure 3. In subsequent measurements on this system by Alfano and co-workers [Sha et al., 1994; Siddique et al., 1995b], emission pulses with a duration as short as 12 ps were measured. The emission in a colloidal sample which is excited at a power which is 1.23 times higher than the threshold power for lasing is shown in Figure 4. A review of measurements on colloidal systems is given in the section on experimental observations.

Amplification of a probe light source by an amplifying random slab has been considered by Zyuzin [1994]. He calculated the coherent backscattering peak in slabs with fixed gain. In a passive medium,

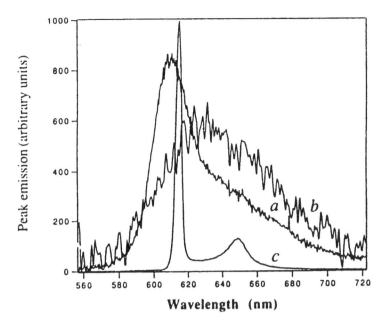

Figure 3 *a*, Emission spectrum of a 2.5 × 10⁻³ M solution of R640 perchlorate in methanol pumped by 3-mJ (7-ns) pulses at 532 nm. *b* and *c*, Emission spectra of the TiO_2 nanoparticle (2.8 × 10¹⁰ cm⁻³) colloidal dye solution pumped by 2.2-μJ and 3.3-mJ (7-ns) pulses, respectively. Emission: *b*: scaled up 10 times, *c*: scaled down 20 times. (From Lawandy, N. M. et al., *Nature,* 368, 436, 1994. With permission.)

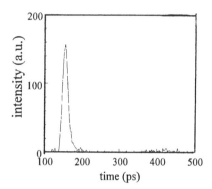

Figure 4 Emitted pulse from R640 dye in methanol (10⁻³ M, ℓ_a = 140 mm) with TiO_2 particles (5 × 10¹¹ cm⁻³, ℓ = 16 μm) pumped by 16-μJ pulses of 10-ps duration at 527 nm. (From Siddique, M. et al., *Opt. Lett.,* 21(7), 450, 1996. With permission.)

the enhancement of reflection from a random sample in the retroreflection direction is a consequence of the constructive interference of counterpropagating waves in the medium. This gives rise to an enhancement of backward scattered light by a factor of two compared with the diffuse background [van Albada and Lagendijk, 1985; Wolf and Maret, 1985; Akkermans et al., 1986]. The angular dependence of the enhancement is given by the Fourier transform of the spatial intensity distribution at the sample surface. This distribution can be calculated using diffusion theory [Akkermans et al., 1986]. Zyuzin calculated the shape of the coherent backscattered cone using a gain term in the diffusion equation. He found that the width of the backscattering peak collapsed at a critical thickness. This corresponds to the point at which the intensity of the light at the frequency of the probe is uniform on the surface of the slab. It will occur when the usual falloff of intensity with transverse position from the center of the incident beam is counterbalanced by the gain. This requires that the intensity of light at the surface is enhanced to a greater degree at points farther from the center of the beam. This occurs because light reaching these points follows longer paths on average and the exponential amplification along the paths therefore enhances the intensity.

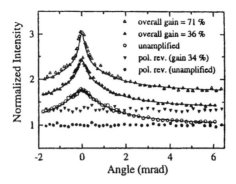

Figure 5 Three backscattering cones in the polarization conserving channel and two backscattered intensities in the polarization reversing (pol. rev.) channel from the same sample at different pump power. Sample: 30% Ti:sapphire particles (10 µm) in water, transport mean free path: 40 µm. The overall gain of 0%, 36% (34% in the pol. rev. channel), and 71% corresponds to pump fluences of 0, 840, and 968 mJ/cm². The intensity is normalized to the diffuse background at zero gain. Solid lines: calculated curves based on diffusion theory. Upon increasing gain, the overall intensity increases and the top of the cones sharpens. (From Wiersma, D. S. et al., *Phys. Rev. Lett.*, 75(9), 1739, 1995. With permission.)

Wiersma et al. [1995b] have carried out coherent backscattering measurements from amplifying random media using optically pumped Ti:sapphire powders. They find that the top of the coherent backscattering cone sharpens with increasing gain. This reflects a broadening of the transverse intensity distribution on the surface in accord with calculations of the intensity distribution based on diffusion theory. The narrowing of the coherent backscattering peak as the gain increases is shown in Figure 5.

An important question is the interaction of localization and gain. Pradhan and Kumar [1994] have considered reflection in a random system of dielectric layers with constant gain. They use a stochastic noise model for the fluctuations in the real part of the dielectric constant. In the absence of gain, all random-layered systems with disorder only in one dimension are localized. The localization length ξ_0 is a few times ℓ. This length describes the exponential decay length of the average transmission with increasing thickness over an ensemble of samples with statistically equivalent disorder. Pradhan and Kumar consider the statistics of fluctuations of the reflection coefficient in a random one-dimensional system. In this system, wave interference is an essential aspect of the problem. In the limit of infinite thickness for samples with intensity gain length ℓ_g, Pradhan and Kumar find a distribution of reflection coefficients $P_\infty(r)$ for various realizations of the disorder which depends only upon the ratio $d = \xi_0/\ell$. The distribution obtained by solving the Riccati equation is $P_\infty(r) = (d/(r-1)^2) \exp(-d/(r-1))$, for $r > 1$ and 0 for $r = 1$. This gives a weak logarithmic divergence for the mean reflection coefficient $\langle r \rangle$. The reflected light appears to be amplified mostly within a localization length of the interface, with the tail of the distribution associated with amplification from regions deep within the sample. This result is quite different from what would be obtained here using an incoherent model. In such a model, longer paths would be represented more strongly and the decay of $P_\infty(r)$ would be less rapid.

Zhang [1995] used the transfer-matrix method to explore the interaction between gain and localization in samples of arbitrary scattering strength. He found the paradoxical result that the tendency toward localization is enhanced by the presence of gain, yielding an even more rapid exponential decay of transmission vs. thickness of a random sample. The inverse localization length is given by $1/\xi = 1/\xi_0 + 1/\ell_g$, where ξ_0 is the localization length in the absence of gain. Apparently, the presence of gain enhances the tendency for constructive interference of waves returning to a plane and, hence, leads to a more rapid decay of the wave with depth into the sample. Zhang finds that generally for thicknesses of several times ξ_0, the reflection probability function approaches the result for $P_\infty(r)$ calculated by Pradhan and Kumar. These results suggest that large amplification of the incident wave is suppressed by localization. This is confirmed in analytic results of Freilikher [personal communication] which suggest that "in the competition between localization and gain the first wins." Although this may be the case for the average over an ensemble of samples, large fluctuations may lead to very different results in particular configurations. It is important then to find a means for introducing the wave to be amplified at some depth into the medium.

The rest of this chapter is concerned with the nature of propagation in random amplifying three-dimensional systems far from the localization limit.

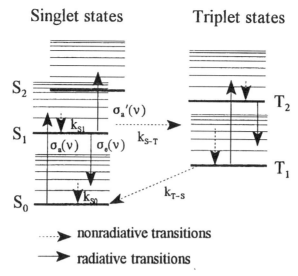

Figure 6 Diagram illustrating the energy levels of a dye. Nonradiative transitions are characterized by transition rates, radiative transitions by cross sections.

3. LASING ACTION IN RANDOM MEDIA

Stimulated emission of light arises from the interaction of photons with atoms or molecules in a host medium which is excited by external pumping. For most purposes, this interaction can be modeled by rate equations; see, for instance, Saleh and Teich [1991]. The rate equation approximation is appropriate for describing interactions on a time scale that is longer than the phase relaxation time of the oscillations of the induced dipole moment. The states of the excited species can then be adequately described by the occupation number densities of their energy states. A schematic energy-level diagram for dyes is shown in Figure 6. It consists of electronic levels with total spin zero (singlet states S_0, S_1,...) and total spin one (triplet states T_0, T_1,...). Each electronic level has a manifold of vibrational states due to the many internal degrees of freedom of a complex molecule. Even more finely spaced are the rotational sublevels of each vibrational state. The total manifold of states is so dense that the absorption/emission spectrum of a dye forms a continuum. The transition rates $k = 1/\tau$ between two energy levels depend on the associated dipole moment and the interaction with the environment which acts as a thermal bath. The decay rate of a population in a particular energy level is determined by the transition rates to all energy levels $1/\tau_i = \Sigma 1/\tau_{ij}$. Normally, the lasing transition is between the lowest level of the S_1 manifold and various levels of the S_0 manifold. The lifetime of the spontaneous emission of the upper lasing level is typically a few nanoseconds for neat dye solutions. The cross section for stimulated emission $\sigma_e(\nu)$ is of the order of 10^{-16} cm^2. In thermal equilibrium, at room temperature, only the S_0 states are occupied because the energy difference between the S_0 and S_1 levels is much larger than kT. Optical pumping brings the molecules from the S_0 state to vibrational/rotationally excited states of S_1. The cross section for this absorption $\sigma_a(\nu_p)$ at the frequency of the pump light ν_p is typically also of order 10^{-16} cm^2. The lifetime of the excited vibrational/rotational levels of S_1 is short, of order 10^{-13} s, because of the rapid nonradiative decay with rate k_{S1} to the lowest level of S_1. Intersystem crossing between the singlet and triplet states is spin forbidden and occurs at a much lower rate $k_{S \to T} \ll k_{S1}$, $k_{T \to S} \ll k_{T1}$. Because the T_1 states can only decay to the S_0 states via such a transition, they are long-lived with lifetimes of the order of 10^{-6} s. Therefore, once in a triplet state, molecules can only participate in the lasing action by transitions within the triplet manifold on time-scales less than the time for intersystem crossing. If there is no possibility of several cycles of excitation, as would be the case, for example, if the pulse duration is short (picoseconds) and the repetition rate is low (several Hertz), the triplet state occupation is negligible. If the pump pulses have a duration of nanoseconds, however, the triplet state occupation can become significant at energy densities above the saturation level. The observed bichromatic spectrum of R640 dye at high pumping levels has been attributed to optical transition between levels other than $S_1 \to S_0$ [John and Pang, 1995]. There is also the possibility of further excitation from the S_1 manifold into higher-lying singlet states. For many dyes, however, the cross section $\sigma_a'(\nu)$ for this excited state absorption is more than a factor of 100 times smaller than for ground state absorption. For an overview of properties of laser dyes, see, for instance, Schäfer [1979].

For our simulations we model the dye as a four-level system. This neglects intersystem crossing as well as excited state absorption. The schematic level diagram is shown in Figure 7. Because of the short lifetime of the excited states of the S_1 and S_0 manifolds, the molecular population in levels 2 and 4 are negligible and the total number density of molecules is, therefore, taken to be $N = N_1 + N_3$, where N_1 and N_3 are number densities of the molecules in the first and the third energy level, respectively.

Transport in a random amplifying medium can be described using a diffusion equation with nonlinear gain and absorption coefficients. The diffusion coefficient is $D = v\ell/3$ where v is the phase velocity in the medium in the limit of low scatterer density. The transport mean free path ℓ of light propagating diffusively in a random medium is related to the scattering mean free path ℓ_s by $\ell = \ell_s/(1 - \langle\cos\vartheta\rangle)$ where $\langle\cos\vartheta\rangle$ is the average cosine of the scattering angle. The scattering mean free path ℓ_s has a simple interpretation as the average distance between two scattering events. In the case of a homogeneous medium with a low concentration of scatterers, ℓ_s is related to the scattering cross section σ_s and the concentration of the scatterers N_s by $\ell_s = 1/\sigma_s N_s$. If the scattering is anisotropic, ℓ is larger than ℓ_s, since several scattering events may be required to randomize the propagating direction.

In order to develop an understanding of the consequences of an incoherent approach to lasing from homogeneous random amplifying media in the reflection geometry, we consider, for concreteness, the multiply scattering limit, in which $\ell \ll \ell_a$, d, L for the pump and emitted light, where d is the diameter of the area illuminated by the pump beam and L is the sample length. We may treat this problem within the diffusion approximation by assuming that the incident pump radiation of intensity $\tilde{I}_p$ is randomized inside the medium at a depth z_p [Garcia et al., 1992; Li et al., 1993]. We represent the incident light by a source at this depth in the medium with a strength equal to that of the incident radiation. The propagation of the randomized pump radiation with photon number density n_p in the medium is then given by,

$$\frac{\partial n_p(r,t)}{\partial t} - D_p \nabla^2 n_p(r,t) + \frac{n_p(r,t)}{\tau_p(r,t)} = \delta(z - z_p)\tilde{I}_p(x,y,t)/h\nu_p \tag{1}$$

where D_p is the diffusion coefficient for the pump radiation and $1/\tau_p$ is the pump absorption rate. The latter is given as $1/\tau_p = v/\ell_a = v\sigma_a(\nu_p)N_1$. Equation 1 is nonlinear because the absorption rate depends upon the intensity at the pump frequency via N_1. Leakage of light out of the sample is determined by Equation 1 and by the boundary conditions. Because the laser pumping is mimicked by a diffusive source at z_p, the only light entering the medium from the boundary in this model is the reflected part of the outgoing flux. As a result of these boundary conditions, the intensity inside the medium extrapolates to zero at a distance beyond the boundary of $z_b = \frac{2}{3}\ell(1 + R)/(1 - R)$ where R is the **internal reflectivity** for diffusively propagating photons as illustrated in Figure 8 [Lagendijk et al., 1989; Garcia et al., 1992].

The propagation of emitted photons of number density $n(r, t; \nu)$ per unit frequency per unit is described by the nonlinear diffusion equation,

$$\frac{\partial \bar{n}(r,t;\nu)}{\partial t} - D\nabla^2\bar{n}(r,t;\nu) + \frac{\bar{n}(r,t;\nu)}{\tau_a(r,t;\nu)} = \frac{N_3(r,t)h\nu}{\tau(\nu)} + \frac{\bar{n}(r,t;\nu)}{\tau_g(r,t;\nu)} \tag{2}$$

The diffusion coefficient at the emission frequency is D. The sources of emitted spontaneous radiation on the right-hand side of Equation 2 are emissions from the lowest level of S_1 and stimulated emission. The rates in Equation 2 are the spontaneous emission rate $1/\tau$, the stimulated emission rate $1/\tau_g = v/\ell_g = v\sigma_e(\nu)N_3$, and the rate of absorption of emitted radiation in the tail of the pump transition $1/\tau_a(\nu) = v/\ell_a(\nu) = v\sigma_a(\nu)N_1$. $g(\nu)$ is the lineshape function. The boundary conditions are the same as for the pump radiation.

The dynamics of the level populations, as described by rate equations, determines the rates of emission and absorption in Equations 1 and 2. Because $N_2 = N_4 = 0$ and $N_1 = N - N_3$, it is sufficient to give the rate equation for the population density N_3,

$$\frac{\partial N_3(r,t)}{\partial t} - \frac{n_p(r,t)/h\nu_p}{\tau_p(r,t)} + \int_{\infty}^{\infty} d\nu \left[\bar{n}(r,t;\nu)\left(\frac{1}{\tau_a(r,t;\nu)} - \frac{1}{\tau_g(r,t;\nu)}\right) - \frac{N_3(r,t)}{\tau} \right]. \tag{3}$$

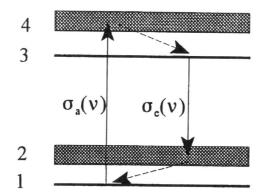

Figure 7 Simplified four-level system used for the rate-equation calculations.

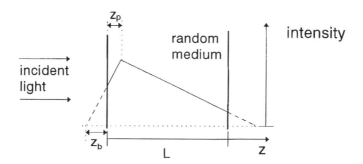

Figure 8 Intensity distribution of diffuse photons within a random slab.

In the absence of gain, the density of emitted photons $\bar{n}\,(\boldsymbol{r},\,t;\,\nu)$ can be expressed as the integral over the sample volume of the product of the source density of spontaneously emitted photons $N_3(\boldsymbol{r}',\,t')\,g(\nu)/\tau$ and the probability that these photon will arive at the point of detection,

$$\bar{n}(\boldsymbol{r},t;\nu) = \int d\boldsymbol{r}'dt' \sum_{\alpha} P_{\alpha}(\boldsymbol{r},\boldsymbol{r}';\nu) \frac{N_3(\boldsymbol{r}',t')g(\nu)}{\tau} \delta\left(t - t' - \frac{S_{\alpha}}{\nu}\right). \tag{4}$$

Here the probability is expressed as the sum of probabilities over all possible paths leading from $\boldsymbol{r}'$, to $\boldsymbol{r}$ for light at frequency ν, S_{α} is the path length, and S_{α}/ν is the travel time $t-t'$ along the path. This corresponds to a formal solution of the ordinary diffusion. In the presence of gain, Equation 4 remains valid once the probability is replaced by the effective probability,

$$P_{\alpha}(\boldsymbol{r},\boldsymbol{r}';\nu) - P_{\alpha}^{(0)}(\boldsymbol{r},\boldsymbol{r}',\nu) \exp\left(\int_0^{S_{\alpha}} dS_{\alpha}\left(\frac{1}{\ell_g(\nu)} - \frac{1}{\ell_a(\nu)}\right)\right) \tag{5}$$

where S_{α} is the length along the path.

Equations 4 and 5 are consistent with stimulated photons continuing along the same trajectory as the spontaneously emitted photon which stimulated their emission as schematically shown in Figure 9. In contrast, Equations 2 and 3 specify only the density of photons and not their direction. The two pictures are consistent, however, because the distribution of paths from any point for both the spontaneously emitted and stimulated photons is found by randomly choosing the direction of the next step. Thus the spontaneous and stimulated photons have the same distribution of trajectories between any tow points.

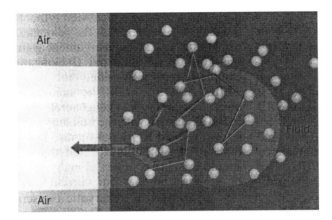

Figure 9 Possible path for a photon emitted and amplified within a random amplifying medium. The lighter region indicates the volume strongly pumped by the laser. (From Genack, A. Z. and Drake, J. M., *Nature,* 368, 400, 1994. With permission.)

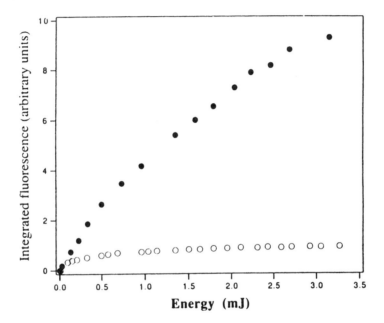

Figure 10 Wavelength-integrated emission as a function of the pump pulse energy for the neat dye (open circles) and the TiO$_2$ nanoparticle colloidal dye solutions (2.8 × 10^{10} cm^{-3}) (closed circles). The dye concentration is 2.5 × 10^{-3} M in both cases. (Lawandy, N. M. et al., *Nature,* 368, 436, 1994. With permission.)

4. MEASUREMENTS OF EMISSION IN COLLOIDAL DYE SUSPENSIONS

Experimental work on lasing in random materials has been performed with a variety of samples including randomly shaped single microcrystals [Fork and Taylor, 1979], powders [Markushev et al., 1986; Gouedard et al., 1993], and, more recently, with solid polymeric dye media [Balachandran and Lawandy, 1995b], dye-treated animal tissue [Siddique et al., 1995a], and colloidal suspensions with a dissolved dye as active medium. In this section we deal exclusively with the dye/colloid system. This system has been most extensively studied, because it allows one to vary the various parameters that are involved in lasing in random media. Most of the experimental work has been performed using R640 dye in methanol solutions. The scattering in the sample is introduced by adding titania particles of submicron size.

   In the first report of lasing in this system, the solution was placed in a cuvette and pumped by a frequency-doubled Nd:YAG laser at 532 nm [Lawandy et al., 1994]. The absorption length in these experiments was 50 μm and the transport mean free path was typically 200 μm. Measurements were

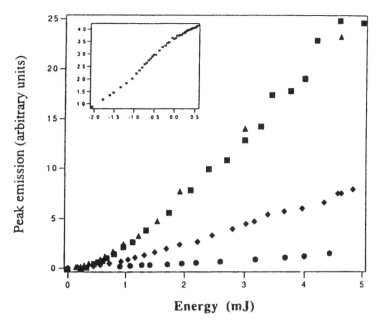

Figure 11 The emission at the peak wavelength as a function of the pump pulse energy for four different TiO$_2$ nanoparticle densities. Nanoparticle densities of 1.4×10^9 cm^{-3}, 7.2×10^9 cm^{-3}, 2.8×10^{10} cm^{-3}, and 8.6×10^{11} cm^{-3} are shown by solid circles, diamonds, squares, and triangles, respectively. The inset shows the data on a log–log plot (same quantities on the x and y axes) for a nanoparticle density of 2.8×10^{10} cm^{-3}. (From Lawandy, N. M. et al., *Nature,* 368, 436, 1994. With permission.)

made using a Q-switched laser producing 7-ns pulses. The repetition rate was less than 25 Hz. The light was focused to a 1.8-mm spot. The emission from the front face of the cell was collected by a lens, and its energy spectrum was analyzed. In the neat dye, the emitted wavelength-integrated fluorescence saturated at pump fluences of about 10 mJ/cm^2. In contrast, the emitted intensity from the dye with scatterers is nearly linear for pump energies which were up to ten times greater, as seen in Figure 10. At the same time, the emission line width that broadened for weak pumping, as compared with the neat dye, narrowed dramatically for pump energies for which saturation in the neat dye was observed. Even more striking is the fact that the emission at the peak wavelength (617 nm) increases considerably faster with the pump energy when the scattering strength of the sample increases (Figure 11). This points to a redistribution of emission in favor of the wavelengths with the strongest amplification. The spectral narrowing of the emission, shown in Figure 12, from around 40 nm associated with spontaneous emission to 4 nm exhibits a clear threshold behavior. For the neat dye, narrowing of the emission was not observed. In this, and in many subsequent experiments, the narrowing of the spectrum was used to specify a threshold. The pump fluence at threshold was approximately 8 mJ/cm^2.

The emission from the random system does not saturate because stimulated processes dominate the deactivation of the excited molecules. The associated spectral line width collapse is responsible for the change in the slope of the peak emission vs. pump energy. After the spectrum has narrowed to its minimum value, the slope returns to its value for weak pumping as can be seen from the inset in Figure 11.

In the experiments by Lawandy et al. [1994], the gain region was in the form of a flat disk with a diameter of about 1.8 mm and a thickness of less than 0.1 mm. One might expect, therefore, that the migration of photons in the transverse direction plays a significant role. This issue was addressed by Wiersma et al. [1995a]. The authors observe spectral narrowing of the emission emerging from the side of the cuvette even in a neat dye and argue that scatterers simply redirect radiation which has been amplified by propogation in the transverse direction so that it emerges from the front surface. Lawandy and Balachandran [1995] studied the emission from a random medium out of the side of the cuvette. They report that increasing the scatterer concentration results in spectral narrowing of the side emission (from 10 to 5 nm) at the same point at which the emission from the front of the cuvette narrowed (from 40 to 8 nm). They state that these results, as well as the threshold dependence on the scatterer concentration, demonstrate that the characteristics of emission cannot be explained by amplified spontaneous

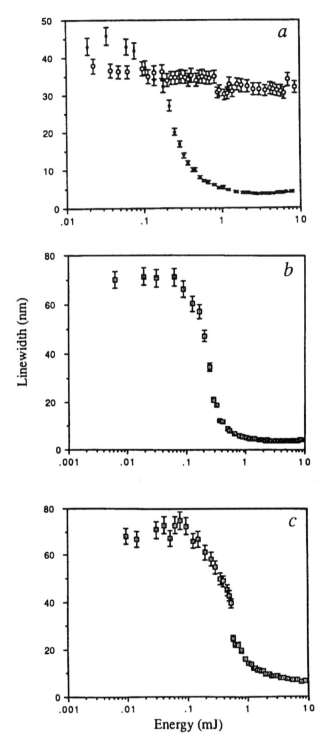

Figure 12 Emission line width as a function of the pump pulse energy for three different TiO_2 nanoparticle densities. *a* to *c* correspond to nanoparticle densities of 5.7×10^9 cm^{-3}, 2.8×10^{10} cm^{-3}, and 1.4×10^{11} cm^{-3}, respectively. The open circles in *a* represent the emission line width of the pure dye as a function of the pump pulse energy. (From Lawandy, N. M. et al., *Nature,* 368, 436, 1994. With permission.)

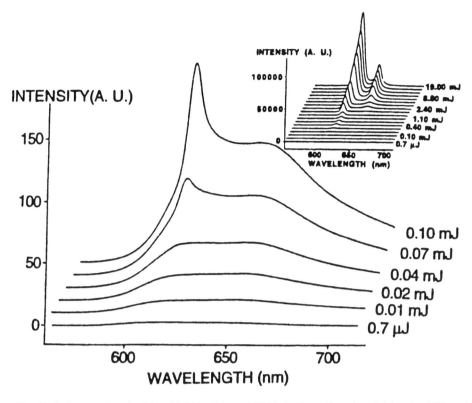

Figure 13 Emission spectra of a 2.5 × 10^{-2} M solution of R640 dye in methanol containing 5 × 10^{11} cm^{-3} TiO$_2$ nanoparticles at different pump energies up to 19 mJ (inset). (From Sha, W. L. et al., *Opt. Lett.,* 19(23), 1922, 1994. With permission.)

emission of light propogating in the transverse direction. The narrow emission from the side of the neat dye is an indication of the importance of long paths in the medium. However, redirection of emission toward the front surface is just one of the many scattering channels in a random system.

Sha et al. [1994] investigated the spectral evolution of the emission and the threshold more closely. The experiments were performed with 3-ns laser pulses from a frequency-doubled Nd:YAG laser at a repetition rate of 20 Hz. The laser spot size on the 1-cm cuvette was about 1 mm. The measured spectrum is shown in Figure 13. Above the threshold of about 0.1 mJ (corresponding to 13 mJ/cm^2), the peak emission at 620 nm was observed to grow rapidly. At even higher pump energies, a second peak at about 650 nm appeared that had also been reported by Lawandy et al. [1994]. Interestingly, Alfano and co-workers report narrowing of the spectrum in the neat dye with a threshold as low as about 9 mJ/cm^2 [Sha et al., 1994]. This pump fluence was even lower than the thresholds that were typically observed for the random systems. The threshold fluences were found to vary slowly with scatterer and dye concentration. Increasing the concentration of the dye by two orders of magnitude lowered the threshold by a factor of less than three, while an increase of the concentration of the scatterers by three orders of magnitude lowered the threshold by a factor of two to six, depending on the dye concentration. The emission peak at 650 nm that became prominent at high pump energies can be associated with an independent, weaker optical transition [John and Pang, 1995].

The temporal behavior of the emission was studied for various values of ℓ and ℓ_a by Siddique et al., [1995b]. There, the system was pumped by single 10-ps pulses at 527 nm using a frequency-doubled Nd:glass laser. At threshold, pulsed emission is observed with a duration comparable to the pump pulse. The shortest emission obtained lasted about 12 ps, taking into account the resolution of the streak camera of 10 ps. The change in the measured emission near threshold is shown in Figure 14. The threshold, defined as the point at which the temporal profile of emission narrowed to approximately 100 ps, changed

316

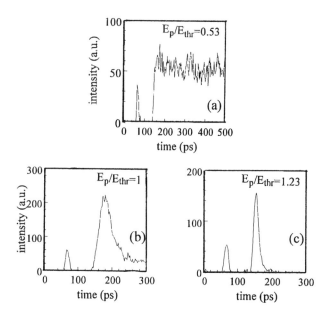

Figure 14 Time-resolved emission from R640 dye in methanol (10^{-3} M, ℓ_a = 140 μm) with titania particles (5×10^{11} cm^{-3}, ℓ = 16 μm). (a) Spontaneous emission after excitation with 7 μJ of pump light. (b) Emission at threshold of approximately 13 μJ pump energy. (c) Short pulse at a pump energy above threshold (16 μJ). The reference pulse shown in each frame indicates the input pulse duration and the time resolution of the detection system. Its timing with respect to the onset of the emission varies from (a) to (c). E_p/E_{thr} is the ratio of the pump to threshold energies. (From Siddique, M. et al., *Opt. Lett.*, 21(7), 450, 1996. With permission.)

by less than a factor of three when the concentration of scatterers was varied over two orders of magnitude. The dependence of the threshold upon ℓ_a was more pronounced. Increasing the concentration of the dye by a factor of 25 led to a threshold reduction by about an order of magnitude. The lowest threshold of 3 mJ/cm^2 was found for the shortest absorption length (ℓ_a = 50 μm) and transport mean free path for the emitted radiation (ℓ = 16 μm).

Let us estimate the threshold for lasing action in a simple case. We consider the diffusion limit of photon migration in which bleaching of the pump transition and the transverse diffusion out of the incident beam are negligible. In this case, the penetration depth of the pump photons is given by the intensity attenuation length $L_a = (\ell\ell_a/3)^{1/2}$ [Genack, 1987]. If we require that the average spontaneously emitted photon on its way through the sample stimulates the emission of another photon, we have $\exp[\gamma_0\langle s\rangle] = 2$. The simulations presented later will show that this point corresponds to the threshold at which significant changes in the spectral and temporal properties of the emission occur. Thus we obtain $\gamma_0\langle s\rangle \approx 1$ as the threshold condition. In the absence of gain or loss for the emitted photons, the average path length of photons which are spontaneously emitted in the excited region of depth L_a is $\langle s\rangle \approx L_a^2/\ell = \ell_a/3$. The small signal gain in the medium is $\gamma_0 = \sigma_e N_3$ where σ_e is the emission cross section of the lasing transition and N_3 is the number density of the excited molecules. For an incident pulse with fluence $\varepsilon_p = E_p/A$, we have $N_3 = N\, \varepsilon_p/\varepsilon_{sat}$ with the saturation fluence $\varepsilon_{sat} = h\nu_p/\sigma_a$. Thus we can write $N_3 = (\varepsilon_p/h\nu_p)/\ell_a$. Using the expressions above, the threshold condition becomes $\gamma_0\langle s\rangle \approx \sigma_e\varepsilon_p/3h\nu_p \approx 1$, which is independent of ℓ and ℓ_a. The threshold pump fluence is then $\varepsilon_p^{thres} \approx 3\sigma_e/\sigma_e\varepsilon_{sat} = h\nu_p/\sigma_e$. For the R640 dye pumped at 532 nm, we obtain $\varepsilon_p^{thres} \approx 0.2\ \varepsilon_{sat} \approx 3$ mJ/cm^2 in agreement with the thresholds observed.

Apart from the neglect of bleaching and transverse diffusion out of the beam, this estimate does not fully represent the situation because it treats the average path length rather than the full distribution of path. As discussed in the previous section, longer paths within the path length distribution are of increasing importance as stimulated emission increases. Furthermore, internal reflection was not taken into account. Nonetheless, the model gives an estimate of what one might expect in the reflection geometry.

In order to explain lasing action in the limit of very weak scattering, Genack and Drake [1994] suggested that some of the emitted light might be confined within the gain region for extended periods as a result of scattering and internal reflection at the boundaries. This question has been investigated by

Balachandran and Lawandy [1995b]. The gain region was confined in all directions to a size comparable to the transport mean free path of about 300 μm. Increasing the boundary reflectivity in this case did not further narrow the emission spectrum near threshold. Indeed, the emission broadened slightly when the back side of the cuvette was replaced by a mirror. This was attributed to enhanced absorption and spontaneous reemission.

We would like to note that absorption followed by spontaneous emission are incoherent effects that do not require some set of selected paths. Thus the spectral broadening when internal reflectivity at the sample boundary is increased is indeed a consequence of the interaction of the photons during their extended stay in the sample and does not allow one to choose between models for lasing in the weak scattering limit. In a more strongly scattering medium, the back-side reflectivity should be negligible as compared with the front-side reflectivity. Simulations described below indicate that higher internal reflectivity of the front side enhances lasing in this regime.

Line width broadening in a more strongly scattering, amplifying medium was reported by Zhang et al. [1995] using Rhodamine B dye. The incident pump fluence was about 60 mJ/cm^2 produced by 7-ns pulses from a frequency-doubled Q-switched Nd:YAG laser focused to a spot diameter of about 2.5 mm. When the particle concentration was increased, the spectrum narrowed from 8 nm at low densities to a minimum of about 4 nm and broadened beyond this point to about 11 nm. The transport mean free path at the point of the narrowest spectrum was roughly 500 μm. At the same time, the peak of the spectrum shifts from about 580 nm for the neat dye to about 595 nm at low particle concentrations. Increasing the particle concentration led to a blue shift back to 580 nm.

One can expect that such effects have no significant impact on the lasing and are presumably very specific to the dyes used. The initial red shift of the emission for low concentrations might be due to the stronger absorption of the emitted light in the blue. With increased concentration, the gain region narrows, making reabsorption of photons traveling deep into the sample less likely. Thus, the emission peak shifts toward the frequencies with the strongest gain. The accompanied broadening of the spectrum may be due to the increased loss of pump photons by diffuse reflection. This and other issues that have been addressed here in a qualitative fashion will be dealt with in the simulations.

A number of attempts have been undertaken to develop theoretical models which are relevant to lasing in the colloid/dye system. One can differentiate these models according to whether selected paths do or do not play a key role. The ring laser model [Balachandran and Lawandy, 1995a] and the randomly distributed feedback laser model [Herrmann and Wilhelmi, 1995] belong in the first category. Models based on the random walk of photons [Siddique et al., 1995b; Berger et al., 1996] and on the diffusion theory [John and Pang, 1995] take the multitude of possible paths into account without giving preference to specific paths.

Balachandran and Lawandy [1995a] consider an ensemble of diffusive ring lasers created by closed paths of length L of the diffusive photons. The probability P of return to a volume V within the gain medium serves as an effective reflection coefficient and is obtained from diffusion theory. The threshold gain then becomes $\gamma_{th} = \ln(1/P)/2L$. The resulting threshold condition is used in a numerical treatment of a two-level model of the dye.

In a diffusive medium, closed loops are one of many possible actual paths of the photons. Thus, in general, $P \ll 1$. The likelihood of a buildup of lasing by many round-trips is, therefore, extremely small. We note that the above model does not appear to apply to the limit of transparent samples.

Herrmann and Wilhelmi [1995] approach the problem with a treatment based on distributed feedback lasers in which feedback is provided by randomly distributed scatterers. The authors argue that "from all scattering partial waves that component which is precisely backscattered in the opposite direction and satisfying the Bragg condition is especially important and yields a reverse coupling for the building-up of the laser regime." This approach clearly relies on the selective action of wave interference on a subset of paths that have small probability in a medium without gain.

Diffusion theory provides the basis for an entirely different approach [Genack and Drake, 1994; Zyuzin, 1994; Siddique et al., 1995b; John and Pang, 1995; Berger et al., 1996]. John and Pang [1995] solved the diffusion equation with nonlinear gain and loss. The interaction between the photons and the molecules was treated by rate equations using a four-level system for the singlet state transitions and a separate emission/absorption cycle between triplet states. The key simplifications are (1) the whole system is homogeneous in the transverse direction, (2) the rate equations are solved under steady state conditions, and (3) bleaching of the ground state is not considered. The nonlinear system of differential equations is solved numerically. Even though the experiments were carried out with excitation pulses comparable to or much shorter than the spontaneous emission time, the steady state solutions account

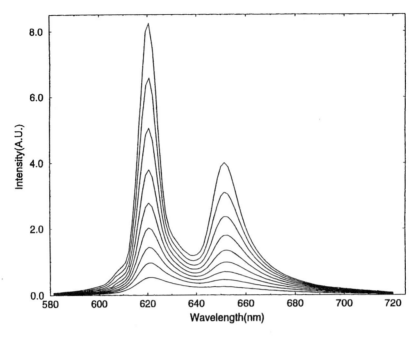

Figure 15 Emission spectra at nine different pump intensities for a dye concentration corresponding to an absorption length $\ell_a = 5.0 \times 10^{-4}$ cm. The transport mean free path is $\ell = 1.95 \times 10^{-3}$ cm. The pump intensities vary from 0.07×10^7 Jcm^{-2} s^{-1} for the innermost curve to 0.63×10^7 Jcm^{-2} s^{-1} for the outermost curve. (From John, S. and Pang, G., 1995, preprint. With permission.)

for many essential features of the experimental observations. An example is the calculation of the emission spectrum shown in Figure 15.*

John and Pang find that, in the reflection geometry and uniform gain, emission at the peak frequency of the spectrum increases according to a power law $I_{peak} \sim (\lambda_{peak}/\ell)^{\alpha}$, where α is a constant which may depend on the dye concentration and the sample thickness, but not the pumping intensity. They suggest that this points to a synergy between laser activity and photon localization which might substantially lower the lasing threshold in the region of the localization transition.

The random walk model by Berger et al. [Siddique et al., 1995b; Berger et al., 1996] is equivalent to solving the nonlinear diffusion equation. This model will be explained in detail below. It makes possible the study of the dynamics of the system under pump conditions that have been realized in the experiments and treats the bleaching of the ground state.

5. MONTE CARLO SIMULATION

In order to investigate the appropriateness of the random walk model of lasing in random media, we carried out a Monte Carlo simulation of this model. As an initial approximation, we assumed uniform excitation in the transverse dimensions. In this case, all variables are functions only of the time t and the longitudinal coordinate z. The average distance traveled by a photon between randomizations is ℓ, the projection of which in the z direction is $\ell/2$. The "sample" in the simulation was thus divided into discrete bins of width $\ell/2$. Each time a photon traversed one bin, its direction was randomized; see Figure 16. The development of the system was followed in discrete time intervals of one mean free time. This is the time $\Delta t = \ell/v$ in which the photons are assumed to travel one bin to the left or right. The population densities of the excited state (N_3) of the dye molecules were tracked in each bin, and three sets of photons were followed separately — pump, spontaneously emitted, and stimulated photons.

---

* A steady-state solution of the diffusion equation was also discussed by Noginov et al. [1995].

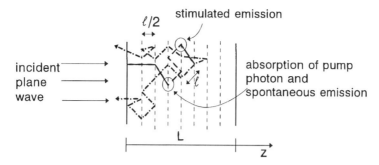

Figure 16 Sample geometry used for the simulations.

Since thermal excitation is negligible, N_3 was initially set to zero everywhere, so that $N_1 = N$. A pulse energy and duration was chosen, and, the number of photons incident in each time interval was calculated assuming a flat pulse.

During each time interval, the appropriate number of photons were injected into the system from the input pulse. From the sample boundary at $z = 0$, they traveled a distance ℓ before being scattered for the first time. Subsequently, the photons of all three types currently found at each site throughout the sample were allowed to move through one bin per mean free-time, their direction of motion being determined by a random number between zero and one. The direction of motion once the photon reached the surface was determined by the internal reflection coefficient. During the time a photon traverses a bin, there is a certain probability, based on the populations of the states of the dye molecules in the bin and on the cross section, that it will be absorbed, or, if it is an emitted photon, that it will stimulate a molecule to emit another photon. The probability for absorption of a photon is given by $1 - \exp[-\sigma_a(\nu)N_1\ell]$, and the probability for stimulated emission is $1 - \exp[-\sigma_e(\nu)N_3\ell]$. Whether or not such an event actually occurred was determined by another random number. If absorption or stimulation occurred, the values of N_1 and N_3 were updated. This process was followed for 1000 photons of each type at each site in the sample, each one of the thousand representing 1/1000 of the actual number of photons found there. We found that scaling up from 1000 photons provided statistically adequate data. Finally, the appropriate number of molecules in each bin was allowed to spontaneously emit, based upon the length of the time interval, the decay rate of the excited state, and the value of N_3 in the bin. All spontaneously emitted and stimulated emitted photons which escaped through the input face of the sample were collected. Only a fraction of this number could be detected in the experiments. Considering the random nature of the emerging photons, however, any substantial fraction of the totally emitted photons is representative of the total emission. This entire process was followed for each time interval.

Note that the Monte Carlo simulation naturally accounts for the nonlinearities of the system. The populations of the different states are continually updated, so that each time a new probabilistic event is considered, the parameters have already been reset based on the immediately preceding events.

For the emission and absorption processes, the spectral distribution was taken into account by dividing the emission line shape of Rhodamine perchlorate into 20 sections of equal areas $\sigma_e(\nu)\Delta\nu$ and treating the photons at each frequency separately.

6. RESULTS OF THE SIMULATIONS

In order to test the random walk model, we performed simulations using parameters from time resolved experiments which were carried out in the much-studied R640 dye/titania scatterer system [Siddique et al., 1995b]. In those experiments, the absorption length was varied from 50 μm to 1.4 mm and the transport mean free path for the emitted light was varied from 16 μm to 1.6 mm. The gain was produced by focusing individual 10-ps pulses at 527 nm to a 0.5-mm-diameter spot on the suspension surface.

The calculation was carried out for the following conditions: (1) The incident light is a plane wave with a fluence corresponding to the total energy in the pulse divided by the area of the spot. (2) The incident pump radiation is a 10-ps-wide square pulse. (3) The absorption cross section σ_a is 1.2×10^{-16} cm² at 527 nm as measured by Siddique et al. [1995b] and the emission has a spectral distribution and a cross section at the emission peak of 5.3×10^{-16} cm² as deduced from the experiments in Lawandy et al. [1994]. (4) The absorption of the emitted light is determined from the absorption spectrum given in

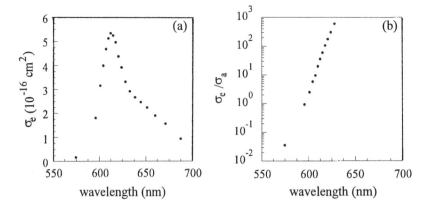

Figure 17   Cross sections of R640 used in the simulations. (a) Emission cross section. The wavelength spacing between the depicted points corresponds to a division of the spectrum into 20 segments of equal area. (b) Ratio of the emission and the absorption cross sections.

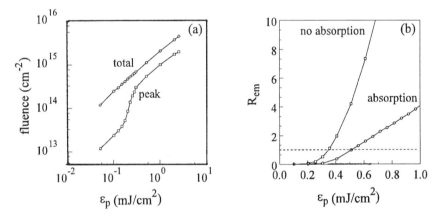

Figure 18   Emission characteristics as a function of pump fluence for a sample with $\ell = 60$ µm, $\ell_a = 300$ µm, and $R = 0.5$. (a) Total emission and the emission around 616 nm. (b) Ratio of stimulated to spontaneously emitted photons, with and without absorption of the emitted light in the sample.

Brackmann [1994]. The value of the emission cross section and its size relative to the absorption cross section are shown in Figure 17. (5) The transport mean free paths for pump and emitted photons are the same. (6) The sample thickness is 1 mm. The fact that this is thinner than the cuvette used in many experiments should not influence the results, because it is much greater than the intensity attenuation length L_a for the range of parameters considered.

Assumption 1 corresponds to neglecting the transverse excitation profile. This is an accurate reflection of the physical situation only if the transport mean free path is much smaller than the spot diameter. Even for the strongest scattering for which measurements were reported by Siddique et al. [1995b], $\ell = 16$ µm, transverse diffusion has a significant impact on lasing. In this case, photons emitted at the center of the incident beam diffuse a transverse distance nearly equal to the beam radius in a time equal to the 10-ps excitation pulse. As discussed below the main consequence of neglecting transverse diffusion out of the excitation region is a lowering of the lasing threshold. Assumption 5 allows us to simplify the simulations considerably. In general, the transport mean free path is different for pump and emitted photons. A much greater transport mean free path for pump photons as compared with emitted photons allows deeper penetration into the medium and might, therefore, be desirable with respect to lowering the threshold. In the experiments reported by Siddique et al. [1995b], the transport mean free path for the pump light was found to be 1.6 times smaller than the transport mean free path for the emitted light.

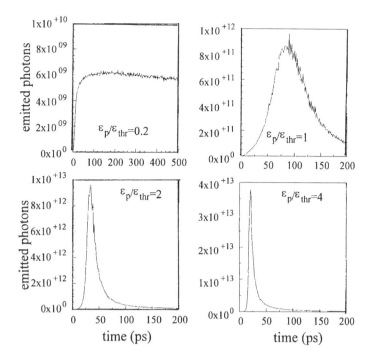

Figure 19 Temporal emission profiles for the same sample as in Figure 18 for various pump energy densities ε_p. The threshold fluence is $\varepsilon_{thr} = 0.255$ mJ/cm². Depicted is the number of emitted photons per mean free time (~0.26 ps).

We will first consider the change of the emission properties with increasing pump energy for a system with $\ell = 60$ µm and $\ell_a = 300$ µm. The number of emitted photons at all wavelengths and around the peak wavelength of 616 nm, as well as the ratio of the numbers of stimulated to spontaneously emitted photons R_{em}, is shown in Figure 18. A detection bandwidth of 5 nm around 616 nm was chosen to find the peak emission in order to encompass shifts of the actual emission peak as a function of pump fluence (see Figure 20). The peak emission shows a behavior that is similar to what was observed by Lawandy et al. [1994], compare Figure 11. The large slope is associated with the line width narrowing near the threshold. Above and below this transition region, the input–output relation is linear with about the same slope. In the simulations, we define the threshold as the pump energy at which the number of stimulated photons equals the number of spontaneously emitted photons ($R_{em} = 1$). We find, as will be seen below, that transitions in the spectral and temporal characteristics of the emission occur at around this point, in agreement with the experimental observations. Figure 19 shows the emitted pulse below, at, and above threshold. The short time required for the buildup of stimulated emission and its short duration are particularly striking. With a shorter mean free path ($\ell = 10$ µm), emission with a duration as short as 4 ps is found with a peak which is delayed by 14 ps from the arrival of the leading edge of the pump pulse, compare Figure 22b. In Figure 20, we show the duration of emission, the delay of the emission peak with respect to the arrival of the pump pulse, the emission line width, and the peak wavelength as functions of pump fluence. The results on the emission line width are consistent with experimental findings [Lawandy et al., 1994]. However, the threshold is about an order of magnitude lower than the experimental values. We believe the largest contributor to this discrepancy is the neglect of the variation of the degree of excitation in transverse directions. Since the transverse gain profile is assumed to be flat, emitted photons cannot escape the gain region by transverse diffusion. The difference between experiments and simulations should diminish as ℓ and ℓ_a become significantly smaller than the beam diameter. Indeed, Siddique et al. [Siddique, personal communication] have observed thresholds as low as 0.8 mJ/cm² in strongly scattering samples where the above condition is met.

The simulations allow us to study separately the influence of various factors upon lasing action in random systems. We will consider here the following factors: (1) the absorption of the emitted photons, (2) the transport mean free path and the absorption length for the pump photons, (3) the internal

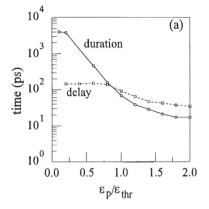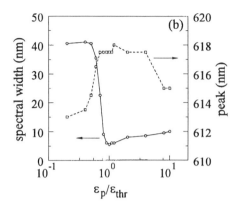

Figure 20 Temporal and spectral emission properties as function of pump fluence. The sample parameters used are the same as in Figure18. (a) Duration of the emission and time delay between arrival of the leading edge of the pump pulse and the peak of the emitted pulse. (b) Spectral width and peak wavelength of the emission.

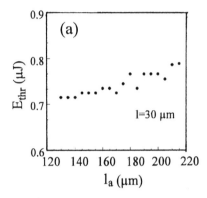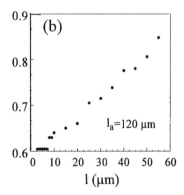

Figure 21 Threshold pump energy (a) vs. ℓ_a for fixed ℓ = 30 μm and (b) vs. ℓ for fixed ℓ_a = 120 μm. The simulation was performed for R = 0.15. Absorption of the emission in the sample is neglected. A pump energy of 0.75 μJ corresponds to a fluence of 0.38 mJ/cm².

reflectivity at the boundary of the medium, and (4) the location at which the pump photons enter the sample. We consider each of these factors in turn.

(1) Absorption of the Emitted Radiation
In the experiments with R640, the dye was pumped relatively far from its absorption maximum at about 570 nm. As a result, the absorption cross sections for the pump and emitted photons are not very different. Absorption of the emission might, therefore, be expected to influence the lasing process significantly. Indeed, we find a considerably faster rise of R_{em} with increased pump fluence when absorption is absent (see Figure 18). However, the threshold drops by a factor of less than two. We find no significant change in the temporal and spectral properties of the emission above threshold. This is in agreement with the previous observation that these properties do not depend significantly on R_{em} above threshold, see Figure 20. Among the experimentally observable properties of the emission only the emission peak wavelength is significantly affected by absorption. The initial red shift of the emission peak with increased pump fluence, as seen in Figure 20(b), is due to the stronger absorption in the blue. For stronger pumping, the increased gain for transitions with the largest emission cross-section offsets this effect, pulling the emission peak back to the blue. Such shifts have been observed by Noginov et al. [1995].

(2) Transport Mean Free Path ℓ and Absorption Length ℓ_a
In Figure 21, we show the results of the computation of the threshold energy a function of ℓ with ℓ_a held constant at 120 μm and as a function of ℓ_a with ℓ held constant at 30 μm. In these calculations, the spectral distribution of the spontaneous emission was ignored. An effective emission cross section

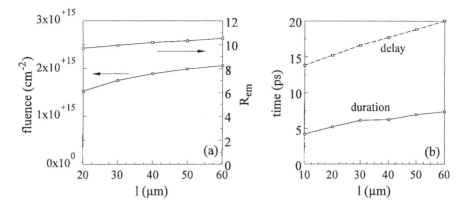

Figure 22 Emission characteristics as a function of the transport mean free path ℓ for a fixed pump fluence of 1 mJ/cm². (a) Total emission and ratio of stimulated to spontaneously emitted photons, and (b) emission duration and delay of emission pulse peak. The pump absorption length is $\ell_a = 300$ µm.

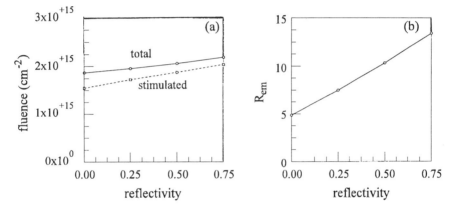

Figure 23 Emission characteristics as a function of the internal reflectivity at the sample boundaries, R, for a fixed pump fluence of 1 mJ/cm². (a) Total and stimulated emission, and (b) ratio of stimulated to spontaneously emitted photons. The sample parameters are $\ell = 60$ µm and $\ell_a = 300$ µm.

of $\sigma_e = 4.2 \times 10^{-16}$ cm² was used which was calculated from the line shape and spontaneous lifetime of the upper lasing level due to transitions to the S_o manifold. The variation of threshold with ℓ and ℓ_a is small, as anticipated from the simple considerations presented previously, and is in accord with experiments [Sha et al., 1994; Siddique et al., 1995b]. For certain parameter ranges, the threshold in the experiments rose when the concentration of scattering particles is increased, see, for instance, Figure 12. These results show that the threshold is so weakly dependent on ℓ and ℓ_a that, under certain conditions, otherwise unimportant factors may determine whether or not the threshold decreases when ℓ and ℓ_a are decreased. One of these factors might be that a larger fraction of pump photons is reflected from the sample before being absorbed as the scattering strength increases. Figure 22 shows this effect as a reduced total emission with decreased ℓ. The ratio R_{em} is also shown in Figure 22.

We conclude that a decrease of the transport mean free path and the absorption length within achievable limits does not seem to provide a substantial lowering of the threshold.

(3) Internal Reflectivity R

Internal reflectivity has been shown to influence the diffuse transport in random media considerably [Lagendijk et al., 1989; Li et al., 1993]. All previously mentioned simulations were performed assuming an internal reflectivity of 0.5. This is the reflection that one expects from the methanol-glass-air boundary for diffuse photons from the Fresnel equations by integration over all angles and both polarizations. Internal reflection affects the emission in two ways. First, it determines the probability for pump photons to leave the sample before being absorbed. This is shown in Figure 23 as an increase of the total emission

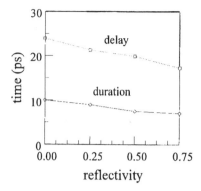

Figure 24   Emission duration and delay of the emission pulse peak as a function of internal reflectivity R. The sample and pump parameters are as in Figure 23.

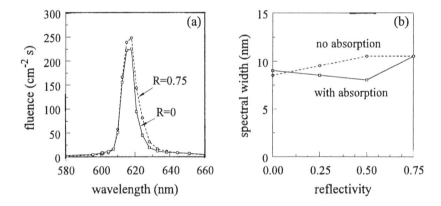

Figure 25   (a) Emission spectrum for two values of internal reflectivity. (b) Width of the spectrum as a function of R with and without absorption of the emitted light. The sample parameters were $\ell = 60$ μm and $\ell_a = 300$ μm, and the pump fluence was 1 mJ/cm².

as R increases. Second, there is an increase in path length for emitted photons as R increases. This increases both the gain and the reabsorption [Balachandran and Lawandy, 1995b] of the photons. Because the photons cannot escape the gain region in the transverse direction in our model, the role of reabsorption should be small. Indeed, we find that the dominant effect of R is the enhancement of the stimulated processes, as can be seen from the increase of R_{em} in Figure 23. The consequences are striking especially in terms of the temporal characteristics of the emission that are shown in Figure 24. In contrast to what one might expect, we find a shorter emission with an earlier onset when R is increased. Only for very large internal reflectivity do we observe a broadening of the emission, as reported by Balachandran and Lawandy [1995b] for the weakly scattering case, see Figure 25.

In experiments, a change of internal reflectivity is necessarily accompanied by a change of the external reflectivity for the incident pump photons. In the above simulations, the number of incident pump photons was kept constant while varying R.

(4) Location of Pump Source

From the above it becomes clear that a substantial lowering of the pump energy threshold might be difficult to achieve in the dye system pumped from the outside. One alternative is to deposit the energy into the medium at a point removed from the surface. The probability of return of emitted photons to the region of high excitation in which they were emitted is then larger than it would be at the boundary. This is manifested in a larger energy density for the case that the source of light is deep inside the sample than when the energy is deposited at the surface. Thus, one expects that the emitted photons experience a larger gain when the pump energy is deposited deeper into the sample. For a sample with

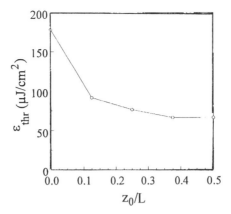

Figure 26 Pump energy threshold as a function of the source location within the sample normalized by the sample thickness, z_0/L. Sample: $\ell = 60$ μm and $\ell_a = 300$ μm, $R = 0.5$. The absorption of the emitted light has been neglected.

$$\varepsilon_p/\varepsilon_{thr}=1$$

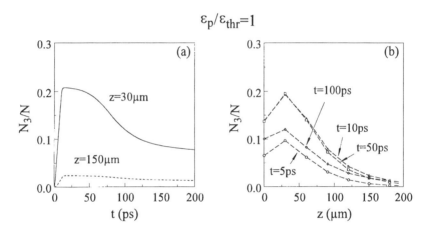

Figure 27 The fraction of the molecules in the excited state at threshold (a) as function of time for two locations within the sample and (b) as function of depth z within the sample for various times t after the arrival of the leading edge of the pump pulse. The pump fluence was 0.255 mJ/cm² and the sample parameters are $\ell = 60$ μm, $\ell_a = 300$ μm, and $R = 0.5$.

$\ell = 60$ μm and $\ell_a = 300$ μm, the lowering of the threshold by this means is shown in Figure 26. A somewhat larger reduction by a factor of five was obtained for $\ell = 20$ μm. When we include absorption in the simulations, the lowering of the threshold is even smaller.

The simulations presented and the experimental observations indicate that the dye/colloid systems require a pump fluence at least close to the saturation fluence to permit lasing action. Figures 27 and 28 show the bleaching of the ground state that is associated with such high pump levels. Here, the system was again pumped from the front surface. The normalized population in the ground state, N_1/N, is related to the normalized population density in the upper lasing level, N_3/N, which is shown in the figures by $N_1/N = 1 - N_3/N$. A threshold the gain region extends somewhat beyond the small-signal intensity penetration depth $L_a \approx 80$ μm (Figure 27). Ten times above threshold, there is considerable bleaching of the ground state at the sample surface at the end of the pump pulse and the gain extends nearly 200 μm in the longitudinal direction (Figure 28). Photons emerging from a depth of $z = 200$ μm in the sample have an average path length of $z^2/\ell \approx 600$ μm and leave the gain region in less than 3 ps. The fast dynamics of the system above threshold results from the fact that virtually all stimulated photons originate from a depth ≤ 200 μm in the above sample.

We have recently performed simulations which include transverse diffusion in a sample excited by a gaussian beam [Berger et al., 1996]. For a 500-μm beam illuminating a sample with $\ell = 100$ μm and

$\varepsilon_p/\varepsilon_{thr}=10$

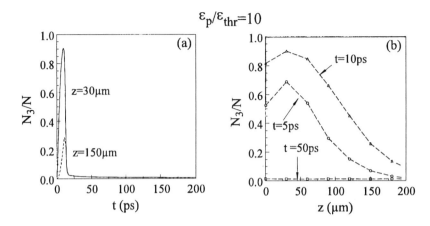

Figure 28 The same as in Figure 25 for a pump fluence of 2.55 mJ/cm², ten times above threshold.

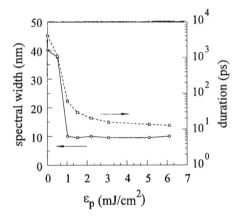

Figure 29 Linewidth and duration of the emission as a function of pump fluence obtained with gaussian beam excitation. The sample parameters are $\ell = 100$ µm, $\ell_a = 500$ µm, and R = 0.15.

$\ell_a = 500$ µm, the minimum width of the emission line is reached when the pump fluence is a factor of three higher than found in the simulations with plane wave excitation. Preliminary calculations in which the transport mean free path of the emission frequency is twice that at the pump frequency indicate further substantial increase in the threshold. These results bring the calculated threshold close to that observed experimentally. The line width and pulse shapes observed for gaussian beam and plane wave excitation are not substantially different above threshold. Figure 29 shows results of the simulations with gaussian beam excitation.

Our simulations show that the random walk model of photons in an amplifying disordered medium can explain all the experimentally observed features of lasing. Figure 30 summarizes the essence of the random walk.

7. CONCLUSIONS

The strong analogy between classic wave propagation in random media and electronic transport has resulted in a rich interplay between these fields. Such an interplay is always enriched when one of the fields exhibits properties that cannot be fully explored in the other. So, for example, the richness of the electron interaction with a magnetic field has stimulated optical studies on optical propagation in the presence of perturbations that break time reversal symmetry, such as the Faraday effect [Erbacher et al.,

Figure 30 Drunkard's walk. (From Gamow, G., *One Two Three ... Infinity,* Viking Press, New York, 1947. With permission.)

1993]. Here we have considered the emerging understanding of the conjunction of wave propagation and stimulated emission which is possible only for bosons and cannot be achieved for electrons.

Stimulated emission completely alters the properties of emission in random media above the threshold for lasing action. In optically pumped powdered and colloidal samples it leads to a shortening of the emitted pulse and a narrowing of the emission spectrum. The pumping thresholds, above which lasing action has been reported in colloidal suspensions in dye solutions, are not far from the saturation fluence of the neat dye solution. Nonetheless, there has been considerable consideration given to potential applications of this phenomenon. Lawandy [1994] has considered commercial applications for search-and-rescue operations, robotic vision, and removing skin discolorations.

In the multiply scattering regime, these observations can be explained using a random walk or diffusion model of wave propagation in which nonlinear interactions are taken into account. A detailed comparison of models for lasing in nearly transparent samples has not been carried out as yet.

Substantial lowering of the threshold seems to be feasible only if the pump energy can be deposited more deeply into the sample than is possible for diffusing waves in the reflection geometry. Strong scattering then leads to a lengthening of the paths of emitted photons in the gain region and can lower the lasing threshold. When the sample is pumped in the reflection geometry, however, stronger scattering reduces the penetration of the pump radiation into the medium.

One way of penetrating more deeply into the medium may be to couple to a localized state which is created in a random medium or in a nearly periodic structure possessing a photonic pseudogap. Since light is always localized in random one-dimensional structures, the emission from localizing media could be studied. Theoretical studies of the interaction of gain and lasing in random one-dimensional systems indicate that enhanced reflection is observed in the situation that uniform gain is maintained in the structure by some external mechanism.

At present, the applications of emission from random media seem to suffer from the high pumping levels required. Lowering of the lasing threshold would require novel optical and electronic pumping configurations. No doubt, attempts in this direction will lead to still new discoveries and to a deepening of our understanding of wave propagation in random media.

8. DEFINING TERMS

Internal reflectivity: The average reflection coefficient for photons reaching the internal surface of a medium from within the bulk is the internal reflectivity. Because the angle of incidence that the randomized light makes with the normal as it approaches the sample boundary may be large, the internal reflection coefficient may be large.

Lasing threshold: The lasing threshold denotes the minimum gain in a medium that is required for lasing. Because the gain is determined by pumping, it is at the same time a pumping threshold. Usually, it is determined from the requirement that in a round-trip of a photon the gain exceeds the losses. This corresponds to the situation that a typical photon stimulates the emission of another photon before being lost as a result of absorption or by leaving the gain region.

Localization: Localized waves cannot propagate in a random medium. The ensemble average of transmission of a localized wave falls exponentially with increasing thickness. Localization results from the constructive interference of waves returning to a point in the medium. In static samples, all waves in random media with dimension less than two are localized. In one dimension, the localization length is somewhat larger than the transport mean free path. In three dimensions, delocalized waves can exist. These waves are separated from the localized waves at the mobility edge, which occurs at the point where $k\ell \approx 1$. Localization in three dimensions can be created by the presence of sufficiently strong disorder that the condition above obtains or in a quasi-periodic structure in which the disorder induces a localized state to fall within the band gap. All such states in the pseudogap are localized.

Random systems: A random medium or system is taken to be a material in which the dielectric function varies randomly in space with statistical properties which are homogeneous throughout the sample. Because of large fluctuations in transport properties that arise as a result of wave interference in the sample, a full statistical picture can often not be obtained by studying a single sample. Rather the properties of an ensemble of different realizations of samples with equivalent inhomogeneity are examined both theoretically and experimentally.

Random walk: Transport as a succession of straight-line paths which follow interactions with isolated scattering centers in the medium.

FOR FURTHER INFORMATION

One can expect continued strong interest in stimulated emission from random media.

The Conference on Lasers and Electro-Optics (CLEO®) and the Quantum Electronics and Laser Science Conference (QELS) provide fora for an update and review of the latest research. For information contact: Optical Society of America, 2010 Massachusetts Ave., Washington, D.C. 20036–1023.

For more information on wave propagation in random media, see *The Scattering and Localization of Classical Waves*, P. Sheng, ed., World Scientific Press, Singapore (1990) and *Photonic Band Gaps and Localization*, C. M. Soukoulis, ed., Plenum Press, New York (1993). More recent research is presented in *OSA Proceeding on Advances in Optical Imaging and Photon Migration*, 1994, Vol. 21, R. R. Alfano, ed. Optical Society of America, Washington, D.C. (1994).

ACKNOWLEDGMENTS

This research is supported by NSF grants #DMR9311605 and #DMR9632789 and by an institutional NASA grant. We thank W. Rudolph for helpful suggestions and comments.

REFERENCES

Akkermans, E., Wolf, P. E., and Maynard, R. 1986. Coherent backscattering of light by disordered media: analysis of the peak line shape. *Phys. Rev. Lett.* 56(14):1471–1474.

Ambartsumyan, R. V., Basov, N. G., Kryukov, P. G., and Letokhov, V. S. 1970. Non-resonant feedback in lasers, in *Progress in Quantum Electronics,* Sanders, J. H., and Stevens, K. W. H., Eds., Pergamon, Oxford, 107–185.

Anderson, P. W. 1958. Absence of diffusion in certain random lattices. *Phys. Rev.* 109(5):1492–1505.

Anderson, P. W. 1985. The question of classical localization: a theory of white paint? *Philos. Mag.* 52(3):505–509.

Arnold, S., Liu, C. T., Whitten, W. B., and Ramsey, J. M. 1991. Room-temperature microparticle-based persistent spectral hole burning memory. *Opt. Lett.* 16(6):420–422.

Balachandran, R. M. and Lawandy, N. M. 1995a. Theory of photonic paint. *QELS'95.* Paper QPD15-1. Baltimore.

Balachandran, R. M. and Lawandy, N. M. 1995b. Interface reflection effects in photonic paint. *Opt. Lett.* 20(11):1271–1273.

Berger, G. A., Kempe, M., and Genack, A. Z., 1996. Ultrafast dynamics of stimulated emission from random media. Preprint.

Brackmann, U. 1994. *Lambdachrome® Laser Dyes*. Lambda Physik, Inc., Göttingen.

Elitzur, M. 1995. Masers in the sky. *Sci. Am.* 272(2):68–74.

Erbacher, F., Lenke, R., and Maret, G. 1993. Optical speckle patterns and coherent backscattering in strong magnetic fields, in *Photonic Band Gaps and Localization*, C. M. Soukoulis, Ed., Plenum, New York, 81–97.

Folan, L. M. and Arnold, S. 1988. Determination of molecular orientation at the surface of an aerosol particle by morphology-dependent photoselection. *Opt. Lett.* 13(1):1–3.

Fork, R. L. and Taylor, D. W. 1979. Unusual optical emission from microcrystals containing Eu^{2+}: experiment. *Phys. Rev. B* 19(7):3365–3398.

Gamow, G. 1947. *One Two Three … Infinity*. Viking Press, New York.

Garcia, N. and Genack, A. Z. 1991. Anomalous photon diffusion at the threshold of the Anderson localization transition. *Phys. Rev. Lett.* 60(14):1850–1853.

Garcia, N., Genack, A. Z., and Lisyansky, A. A. 1992. Measurements of the transport mean free path of diffusing photons. *Phys. Rev. B* 46(22):14475–14479.

Genack, A. Z. 1986. Light localization laser. Patent memorandum 86CRL087. Exxon Research and Engineering Company.

Genack, A. Z. 1987. Optical transmission in disordered media. *Phys. Rev. Lett.* 58(20):2043–2046.

Genack, A. Z. and Drake, J. M. 1994. Scattering for super-radiation. *Nature* 368:400–401.

Golubentsev, A. A. 1984. Suppression of interference effects in multiple scattering of light. *Sov. Phys. JETP* 59(1):26–32.

Gouedard, C., Huusson, D., Sauteret, F., Auzel, F., and Migus, A. 1993. Generation of specially incoherent short pulses in laser-pumped neodymium stroichlometric crystals and powders. *J. Opt. Soc. Am. B* 10:2358.

Herrmann, J. and Wilhelmi, B. 1995. Randomly-distributed-feedback lasers with dispersed scattering nano particles. Preprint.

John, S. 1984. Electromagnetic absorption in a disordered medium near a photon mobility edge. *Phys. Rev. Lett.* 53(22):2169–2172.

John, S. 1987. Strong localization of photons in certain disordered dielectric superlattices. *Phys. Rev. Lett.* 58(23):2486–2489.

John, S. and Pang, G. 1995. Theory of paint-on lasers. Preprint.

Lagendijk, A., Vreeker R., and de Vries, P. 1989. Influence of internal reflection on diffusive transport in strongly scattering media. *Phys. Rev. A* 136A:81–88.

Lawandy, N. M. 1994. "Paint-On Lasers" light the way for new technologies. *Photonics Spectra* 28(7):119–124.

Lawandy, N. M., Balachandran, R. M., Gomes, A. S. L., and Sauvain, E. 1994. Laser action in strongly scattering media. *Nature* 368:436–438.

Lawandy, N. M. and Balachandran, R. M. 1995. Random laser? — Reply. *Nature* 373:204.

Letokhov, V. S. 1968. Generation of light by a scattering medium with negative resonance. *Sov. Phys. JETP* 26(4):835–840.

Li, J. H., Lisyansky, A. A., Cheung, T. D., Livdan, D., and Genack, A. Z. 1993. Transmission and surface intensity profiles in random media. *Europhys. Lett.* 22(9):675–680.

Maret, G. and Wolf, P. E. 1987. Multiple light scattering from disordered media: the effect of Brownian motion of scatterers. *Z. Phys. B* 65:409–413.

Markushev, V. M., Zolin, V. F., and Briskina, Ch. M. 1986. Luminescence and stimulated emission of neodymium in sodium lanthanum molybdate powders. *Sov. J. Quantum Electron.* 16(2):281–283.

Martorell, J. and Lawandy, N. M. 1990. Observation of inhibited spontaneous emission in a periodic dieleectric structure. *Phys. Rev. Lett.* 65(15):1877–1880.

Noginov, M. A., Caulfield, H. J., Noginova, N. E., and Venkateswarlu, P. 1995. Line narrowing in the dye solution with scattering centers. *Opt. Comm.* 118: 430–437.

Pine, D. J., Weitz, D. A., Maret, G., Wolf, P. E., Herbolzheimer, E., and Chaikin, P. M. 1990. Dynamical correlations of multiply scattered light, in *Scattering and Localization of Classical Waves in Random Media*, Sheng, P., Ed., World Scientific, Singapore, 312–372.

Pradhan, P. and Kumar N. 1994. Localization of light in coherently amplifying random media. *Phys. Rev. B* 50(13):9644–9647.

Saleh, B. E. A. and Teich, M. C. 1991. *Fundamentals of Photonics*. John Wiley & Sons, New York.

Schäfer, F. P., Ed. 1979. *Dye Lasers*. Springer, Berlin.

Shapiro, B. 1986. Large intensity fluctuation for wave propagation in random media. *Phys. Rev. Lett.* 51(17):2168–2171.

Sha, W. L., Liu, C.-H., and Alfano, R. R. 1994. Spectral and temporal measurements of laser action of Rhodamine 640 dye in strongly scattering media. *Opt. Lett.* 19(23):1922–1924.

Sheng, P., Ed. 1990. *Scattering and Localization of Classical Waves in Random Media*, World Scientific, Singapore.

Siddique, M., Yang, L., Wang, Q. Z., and Alfano, R. R. 1995a. Mirrorless laser action from optically pumped dye-treated animal tissues. *Opt. Common.* 117:475–479.

Siddique, M., Alfano, R. R., Berger, G. A., Kempe, M., and Genack, A. Z. 1995b. Time-resolved studies of stimulated emission from colloidal dye solutions. *Opt. Lett.*, 21(7):450–452.

Soukoulis, C. M., Ed. 1993. *Photonic Band Gaps and Localization*, Plenum Press, New York.

Stephen, M. J. 1988. Temporal fluctuations in wave propagation in random media. *Phys. Rev. B* 37(1):1–5.

Thouless, D. J. 1977. Maximum metallic resistance in thin wires. *Phys. Rev. Lett.* 39(18):1167–1169.

van Albada, M. P. and Lagendijk, A. 1985. Observation of weak localization of light in a random medium. *Phys. Rev. Lett.* 55(24):2692–2695.

Weinberg, A. M. and Wigner, E. P. 1958. *Physical Theory of Neutron Chain Reactors*. University of Chicago, Chicago.

Wiersma, D. S., van Albada, M. P., and Lagendijk, A. 1995a. Random laser? *Nature* 373:203–204.

Wiersma, D. S., van Albada, M. P., and Lagendijk, A. 1995b. Coherent backscattering of light from amplifying random media. *Phys. Rev. Lett.* 75(9):1739–1742.

Wolf, P. E. and Maret, G. 1985. Weak localization and coherent backscattering of photons in disordered media. *Phys. Rev. Lett.* 55(24):2696–2699.

Yablonovitch, E. 1987. Inhibited spontaneous emission in solid-state physics and electronics. *Phys. Rev. Lett.* 58(20):2059–2062.

Yablonovitch, E., Gmitter, T. J., Meade, R. D., Rappe, A. M., Brommer, K. D., and Joannopoulos, J. D. 1991. Donor and aceptor modes in photonic band structure. *Phys. Rev. Lett.* 67(24):3380–3383.

Zhang, W., Cue, N., and Yoo, K. M. 1995. Linewidth broadening of laser action in random gain media. *Opt. Lett.* 20: 961–963.

Zhang, Z.-Q. 1995. Light amplification and localization in randomly layered media with gain. *Phys. Rev. B* 52(11):7960–7965.

Zyuzin, A. Yu. 1994. Weak localization in backscattering from an amplifying medium. *Europhys. Lett.* 26:517–520.

Chapter 12

Optical Properties of Diamond Films and Particles

Leah Bergman and Robert J. Nemanich

Reviewed by Alan T. Collins

TABLE OF CONTENTS

1. INTRODUCTION

The unique properties of diamond, such as hardness, UV optical transparency, high thermal conductivity, and its semiconducting behavior, have led to the extensive development of new growth methods and to innovative research in the field of diamond devices and applications.[1-3] Diamond films differ in many ways from natural diamond crystals because of the synthetic method of growth; in particular, they have their own characteristic defects and impurities.[3] In order to achieve the goal of producing diamond films of sufficiently high quality to be viable as high-performance semiconductor devices, it is crucial to gain knowledge of the characteristic impurity and defect states.

There are various methods of characterizing semiconductor materials that, in general, can be categorized as surface, imaging, and bulk analysis techniques.[4] Surface analysis techniques include Auger and photoelectron spectroscopy. Imaging analysis techniques consist of scanning electron microscopy (SEM) and transmission electron microscopy (TEM), while the bulk analysis techniques are cathodoluminescence (CL), photoluminescence (PL), and Raman spectroscopy. The latter two spectroscopy techniques have been proved useful in analyzing crystalline diamond and, more recently, diamond films.[5-10] Raman and PL, moreover, are nondestructive methods of spectroscopy that can be applied to a wide range of diamond characterizations and analyses.

Important applications of Raman spectroscopy to diamond films are to identify the diamond structure and the amorphous structure of carbon (graphitic phase) that coexist in many of the diamond films,[8,9] as well as to obtain a quantitative measure of the stress in the film and to analyze its possible sources.[11-15] The analysis of PL spectra, on the other hand, can be used to provide information about the impurities and defects that are present in the diamond film and may give an insight into the mechanism of impurity incorporation and doping.[12,15,16] These two complementary optical techniques can provide crucial initial insight into the diamond film quality.

This chapter presents a review of investigations into the structure, substrate–film interface, and impurities of diamond thin films grown on silicon substrates utilizing Raman and PL spectroscopy. The remainder of this chapter is organized as follows. In Section 2 the background theory of Raman scattering and PL is presented. The issue of the graphitic phase, as a defect in diamond film, is discussed in Section 3. Lastly, Sections 4 and 5 focus on the nitrogen and silicon impurities in diamond films, respectively.

2. RAMAN SCATTERING AND PHOTOLUMINESCENCE: GENERAL ASPECTS

The Raman effect in solids[17-23] is an inelastic scattering of the incident photons by the crystal electrons distributed around the lattice nuclei. In the photon–electron interaction, the incoming photons exchange a quantum of energy with the electrons via the creation or annihilation of lattice vibrations (phonons). As a result, the scattered photons lose or gain an energy quantum depending on whether a phonon was created (Stokes process) or annihilated (anti-Stokes process). In Raman spectroscopy, the energy of the scattered photons is measured; thus, a characteristic value of the vibration energy of a specific material may be obtained.

Figure 1 illustrates the Raman scattering process. The sample is irradiated by a monochromatic laser beam of frequency v_i chosen to be in the visible region. The incident photons create or annihilate crystal vibrations of frequency v_p, and the observed scattered beam consists of light of frequency $v_i - v_p$ and $v_i + v_p$ due to the Stokes and anti-Stokes mechanisms, respectively. The scattered beam has an additional light component of the same frequency as the incident light (v_i) that arises from the elastic Rayleigh scattering. A typical Raman spectrum of a crystal exhibits sharp bands at frequencies $\pm v_p$, which are measured as a shift, called the Raman shift, from the incident beam frequency v_i.

The spectral unit of the wave number, cm^{-1}, is conventionally employed in Raman spectroscopy. The wave number $\tilde{v}$ is defined by

$$\tilde{v} = \frac{v}{c} = \frac{1}{\lambda} cm^{-1} \tag{1}$$

where c is the velocity of light (3×10^{10} cm/s), v is the frequency of the scattered light in units of $1/s$ (Hz), and λ is the wavelength in centimeters.

The classic theory of the Raman effect treats the light scattering event in terms of an oscillating dipole.[24-26] In a material of polarizability tensor α, the induced dipole moment, $\mathbf{p}$, is given by

$$\mathbf{p} = \alpha \mathbf{E} \tag{2}$$

where $\mathbf{E}$ is due to the oscillating incident electric field of a monochromatic source (laser) of frequency v_i represented by

$$\mathbf{E} = \mathbf{E_0} \cos(2\pi v_i t) \tag{3}$$

The vibrational modes of a solid are also of an oscillating nature that can be described by the normal coordinate of the vibrations, Q:

$$Q = Q_0 \cos(2\pi v_p t) \tag{4}$$

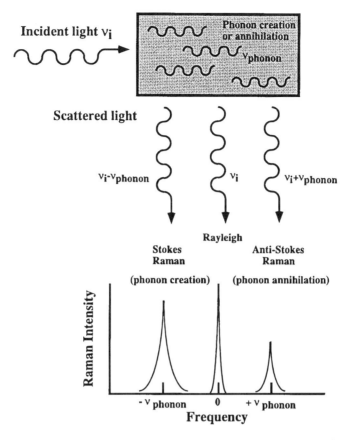

Figure 1   Schematic representation of the Raman scattering effect in a solid and the resulting Raman spectrum of the Stokes and the anti-Stokes bands.

The oscillation presented in Equation 4 may affect the polarizability of the material; this effect for small vibrational amplitude can be expressed as

$$\alpha = \alpha_0 + \left(\frac{\partial \alpha}{\partial Q}\right)_0 Q + \text{higher-order terms} \tag{5}$$

Substituting Equations 3 to 5 into Equation 2 yields what is known as the Raman scattering equation:

$$\mathbf{p} = \alpha_0 \mathbf{E_0} \cos(2\pi\nu_i) + \frac{1}{2}\mathbf{E_0}Q_0\left(\frac{\partial \alpha}{\partial Q}\right)_0 \left[\cos\{2\pi(\nu_i + \nu_p)t\} + \cos\{2\pi(\nu_i - \nu_p)t\}\right] \tag{6}$$

The first term in Equation 6 represents an oscillating dipole that radiates light of frequency ν_i corresponding to the elastic Rayleigh scattering, while the other two terms correspond to the Raman inelastic scattering process of frequencies $\nu_i \pm \nu_p$ (anti-Stokes and Stokes). When the polarizability is independent of a change of the normal coordinates, i.e., $\partial \alpha / \partial Q = 0$, the vibration of frequency ν_p is not Raman active.

Equation 2 has important practical applications to determining the phonon symmetry in a given material. Since the polarizability α (a symmetric 3×3 tensor) exhibits the same symmetry as the phonon vibrations, and the polarization states of the incident ($\mathbf{E_0}$) and scattered ($\mathbf{p}$) light can experimentally be measured, the components of α can readily be determined.

One of the most important aspects of Raman scattering in solids arises from conservation of energy and wave vector, as described by the conservation laws:

$$h\nu_{scattering} = h\nu_i \pm h\nu_p \qquad (7a)$$

for energy conservation, and

$$\mathbf{k}_{scattering} = \mathbf{k}_i \pm \mathbf{k}_p \qquad (7b)$$

for conservation of the wave vector $\mathbf{k}_p$ of the crystal momentum, where the phonon energy and wave vector are related via the dispersion relation of the material. In a material consisting of a long range periodic lattice, all of the vibrational modes can be described with a well-defined $\mathbf{k}_p$; the above conservation laws thus hold, and the Raman spectrum exhibits a sharp band at the characteristic ν_p. The resulting Raman line shape can be described as a Lorentzian function of a parameter Γ that specifies the natural line width of the function.[27,28] The physical interpretation of the natural line width may be derived from the uncertainty principle

$$\tau\Delta E = h / 2\pi \qquad (8)$$

where ΔE is taken to be the line width Γ (in units of energy) and τ is the characteristic phonon lifetime. In a high-quality diamond crystal the natural Raman line width is determined by the time it takes the optical phonon to decay into the acoustical phonons.[28] Any value of line width that is larger than the one expected from a good quality crystal implies the existence of lifetime shortening mechanisms in the material.

If the material does not exhibit long-range order, the vibrational modes cannot be described with a single charcteristic $\mathbf{k}_p$ (as expressed in Equation 7). Each individual mode may be represented by a distribution of phonon wave vectors. Thus, all modes with wave vector components that satisfy the scattering relations will be observed. The Raman spectrum in this case is very broad and often reflects the phonon density of states of the material. Another broadening mechanism that may affect the Raman line shape arises from the confinement of the phonons in small domains.[29] The domain size, d, is related to the phonon wave vector, $\mathbf{k}$, via the uncertainty relation $\Delta k \sim 2\pi/d$; thus, the domain size determines the range of the wave vectors that are allowed to participate in the Raman process. The resulting Raman line shape reflects the dispersion relation for a given material.

As opposed to the Raman process, where the light scatters through interaction with the continuum of the crystal vibrations, PL is due to a radiative recombination mechanism.[27,30-32] The most common origins of PL are from recombination of carriers at or near the band edge or from defect or impurity centers within the band gap. The optically active defect and impurity centers, termed *optical centers*, can often be excited with both above band gap and sub-band gap light. Radiative recombination is one way an optical center returns to its equilibrium state after being perturbed by the excitation.

The basic mechanism of the PL process is presented in Figure 2. In the figure, process **1** indicates the absorption of light by an optical center which results in the up-transition of the optical center into its excited state. Process **2** is the spontaneous emission of a quantum of light by the optical center in order to reach the ground state. These two processes are the most common ones that an optical center may undergo. A less common route to the ground state is the induced emission which requires a quantum of radiation (light) to be absorbed by the optical center in its excited state in order to undergo the transition, shown by process **3** in Figure 2.

The dynamics of the transitions are determined by the recombination mechanisms that the optical systems undergo and can be described as probabilistic events.[33] Let N_G and N_E be the concentrations of the optical systems in the ground and excited state, respectively. Then, the rate of change of population N_E due to the absorption is

$$dN_E / dt = N_G B_1 \rho(\tilde{\nu}) \qquad (9)$$

where $\rho(\tilde{\nu}) \sim (\tilde{\nu})^3/(\exp(hc\,\tilde{\nu}/kT) - 1)$ is the spectral radiation density and $\tilde{\nu}$ is the wave number. The constant B_1 is the Einstein coefficient which is directly related, as will be explained in the next paragraph, to the coupling process of the light with the optical center. Similarly, the rate of change of N_E due to the induced emission is given by

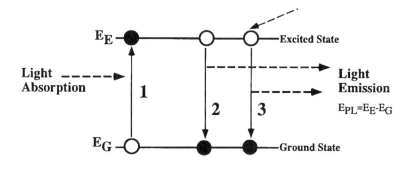

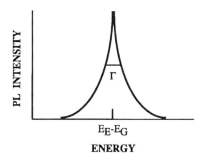

Figure 2   The PL process in an optical center involves an excitation via light absorption and a recombination mechanism along with light emission.

$$dN_E / dt = -N_E B_1 \rho(\tilde{v}) \tag{10}$$

and for the spontaneous emission

$$dN_E / dt = -N_E B_2 \tag{11}$$

where B_2 is the Einstein coefficient for this process. Upon shining a light source on the optical systems all of the above processes are initiated concurrently, and the net rate of population change can be analytically described by

$$dN_E / dt = N_G B_1 \rho(\tilde{v}) - N_E B_1 \rho(\tilde{v}) - N_E B_2 \tag{12}$$

When no nonradiative channels are present, the population change as expressed in Equation 12 is proportional to the luminescence intensity. From the thermal equilibrium conditions $dN_E/dt = 0$ and $N_E/N_G \sim \exp\{-(E_E - E_G)/kT)\}$ the relation $B_2 \sim \tilde{v}^3 B_1$ between the Einstein coefficients is obtained.

The physics of luminescence, which specifically concerns the interaction of the light with the optical center, is conveyed in the Einstein coefficients. The absorbed light induces an electric dipole moment μ in the optical center which couples the ground state ψ_G to the excited state ψ_E. The transition probability $|R|^2$ analytically describes this process:

$$|R|^2 = \left| \int \psi_E^* \mu \psi_G d\tau \right|^2 \propto B_i \tag{13}$$

The transition probability function (and thus the Einstein coefficients B_i) determines the selection rules of a transition: it is zero for a forbidden transition and nonzero for an allowed transition. Furthermore,

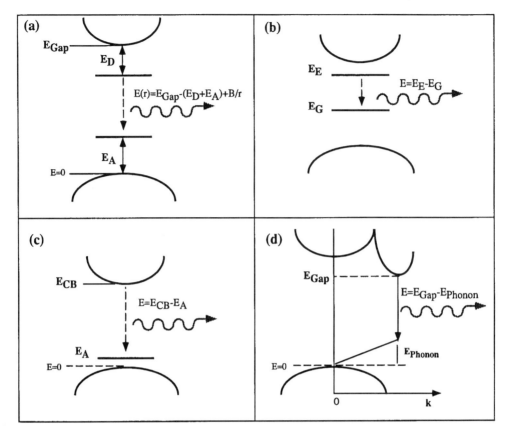

Figure 3 Various PL mechanisms due to: (a) donor–acceptor recombination, (b) transition between impurity energy levels, (c) conduction band and acceptor (or impurity) transition, (d) conduction to valence band indirect recombination.

when spontaneous emission is the only channel of decay to the ground state, the lifetime of the transition is inversely proportional to B_2.

In the above discussion the physical system of the states E_G and E_E in Figure 2 was not specified. Various PL mechanisms in semiconductors may arise depending on the exact physical system. For example, in semiconductor materials these two energy states can be associated with the acceptor and donor levels which lie in the band gap of the material. Upon an electron transition from the donor level to the acceptor level a quantum of energy may be dissipated into light and give rise to luminescence. The energy released via the donor–acceptor transitions can be expressed as a Coulomb interaction of the form: $E(r) = E_G - (E_D + E_A) + B/r$ where E_G, E_D, and E_A are the gap, donor, and acceptor ionization energies, respectively, r is the separation of the donor and the acceptor in the matrix, and B is a constant. Figure 3a to d schematically represents four possible routes for luminescence to occur. As in Raman scattering, the resulting line shape of the PL process in a high-quality crystalline material is a Lorentzian the width of which is in this case related to the lifetime of the transition.

3. THE DIAMOND AND GRAPHITIC PHASES IN DIAMOND FILMS

3.1. Diamond and Graphitic Bonding in Carbon Films

Carbon films produced by vapor deposition techniques have been characterized as diamond[34-36] or diamond-like.[37-40] The difference between the two categories lies in the types of atomic bonding configurations that the carbon atoms possess as a result of the growth conditions.

Raman spectroscopy has been used to investigate the various bonding configurations in the carbon films and to classify the composite nature found in a given film.[8] The latter is of great interest in semiconductor applications since the various types of microstructures and carbon forms in a particular

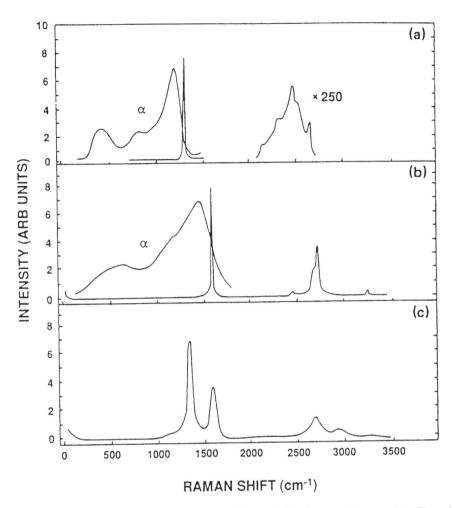

Figure 4 The Raman spectra of (a) diamond, (b) graphite, and (c) microcrystalline graphite. The solid lines in (a) and (b) represent the first- and second-order spectra of crystalline diamond and graphite, respectively. The line in (a) labeled α is the spectrum of amorphous Si scaled to the diamond frequency to represent the spectrum of amorphous sp³-carbon, while the line in (b) labeled α is that due to amorphous sp²-carbon. (From Nemanich, R. J. et al., *J. Vac. Sci. Technol. A,* 6, 1783, 1988. With permission.)

film may be a controlling factor in electronic transport as well as in the electronic emission properties. According to this work, the carbon films can be in four different possible forms (that may coexist in a given film as a composite): sp³-carbon in either amorphous or crystalline (diamond) form, and sp²-carbon also in either amorphous or crystalline (graphite) structures. In the study four possible types of carbon bonding were investigated; in addition, the influence of possible structures on the spectra was analyzed, and in light of the results the bonding types in diamond and diamond-like films were identified. The following presents a more detailed review of the Raman investigations carried out in these studies.[8] Related studies that have supported these findings have been reported by Hyer et al.,[41] Knight and White,[42] and Kobashi et al.[43]

The first- and second-order Raman spectra of diamond crystal[44] and graphite crystal[45] are compared in Figure 4a and b. The sharp features at 1332 and 1580 cm⁻¹ are the first-order modes of diamond and graphite crystals, respectively. These frequencies are indicative of the different bond strengths for which the graphite bonding is the strongest of the two. The bonding in graphite exhibits one of the largest anisotropies of any solid. The nearest-neighbor C–C bonding in graphite is considerably stronger than the C–C bond in diamond.[46] In contrast, the bonding between the graphite planes is very weak and exhibits a van der Waals character.[46] The very weak and very strong bonding in graphite yields both unusually low (42 cm⁻¹) and high (1580 cm⁻¹) frequency contributions to the lattice vibration spectrum.[45]

The second-order Raman feature is due to the scattering mechanism which involves two phonons with opposite crystal momentum vectors of nearly equal magnitudes. A detailed investigation of the second-order Raman scattering of crystalline and microcrystalline graphite is presented in Reference 45.

Samples considered to be amorphous sp^2-carbon have been reported by Solin and Kobliska.[47] The spectrum (labeled α in Figure 4b) of the film, overlaid with that of the crystalline graphite, is shown in the figure. The spectra of amorphous films consisting of pure disordered sp^3-carbon structure have not been reported, and the Raman spectra of such material could only be speculated to be similar to amorphous silicon. This is a reasonable approximation since amorphous silicon consists of a disordered state of tetrahedral sp^3-type bonding. The inferred spectra of amorphous sp^3-carbon is shown in Figure 4a.

3.1.1. Composite Properties

It has been suggested that three different composite structures might exist in the carbon films: the micron-scale, the microcrystalline, and the atomically disordered network. The characteristics of the Raman scattering from each structure have been investigated in Reference 8. The following discussion outlines some aspects of the structural properties of the films and the related vibrational excitations.

Micron-scale composite is defined as a film consisting of ~1 μm domains of diamond as well as of graphite. The decay length of the phonons is in general <1 μm; therefore, boundary scattering will not significantly affect the phonon lifetime. This means that the vibrational excitations exhibit the same spectral response as the bulk materials, i.e., the same Raman frequencies. However, the optical properties of graphite (which is absorbing material) and of diamond (which is transparent material) are different and as such will affect the Raman scattering intensities. In general, the Raman cross section, and hence the intensity of absorbing materials, is enhanced due to the resonance effects. It has previously been reported that the cross section of the 1580 cm^{-1} mode of graphite is ~75 times larger than that of diamond when normalized for carbon sites.[48] Thus, Raman scattering from similar volumes of graphite and diamond will emphasize the graphite structure.

Microcrystalline (or nanocrystalline) samples can exhibit domains of various sizes which are usually much less than 1 μm; the domains that are in this nanometer scale have a strong effect on the Raman scattering, as is discussed next. The phonon decay length is often larger than the domain size dimension; thus, the boundary scattering causes the vibrational excitations to exhibit lifetime broadened peak widths. In addition, from the Heisenberg uncertainty principle, the wave vector of the excitation is uncertain ($\Delta k \sim 2\pi/d$, where d is the domain size), and the momentum selection rules of the Raman process are relaxed.[29] The domain size effects are pronounced on microcrystalline graphite as is shown in Figure 4c. The Raman lines of the microcrystalline graphite exhibit broader peaks than those of the crystalline graphite. Furthermore, a new band in the first-order spectrum can be observed at 1355 cm^{-1} which is attributed to the uncertainty of the wave vector of the vibrations due to the finite size of the domains. An additional band can also be observed at 2940 cm^{-1} in the second-order spectrum. As in the micron-scale composite, the Raman cross section of the graphite component in the microcrystalline domain is ~75 larger than that of diamond.

The last structure to be considered is the atomically disordered (amorphous) network. In the case of carbon material, the amorphous network may possibly be composed of sp^2 and sp^3 bonding.[49] The vibrational excitations would not be confined to a single atomic site, and thus would represent an average of the network possibilities. The optical properties would not strongly favor enhancement of the sp^2 over the sp^3 bonding, and the Raman cross sections are speculated to be similar. The anticipated Raman spectrum should be very broad extending from ~1100 to 1600 cm^{-1}. Extensive studies on amorphous carbon networks have been conducted and are ongoing,[49-51] but at present there is yet to be achieved a complete understanding of its complex nature.

3.1.2. Diamond and Diamond-Like Films

Lastly, the properties of diamond-like and diamond film have also been investigated.[8] The Raman spectra of two diamond-like films produced under different deposition conditions are shown in Figure 5. The spectra presented in the figure are not similar to any of the spectra previously discussed as characteristic of amorphous or crystalline sp^2-carbon or sp^3-carbon (see Figure 4a and b). Although the 1590-cm^{-1} band can be attributed to graphite, the 1355-cm^{-1} band has no apparent origin.

One argument against the assignment of the 1355-cm^{-1} line to diamond is that its frequency is higher than any of the vibration frequencies of the diamond lattice. While effects such as strain are known to shift frequencies to a higher wavelength, the observed large shift would indicate an infeasible compressive stress of ~12 GPa. The other argument that this peak is most likely not due to the diamond structure is

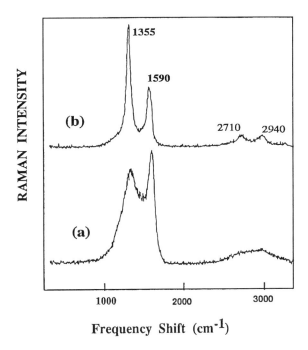

Figure 5 The Raman spectra of diamond-like films produced with different deposition conditions. Both spectra exhibit the 1355- and the 1590-cm⁻¹ bands attributed to the disordered and to the ordered mode of graphite, respectively. (From Nemanich, R. J. et al., *J. Vac. Sci. Technol A*, 6, 1783, 1988. With permission.)

based on the interpretation of the second-order spectrum of graphite and the spectrum of microcrystalline graphite samples. The strongest band in the second-order spectrum of crystalline graphite is at 2710 cm⁻¹ which is also present in the spectra of the diamond-like films. This band at twice the frequency of 1355-cm⁻¹ indicates that it is a first order Raman band related to graphite. Since the 1355-cm⁻¹ band is also observed in the spectrum of the microcrystalline graphite (due to small-domain-size activation), it may be concluded that the 1355-cm⁻¹ band in the diamond-like sample is due to graphite present in the small domains of the microcrystalline sample.

The conclusion that the Raman bands of diamond-like films at 1355 and 1590 cm⁻¹ are due to graphite bonding does not exclude the presence of regions of diamond structure. Electron energy loss experiments have shown features which can be attributed to diamond structures. The origin of the inconsistency between the two spectroscopies may possibly lie in the Raman cross section which may limit the detection of diamond forms.

The bond types of diamond thin films are substantially different from those of diamond-like films; these intrinsic differences are manifested in the Raman spectra. A Raman spectrum of microcrystalline diamond film is presented in Figure 6. The spectrum shows a sharp Raman band at 1332 cm⁻¹ and a broader Raman band at ~ 1500 cm⁻¹. The spectral properties of the 1332-cm⁻¹ band indicate that this band is due to micron-scale domains of diamond or large regions of diamond present in the microcrystal. The 1500-cm⁻¹ band, which is often referred to as the graphitic band, is ascribed to the amorphous network of sp^2 bonding (possibly combined with a smaller component of sp^3 bonding). The sp^2 amorphous network (the graphitic phase) may exist between the diamond crystalline domains on grain boundaries or may exist as inclusions.

3.2. Analytical Raman Spectroscopy of Diamond Films

The Raman intensity may be a quantitative measure of the relative concentrations of the constituents of a composite since the intensity depends on the absorption coefficient, a quantity which is related to dimension. Shroder et al.[9] have developed a model in which the concentrations of diamond and graphitic components (a transparent material and an absorbing material, respectively) in diamond films may be obtained from the Raman intensities. The following section outlines the principal results.

The Raman intensity, I, from a material (in a backscattering geometry) has been shown by Loudon[26] to follow the relation:

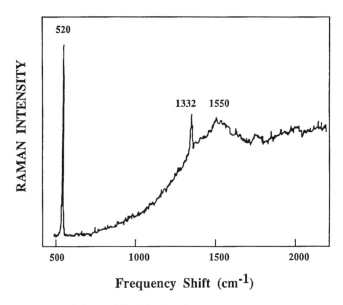

Figure 6 The Raman spectra of diamond film. The band at 1332 cm⁻¹ is due to the diamond bonding and the band at ~1550 cm⁻¹ is due to graphitic bonding. The sharp feature at 520 cm⁻¹ is due to the silicon substrate. (From Nemanich, R. J. et al., *J. Vac. Sci. Technol. A*, 6, 1783, 1988. With permission.)

$$I = \frac{I_0 S}{S + \alpha_1 + \alpha_2} \left\{ 1 - \exp\left[-(S + \alpha_1 + \alpha_2)L \right] \right\} \tag{14}$$

where S is the scattering efficiency, I_0 is the incident intensity, L is the sample thickness in the direction of the incident laser light, and α_1 and α_2 are the absorption coefficients at the frequencies of the incident and scattered light, respectively. Wada and Solin[48] showed that the equation could be modified to give the ratio of the Raman intensities of two different materials:

$$\frac{I_D}{I} = \frac{I_{0D}}{I_0} \left[\frac{A_D}{A} \right] \left[\frac{L_D(\alpha_1 + \alpha_2)}{1} \right] \left[\frac{\Delta\Omega_D}{\Delta\Omega} \right] \left[\frac{1 - R_D}{1 - R} \right]^2 \left[\frac{\left\{ \sum_j \left(\hat{e}_2 \cdot R_j \cdot \hat{e}_1 \right)^2 \right\}_D}{\left\{ \sum_j \left(\hat{e}_2 \cdot R_j \cdot \hat{e}_1 \right)^2 \right\}} \right] \tag{15}$$

where S has been redefined in terms of a scattering efficiency (A), and a summation over the inner product of the Raman tensor (R_j) and the polarization unit vectors of the incident and scattered light, $\hat{e}_1$ and $\hat{e}_2$. I is the scattering intensity of the material being compared; I_D represents the scattered intensity from the diamond, and I_0 is the incident intensity. $\Delta\Omega$ is the solid angle into which light is scattered, and the term involving R is a correction term for reflection of the scattered light at the sample surface and multiple reflections within the sample. Here, α_1 and α_2 are the previously defined absorption coefficients of the material to be compared with diamond since it has been assumed that diamond is transparent to the visible laser radiation.

In order to investigate the Raman intensity from a composite, Shroder et al.[9] prepared samples consisting of compressed powder of ~1 μm diamond and ~40 μm graphite particles for which the concentrations in each sample were known. In order to apply Equation 15 to the diamond–graphite composite samples several approximations relevant to the experimental conditions had to be made. The first approximation was that the values of L_D and $\Delta\Omega$ were the same for both materials since the Raman signal was being collected from a region of discrete particles. Moreover, the reflection losses due to light scattering between the graphite and diamond particles were assumed to be minimal and thus were disregarded. Finally, because of the random orientations of the particles, an angle-averaged value of the summation over all possible polarization directions was taken.

In light of these approximations, the ratio of the Raman scattering intensities of diamond to graphite may be given as:

$$\frac{I_D}{I_G} = \frac{4A_D'N_DV_D}{3A_G'N_GV_G'} = \frac{4A_D'N_DV_D}{3A_G'N_GV_G}\left[\frac{V_G}{V_G'}\right] \tag{16}$$

where A' is the angle- and polarization-averaged scattering efficiency per nearest-neighbor bond, N is the atomic density, and V_D and V_G' are the volumes of the diamond and graphite, respectively, which are sampled by the Raman scattering. The absorption factor of graphite is accounted for in the factor V_G'/V_G which represents the fraction of each graphite particle sampled in the Raman process. Such a discriminating proportionment of volume is not applicable to diamond since it is essentially transparent to the laser light, so the entire volume of diamond is sampled by the Raman scattering. The factor of 4/3 accounts for diamond having four nearest neighbors and graphite having three nearest neighbors at each site. Equation 16 can be written in terms of the percentage of diamond in the composites, P_D, as follows:

$$\frac{I_D}{I_G} \sim \frac{4A_D'}{3A_G'}\left[\frac{P_D}{1-P_D}\right]\left[\frac{V_G}{V_G'}\right] \tag{17}$$

At 514.5-nm laser excitation, graphite has an absorption depth of ~30 nm.[48] However, because the scattered light must also exit the absorption region, an absorption length of ~15 nm was considered instead (see Figure 7A). The ratio of the scattering efficiencies, A_D'/A_G', was taken to be ~1/75.

Equation 17 states that for a given composite the relative Raman intensities are modulated not only by the relative Raman scattering efficiency but also by the volume of the absorbing component, which is actually sampled by Raman scattering. The effect of the absorption on the Raman spectra can be seen qualitatively in the spectra presented in Figure 7B. In the figure, the spectra of diamond–graphite composites (of ~1 μm diamond and ~40 μm graphite particles) are shown for which the relative concentration of diamond in the samples ranges from ~1% up to 50%. An interesting aspect of these spectra is that the ~1% diamond composite displays a 1:1 ratio between the peak intensities of diamond and graphite. At 50% diamond concentration, it can be seen that the peak due to graphite has practically disappeared even though the Raman cross section is 75 times larger in graphite than in diamond. Thus, the absorption of graphite has a significant effect on the Raman spectra of the composites, as is shown in the large disparity of the measured intensities of the samples.

The quantitative predictions of the model, as given in Equation 17, are evidenced in Figure 8, where the ratio of the two peaks vs. concentration of diamond is plotted for various particle sizes of graphite. It can be seen in the figure that when the particle size was taken to be 42 μm, the model and the experimental data are in agreement. Within the model, when the graphite domains are smaller than the absorption length, the value of V_G'/V_G becomes 1 and the effect of absorption can be ignored. The only intensity-modulation factor in that case is the relative Raman scattering efficiency. In high-quality diamond film, the graphitic domains are very small, so in applying the model in most cases the absorption effect may similarly be ignored.

3.3. Diamond Raman Line Shape of Diamond Films

The Raman frequency of a high-quality material yields information about the vibrational energy of the phonons whereas the Raman line width is a measure of the phonon lifetime. When the quality of a material is degraded because of the presence of defects, the resulting Raman line shape reflects the effect of the defects on the phonon characteristics. The focus of this section is recent investigations of Raman line shape as it relates to the various defects which are present in diamond thin films.

The Raman line width in general can be broadened via several mechanisms: the main mechanisms applicable to the diamond line are homogeneous broadening[28] and broadening due to the size effect of

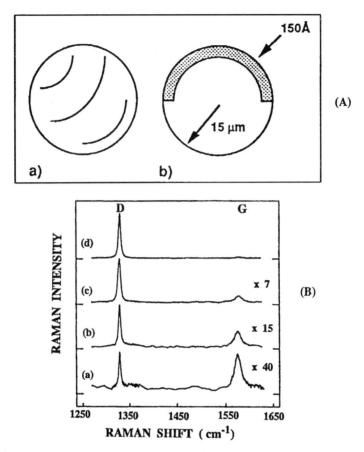

Figure 7 (A) Model of the unit volumes in the Raman scattering from composites: (a) fully illuminated diamond particle of ~30 μm size and (b) graphite particle ~30 μm size, partially illuminated in a 15-nm surface layer. (B) The Raman spectra of the composites of diamond and graphite powders. The relative concentrations of diamond in the samples are (a) 1.3%, (b) 6.6%, (c) 21.5%, and (d) 50%. The diamond band (D) is at 1332 cm⁻¹ and the graphite band (G) is at 1580 cm⁻¹. (From Shroder, R. E. et al., *Phys. Rev. B,* 41, 3738, 1990. With permission.)

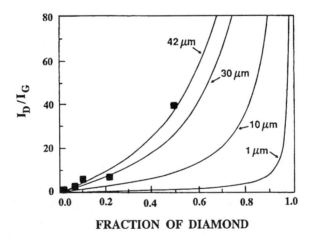

Figure 8 Ratio of peak intensities (I_D/I_G) vs. relative concentration of diamond in the composite samples. The solid lines are derived from Equation 17 assuming an average graphite particle size of 42, 30, 10, and 1 μm, respectively. (From Shroder, R. E. et al., *Phys. Rev. B,* 41, 3738, 1990. With permission.)

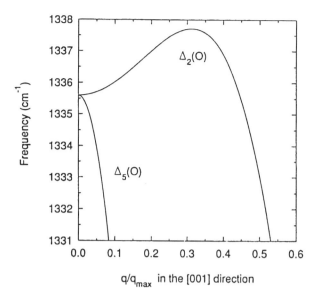

Figure 9 Detail of the (001) phonon-dispersion curves for diamond.[55] The $\Delta_5(O)$ has a maximum at Γ point. The $\Delta_2(O)$ curve exhibits a shallow maximum away from the Γ point. (From Ager, J. W. et al., *Phys. Rev. B*, 43, 6491, 1991. With permission.)

the crystal, a theory which has been developed to explain the line shape of boron-nitride material.[29] Homogeneous broadening arises from a decrease of the lifetime of the crystal phonons. The theory of homogeneous spectral line shape predicts that the line width is inversely proportional to the phonon lifetime and that the line shape is expected to be a Lorentzian.[27] The other possible mechanism which results in Raman line broadening is phonon confinement in a small domain size.[29]

The well-established confinement model is based on the uncertainty principle, $\Delta k \sim 2\pi/d$, which states that the smaller the domain size d, the larger the range of different phonons (with different q vector and different energy) that are allowed to participate in the Raman process. Hence, the broadening of the Raman line in this case is due to the spread in phonon energy, and the line shape reflects the shape of the phonon-dispersion curve. The Raman line shape in the phonon-confinement model is given by

$$I(\omega) \cong \int_0^1 \frac{dq\, \exp\!\left(-q^2 L^2/4\right) 4\pi q^2}{\left[\omega - \omega(q)\right]^2 + \left(W_0/2\right)^2} \tag{18}$$

where L is the confinement size, W_0 is the diamond natural line width (~ 2 cm^{-1}), and $\omega(q)$ is the phonon-dispersion curve of the form $A + B \cos(q\pi)$.[29,52-54] In general, the width, shape, and peak position are dependent on phonon-dispersion curves. In particular for silicon, the above model predicts that, as the Raman line gets broader, the peak of the line shifts to a lower frequency and the line shape becomes asymmetric.[53]

Ager et al.[52] have investigated the application of the confinement model to the Raman line shape of diamond thin films. For each of the relevant phonon-dispersion relations, Ager et al. modeled the expected line shape using the confinement model (Equation 18) and compared the results with the Raman line shape of diamond films. The study of the expected Raman line involved three sets of line shape calculations in which different one-dimensional dispersion curves were used. First, in case (a), a dispersion relation was used for which the phonon frequency decreases away from the Γ point. The dispersion relation has the form $\omega(q) = A + B \cos(q\pi)$, where A = 1241.25 cm^{-1} and B = 91.25 cm^{-1}; the shape of this curve is similar to that used in Si and GaAs phonon-confinement calculations. For case (b), the $\Delta_2(O)$ dispersion curve was used; this curve in diamond has a shallow maximum which is farther from the Γ point. Figure 9 depicts the characteristics of the first two dispersion relations.[55] Lastly, in case (c), the weighted averaged three-dimensional dispersion curve was used in the calculations (see Reference 52 for a more detailed analysis).

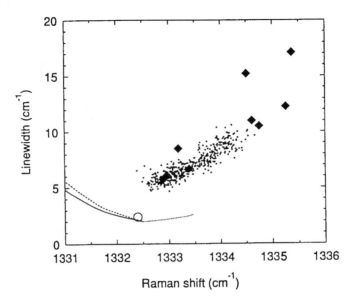

Figure 10 Raman line width vs. Raman frequency for diamond: (circle) single crystal; (diamonds) single-point measurements from ten diamond films; (dots) 500 spatially resolved measurements from one sample; (solid line) phonon-confinement theory as predicted from case (a); (dotted line) the predicted model case (b), and (long dashes) of case (c). The experimental line width is an increasing function of the Raman peak position, an opposite behavior to the one expected from the confinement model. (From Ager, J. W. et al., *Phys. Rev. B*, 43, 6491, 1991. With permission.)

Figure 10 presents the Raman line width vs. the Raman frequency, as obtained from Equation 18, for the three cases of dispersion relations. The Raman line width vs. frequency correlation which was obtained from diamond films is also presented in the figure. Figure 11 shows the predicted Raman line shapes as calculated from the phonon-confinement theory for various values of domain size, L, and for dispersion relations (a) and (b). In Figure 12 the Raman spectra of diamond films and crystal are presented. As can be seen in the figures, there is no agreement between the Raman diamond line shapes obtained from diamond films (which are symmetric) and those predicted from the confinement model (which are asymmetric). From these results Ager et al. have concluded that the size effect in diamond films is not a dominant factor in determining the line shape characteristics of diamond films. They suggested that the internal compressive stress in diamond is a more likely mechanism in determining the Raman line shape. Results similar to those obtained by Ager et al. also have been reported by Bergman and Nemanich;[15] moreover, in the latter study the Raman line shape was found to be correlated to the internal compressive stress imposed, for most part, by the graphitic phase present in the diamond thin films.

3.4. Effect of the Graphitic Phase on the Luminescence of Diamond Films

This section presents a study of the broadband and the nitrogen-related luminescence in chemical vapor deposition (CVD) diamond films.[56] A broadband luminescence extending from approximately 1.5 to 2.5 eV and centered at ~2 eV has been observed in various PL studies of diamond films.[57,58] A complete model has yet to be formulated to explain the origin of this broadband PL. Studies utilizing CL and absorption spectroscopy of crystal diamonds of types Ia and Ib which contain nitrogen have shown that similar luminescence has its origin in the electron–lattice coupling (vibronic interaction) of nitrogen-related centers with zero phonon lines (ZPL) at 1.945 and 2.154 eV.[59,60] Luminescence studies on natural brown diamonds[61] have shown that the brown diamonds luminesce in the yellow and in the red region of the spectrum. The luminescence appears in the optical spectra as wide bands centered at ~2.2 and ~1.8 eV and is very similar to the one observed in the spectra of the CVD diamond films. The origin of the PL bands of the brown diamonds has also been determined to be of vibronic nature with numerous ZPL, the principal ones at 2.721 and at 2.145 eV.[61,62]

An alternative mechanism which could give rise to the broadband PL in the CVD diamond films is the amorphous phase of the sp^2-bonded carbon (also called the graphitic phase), the presence of which

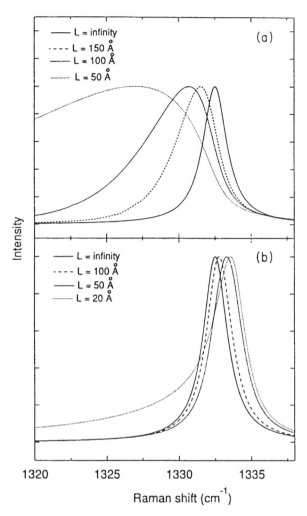

Figure 11 Calculated Raman line shapes from the phonon-confinement theory for different values of domain size L. (a) Results when using dispersion relation described in case (a); (b) the results of case (b). (From Ager, J. W. et al., *Phys. Rev. B,* 43, 6491, 1991. With permission.)

has been widely confirmed. The PL of amorphous carbon films exhibits an emission centered at ~1.8 to 2 eV which is of similar line shape to that observed in the diamond films.[49,63] According to the general model of the state distribution of amorphous materials,[64] the distortions of bond angles and of bond lengths which constitute the amorphous phase introduce a continuous state distribution in the optical band gap of the material. The PL of amorphous carbon films has been suggested to originate in the optical transitions of an in-gap state distribution related to the disordered forms of the sp²-carbon bonding.[49]

In a study, which is outlined here, the presence of an in-gap state distribution due to the sp² bonding has been established, and this was suggested to be the likely cause of the broadband luminescence.[56] The first part of the study focused on obtaining the PL spectra of nitrogen-doped and undoped diamond films, identifying the nitrogen-related PL bands, and examining the influence of the nitrogen on the broadband PL. In Figure 13, the PL spectra of the nitrogen-doped and undoped diamond films are shown. Both spectra were obtained utilizing the 514.5-nm green line of the argon laser. The PL spectra are shown on an absolute energy scale. The PL spectrum of the undoped diamond film exhibits the fairly smooth broadband line shape centered at ~2.05 eV and also exhibits the 1.681-eV band which has been attributed to an optical transition in an Si complex center.[65] However, the spectrum of the nitrogen-doped film indicates a red shift of the broadband luminescence as well as a line shape change. Furthermore, the nitrogen-related bands at 2.154 and 1.945 eV are present. Studies carried out by Davies and Hamer[60]

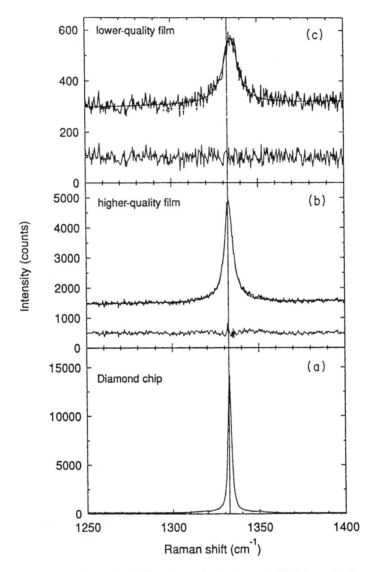

Figure 12 Raman spectra of diamonds. (a) Type IIa synthetic diamond; (b) high quality diamond film, and (c) lower-quality diamond film. The lines exhibit a Lorentzian line shape. (From Ager, J. W. et al., *Phys. Rev. B, 43,* 6491, 1991. With permission.)

have suggested that the 1.945-eV band is due to the substitutional nitrogen-vacancy optical center. Collins and Lawson[59] proposed that the 2.154-eV band is the result of a transition in a center consisting of a single substitutional nitrogen atom with one or more vacancies. Yet another PL band at 1.967 eV is also present in the spectrum (barely distinguishable from the 1.945-eV band), which might also be due to a nitrogen-related center.

In order to examine in further detail the line shape of the broadband PL, the 457.9-nm blue laser line was used for excitation. Figure 14 shows the spectra of the nitrogen-doped and undoped diamond films for this laser frequency. The broadband PL of the undoped diamond film retained its relatively unstructured line shape; however, the maximum intensity is shifted toward higher energy and is centered at ~2.2 eV. The spectrum of the nitrogen-doped diamond film exhibits the nitrogen-related bands at 2.154 and at 1.967 eV. The 1.945-eV band, which appeared with the 1.967-eV band as a doublet in the spectrum obtained using the green laser line, cannot be clearly distinguished in the spectrum taken using the blue line. The overlapping of the two bands is a resolution artifact of the scaling of the spectrum taken using the blue laser line. A relatively wide band with line width ~0.3 eV centered at 2.46 eV is also present. Similar wideband CL has been observed in both natural and synthetic diamonds and is commonly referred

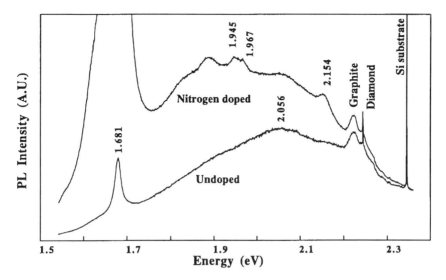

Figure 13   The PL spectra of the nitrogen-doped and of the undoped diamond films employing the 514.5-nm laser line. Raman bands are labeled as to origin and the peak energies of the PL bands are indicated. The PL nitrogen-related centers at 1.945, 1.967, and 2.154 eV are present in the spectrum of the nitrogen-doped sample. (From Bergman, L. et al., *J. Appl. Phys.,* 76, 3020, 1994. With permission.)

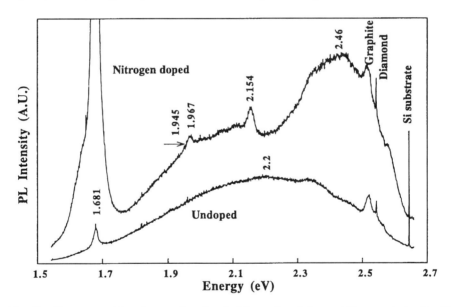

Figure 14   The PL spectra of the nitrogen-doped and of the undoped diamond films employing the 457.9-nm blue laser line. The PL nitrogen-related centers at 1.945, 1.967, 2.154, and at 2.46 are present in the spectrum of the nitrogen-doped sample. The spectrum of the undoped sample consists of broadband luminescence. (From Bergman, L. et al., *J. Appl. Phys.,* 76, 3020, 1994. With permission.)

to as *green band A* luminescence.[66] It is evident from the spectra in Figures 13 and 14 that the incorporation of nitrogen caused a distortion in the line shape of the underlying broadband luminescence. If the broadband PL had been due to a nitrogen–lattice interaction, the line shape would have been invariant, and a change in the intensity would have been anticipated.

3.4.1. The Temperature Characteristics of the Broadband PL

A second series of experiments and analysis was also conducted to further rule out the possibility of the broadband PL being of vibronic origin.[56] According to the theoretical model of the electron–lattice interaction,[32,67] the total band intensity which includes the ZPL and its vibronic sideband is expected to

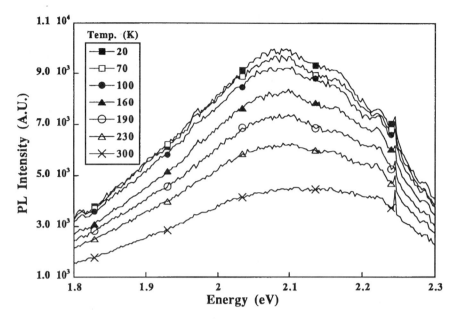

Figure 15 The spectra of the broadband PL of the undoped diamond sample at various temperatures. The spectra do not exhibit any phonon lines. (From Bergman, L. et al., *J. Appl. Phys.,* 76, 3020, 1994. With permission.)

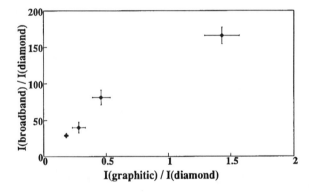

Figure 16 Correlation between the broadband PL intensity and the graphitic Raman intensity. Intensities of PL were normalized to the diamond Raman intensities. (From Bergman, L. et al., *J. Appl. Phys.,* 76, 3020, 1994. With permission.)

be independent of temperature. As the temperature increases the ZPL intensity decreases, and the vibronic band intensity is expected to increase so as to keep the total intensity constant with temperature (where the ZPL and sideband intensities are taken relative to the total band intensity). The width of the vibronic band is also expected to increase with temperature. However, as shown in Figure 15 it was found instead that the broadband PL intensity of the undoped sample exhibits an ~60% decrease with increasing temperature without any significant change in the bandwidth. Furthermore, no ZPL were present at the low-temperature spectra. However, the ZPL responsible for the vibronic band in brown diamonds as well as the ZPL of nitrogen centers are known to be sharp and well pronounced in low-temperature spectra.[66] The temperature dependence observed is thus not characteristic of a vibronic interaction.

Figure 16 shows the correlation between the Raman intensity of the graphitic phase and the intensity of the broadband PL. This correlation was found and described in detail in a previous study.[57] It was found that, as a function of growth time, as the graphitic phase increases so does the intensity of the broadband PL. It has been suggested that the amorphic graphitic phase introduces a state distribution in the band gap which provides transition centers for the photoexcited carriers, thus resulting in the broadband PL.

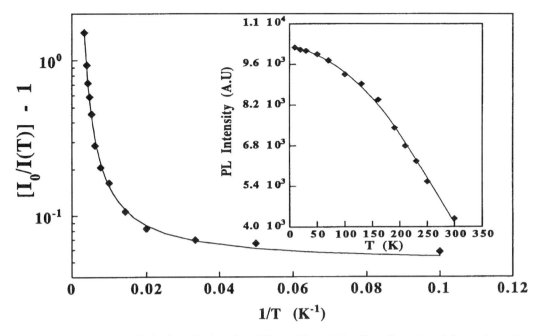

Figure 17   The functional behavior of the broadband PL vs. 1/T as obtained from the undoped diamond sample. (From Bergman, L. et al., *J. Appl. Phys.,* 76, 3020, 1994. With permission.)

In order to prove the above postulate, a further investigation of the broadband PL had to be carried out. In general, the PL process can be expressed by the following equation:

$$I / I_0 = P_R / \left(P_R + P_{NR}\right) \tag{19}$$

where P_R and P_{NR} are the probabilities for the radiative and the nonradiative recombination, respectively,[31] I is the PL intensity, and I_0 is the PL intensity for the temperature approaching absolute zero. If there exists a single activation energy E_A for P_{NR} for which the thermal quenching of the PL is of the form of a Boltzmann activated process then Equation 19 becomes:

$$P_{NR} / P_R = \left[I_0 / I(T)\right] - 1 \sim \exp\left(-E_A / k_B T\right) \tag{20}$$

By plotting $\log\{[I_0/I(T) - 1]\}$ vs. 1/T a straight line should be obtained from which E_A can be evaluated. Figure 17 shows this plot for the experimental data, I(T); the inset in the figure shows I(T) vs. temperature. The continuous curve of Figure 17 indicates the existence of a continuous distribution of activation energies E(T) rather than a single E_A associated with one energy level of a specific defect. Such a continuous distribution of activation energies E(T) indicates in turn a corresponding distribution of localized energy states in the band gap of the diamond film. E(T) may thus be viewed as corresponding to the binding energies of these localized states. The data in Figure 17 can be fitted by the equation

$$\left[I_0 / I(T)\right] - 1 \sim \exp\left(T / T_0\right) \tag{21}$$

for which T_0 is a constant to be determined. This form of quenching of the PL has also been observed and its theory developed by Street [68] in his extensive work on amorphous Si:H (a-Si:H). It should be noted that the broadband PL intensity in the diamond films exhibits a much slower decrease with increasing temperature than the PL intensity reported for a-Si:H. A smaller dependence on temperature was also reported for amorphous C:H (a-C:H)[63] with temperature dependence of the form of Equation

21, consistent with the findings presented here. According to the model developed by Street for amorphous materials, T_0 is a measure of the width of an exponential in-gap state distribution from which optical transitions can occur.[68] More extensive experiments need to be carried out to further quantify and model the state distribution and to determine the bands involved in the optical transitions. In amorphous carbon material the sp^2 bonding creates sigma-bands (σ, σ^*) and pi-bands (π, π^*) for which optical transitions can occur.[49] At present, it is hypothesized that the $\pi–\pi^*$ band transitions are responsible for the broadband PL; these bands constitute allowable optical transitions and are in the energy range closest to the laser excitation energy.

In summary, the spectra of both nitrogen-doped and undoped films exhibited the broadband PL; the nitrogen-doped sample, however, had a distortion of the line shape of the underlying broadband PL due to the vibronic interaction of the nitrogen centers. The nitrogen optical centers at 2.154, 1.945, and at 2.46 eV (the green band A) were observed, as well as a new, possibly nitrogen-related, center at 1.967 eV. The temperature behavior of the broadband PL indicates that the band does not originate from a vibronic interaction. Moreover, the intensity of the broadband PL was found to exhibit a temperature dependence characteristic of optical emission from a continuous distribution of gap states. In light of the above findings and from the correlation of the PL intensity to the graphitic phase, it is suggested that the broadband PL in diamond films is due to the optical transitions in an in-gap state distribution, where the in-gap state distribution is introduced by the amorphous phase of the sp^2 hybrid bonding.

Complementary studies of the origin of the broadband PL were recently carried out by Dallas et al.[69] and by Gangopadhyay et al.[70] utilizing time-resolved PL (TRPL) spectroscopy. In these studies the researchers measured the PL decay times for undoped and nitrogen-doped diamond films as well as the decay time of a hydrogenated amorphous carbon (a-C:H) film. The decay times of the undoped diamond film were measured at several energies in the 1.55 to 2.07 eV range of the broadband PL. The analysis indicated that the decay times at the various energies of the broadband PL are of the same magnitude, ~1.62 ns. A similar decay time (1.69 ns) was measured for the PL band of the a-C:H film. The nitrogen-doped decay time, on the other hand, was found to vary with the PL energy and was dominated by the decay time of the nitrogen centers at the specific energy; at emission energies 2.15, 2.07, 1.95, and 1.55 eV the measured lifetimes are 25.6, 24.9, 14.9, and 11.4 ns, respectively. The decay times of the nitrogen-doped film were much longer than those of the undoped diamond film.

The comparable decay times of the undoped diamond film and the a-C:H film led the researchers to conclude that both materials exhibit similar radiative recombination centers.[69] Hence, the broadband PL in the undoped diamond film was concluded to be of similar origin as the PL of the a-C:H film.

4. NITROGEN OPTICAL CENTERS

4.1. Forms of Nitrogen Impurities and Optical Centers in Diamond

The two main forms that nitrogen impurities assume in the diamond lattice are the single substitutional nitrogen and the nitrogen A aggregates.[66,71] Natural and synthetic diamonds containing the A aggregate are referred to as type Ia diamonds while those containing the single substitutional are type Ib diamonds. The energy level of the single substitutional nitrogen lies ~1.7 eV from the conduction band; the center is not luminescence active, and its presence in the diamond can be detected via absorption and electron paramagnetic resonance spectroscopy.[66] Moreover, because of the large ionization energy of the single substitutional nitrogen the center is not an active donor.

The A aggregate, which consists of two nearest-neighbor nitrogen atoms and has trigonal symmetry, is a very common form of nitrogen in natural diamond.[72-74] The trigonal symmetry of the A aggregate is depicted in Figure 18, where X and Y represent the nitrogen atoms. The energy level of the A aggregate lies ~4 eV from the conduction band, and, like the single substitutional impurity center, the A aggregate is not a luminescence-active center but can be detected via absorption spectroscopy.

In general, when nitrogen atom(s) from the single substitutional or from the A aggregate are combined with vacancies, luminescence centers are created in the diamond. There are many nitrogen-related optical centers in diamond with emission energies spanning the visible spectrum.[66,71,75] A short summary of optical centers relevant to this review is given below.

The 1.945-eV PL band arises from transitions in optical centers which consist of a substitutional nitrogen atom and a vacancy at a nearest-neighbor site (N-V); the symmetry of the center has been suggested to be trigonal C_{3v}.[66,76] Figure 18 presents the structure of the 1.945-eV optical center; the C labels represent the carbon atoms and the X and Y represent the nitrogen atom and vacancy, respectively.

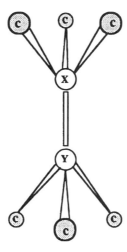

Figure 18 The trigonal symmetry. The C labels represent the carbon atoms. For the A aggregate the X and Y represent the nitrogen atoms, while for the 1.945-eV optical center the X and Y represent the nitrogen atom and vacancy, respectively.

The 1.945-eV center has been observed in crystal diamond as well as in nitrogen-doped diamond films. The actual energy level of the 1.945-eV center in the diamond band gap is as yet unknown. However, it has been established that the transitions of this optical center are between excited and ground states of a deep state that lies in the diamond gap.[66]

The other optical center which is common to both diamond crystals and nitrogen-doped films is the 2.154 eV. The symmetry of this center has been suggested to belong to one of the trigonal subgroups and the structure to consist of a single nitrogen atom and one or more vacancies.[59,77] Like the 1.945-eV center, the 2.154 eV is considered to be a deep-level optical center of intraimpurity transitions.

The 2.462-eV (H3) optical center is one of the most commonly observed centers in natural and synthetic diamond.[66] The structure of the H3 optical center consists of two nitrogen atoms from the A aggregate and one vacancy (N-V-N). Recently, Collins[71] has detected the H3 CL line from a single particle of CVD diamond, implying that small concentrations of A nitrogen can be present in this material.[71]

The band A luminescence has been observed to be present in the spectra of crystalline diamonds, as well as in those of nitrogen-containing diamond films. The band A luminescence is relatively wide (~0.5 eV), and its peak position has been reported to vary from ~2.3 to 2.7 eV, depending on the diamond type.[1,75,78-81] It has been proposed that the band A PL could be a result of donor–acceptor pair recombination transitions.[82] In the proposed mechanism the donor has been suggested to be one nitrogen atom of an A aggregate and the acceptor to be the boron impurity.

The energy released via the donor–acceptor transitions can be expressed as a Coulomb interaction of the form

$$E(r) = E_G - \left(E_D + E_A\right) + B/r \tag{22}$$

where E_G, E_D, and E_A are the diamond gap, the donor, and acceptor ionization energies, respectively, r is the separation of the donor and the acceptor in the diamond matrix, and B is a constant. Hence, the resulting luminescence line shape reflects the distribution of the donor–acceptor separation in the crystal. The energy level E_D of the A aggregate is positioned ~1.5 eV from the valence band, and the energy level of the boron E_A is ~0.37 eV from the valence band.[66] Figure 19 schematically presents the energy levels of the donor and the acceptor in the diamond gap. It should be noted that the donor–acceptor model is widely regarded as the most likely mechanism responsible for the band A luminescence, although further experimental evidence is required to establish its certainty. The only strong correlation that has been established is between the existence of the nitrogen A aggregates in the diamond and the presence of the band A in the PL spectra.[66]

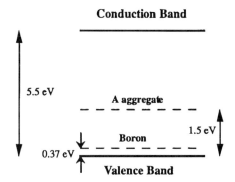

Figure 19 The energy levels of the A aggregate and boron donors in the band gap of the diamond.

4.2. Correlation of the Band A Luminescence to Structural Defects

Two luminescence broadbands have often been observed in the spectra of crystalline diamonds as well as in diamond films: a blue band with peak in the range 2.6 to 3 eV and a green band with peak in the range 2.2 to 2.5 eV, referred to as the blue band A and the green band A, respectively. Both bands have been suggested to originate from the donor–acceptor radiative recombination mechanism. The occurrence of the blue and the green bands, according to the model, depends on the mean separation of the donor–acceptor pairs: the blue emission corresponds to the closely spaced pairs, while the green emission is due to transitions between the more widely spaced pairs. It has been argued that in natural diamond the donor–acceptor close pairs resulted from diffusion of the impurities at elevated temperatures over periods of millions of years. As has been noted by Collins et al.,[65,83] it is not immediately evident how such close pairs can be produced during the growth of CVD diamond.

The blue band A is of a great interest because of the established correlation between its emission line and the presence of dislocations in the diamond.[84-86] Yamamoto et al.[86] have developed a technique which allows both high-resolution transmission electron images and optical emission (CL) to be acquired from the same isolated dislocation. This type of spectroscopy is advantageous in the investigation of a nanoscale structure since it also conveys information about the optical properties of the structure. In the experiments by Yamamoto et al. it was observed that dislocations in natural diamonds emit the blue band A CL at ~2.8 eV. The luminescence line shape characteristic was found to be similar for screw and edge dislocations, and the luminescence exhibited polarization along the dislocation line. It was suggested that the blue emission from the dislocations may be due to recombination of donor–acceptor close pairs which are arranged with variable spacing along the dislocation cores.

Graham et al.[80,87] found a direct correlation between the blue band A CL ~2.9 eV and the presence of dislocations in CVD diamond films. In their work they obtained TEM images of defect-free grains and of grains with dislocations, and the corresponding CL spectra of each grain. The band A CL was found to be prominent only in the spectra of the grains which contained the dislocations, an observation which is consistent with that reported for bulk diamond. Although the aforementioned experimental findings clearly indicate the existence of luminescent dislocations, the actual mechanism of the luminescence is as yet uncertain; the suggested donor–acceptor model, in particular, bears further investigation.

4.3. Optical Analysis of Stress Sources in Diamond
4.3.1. Luminescence and Crystal Stress: General Aspects

In diamond crystal at low temperatures the PL line shape is determined almost entirely by the crystal strain inhomogeneities. The inhomogeneous broadening is at least 1000 times greater than the homogeneous broadening,[72] the latter of which arises from the lifetime characteristics of the transition rate. In general, the principal mechanism of the inhomogeneous broadening at low temperatures is the strain broadening that arises from the presence of dislocation-type defects and/or point defects in the crystal.[72,88-90] The defects introduce strain fields throughout the crystal that interact with and perturb the energies of the optical transitions. The statistical distribution and density of the optical centers and defects in the crystal as well as the defect type, determine the variations in the transition energies of the optical centers and the respective PL line shape.

The general theory of the strain line broadening is discussed in detail by Stoneham.[90] A brief summary of Stoneham's relevant results is presented below.

1. For a symmetric lattice, in which the lattice sites accommodate at random the optical centers and defects, the line shape is expected to be symmetric.
2. When the strain in the crystal arises solely from uniformly distributed point defects, the luminescence line shape, $I(\omega)$, is expected to be a Lorentzian of the form

$$I(\omega) = I_0 / \left\{ 1 + \left[A(\omega - \omega_0) / W_L \right]^2 \right\} \tag{23}$$

where W_L is the Lorentzian line width, ω_0 is the center frequency, and A is a constant.

3. When the sources of strain in the crystal are uniformly distributed dislocations (line-type defects) the line shape is expected to be a Gaussian with line width W_G:

$$I(\omega) \sim I_0 \exp \left\{ -\left[B(\omega - \omega_0) / W_G \right]^2 \right\} \tag{24}$$

4. In the case when both types of defects are present in the crystal, the resulting line is the convoluted line shape of the Gaussian and the Lorentzian: this line shape is known as the Voigt profile and has no closed-form mathematical expression. The relative line widths of the Gaussian component and the Lorentzian component in the Voigt profile, W_L/W_G, reflect which defect is the source of dominant stress in the crystal.

The above results of Stoneham assumed an ideal continuous single crystal for which linear elasticity applies. Experimental evidence supporting Stoneham's work has been found which correlates the line shape of the luminescence to the defect type, to the defect concentration, and to the distribution of the defects (see References 88 and 90 and references within). The above optical analysis of crystal stress and strain has proved to be useful, particularly in the cases when an appreciable amount of point defects as well as dislocations are present in the crystal (see Reference 88 and references within).

4.3.2. A Aggregate as Source of Stress in Natural Crystalline Diamond

Davies[72] has investigated the 503-nm (2.462-eV) PL line shape of natural diamond crystals containing different concentration of nitrogen. According to Stoneham's theory, the expected PL line shape should be a Voigt profile for which the Lorentzian line width, W_L, would be different for each sample, reflecting the amount of nitrogen (point defects) in a given crystal. As a result, the ratio W_L/W_G for each of the PL lines is expected to have a different value. However, Davies found that the spectral line shape of many of the diamonds is a bi-Lorentzian function with line width W_{BL}:

$$I(\omega) = I_0 / \left\{ 1 + \left[A(\omega - \omega_0) / W_{BL} \right]^2 \right\}^2 \tag{25}$$

This line shape lies midway between the Lorentzian and Gaussian shapes and is close to a Voigt profile with Lorentzian and Gaussian components whose ratio $W_L/W_G \sim 0.5$ is independent of the nitrogen concentration. Moreover, the bi-Lorentzian line width W_{BL} was found to be directly correlated with the nitrogen concentration.

The results of Davies imply that the stress contribution of line defects is negligible and that the broadening mechanism is controlled entirely by the presence of nitrogen. In order to explain the effect of the stress due to nitrogen impurities on the optical centers, as manifested in the observed bi-Lorentzian line shape, Davies suggested a new model which modifies Stoneham's original theory. The model basically consists of imposing a restriction on the continuum elasticity assumed in Stoneham's theory, as well as postulating on the basis of strong experimental evidence that the nitrogen is in the form of the A aggregate. The former restriction consists of requiring a critical distance between the optical center and the nitrogen defect which exerts the strain. The main conclusions of Davies are that, in the case

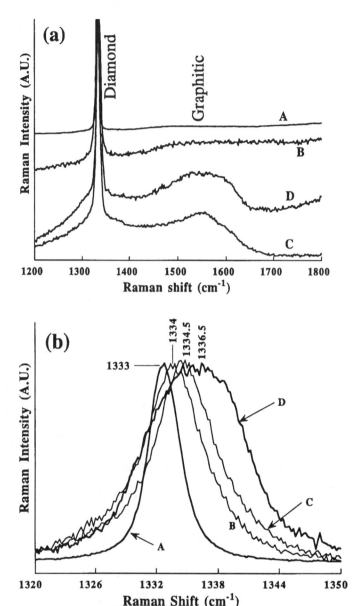

Figure 20 The Raman spectra of the diamond films: (a) showing the diamond and the graphitic signals, (b) the normalized high-resolution Raman spectra of the diamond signal of the films. Sample A was grown in the controlled combustion chamber and is ~3 μm thick, sample B was grown via the CVD method and is ~2 μm thick, samples C and D were prepared utilizing the hot filament growth method, and the substrate coverage is ~80%. (From Bergman, L. and Nemanich, R. J., *J. Appl. Phys.*, 78, 6709, 1995. With permission.)

when the strain field in crystalline diamond is predominantly due to the aggregates of nitrogen, the discrete nature of the crystal has to be taken into account. The resulting PL line shape is then a bi-Lorentzian function.

4.3.3. Optical Analysis of Stress in Diamond Thin Films
4.3.3.1. Raman Analysis
In contrast to the high-quality crystalline diamonds investigated by Davies, diamond films are known to contain a variety of defect types whose relative contribution to the stress needs to be determined. In addition to the silicon and nitrogen impurities present in diamond films (which are point-type defects),

a high concentration of line and extended defects was found utilizing TEM spectroscopy.[79,87,91,92] Yet another defect, the graphitic phase, appears as the most common defect in diamond films, although it has not yet been definitely classified as either a point or a line defect. Hence, an optical analysis as proposed by Stoneham is a potentially useful technique to classify the graphitic phase and to differentiate the contributions to the strain from each defect type.

Recently, the stress state in diamond thin films containing various concentrations of defects and impurities was investigated utilizing Raman as well as PL spectroscopy.[15] The following presents a summary of the analysis. Figure 20a shows the diamond and the graphitic Raman signals for four samples; Figure 20b shows the high-resolution normalized diamond Raman spectra. From the dependence of the Raman shift Δv on the stress σ[10,93], given by

Table 1 The Raman characteristics of the four diamond thin films

Sample	Diamond position [cm^{-1}]	$I_{DIAMOND}/I_{GRAPHITIC}$	Δv [cm^{-1}]	σ_{net} (GPa)
A	1333.0	4.6	0.7	−0.37
B	1334.0	1.8	1.7	−0.90
C	1334.5	0.7	2.2	−1.16
D	1336.5	0.6	4.2	−2.21

From Bergman, L. and Nemanich, R. J., *J. Appl. Phys.*, 78, 6709, 1995. With permission.

$$\Delta v = v - v_0 = -\alpha\sigma \tag{26}$$

where v_0 (~1332.3) is the Raman peak position of an unstressed diamond and α (~1.9 cm^{-1}/GPa) is the pressure coefficient, it was found that all of the diamond films exhibit a net compressive stress (σ_{net}). Table 1 lists the calculated stress values as well as the Raman characteristics of the films obtained from Figure 20.

Next, the sources of stress in the diamond films were analyzed by identifying the stress components and estimating the contributions to the net stress, σ_{net}, in each of the diamond films. The observable σ_{net} is given by Equation 27 in terms of the thermal stress, σ_{TH}, and the sum of the internal stresses σ_{IN}:

$$\sigma_{net} = \sigma_{TH} + \Sigma \sigma_{IN} \tag{27}$$

The thermal stress which arises from the difference in the thermal expansion coefficients of silicon and diamond is expected to be compressive in the growth temperature range of the films studied here.[94] The values of σ_{TH} were determined from the growth temperature and are listed in Table 2. The total internal stress component $\Sigma\sigma_{IN}$ in the diamond film may be due to various sources — impurities, structural defects such as dislocations, and interactions across grain boundaries. The interactions across grain boundaries, due to atomic attractive forces, have been reported to be the possible origin of the main intrinsic tensile stress, $\sigma_{IN,GB}$, in the diamond films.[94,95] The $\sigma_{IN,GB}$, has been found to be inversely proportional to the average grain diameter, d, in a film:[96]

$$\sigma_{IN,GB} = \left[E(1-v)\right](\delta/d) \tag{28}$$

where $\delta = 0.077$ nm is the constrained relaxation of the lattice constant and $E(1-v) = 1345$ GPa is the biaxial Young's modulus of diamond. The average grain sizes for the A, B, C, and D samples were ~3.5, 1, 1.5, and 2.5 μm, respectively. The values of $\sigma_{IN,GB}$ as well as the values of the previous stress components found so far are listed in Table 2. From the results listed in Table 2 it may be concluded that the thermal stress and the stress due to the grain boundaries are not very significant, and that, after compensating for their contributions ($\sigma_{calculated}$), the samples still exhibit an appreciable excess of internal compressive stress.

Table 2 The experimental and the calculated stress components of the diamond films

Sample	σ_{net} (GPa)	σ_{TH} (GPa)	$\sigma_{IN,GB}$ (GPa)	$\sigma_{calculated}$ = σ_{TH} + $\sigma_{IN,GB}$	Internal stress (GPa) = σ_{net} − $\sigma_{calculated}$
A	−0.37	−0.155 (1000°C)	+0.03	−0.125	−0.25
B	−0.90	−0.25 (750°C)	+0.104	−0.146	−0.75
C	−1.16	−0.23 (850°C)	+0.07	−0.16	−1.00
D	−2.21	−0.23 (850°C)	+0.042	−0.188	−2.02

From Bergman, L. and Nemanich, R. J., *J. Appl. Phys.,* 78, 6709, 1995. With permission.

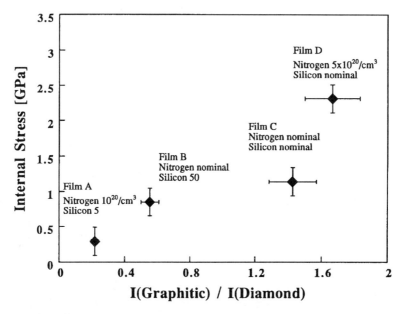

Figure 21   The correlation between the internal stress Δ and the graphitic phase Raman signal. Concentrations of the various impurities are also listed. The nitrogen concentration was obtained via SIMS while that of silicon in arbitrary units was estimated from the ratio of the intensities of the silicon PL and the diamond Raman. (From Bergman, L. and Nemanich, R. J., *J. Appl. Phys.,* 78, 6709, 1995. With permission.)

To investigate the role of the graphitic phase as a source of internal stress in the thin diamond films, the relative graphitic concentration vs. the internal stress was plotted. The graph is shown in Figure 21, and for each sample the other observed nongraphitic impurities are listed. The correlation shown in Figure 21 indicates that the graphitic phase may be a major contributor to the internal compressive stress but may not be the sole cause of the stress. The deviation from linearity is attributed to the internal stress exerted by the silicon and nitrogen impurities as well as possibly by structural defects.

4.3.3.2. PL Analysis
In order to further investigate the stress state in the diamond films, PL line shape analysis was carried out.[15] Figure 22a and b show the 2.154-eV PL bands of samples D and A, respectively, as well as the Lorentzian and Gaussian line shape (of the same widths as the PL line widths). It can be seen from the figures that both PL line shapes exhibit a high degree of symmetry. In this respect, the PL bands of the films are similar to the symmetric 2.154-eV PL line shape of Ia crystalline diamond.[72] It can also be seen in Figure 22a and b that the PL line shape of sample D is mainly a Gaussian with a very small Lorentzian component and that the line shape of sample A lies between a Lorentzian and a Gaussian function. The relative line widths of the Gaussian and Lorentzian components of each of the PL bands may be found by a deconvolution procedure and can be used to determine the respective stress contributions from line- and point-type defects. The following paragraphs outline the procedure. It has been demonstrated that the Voigt profile can be approximated by a linear combination of the Gaussian and the Lorentzian functions:[97]

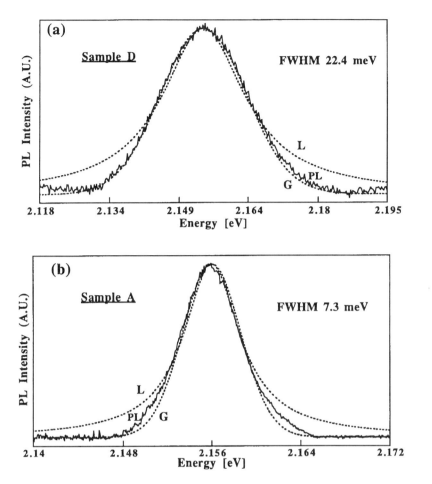

Figure 22 The Lorentzian (L) and the Gaussian (G) fits to the 2.154-eV nitrogen PL band of (a) sample D and (b) sample A. (From Bergman, L. and Nemanich, R. J., *J. Appl. Phys.*, 78, 6709, 1995. With permission.)

$$I(\omega)/I_0 = \left[1 - W_L/W_V\right] \exp\left[-2.772\left\{(\omega - \omega_0)/W_V\right\}^2\right]$$
$$+\left[W_L/W_V\right]\left\{1/\left(1 + 4\left((\omega - \omega_0)/W_V\right)^2\right)\right\}$$

(29)

where W_L and W_V are the Lorentzian and the Voigt line widths, respectively (to be determined). Once W_L and W_V have been determined from the curve fit of the given PL line to Equation 29, the following relation between the Lorentzian and Gaussian line widths:[97]

$$W_V = \left(\frac{W_L}{2}\right) + \sqrt{\left(W_L^2/4\right) + W_G^2}$$

(30)

is used to find the Gaussian line width component W_G.

Figure 23 shows that the PL line of the sample A (obtained from a high-resolution PL spectrum) can be approximated well by Equation 29; from the curve fit and Equation 30 the values of W_L and W_G were calculated to be 2.3 and 6.0 meV, respectively. The ratio of $W_L/W_G = 0.38$ implies that the Gaussian stress, which will be referred to as S_G, is approximately 2.6 times greater than the Lorentzian stress S_L. The same analytical procedure was performed on the PL band of sample D, and its results are presented

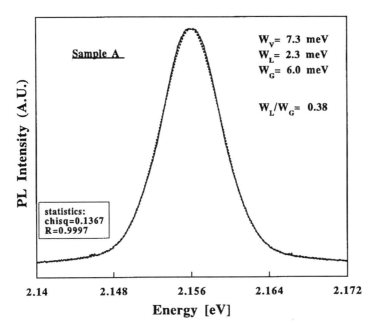

Figure 23 The high-resolution spectrum of the 2.154-eV ZPL of sample A. The curve fit to the Voigt profile (Equation 29) is represented by the dashed line. In order to get a meaningful fit, a bilateral symmetric spectrum was derived by mirroring the side which exhibits the lowest background noise. (From Bergman, L. and Nemanich, R. J., *J. Appl. Phys.*, 78, 6709, 1995. With permission.)

in Figure 24. The values of W_L and W_G shown there are 3.2 and 20.8 meV, respectively; hence, the Gaussian stress in this sample is approximately seven times greater than the Lorentzian stress.

Next, the assignment of stress values to the observed Gaussian and Lorentzian line-broadening components was performed as follows. It has been demonstrated that the total internal stress, S (= S_L + S_G), in a given diamond sample may be obtained from the line width W of a nitrogen PL band via the following approximation:[12,98]

$$S \cong W / 10 \tag{31}$$

where S is in units of GPa and the line width W is in meV. Hence, the stresses of samples A and D as obtained from the PL line widths are ~0.73 and 2.24 GPa, respectively. These stress values are of a similar order of magnitude and consistent with the compressive stress values obtained previously from the Raman shift, implying that the Raman shift and the PL line broadening result from the same compressive stress sources. By combining the results calculated via Equation 31 for the S values with the previously calculated ratios W_L/W_G ($\equiv S_L/S_G$), it was found that for sample A the Lorentzian stress S_L is ~0.20 GPa and the Gaussian stress S_G is ~0.53 GPa, while for sample D S_L ~ 0.29 GPa and S_G ~ 1.95 GPa. Table 3 summarizes the above results for both samples.

According to the results listed in Table 3, it was concluded that the samples exhibited mainly a Gaussian stress and to a much smaller degree a Lorentzian stress. Therefore, the PL investigation indicated that line-type defects contribute far more to the stress existing in the diamond thin films than do the point defects.

Based on the result obtained from the Raman analysis, that the graphitic phase is a major contributor to the internal compressive stress, it was further concluded that the dominant Gaussian stress is mainly due to the graphitic phase. One implication of this finding is that the sp^2 bonding does not act as a point defect stress source but rather as a line- or extended-type defect.

4.3.4. Stress in Diamond Grit

Raman and PL spectroscopy have also been used to investigate the stress state in diamond grit. Synthetic diamond grit consists of 100 to 1000 μm size diamond crystallites which are grown by high-pressure high-temperature (HPHT) processes. In the studies by McCormick et al.[16,99] the optical characteristics

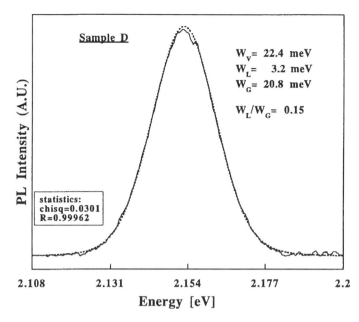

Figure 24 The high-resolution spectrum of the 2.154-eV ZPL of sample D. The curve fit of the Voigt profile (Equation 29) is represented by the dashed line. (From Bergman, L. and Nemanich, R. J., *J. Appl. Phys.*, 78, 6709, 1995. With permission.)

Table 3 The line width of the 2.154-eV PL nitrogen bands and the Gaussian and Lorentzian stress components

Sample	FWHM (meV)	Total stress S (GPa)	S_G (GPa)	S_L (GPa)
A	7.3	0.73	0.53	0.20
D	22.4	2.24	1.95	0.29

From Bergman, L. and Nemanich, R. J., *J. Appl. Phys.*, 78, 6709, 1995. With permission.

of different types of grit (i.e., prepared with different growth conditions) were correlated with the mechanical strength of the grit. The latter was measured in terms of the amount of force applied on the grit in order to initiate cracking, referred to as the crushing force. The following presents a short summary of the optical analyses of the diamond grit as described in References 16 and 99.

One of the main issues of the study was to ascertain whether or not there was a correlation between the optical centers and the strength of the crystals. The measurements attempted to correlate microscopic properties relating to the defect and impurity concentrations and configurations with macroscopic properties, namely, strength. The issue is of relevance to synthetic diamond grit since impurities like nitrogen can readily be incorporated (due to the extreme growth conditions) and strongly affect the mechanical properties of the small-volume crystallites.

In order to examine the influence of the nitrogen impurity on the strength of the grit, the relative abundance of the nitrogen optical centers of several different groups of grit were qualitatively inferred from the PL measurements. The measurements were performed with a microfocus system such that (111) faces of individual particles (crystals) could be examined. All measurements were obtained from surfaces in an unstrained area that was free of inclusions, cracks, or other visible defects. The intensity of the 1.945-eV (N-V) PL spectra of several samples is presented in Figure 25. Here, it should be noted that the PL experiments will only measure the optically active nitrogen centers (i.e., see Section 4.1), and much of the nitrogen may actually be bonded in optically inactive configurations. Figure 26 presents a correlation between the PL intensity of the 1.945-eV PL (N-V) band and the crushing force. The correlation presented in the figure indicates a negative effect of the N-V centers on the strength of the diamond grit. However, when the 2.46-eV (H3) optical center, which consists of an N-V-N configuration, was correlated with the crushing force, it was found that the H3 optical centers exert a positive influence on the strength of the diamond grit (Figure 27).

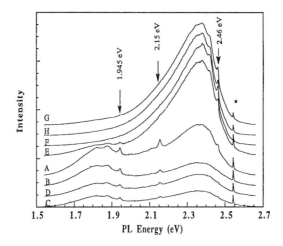

Figure 25 PL spectra of the various diamond grit samples. (From McCormick, T. L. et al., Eds., *MRS Symp. Proc. Novel Forms of Carbon II,* 349, 445, 1994. With permission.)

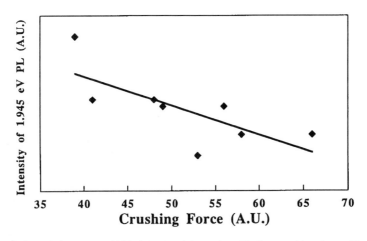

Figure 26 Correlation of the 1.945-eV PL integrated intensity with the crushing force. The PL intensity is normalized to the diamond Raman intensity. (From McCormick, T. L., Master thesis, North Carolina State University, 1994. With permission.)

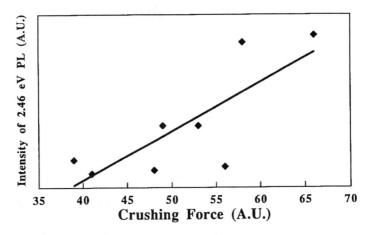

Figure 27 The correlation of the 2.46-eV PL normalized intensity with the crushing force. (From McCormick, T. L., Master thesis, North Carolina State University, 1994. With permission.)

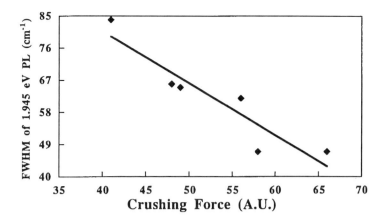

Figure 28   The internal stress (expressed in terms of the FWHM of the 1.945-eV ZPL in units of cm⁻¹) of the diamond grit in correlation with the crushing force. (From McCormick, T. L., Master thesis, North Carolina State University, 1994. With permission.)

There have been many conflicting reports on the strengthening role of nitrogen; however, the above results are in accord with the suggestion of Bokii et al.[100] that a balance of the nitrogen forms and concentrations must be achieved in order to maximize the strength. Further investigation should be initiated to explain why some forms of nitrogen act as a strengthening agent while other forms do not.

The magnitude of the internal stress in the various grit samples was estimated using Equation 31 and the line width of the 1.945-eV PL band. It was found that the internal stress ranges between 0.5 and 1 GPa and exhibits a correlation with the crushing force, as can be seen in Figure 28. The internal stress was attributed to the combined (and possibly correlated) effect of the various point and line defects, as well as to macroscopic inclusions in the grit. Figure 29 presents photographs of characteristic defect-free and defective grit samples.

Annealing studies of the diamond grit have also been conducted by McCormick[99] and Webb and Jackson.[101] Both studies employed PL in an attempt to characterize the properties of the defect and impurity structures. Figure 30 compares the PL spectra of diamonds before and after annealing. The sample was annealed at HPHT conditions (60 kbar and 1200°C, respectively) for 1 h. It can be seen in the figure that after annealing, the PL intensity of the H3 center and its vibronic sideband was somewhat diminished while the 2.154- and 1.945-eV systems exhibit a considerable increase in intensity. Figure 31 presents the spectra of the grit samples before and after the low-pressure high-temperature (LPHT) anneal (i.e., at 1 atm and 1200°C). As opposed to the HPHT, the LPHT anneal resulted in an increase in intensity of the H3 vibronic system. In addition, the study also noted changes in the strength of the diamonds depending on the annealing conditions.

The large changes in the PL demonstrate the value of this technique in exploring how defects and microstructure are affected in processing environments. The observed PL changes may be indicative of the interactions of defects and impurities within the diamond. However, previous studies have suggested that high temperature annealing may involve plastic deformation and creation of vacancies that then enhance the PL of previously inactive nitrogen sites.[12,98]

5. 1.68 eV OPTICAL CENTER

5.1. Structure

A luminescence band with a ZPL at approximately 1.68 eV, as shown in Figure 32, has been reported to be present in the PL and CL spectra of many diamond films.[15,56,57,102-106] Early studies have speculated that the 1.68-eV center may be attributed to the GR1 optical center which can be produced by radiation damage in natural and synthetic diamond and has been established to consist of a neutral vacancy.[66] This hypothesis originated from the observation that the GR1 in natural diamond gives rise to the closely proximate PL band with a ZPL at 1.673 eV. The small deviation between this PL line in natural diamond and that at 1.68 eV observed in diamond films had been attributed to the stress existing in the film which may interact with the GR1 center and shift the emission energy to 1.68 eV.[58]

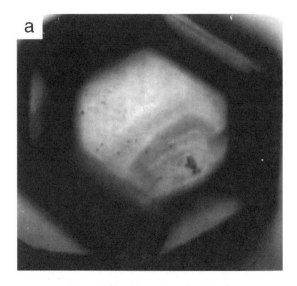

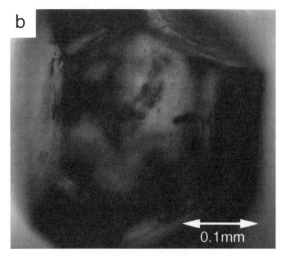

0.1mm

Figure 29 Characteristic pictures of the diamond grit. Sample (a) is relatively free of defects, and sample (b) is filled with crack and inclusion content. (From McCormick, T. L., Master thesis, North Carolina State University, 1994. With permission.)

Studies of silicon ion implantation into natural diamonds have been reported by Vavilov et al.,[103] Clark and Dickerson,[107] and Collins et al.[65] There it was found that, as a result of the silicon ion implantation, the 1.68-eV luminescence band appeared in the spectra. It was suggested by the researchers that the 1.68-eV band is due to a transition of a center involving a silicon atom.[65,103,107] Furthermore, the annealing behavior of the 1.68-eV center was found to be inconsistent with that of the GR1 center.[107] The intensity of the GR1 luminescence band diminishes significantly at ~700°C,[108,109] since the vacancy in diamond becomes mobile at this temperature, while the 1.68-eV band intensity was observed to increase.[107]

In order to examine the influence of silicon incorporation in CVD diamond films, Badzian et al.[102] have conducted experiments in which diamond was deposited on a graphite substrate, as well as others where a piece of a silicon wafer was placed on the graphitic substrate. The PL spectra obtained from the film grown in the latter experiments exhibited a significant increase in intensity of the 1.68-eV band relative to the intensity from the films grown on the bare graphite substrate. Other work[110] showed a direct correlation between the 1.68-eV luminescence intensity and the concentration of silicon in diamond films grown on silicon substrate. By now, it is well established that the 1.68-eV optical band of diamond films is not associated with the neutral vacancy but rather has a silicon atom as one of its constituents.

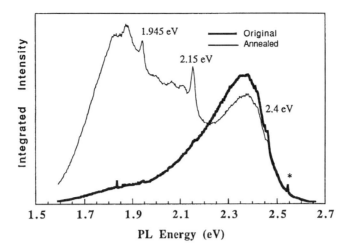

Figure 30 PL spectra of diamond grit before and after annealing at HPHT conditions. (From Webb, S. W. and Jackson, W. E., *J. Mater. Res.,* 10, 1700, 1995. With permission.)

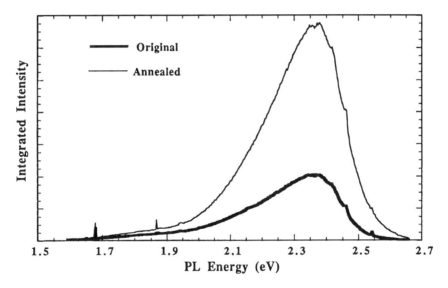

Figure 31 PL spectra of diamond grit before and after annealing at LPHT conditions. (From Webb, S. W. and Jackson, W. E., *J. Mater. Res.,* 10, 1700, 1995. With permission.)

A definitive model of the structure of the 1.68-eV center has yet to be formulated; however, two possible models have been suggested which are supported by experiment. From the observed quadratic dependence of the luminescence intensity on the silicon implantation dose, Vavilov et al.[103] have suggested that the center may contain two silicon atoms. Another suggested possible structure of the center is that of a silicon–vacancy pair:[107,111] this structure was found to be consistent with the annealing behavior reported by Clark and Dickerson.[107] In that work, the CVD diamond film was first irradiated by energetic electrons to induce local disorder and vacancies, and then annealed. The sharp increase in the 1.68-eV intensity band that was observed at the mobile vacancy temperature (700°C) led to the suggestion that the center involves a silicon atom and a vacancy.

5.2. Incorporation of the Silicon Impurity in Diamond Films: Time-Dependence Study
The deposition of diamond films on silicon substrates is a widely used method and as such the silicon substrate becomes a principal potential source of silicon atom impurities in the diamond films.[104,112] Other possible sources of silicon contaminant may be the walls of the growth chamber which in some reactors are made of quartz (SiO_2),[107] as well as residual silicon atoms which exist in the growth

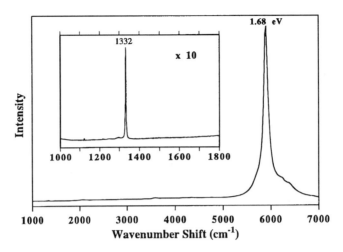

Figure 32 The characteristic room-temperature PL spectra of a diamond film showing the 1.68-eV band. The inset shows the Raman spectra of the film. (From Bergman, L. et al., *J. Appl. Phys.*, 73, 3951, 1993. With permission.)

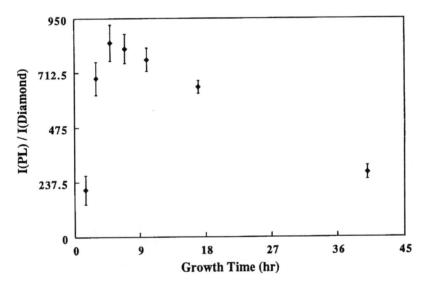

Figure 33 The relative integrated PL intensity of the 1.68 eV band as a function of deposition time. The highest intensity occurs at ~ 7 h. (From Bergman, L. et al., *J. Appl. Phys.*, 73, 3951, 1993. With permission.)

environment of the quartz and stainless-steel chambers. Although the source of silicon contaminant depends, to large extent, on the specific growth system, the general mechanisms involved in the origin of the silicon impurity and its dynamic incorporation into the diamond film is an important topic and has been investigated by several researchers.[57,104,107,112]

The formation of the 1.68-eV centers in CVD diamond film deposited on silicon substrate as a function of growth time has been the subject of investigation.[57] In this study, the PL integrated intensity of the 1.68-eV line was calculated and normalized to the diamond integrated Raman line for the consecutive growth times (1.5, 3, 5, 7, 10, 17, and 40 h) of a redeposited diamond sample. The results shown in Figure 33 indicate an initial increase of the 1.68-eV relative intensity until a maximum is reached at about 8 h; thereafter, the relative PL intensity is seen to decrease with increasing growth time. The SEM images of this sample shown in Figure 34 reveal that initially the film consists of isolated diamond particles; after ~8 h the diamond particles start to coalesce, forming grain boundaries until most of the Si substrate is covered by the growth. A possible mechanism for the effect shown in Figure 33 is that etching of the Si substrate by the plasma releases Si atoms in the gas phase and allows them to become incorporated into the growing diamond film. In the early stages when the nucleation and growth of the isolated particles take place, the probability of creating the 1.68-eV centers is high since

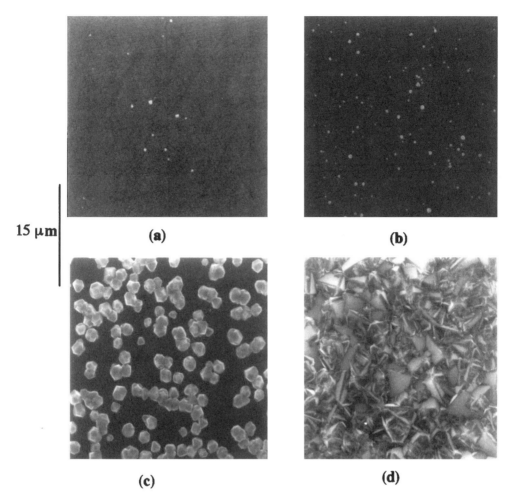

15 μm

(a)

(b)

(c)

(d)

Figure 34 SEM micrographs of the diamond sample at (a) 1.5, (b) 3, (c) 10, and (d) 40 h of deposition time. (From Bergman, L. et al., *J. Appl. Phys.,* 73, 3951, 1993. With permission.)

the Si substrate is almost entirely exposed to the plasma. As the diamond nuclei continue to grow in an isolated fashion, the concentration of the defect centers increases. At deposition times longer than 8 to 10 h, less of the Si substrate is exposed to the plasma, resulting in a reduced concentration of the 1.68-eV defect centers.

CL studies carried out by Robins et al.[112] on the formation of the 1.68-eV center in CVD diamonds have resulted in a time-dependence behavior similar to that presented in Figure 33. The observed time-dependence of the incorporation of the optical centers at early stages of growth implies that the spatial distribution of the centers is concentrated at the interface of the silicon substrate and the diamond film.

5.3. Vibronic System and Temperature Dependence of the 1.68-eV Center

As discussed in the previous sections, optical centers, when excited, can interact with the lattice vibrations through various mechanisms. One type of electron–lattice interaction is energy transfer from a photo-excited optical center to the phonons which results in a vibronic sideband spectra relative to the ZPL band. By calculating the energy difference between the sideband and the ZPL peaks, the energy of the specific phonon involved in the interaction may be obtained. Figure 35 shows the PL emission spectrum of the 1.68-eV optical line and its vibronic sidebands for a CVD diamond film as reported by Feng and Schwartz.[106] Feng and Schwartz interpreted the spectrum presented in Figure 35 as consisting of a ZPL at 1.681 eV (labeled Z) and its vibronic sidebands. Two of the dominant sidebands, the 1.639 eV and the 1.616 eV (labeled A and B, respectively), which correspond to ~42- and ~65-meV phonons, were attributed to a local vibrational mode of the center and to a transverse acoustic (TA) mode of the diamond lattice, respectively.

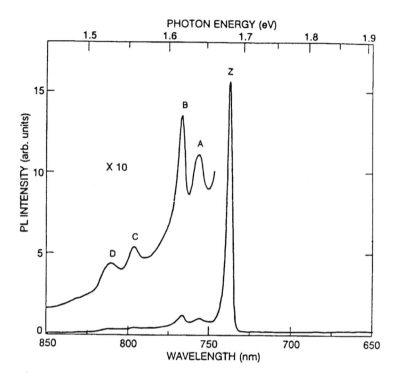

Figure 35 The 1.681-eV PL emission spectrum at 10 K of a diamond film. Peaks A to D are vibronics related to the 1.681-eV ZPL labeled Z. (From Feng, T. and Schwartz, B. D., *J. Appl. Phys.*, 73, 1415, 1993. With permission.)

According to the model of electron–lattice interaction,[32] the total band intensity which includes the ZPL and its vibronic sideband is expected to be independent of temperature. As the temperature increases, the ZPL intensity decreases, and the vibronic band intensity is expected to increase proportionally so as to keep the total intensity constant with temperature. Figure 36 shows the 1.68-eV PL band for various temperatures as reported in Reference 56. It can be observed in the figure that the conservation of the total intensity of the ZPL and its sidebands does not hold as a function of temperature, indicating a weak electron–lattice interaction and the existence of another nonradiative channel of the 1.68-eV luminescence.

Another possible mechanism through which the PL intensity of a ZPL can be quenched is known as the Boltzmann process. In this process there exists an activation energy E_A associated with a certain nonradiative decay channel in the crystal which acts to decrease the PL intensity according to the relation

$$\frac{I_0}{I(T)} - 1 \sim \exp\left(\frac{-E_A}{k_B T}\right) \qquad (32)$$

where $I(T)$ is the PL intensity at a given temperature and I_0 is the intensity for the temperature approaching absolute zero. Figure 37 shows that the sampled 1.68-eV PL intensity can be closely fitted by Equation 32, with an activation energy $E_A = 90$ meV (± 10 meV).[56] The Boltzmann process was also reported to be the dominant luminescence quenching mechanism by Feng and Schwartz[106] with $E_A \sim 70$ meV and by Collins et al.[111] with $E_A = 56$ meV. The different values of the activation energy may be due to experimental error, but one cannot exclude the possibility that different activation energies are due to differences in film quality. At present, the identity of the nonradiative channel is not known; nonetheless, in light of the above-reported findings it can be speculated to be a defect or an impurity rather than a vibronic interaction.

Complementary to PL techniques, the absorption and CL characteristics of the 1.68-eV optical system in CVD diamond films were recently investigated by Collins et al.[113] The spectrum they obtained is shown in Figure 38, and it can be seen that the center absorbs and emits at the same energy. This finding

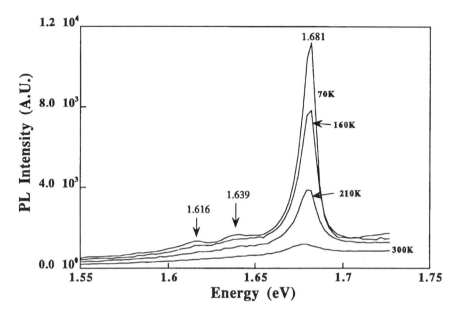

Figure 36 The 1.681-eV PL band and its vibronic sidebands of diamond film at various temperatures. (From Bergman, L. et al., *J. Appl. Phys.,* 76, 3020, 1994. With permission.)

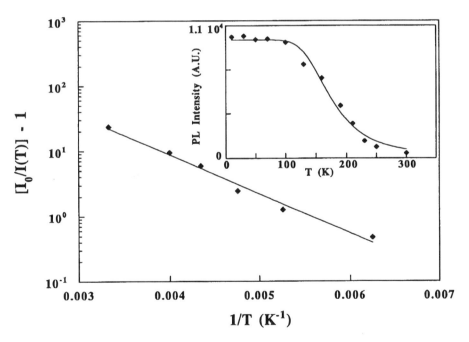

Figure 37 The functional behavior of the 1.681-eV PL intensity vs. 1/T. The inset shows the PL intensity vs. temperature. (From Bergman, L. et al., *J. Appl. Phys.,* 76, 3020, 1994. With permission.)

may be indicative of a relatively small vibronic interaction and is consistent with the lack of conservation of intensity of the vibronic center observed in Reference 56.

6. CONCLUDING REMARKS

Raman and luminescence spectroscopy have become effective and well-established techniques for the characterization of diamond films. This chapter has presented a review of the various Raman and luminescence techniques employed in the study of defects and impurities in diamond films.

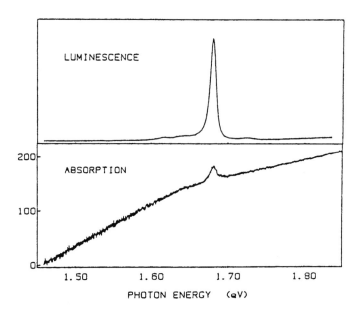

Figure 38 CL and absorption spectra of the 1.681 ZPL of diamond film. The sample temperature was 77 K for both. (From Collins, A. T. et al., *Proc. Mater. Res. Soc. Symp.,* 162, 225, 1990. With permission.)

One of the main applications of Raman spectroscopy is the identification of the sp^2-graphitic and sp^3-diamond phases which may coexist in a diamond film. Moreover, the relative concentrations of the two phases, a measure of the film purity, may be calculated via the Raman intensities.

Raman together with PL spectroscopy may also be utilized as complementary optical characterization techniques in order to investigate the stress state of a diamond film. When both techniques have been employed in the investigation of the stress in thin (<3 μm) films, it was found that the internal stress is compressive and is due mainly to extended and line-type defects, in particular the graphitic phase. The source of the internal stress in high-quality natural diamond crystals, in contrast, is due to point defects consisting of nitrogen impurities of the A aggregate form. In addition to the investigation of pointlike optical centers, CL spectroscopy has been proven to be useful also in the study of line-type defects: the presence of dislocations in diamond, bulk as well as film, may be inferred from the occurrence of the blue band A (~2.8 eV) luminescence.

Other significant results obtained via Raman and luminescence spectroscopy of diamond films are the following. (1) The diamond Raman line shape is symmetric and can be approximated by a Lorentzian function. This result excludes the phonon confinement as the primary mechanism of line broadening. The mechanism proposed for the Raman line broadening is the phonon scattering at impurities and defects. (2) The broadband PL (~2 eV) is due to optical transitions of in-gap states which in turn are attributed to a disordered form of sp^2-type bonding. (3) The 1.681-eV luminescence band is attributed to an optical center containing silicon atom(s) and, possibly, a vacancy. In the case of diamond grown on a silicon substrate, most of the 1.681-eV centers reside near the silicon/diamond interface. This optical center luminesces and absorbs at the same energy, has a very weak vibronic interaction, and exhibits Boltzmann quenching with an activation energy of ~70 meV.

Among the many issues of interest which bear further investigation are the donor–acceptor recombination mechanism as an origin to the band A luminescence, the structure and symmetry of the 1.681-eV optical center, and the spatial configurations (cluster size and dimensionality) of the graphitic phase, in particular the configurations which are responsible for the stress and for the broadband luminescence.

ACKNOWLEDGMENTS

The authors gratefully acknowledge Alan T. Collins, Terri McCormick, and Peter Mills for their contribution. We also acknowledge the support of the Office of Naval Research.

REFERENCES

1. **K. V. Ravi,** *Mater. Sci. Eng. B* 19, 203–27, 1993.
2. **J. E. Field,** Ed., *The Properties of Natural and Synthetic Diamond*, Academic Press, New York, 1992.
3. **G. Davies,** Ed., *Properties and Growth of Diamond*, INSPEC, Institute of Electrical Engineers, London, 1994.
4. **G. E. McGuire,** in *Semiconductor Materials and Process Technology Handbook*, G. E. McGuire, Ed., Noyes, Park Ridge, NJ, 1987.
5. **L. H. Robins, E. N. Farabaugh, and A. Feldman,** *SPIE Vol. 1325 Diamond Optics III*, 1990, 130.
6. **E. S. Etz, E. N. Farabaugh, A. Feldman, and L. H. Robins,** *SPIE Vol. 969 Diamond Optic,* 1988, 86.
7. **D. S. Knight, and W. B. White,** *SPIE Vol. 1055 Raman Scattering, Luminescence, and Spectroscopic Instrumentation in Technology,* 1989, 144.
8. **R. J. Nemanich, J. T. Glass, G. Lucovsky, and R. E. Shroder,** *J. Vac. Sci. Technol. A* 6, 1783, 1988.
9. **R. E. Shroder, R. J. Nemanich, and J. T. Glass,** *Phys. Rev. B* 41, 3738–3745, 1990.
10. **S. K. Sharma, H. K. Mao, P. M. Bell, and J. A. Xu,** *J. Raman Spec.* 16, 350, 1985.
11. **J. W. Ager, and M. D. Drory,** *Phys. Rev. B* 48, 2601, 1993.
12. **T. Evans, S. T. Davey, and S. H. Robertson,** *J. Mater. Sci.* 19, 2405, 1984.
13. **M. Yoshikawa, H. Ishida, H. Ishitani, T. Murakawa, S. Koizumi, T. Inuzuka,** *Appl. Phys. Lett.* 57, 428, 1990.
14. **M. Yoshikawa, G. Katagiri, H. Ishida, A. Ishitani, M. Ono, K. Matsumura,** *Appl. Phys. Lett.* 55, 2608, 1989.
15. **L. Bergman, and R. J. Nemanich,** *J. Appl. Phys* 78, 6709–6719, 1995.
16. **T. L. McCormick, W. E. Jackson, R. J. Nemanich, and C. L. Renschlers,** Eds., *MRS Symp. Proc. Novel Forms of Carbon II*, 349, 445, 1994.
17. **F. H. Pollak,** in *Analytical Raman Spectroscopy*, J. G. Grasselli, and B. J. Bulkin, Eds., John Wiley and Sons, New York, 1991, 137.
18. **B. J. Bulkin,** in *Analytical Raman Spectroscopy*, J. G. Grasselli, and B. J. Bulkin, Eds., John Wiley and Sons, New York, 1991, 1.
19. **M. Cardona,** *SPIE Vol. 822 Raman and Luminescence Spectroscopy in Technology,* 1987, 2.
20. **S. P. S. Porto,** in *Light Scattering Spectra of Solids*, G. B. Wright, Ed., Springer-Verlag, New York, 1969, 1.
21. **R. J. Nemanich,** *Mater. Res. Soc. Symp. Proc.* 69, 23, 1986.
22. **M. Cardona,** in *Light Scattering in Solids I*, M. Cardona, Ed., Springer-Verlag, New York, 1983, 1.
23. **A. Pinczuk, and E. Burstein,** in *Light Scattering in Solids I*, M. Cardona, Ed., Springer-Verlag, New York, 1983, 23.
24. **D. A. Long,** *Raman Spectroscopy,* McGraw-Hill, New York, 1977.
25. **R. Loudon,** *Proc. R. Soc. A* 275, 218, 1963.
26. **R. Loudon,** *J. Phys.* 26, 677, 1964.
27. **B. Di Bartolo,** *Optical Interactions in Solids*, John Wiley & Sons, New York, 1969.
28. **W. J. Borer, S. S. Mitra, and K. V. Namjoshi,** *Solid State Commun.* 9, 1377, 1971.
29. **R. J. Nemanich, S. A. Solin, and R. M. Martin,** *Phys. Rev. B* 23, 6348–6356, 1981.
30. **M. Gershenzon,** in *Semiconductors and Semimetals*, R. K. Willardson, and A. C. Beer, Ed., Academic Press, New York, 1966, Vol. 2, 289.
31. **D. Curie,** *Luminescence in Crystals*, John Wiley and Sons, New York, 1963.
32. **B. Henderson, and G. F. Imbusch,** *Optical Spectroscopy of Inorganic Solids,* Clarendon Press, Oxford, 1989.
33. **J. M. Hollas,** *Modern Spectroscopy*, John Wiley and Sons, New York, 1992.
34. **S. Matsumato, Y. Sato, M. Tsutsumi, and N. Setaka,** *J. Mater. Sci.* 17, 3106, 1982.
35. **A. Sawabe, and T. Inuzuka,** *Thin Solid Films* 137, 89, 1986.
36. **Y. Hirose, and Y. Terasawa,** *Jpn. J. Appl. Phys.* 25, L519, 1986.
37. **C. B. Zarowin, N. Venkataramanan, and R. R. Pool,** *Appl. Phys. Lett.* 48, 759, 1986.
38. **V. Natarajan, J. D. Lamb, J. A. Woollam, D. C. Liu, and D. A. Gulino,** *J. Vac. Sci. Technol. A* 3, 681, 1985.
39. **T. Mori, and Y. Namba,** *J. Vac. Sci. Technol. A* 1, 23, 1983.
40. **J. C. Angus, and F. J. Jansen,** *J. Vac. Sci. Technol. A* 6, 1778, 1988.
41. **R. C. Hyer, M. Green, and S. C. Sharma,** *Phys. Rev. B* 49, 14573, 1994.
42. **D. S. Knight, and W. B. White,** *J. Mater. Res.* 4, 385, 1989.
43. **K. Kobashi, K. Nishimura, Y. Kawate, and T. Horiuchi,** *Phys. Rev. B* 38, 4067, 1988.
44. **S. A. Solin, and A. K. Ramdas,** *Phys. Rev. B* 1, 1687, 1970.
45. **R. J. Nemanich, and S. A. Solin,** *Phys. Rev. B* 20, 392–401, 1979.
46. **A. R. Ubbelohode, and F. A. Lewis,** *Graphite and Its Crystal Compounds,* Clarendon, Oxford, 1960.
47. **S. A. Solin, and R. J. Kobliska,** *Amorphous and Liquid Semiconductors*, J. Stuke, Ed., Taylor Francis, London, 1974.
48. **N. Wada, and S. A. Solin,** *Phys. B* 105, 353–356, 1981.
49. **J. Robertson,** *Adv. Phys.* 35, 317, 1986.
50. **J. Robertson, and E. P. O'Reilly,** *Phys. Rev. B* 35, 2946, 1987.
51. **J. Robertson,** *Diamond Relat. Mater.* 4, 297, 1995.

52. J. W. Ager, D. K. Veirs, and G. M. Rosenblatt, *Phys. Rev. B* 43, 6491–6499, 1991.
53. I. H. Campbell, and P. M. Fauchet, *Solid State Commun.* 58, 739, 1986.
54. P. M. Fauchet, and I. H. Campbell, *Crit. Rev. Solid State Mater. Sci.* 14, S79–101, 1988.
55. R. Tubino, and J. L. Birman, *Phys. Rev. B* 15, 5843, 1977.
56. L. Bergman, M. T. McClure, J. T. Glass, and R. J. Nemanich, *J. Appl. Phys.* 76, 3020–3027, 1994.
57. L. Bergman, B. R. Stoner, K. F. Turner, J. T. Glass, and R. J. Nemanich, *J. Appl. Phys.* 73, 3951–3957, 1993.
58. J. A. Freitas, J. E. Butler, and U. Strom, *J. Mater. Res.* 5, 2502, 1990.
59. A. T. Collins, and S. C. Lawson, *J. Phys. Condens. Mater.* 1, 6929, 1989.
60. G. Davies, and M. F. Hamer, *Proc. R. Soc. Lond. A.* 348, 285, 1976.
61. M. E. Pereira, M. I. B. Jorge, and M. F. Thomaz, *J. Phys. C Solid State Phys.* 19, 1009, 1986.
62. M. H. Nazare, M. I. B. Jorge, and M. F. Thomaz, *J. Phys. C Solid State Phys.* 18, 2371, 1984.
63. R. C. Fang, *J. Lumin.* 48/49, 631, 1991.
64. N. F. Mott, and E. A. Davis, *Electronic Processes in Non-Crystalline Materials,* Clarendon Press, Oxford, 1979.
65. A. T. Collins, M. Kamo, and Y. Sato, *J. Mater. Res.* 5, 2507–2513, 1990.
66. G. Davies, in *Chemistry and Physics of Carbon,* P. L. Walker, and P. A. Thrower, Eds., Marcel Dekker, New York, 1977, Vol. 13, 2.
67. G. Davies, *Rep. Prog. Phys.* 44, 787, 1981.
68. R. A. Street, *Adv. Phys.* 30, 593, 1981.
69. T. Dallas, S. Gangopadhyay, S. Yi, and M. Holtz, *Applications of Diamond Films and Related Materials: Third International Conference,* A. Feldman, Y. Tzeng, W. A. Yarbrough, M. Yoshikawa, and M. Murakawa, Eds., NIST Special Publication 885, 1995.
70. S. Gangopadhyay, private communication, 1995.
71. A. T. Collins, *Diamond Relat Mater.* 1, 457–469, 1992.
72. G. Davies, *J. Phys. C: Solid State Phys.* 3, 2474, 1970.
73. G. Davies, *Nature* 228, 758, 1970.
74. G. Davies, *J. Phys. C: Solid State Phys.* 9, L537, 1976.
75. J. Walker, *Rep. Prog. Phys.* 42, 1605, 1979.
76. C. D. Clark, and C. A. Norris, *J. Phys. C: Solid State Phys.* 4, 2223, 1971.
77. G. Davies, *J. Phys. C: Solid State Phys.* 12, 2551, 1979.
78. L. H. Robins, L. P. Cook, E. N. Farabaugh, and A. Feldman, *Phys. Rev. B* 39, 13367–13377, 1989.
79. R. J. Graham, *Mater. Res. Soc. Symp. Proc.* 242, 97, 1992.
80. R. J. Graham, and K. V. Ravi, *Appl. Phys. Lett.* 60, 1310, 1992.
81. A. T. Collins, *J. Phys. C: Solid State Phys.* 13, 2641, 1980.
82. P. J. Dean, *Phys. Rev. A* 139, 558, 1965.
83. A. T. Collins, M. Kamo, and Y. Sato, *J. Phys. Condens. Matter* 1, 4029, 1989.
84. A. A. Gippius, A. M. Zaitsev, and V. S. Vavilov, *Sov. Phys. Semicond.* 16, 256, 1982.
85. S. J. Pennycook, L. M. Brown, and A. J. Craven, *Philos. Mag. A* 41, 589, 1980.
86. N. Yamamoto, J. C. H. Spence, and D. Fathy, *Philos. Mag. B* 49, 609, 1984.
87. R. J. Graham, T. D. Moustakas, and M. M. Disko, *J. Appl. Phys.* 69, 3212, 1991.
88. A. E. Hughes, *J. Phys. Chem. Solids* 29, 1461, 1968.
89. A. M. Stoneham, *Proc. Phys. Soc.* 89, 909, 1966.
90. A. M. Stoneham, *Rev. Mod. Phys.* 41, 82, 1969.
91. W. Zhu, A. R. Badzian, and R. Messier, *J. Mater. Res.* 4, 659–663, 1989.
92. B. E. Williams, J. T. Glass, R. F. Davis, K. Kobashi, and T. Horiuchi, *J. Vac. Sci. Technol. A* 6, 1819–1820, 1988.
93. M. H. Grimsditch, E. Anastassakis, and M. Cardona, *Phys. Rev. B* 18, 901, 1978.
94. H. Windischmann, G. F. Epps, Y. Cong, and R. W. Collins, *J. Appl. Phys.* 69, 2231–2237, 1991.
95. J. A. Baglio, B. C. Farnsworth, S. Hankin, G. Hamill, and D. O'Neil, *Thin Solid Films* 212, 180, 1992.
96. F. A. Doljack, and R. W. Hoffman, *Thin Solid Films* 12, 71, 1972.
97. E. E. Whiting, J. Quantum. Spectrosc. Radiat. Transfer. 8, 1379, 1968.
98. A. T. Collins, and S. H. Robertson, *J. Mater. Sci. Lett.* 4, 681, 1985.
99. T. L. McCormick, The characterization of strain, impurity content and crush strength of single crystal diamonds, Master thesis, North Carolina State University, Raleigh, 1994.
100. G. B. Bokii, N. F. Kirova, and V. I. Nepsha, *Sov. Phys. Dokl.* 24, 83, 1979.
101. S. W. Webb, and W. E. Jackson, *J. Mater. Res.* 10, 1700, 1995.
102. A. R. Badzian, T. Badzian, R. Roy, R. Messier, and K. E. Spear, *Mater. Res. Bull.* 23, 531–548, 1988.
103. V. S. Vavilov, A. A. Gippius, A. M. Zaitsev, B. V. Deryagin, B. V. Spitsyn, A. E. Aleksenko, *Sov. Phys. Semicond.* 14, 1078, 1980.
104. J. Ruan, W. J. Choyke, and W. D. Partlow, *Appl. Phys. Lett.* 58, 295, 1991.
105. J. Ruan, W. J. Choyke, and W. D. Partlow, *J. Appl. Phys.* 69, 6632–6636, 1991.
106. T. Feng, and B. D. Schwartz, *J. Appl. Phys.* 73, 1415, 1993.
107. C. D. Clark, and C. B. Dickerson, *Surf. Coating Technol.* 47, 336, 1991.

108. **C. D. Clark, E. W. J. Mitchell, J. W. Corbett, and G. D. Watkinss,** Eds., *Proc. 1970 Conf. Radiation Damage in Semiconductors*, Gordon and Breach, London, 1971, 257.

109. **A. T. Collins,** *Inst. Phys. Conf. Ser.* 31, 1977.

110. **B. G. Yacobi, A. R. Badzian, and T. Badzian,** *J. Appl. Phys.* 69, 1643, 1991.

111. **A. T. Collins, L. Allers, C. J. H. Wort, and G. A. Scarsbrook,** *Diamond Relat. Mater.* 3, 932, 1994.

112. **L. H. Robins, E. N. Farabaugh, A. Feldman, and L. P. Cook,** *Phys. Rev. B* 43, 9102, 1991.

113. **A. T. Collins, M. Kamo, and Y. Sato,** *Proc. Mater. Res. Soc. Symp.* 162, 225–230, 1990.

Chapter 13

Guided Wave Optics for the Characterization of Polymeric Thin Films and Interfaces

Wolfgang Knoll

Reviewed by J. D. Swalen

TABLE OF CONTENTS

1. INTRODUCTION

Polymeric thin films are ubiquitous in our daily life and in modern technology. In some cases they just serve as passive coatings for other materials, e.g., as lubricants in tribological applications, adhesives, paints, or protective coatings against corrosion, etc. In many other, more interesting applications, these polymer films also play an active role in processing steps or in device configurations. Examples are photoresists that are required for lithographic processes in microelectronic device fabrication;[1] alignment layers at the interface of a solid support and the liquid-crystalline material of a display cell (in some cases even exhibiting externally controlled aligning properties, switchable by light,[2] heat, or external fields[3]); (planar) waveguides with nonlinear optical properties as basic structures for optoboards in integrated optics and photonics;[4] and, most recently, as materials with light-emitting properties in large-area diodes and displays.[5]

In all of these applications the polymeric materials typically have to meet rather stringent requirements in terms of their optical properties, e.g., absorption, refractive indices, their anisotropy, (lateral) homogeneity, nonlinearity, etc. The characterization and control of all of these parameters constitute a considerable experimental challenge because certain samples range in thickness from a few micrometers to only a few tenths of a nanometer, in some cases only a monomolecular layer thick.

Therefore, the past years have brought a remarkable activity in the field of thin film analysis. Optical techniques have been developed with ever increasing sensitivity and spatial resolution. The "classical" polarization-optical technique for the characterization of ultrathin coatings is ellipsometry.[6] This method has been applied to the analysis of monomolecular polymer layers[7] and can be used also in an imaging mode.[8] Interferometric techniques with monochromatic[9] or with white light[10] have been demonstrated to give optical information about thin films at nanometer thickness resolution and can be used in a microscopic setup.[11]

Another powerful and very promising technique for the characterization of ultrathin polymer films is based on evanescent wave optics.[12] The excitation of surface electromagnetic modes traveling at the interface between two media with dielectric constants of opposite sign[13] is the basis for a wide range of optical techniques that use the interaction of this "surface light" with thin polymer films deposited onto such an interface.

In the following we will give first a brief introduction to the theoretical description of surface polaritons[14] and present a number of examples that illustrate some general features and characteristics of the special case of surface plasmon spectroscopy employed for thin film characterization. Then we will discuss the use of surface plasmons for the optical microscopic imaging of laterally structured thin film samples.[15] We will give examples that demonstrate the high contrast (in thickness and/or refractive index) that can be achieved in these surface plasmon micrographs[16] and discuss the limits of the obtainable lateral (spatial) resolution.[17] For thicker samples (ranging in thickness from d ~ 0.2 μm to a few micrometers) additional resonances, guided optical waves, can be excited in the polymer films.[18] These modes can give very accurate data on the optical properties of thin polymer layers. For example, the anisotropy of the refractive index tensor (the indicatrix)[19] or the Pockels response of electro-optically active polymers[20] can be analyzed with high precision. The last examples that we present concern the use of waveguide modes in a microscopic setup[21] similar to surface plasmon microscopy. This technique is particularly valuable for the characterization of lateral heterogeneities of planar waveguide structures needed in integrated optics.[22]

The polymer systems introduced range from monomolecular layers prepared by a chemisorption process or deposited by the so-called Langmuir–Blodgett–Kuhn (LBK) technique.[23] Another convenient way of preparing thicker polymer films is given by the spin-casting technique well developed in microelectronics for the homogeneous deposition of photoresists.[1] And finally, we will present some examples with polymer films prepared by a so-called "grafting-from" technique recently introduced by Rühe and co-workers:[24,25] initiator molecules covalently bound to a solid support are thermally or optically activated in the presence of a suitable monomer solution such that a monomolecular polymer layer grows at the interface, with each polymer chain being covalently attached to the solid support.[25]

2. SURFACE POLARITONS AT THE INTERFACE BETWEEN TWO HALF-INFINITE SPACES

We consider an interface in the xy-plane between two half-infinite spaces, 1 and 2, of materials the optical properties of which are described by their complex frequency-dependent dielectric functions $\tilde{\varepsilon}_1(\omega)$ and $\tilde{\varepsilon}_2(\omega)$, respectively. We ignore magnetic materials. Surface polaritons can only be excited at such an interface if the dielectric displacement $\vec{D}$ of the electromagnetic mode has a component normal to the surface ($\parallel \vec{z}$) which can induce a surface charge density σ

$$\left(\vec{D}_2 - \vec{D}_1\right) \cdot \vec{z} = 4\pi\sigma \tag{1}$$

S-polarized light propagating along the x-direction possesses only electric field components, $\vec{E}_i$, parallel to the surface ($\parallel$ y-direction), i.e., transversal electric (TE) waves have $\vec{E}_i = (0, E_y, 0)$, and hence are unable to excite surface polaritons. Only p-polarized light (transversal magnetic TM) modes with $E = (E_x, 0, E_z)$, or, equivalently, $\vec{H} = (0, H_y, 0)$, can couple to such modes. The resulting surface electromagnetic wave, therefore, will have the following general form

$$\vec{A}_1 = \vec{A}_{10}\, e^{i\left(\bar{k}_{x1}\bar{x} + \bar{k}_{z1}\bar{z} - \omega t\right)} \quad \text{in medium 1} \tag{2a}$$

and

$$\vec{A}_2 = \vec{A}_{20}\, e^{i\left(\bar{k}_{x2}\bar{x} + \bar{k}_{z2}\bar{z} - \omega t\right)} \quad \text{in medium 2} \tag{2b}$$

where $\vec{A}$ stands for $\vec{E}$ and $\vec{H}$; $\vec{k}_{x1}$, $\vec{k}_{x2}$ are the wave vectors in the x-direction; $\vec{k}_{z1}$, $\vec{k}_{z2}$ those in the z-direction, i.e., normal to the interface; and ω is the angular frequency. Both fields $\vec{E}$ and $\vec{H}$ must fulfill the Maxwell equations:

$$\nabla \cdot \vec{H} = 0 \tag{3}$$

$$\nabla \cdot \vec{E} = 0 \tag{4}$$

$$\nabla \times \vec{E} + \frac{1}{c} \frac{\partial \vec{H}}{\partial t} = 0 \tag{5}$$

$$\nabla \times \vec{H} - \frac{\varepsilon}{c} \frac{\partial \vec{E}}{\partial t} = 0 \tag{6}$$

with c being the speed of light *in vacuo* and ε the dielectric function of the material. The tangential components of $\vec{E}$ and $\vec{H}$ have to be equal at the interface, i.e.,

$$E_{x1} = E_{x2} \tag{7}$$

and

$$H_{y1} = H_{y2} \tag{8}$$

From Equation 7 it follows immediately that $k_{x1} = k_{x2} = k_x$. On the other hand, it follows from Equations 2 and 6 that

$$k_{z1} H_{y1} = \frac{\omega}{c} \varepsilon_1 E_{x1} \tag{9}$$

and

$$k_{z2} H_{y2} = \frac{\omega}{c} \varepsilon_2 E_{x2} \tag{10}$$

This leads to the only nontrivial solution if:

$$\frac{k_{z1}}{k_{z2}} = -\frac{\varepsilon_1}{\varepsilon_2} \tag{11}$$

Equation 11 indicates that surface electromagnetic modes can only be excited at interfaces between two media with dielectric constants of opposite sign. For a material in contact with a dielectric medium the dielectric constant, ε_d, of which is positive, this can be fulfilled for a whole variety of possible elementary excitations provided their oscillator strength is sufficiently strong to result in a negative ε. Within a limited spectral range this can be the case for phonons as well as for excitons. The coupling of these excitations to an electromagnetic field has been shown to produce phonon surface polariton or exciton surface polariton modes, respectively.

We are dealing here with the interface between a metal (with its complex dielectric function ($\varepsilon_m = \varepsilon'_m + i\varepsilon''_m$) and a dielectric material ($\varepsilon_d = \varepsilon'_d + i\varepsilon''_d$), hence, with the coupling of the collective plasma oscillations of the nearly free electron gas in a metal to an electromagnetic field. These excitations are called plasmon surface polaritons (PSP) or surface plasmons, for short. From Equations 5, 6, 9, and 10 we obtain

$$k_x^2 + k_{zd}^2 = \left(\frac{\omega}{c}\right)^2 \varepsilon_d \tag{12}$$

or

$$k_{zd} = \sqrt{\varepsilon_d \left(\frac{\omega}{c}\right)^2 - k_x^2} \tag{13}$$

With Equation 11 this leads to the dispersion relationships (i.e., the energy–momentum relation) for surface plasmons at a metal/dielectric interface:

$$k_x = \frac{\omega}{c} \sqrt{\frac{\varepsilon_m \cdot \varepsilon_d}{(\varepsilon_m + \varepsilon_d)}} \tag{14}$$

A few points are noteworthy. (1) In the usual treatment, ω is taken to be real. Since ε_m is complex, k_x is also complex, i.e., $k_x = k_x' + ik_x''$. As a consequence, PSP modes propagating along a metal/dielectric interface exhibit a finite propagation length, L_x, given by $L_x = 1/k_x''$. This has a strong impact on lateral resolution that we want to obtain in the characterization of laterally structured samples investigated with plasmon light in a microscopic setup (see below). (2) In the frequency (spectral) range of interest we have:

$$\sqrt{\frac{\varepsilon_m \cdot \varepsilon_d}{(\varepsilon_m + \varepsilon_d)}} \geq \sqrt{\varepsilon_d} \tag{15}$$

This has two important consequences. The first can be seen from Equation 13. Inserting Equation 15 shows that in this case the z-component of the PSP wave vector is purely imaginary. From Equation 2 we see that the surface plasmon is a bound, nonradiative evanescent wave with a field amplitude, the maximum of which is at the interface ($z = 0$) and which is decaying exponentially into the dielectric (and into the metal). The mode is propagating as a damped oscillatory wave (Figure 1). All parameters characterizing the properties of PSPs can be quantitatively described on the basis of the dielectric functions of the involved materials, e.g., the exponential decay of the optical field intensity normal to an Ag/air interface. The penetration depth, l_c, of this light into the dielectric medium is found to be a few hundred nanometers only, and it is this surface specificity that makes it such an interesting probe field. The enormous field intensity enhancement can amount to more than two orders of magnitude for these modes and is the source of the very high surface sensitivity, obtainable, e.g., in PSP Raman spectroscopy and microscopy.

The second consequence of Equation 15 is that the momentum of a free photon propagating in a dielectric medium

$$k^{ph} = \frac{\omega}{c} \cdot \sqrt{\varepsilon_d} \tag{16}$$

is always smaller than the momentum of a surface plasmon mode propagating along an interface between that same medium and the metal (Figure 2). The dispersion of photons is described by the light line, $\omega = c_d \cdot k$ (Figure 2a), with $c_d = c/\sqrt{\varepsilon_d}$. For the excitation of surface plasmons only the photon wave vector projection to the x-direction is the relevant parameter. For a simple reflection of photons (with energy $\hbar\omega_L$, e.g., from a laser) at a planar dielectric/metal interface (see Figure 3a) this means that by changing the angle of incidence, ϕ, one can tune $k_x^{ph} = k^{ph} \cdot \sin\phi$ from zero at normal incidence (point 1 in Figure 2) to the full wave vector k^{ph} at grazing incidence (point 2 in Figure 2). Equation 14 or 15, however, tells us that this is not sufficient to fulfill, in addition to the energy conservation, the momentum-matching condition for resonant PSP excitation because, for very low energies, the PSP dispersion curve (Figure 2b) asymptotically reaches the light line (Figure 2a), whereas for higher energies it approaches the cutoff angular frequency ω_{max} determined by the plasma frequency of the employed metal, ω_p:

$$\omega_{max} = \omega_p / \sqrt{1 + \varepsilon_d} \tag{17}$$

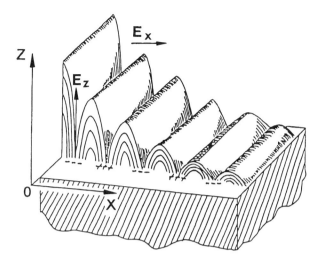

Figure 1 Schematic representation of a PSP as a surface-electromagnetic mode with field components E_x and E_z propagating along the x-direction coupled to a surface charge density wave.

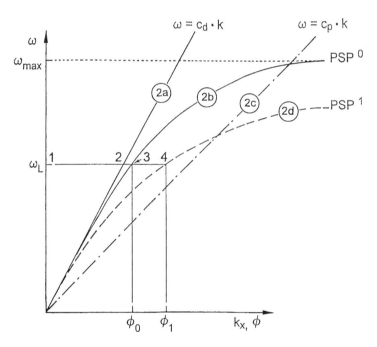

Figure 2 Dispersion relation, angular frequency ω vs. wave vector in the x-direction k_x, of PSPs at a metal–dielectric (e.g., air) interface (PSP[0], curve b) and at a metal-coating–dielectric interface (PSP[1], curve d), respectively. Given also are the light lines of photons propagating in the bulk dielectric material ($\omega = c_d \cdot k$, curve a) and in a prism material ($\omega = c_p \cdot k$, curve c). Laser light of energy $\hbar\omega_L$ couples to PSP states at angles ϕ_0 and ϕ_1 (points 3 and 4, respectively), given by the energy- and momentum-matching condition (see the intersection of the horizontal line at ω_L with the two dispersion curves). All other symbols are given in the text.

3. SURFACE PLASMON SPECTROSCOPY WITH PRISM COUPLER

One way to overcome this problem was introduced by the experimental setup shown schematically in Figure 3b. Photons are not coupled directly to the metal/dielectric interface, but via the evanescent tail of light total internally reflected at the base of a high-index prism (with $\varepsilon_p > \varepsilon_d$).[26] This light is characterized by a larger momentum (Figure 2c as the dash-dotted light line) that for a certain spectral range can exceed the momentum of the PSP to be excited at the metal surface. So, by choosing the appropriate (inner) angle of incidence, ϕ_0 (see point 3 in Figure 2), resonant coupling between evanescent photons and surface plasmons can be obtained.

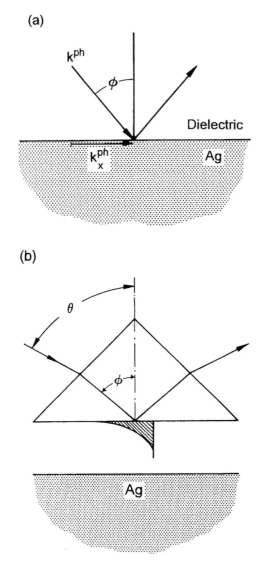

Figure 3   (a) Reflection geometry at an Ag–dielectric interface of photons with momentum k^{ph} incident at an angle ϕ. k_x^{ph} is the photon wave vector projection along the x-direction. (b) The Otto configuration:[26] total internal reflection of a plane wave incident at an (external) angle θ at the base of a prism. The evanescent tail of this inhomogeneous wave can excite PSP states at an Ag–dielectric interface, provided the coupling gap is sufficiently narrow. (c) (Top) Attenuated total internal reflection (ATR) setup for PSP excitation in the Kretschmann geometry:[27] a thin metal film (d ~ 50 nm) is evaporated onto the base of the prism and acts as a resonator driven by the photon field incident at an (external) angle θ ($\hat{=}$ internal angle ϕ). (Bottom) Resonant coupling is observed by a detector as a sharp dip in an angular scan of the reflected intensity, R, at θ_0; θ_c is the critical angle for total internal reflection.

Experimentally, this resonant coupling is observed by monitoring, as a function of the incident angle, the laser light of energy $\hbar\omega_L$ that is reflected by the base of the prism, which shows a sharp minimum. The major technical drawback of this Otto configuration is the need to get the metal surface close enough to the prism base, typically to within ~200 nm. Even a few dust particles can act as spacers, thus preventing efficient coupling. So, despite its potential importance for the optical analysis of polymer-coated bulk metal samples, this version of surface plasmon spectroscopy has not gained any practical interest.

By far the most widespread version of surface plasmon spectroscopy is based on the experimental configuration introduced by Kretschmann and Raether[27] (Figure 3c). Conceptually, this scheme for exciting PSPs is rather similar to the aforementioned technique with the exception that this time the (high-momentum) photons in the prism couple through a very thin metal layer (typically, approximately

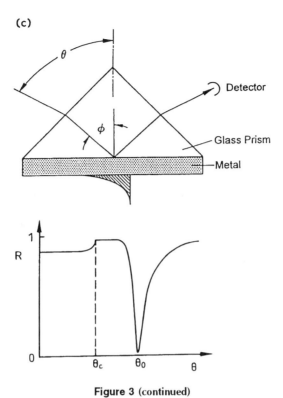

(c)

Figure 3 (continued)

45 to 50 nm thick, evaporated onto the base of the prism) to the PSP states at the other side in contact to the dielectric medium. Qualitatively, the same considerations for energy- and momentum-matching apply as discussed for Figure 2; however, quantitatively one has to take into account that the finite thickness of the metal layer causes some modifications of the dispersion behavior of the PSP modes. In particular, the possibility of coupling out some of the surface plasmon light through the thin metal layer and the prism opens a new, radiative loss channel for PSPs in addition to the intrinsic dissipation in the metal.

What is important for our interest in using surface plasmons for thin film characterization is a qualitative understanding of how this light monitors changes in the optical architecture of the interface. Again, for a qualitative picture only, we just note that depositing an ultrathin layer (with a thickness of $d \ll 2\pi/k_{zd}$) of a material with an index of refraction, $n = \sqrt{\varepsilon}$ larger than that of the ambient dielectric, e.g., air $n = 1$, for a surface plasmon mode, is equivalent to an increase of the overall effective index integrated over the evanescent field. The net effect is a slight shift of the dispersion curve (Figure 2d) corresponding to an increase of k_x for any given ω_L.[28] This is depicted in Figure 2 (dashed curve labeled PSP[1]). As a consequence, the angle of incidence which determines the photon wave vector projection along the PSP propagation direction has to be slightly increased (to ϕ_1, point 4 in Figure 2) in order to again couple resonantly to PSP modes.[29]

For all practical purposes, the quantitative treatment of this problem is based on the Fresnel theory for calculating the overall transmission and reflection of a general multilayer assembly. The latter would, in our case, consist of the prism material, the metal layer, the organic layer(s), and the superstrate, typically air or a transparent liquid, e.g., water. Different algorithms based on either a matrix formalism or a recursion formula procedure for calculating the Fresnel coefficients of the i-th layer for s- and p-polarized light have been treated in the literature. The angular dependence of the overall reflectivity can be computed and compared with the measured curves.[30]

4. ANGULAR SCANS OF REFLECTED LIGHT FOR DETERMINATION OF OPTICAL THICKNESS

The first example for the thickness determination of a thin organic coating by surface plasmon spectroscopy concerns thin polymer films prepared by the grafting-from polymerization procedure.[24,25] The schematics of this surface modification are depicted in Figure 4a. The OH-groups of a suitable substrate

(a)

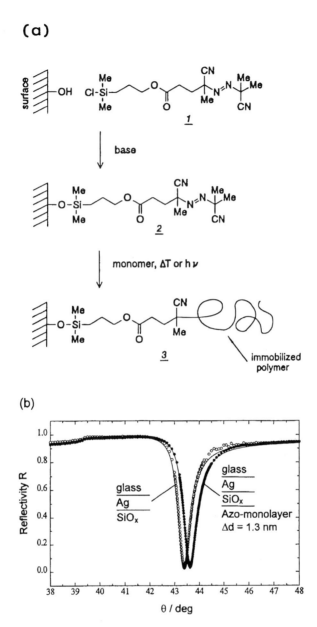

(b)

Figure 4 (a) Formation of covalently bonded polymer films through self-assembled monolayers of azo-initiators. (b) Angular reflectivity scan of the bare substrate; glass/Ag/SiO$_x$: -o-, and after coating with an azo-initiator monolayer: -●-. The obtained shift of the resonance amounts to $\Delta = 0.3°$. (c) Reflectivity scan of a poly(styrene) (PS) monolayer (-o- sample before modification; -■- after deposition of a polymer monolayer). Full lines show reflection curves calculated from Fresnel equations giving the following layer thicknesses: 50.0 nm Ag, 27.9 nm SiO$_x$, 36.1 nm PS.

(typically glass, plasma-treated mica, or evaporated and conditioned SiO$_x$) are used to couple a mono-functional silane group carrying an initiator molecule for radical polymerization (an AIBN analogue in our case). The result is a covalently attached initiator system that can be activated by heat or light in the presence of a monomer solution to initiate the growth of two polymer chains: one in solution, the other covalently bound by one end to the surface.

In order to be able to characterize the resulting thin films by surface plasmon spectroscopy the required Ag- or Au-substrate has to be precoated by an evaporated or sputtered SiO$_x$ film that serves as the oxide surface. The angular reference scan of this sandwich-substrate measured in air is given in Figure 4b (open circles), together with the Fresnel characterization of this resonance curve (full curve). The

(c)

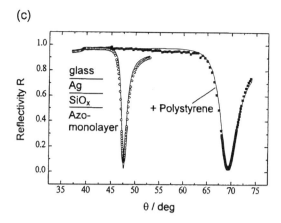

Figure 4 (continued)

formation of a single monolayer of the azo-initiator is easily seen as a clear shift of the resonance curve by about $\Delta\theta = 0.3°$ (closed circles). Assuming an index of refraction of n = 1.5 (a typical value for organic layers, but see also below) the Fresnel fit of this shifted curve yields a thickness of $\Delta d = (1.3 \pm 0.3)$ nm compatible with the expectation of a monomolecular layer of the initiator molecules.

After the thermal initiation of the polymerization reaction in the presence of a styrene solution to grow polystyrene, and a careful removal of all free polymer chains by repeated soxhlet extraction the resulting polymer film measured after drying in air is given in Figure 4c (full squares).[31] The Fresnel fit for this preparation gives a thickness of $\Delta d = (42.3 \pm 1.0)$ nm if the bulk refractive index for polystyrene of n = 1.591 is used (see, however, the index contrast variation experiments below). Very homogeneous coatings can be prepared by this procedure ranging in thickness from a few tens of nanometers up to more than 1 µm, each time only a monomolecular layer thick with each polymer chain being covalently attached to the substrate by one end. More examples for these systems and their characterization by surface plasmon and guided optical waves will be given below.

Another example that demonstrates the potential of surface plasmon spectroscopy is based on the formation of polymeric multilayer assemblies prepared by the LBK technique. Here, a monomolecular layer of an amphiphilic polymer is first spread and compressed at the water/air interface and then transferred to a solid support by either vertical or horizontal dipping of the substrate through the preorganized monolayer. The number of down- and upstroke cycles of a repetitive deposition determines the total thickness of the final multilayer assembly.

The system that we describe is based on the so-called AB deposition, where a suitable double trough system allows for the deposition of one type of monolayer, A, upon the downstroke through the first monolayer, whereas another material, B, is transferred upon the upstroke through the second monolayer. The resulting supramolecular architectures hold potentials, e.g., for electro-optic applications where a stable noncentrosymmetric organization of the nonlinear optically (NLO) active chromophores is essential for the Pockels response.[32]

The employed materials are depicted in Figure 5a. The NLO active azo-chromophore (synthesized by R. Advincula) and the amphiphilic ionene[33] were compressed to a surface pressure of $\pi = 35 - 40$ mN · m^{-1} and transferred under class 100 clean-room conditions. The substrate was a glass slide (after deposition index matched to a prism) evaporation coated with 45 nm of Au and treated with an octadecyl-thiol solution which resulted in the spontaneous formation of a monolayer. Figure 5b shows schematically the final sample architecture with the Au film, the alkylthiol monolayer, and the alternating AB multilayer assemblies.

Figure 5c summarizes angular scans taken at various stages of the sample preparation.[29] The full circles were taken after the Au evaporation; the first full curve indicates the resonance shift induced by the thiol monolayer prepared from solution by a self-assembly process. The next curves, labeled 2, 4, 6, 8, 10, 12 layers, were taken after each of the deposition cycles and demonstrate the consecutive increase of the sample thickness. A Fresnel analysis (not shown as angular fits) based on an index of refraction of n = 1.50 gave the thicknesses as summarized in the inset of Figure 5c. The average bilayer thickness of $\Delta d = 4.0$ nm thus derived agreed well with the results obtained by X-ray reflectometry analysis performed in parallel.

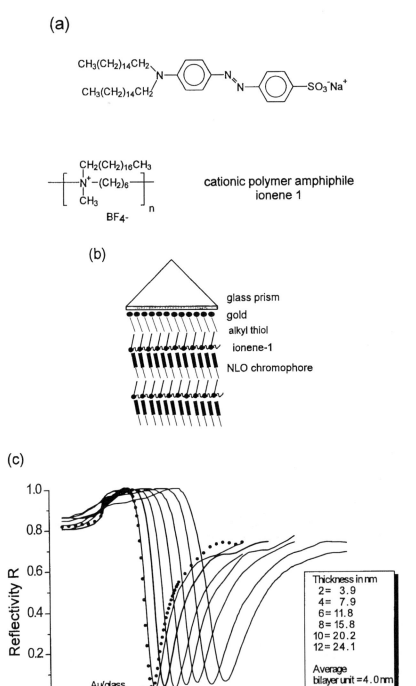

Figure 5 (a) Structure formula of NLO-active azobenzene-amphiphilic chromophore and a cationic polymer amphiphile, ionene-1. (b) Alternating multilayer architecture at the base of an Au/alkylthiol-coated glass prism. (c) Series of ATR scans (R vs. θ) for an increasing number (as indicated) of alternating bilayers of the azobenzene — amphiphile and ionene-1.

The last example that we present concerns the formation of polymeric multilayer assemblies by a protocol recently introduced by Decher and Hong.[34] A charged substrate is sequentially exposed to solutions of cationic and anionic polymers resulting in the adsorption of very stable polymer films the thicknesses of which are controlled by the number of deposition cycles.

In our experiments we used a modified version of the preparation procedure that allowed for the on-line control of the polyelectrolyte deposition in a flow cuvette.[35] Figure 6a gives a schematic representation of the setup: onto the glass (high index, LaSFN9, n = 1.85@λ = 633 nm) /Au /octadecyl-thiol layer sandwich one monolayer of the positively charged amphiphilic ionene-1 (see Figure 5a) was deposited in the Langmuir trough on the downstroke of the LBK process. The sample was then mounted to the flow cell under water so as to keep the charged ammonium head groups of the ionene polymer in constant contact with the electrolyte solution. The angular scan of this reference surface is given in Figure 6c, full circles (note the large coupling angle which is due to the fact that all data were taken with the surface being in contact with electrolyte solution). Now, the first polymer solution, PAZO (structure formula given in Figure 6b), was injected into the cell and the polyelectrolyte allowed to adsorb with its negative charges to the positive groups of the solid interface. A period of 15 min was found to be sufficient for a stable monolayer formation, and the corresponding plasmon resonance scan was taken. After careful rinsing with pure buffer (which had no net effect on the adsorbed polyelectrolyte layer thickness), the positively charged ionene-2 (see Figure 6b) was injected and allowed to adsorb, again for 15 min. The angular scan was recorded, and the procedure repeated several times. As a result, a multilayer assembly of alternating polymers was obtained with opposite charges as sketched in Figure 6b. A sequence of raw reflectivity data taken after consecutive depositions is shown in Figure 6c as full curves each shifted relative to the one taken before. The net angular shift per deposition is relatively small so that some fluctuation in the sequence of reflectivity scans is obvious. Nevertheless, a Fresnel analysis based on a refractive index of the layers of n = 1.50 gives an effective thickness increment per layer which results in a linear increase of the multilayer thickness, as it is shown in Figure 6d. A closer inspection of the data points reveals that the odd-numbered depositions corresponding to the PAZO adsorption gave a larger thickness increase of Δd = (1.55 ± 0.27) nm, whereas the even-numbered ones obtained with the ionene 2 solution gave only Δd = (0.63 ± 0.3) nm. From the slope of the least-squares fit shown in Figure 6d, one obtains an average thickness increase per monolayer deposition of Δd = 1.09 nm.

We should point out that, because of the evanescent character of surface plasmons, the recording of reflectivity scans in the Kretschmann configuration was easily possible also at this solid/solution interface. The setup had to be modified only to allow for the installation of a liquid cell and thus proved to be a very versatile tool in characterizing the optical properties of thin polymer films.

At this stage of the analysis of reflectivity scans by Fresnel simulations we had to assume a (reasonable) value for the refractive index. However, if these investigations are combined with X-ray reflectivity studies on the same thin film architectures probing the electron density profile normal to the surface, a unique discrimination between thickness and refractive index is possible.[36] Moreover, as we will show below, "alt-optical" solutions are also feasible for the problem using either an optical contrast experiment with surface plasmon studies of thin films in contact with media of different bulk refractive indices or for thicker films by exciting, in addition to surface plasmons, waveguide modes.

5. KINETIC STUDIES WITH THIN FILMS

So far, we have discussed angular reflectivity scans of polymeric thin film samples without paying attention to any time-dependent variations of their optical properties. Many technological applications, however, depend crucially on a purposeful modification of their thickness, of their refractive index or its anisotropy, or of any other optics-related property of these layers.

In the following we present, therefore, just a few examples of how surface plasmon optics can also yield valuable information in time-dependent studies of polymer properties.

One obvious way is to simply monitor reflectivity scans at various times. This gives reliable results whenever the spontaneous or induced changes are sufficiently slow, or when they proceed in an incremental way. One example of the latter kind is presented in Figure 7a.[31] Here, a polystyrene film grafted from the oxide surface as described above (see Figure 4) was characterized with respect to its ablation

(a)

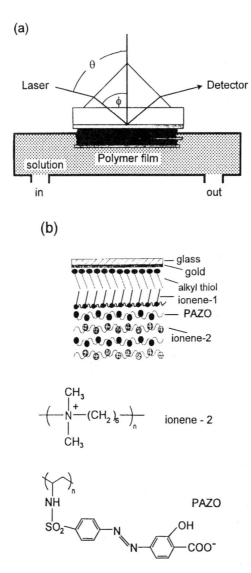

(b)

Figure 6 (a) Experimental setup for on-line surface plasmon optical observation of the alternating deposition of cationic and anionic polymers from solution to an Au substrate precoated by an alkyl thiol and an ionene-1 monolayer. (b) Schematic representation of the final multilayer architecture at the interface after alternating deposition of ionene-2 and PAZO. (c) Series of ATR scans taken after each polyelectrolyte monolayer deposition. (d) Thickness increase as obtained from the ATR scans given in (c).

behavior under the influence of UV photons from a low-pressure Hg lamp of P = 400 µW/cm² power density in an argon atmosphere. The initial thickness as derived from a Fresnel fit calculation with n = 1.591 to the data points (full squares and curve labeled 0 in Figure 7a) was 43.3 nm. After increasing times of irradiation (in time intervals of 30 min), surface plasmon resonance curves were recorded that shifted to smaller angles indicating a loss of thickness by the UV-induced ablation process. A sequence of reflectivity curves is given in Figure 7a, labeled 1 to 8, together with their corresponding Fresnel fits. If the obtained thickness data are plotted as a function of the irradiation time (see Figure 7b) the ablation behavior of this process is derived and can be analyzed in terms of a kinetic model. For comparison, the data obtained by the same sample treatment, but performed in air is also given in Figure7b. Remarkably, the kinetic behavior is very similar (indicating that photo-oxidative processes play only a minor role) and results in a complete removal of the film after ~4-h irradiation time.

For continuous changes of thin film properties that are significantly faster (compared with the time scale of an angular scan which is of the order of minutes), one could use a different mode for recording

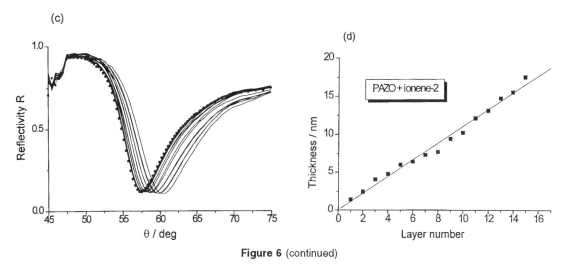

(c)

(d)

PAZO + ionene-2

Figure 6 (continued)

angular scans: if an expanded laser beam is focused onto the sample at the base of the prism by a cylindrical lens, a whole range of incident angles probes the resonance behavior of the interface simultaneously. The reflected divergent light beam contains the same information as an angular scan and can be read out by a linear diode array or a CCD camera.[37] The time resolution is then given by the data transfer rate which amounts for the fastest CCD systems to about 100 μs.

For only small thickness (or index) changes which cause only minor angular shifts of the surface plasmon resonance curve, a much simpler procedure can give kinetic information very easily and with high sensitivity. This is demonstrated for the time-dependent formation of a self-assembled monolayer from a binary mixture of functionalized water-soluble thiol molecules.[38] The structure formulas of the two employed species are given in Figure 8a: one is a biotinylated system designed for specific binding to the protein streptavidin; the other molecule is OH-terminated and is used to laterally dilute the biotin ligands in order to ensure maximum binding.

Prior to the experiment the bare Au surface in contact with the aqueous phase is characterized by recording an angular scan. At a fixed angle of incidence at or near the steep slope of the resonance (see, e.g., the arrow in Figure 4b) the reflected intensity is now continuously recorded as a function of time. If the pure buffer, in our case, is exchanged (within ~2 s) against the thiol solution (see arrow in Figure 8b), the onset of the monolayer formation causes the resonance shifting to higher angles. Hence, the reflected intensity (at this angle of incidence) increases. As can be seen from Figure 8b a time resolution of better than seconds can easily be achieved and allows here the on-line observation of the self-assembly process. From the signal-to-noise ratio of this intensity increase one can derive a thickness sensitivity better than 0.1 nm.

Another example of kinetic studies in thin polymer films will be given below for the light-induced trans-cis isomerization reaction in azobenzene-derivatized liquid-crystalline polymer films prepared by the LBK technique.[39] In these studies it is not the thickness primarily that is changed, but rather the (anisotropy of the) index of refraction.

6. CONTRAST VARIATION EXPERIMENTS

As has been pointed out already, all resonance shifts induced by the thin films deposited to the surface plasmon–carrying metal surface could be converted to (effective) film thickness data only if the index of refraction was either known or estimated for the Fresnel calculation based on a reasonable assumption. The reason for this ambiguity is that for each angular shift an infinite number of pairs (n, d) of film index and thickness, respectively, are compatible with the measured $\Delta\theta$. However, since $\Delta\theta$ depends not on the absolute value of n (or ε) but rather on the contrast, i.e., the index difference of the film to the surrounding bulk medium, a so-called contrast variation, in principle, can solve the problem.[40] To this end, one has to measure reflectivity scans of one polymer film in at least two dielectric media of different refractive indices n_1, n_2. For many systems this could be air and a liquid, e.g., water. The measured angular shifts, $\Delta\theta_1$ and $\Delta\theta_2$, are both known functions of the thickness and the contrast, i.e.,

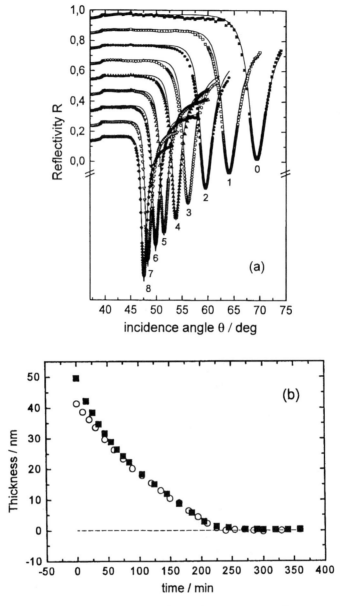

Figure 7 (a) Photoablation of poly(styrene) monolayer under argon atmosphere as measured by surface plasmon spectroscopy. Reflectance curves recorded after different times of exposure to UV light; 0 = 0, 1 = 30, 2 = 60, 3 = 90, 4 = 120, 5 = 150, 6 = 180, 7 = 210, 8 = 240, min; initial film thickness 42.3 nm, low pressure Hg lamp (pen ray, L.O.T.; 17 mA) 2.5 cm distance; solid lines are Fresnel calculations. (b) Thicknesses of the PS films as a function of irradiation time according to Fresnel calculations; n = 1.591; conditions as in (a); (■) ablation in air, initial thickness 49.6 nm; (○) argon atmosphere, initial film thickness 41.4 nm; dashed line represents sample before modification.

$$\Delta\theta_1 = f(d, n - n_1)$$
$$\Delta\theta_2 = f(d, n - n_2)$$

(18)

If more contrasts are measured, then the corresponding plots of n vs. d compatible with each of the measured $\Delta\theta_i$ should intersect at one point, which is then the only pair of n and d compatible with all measurements.

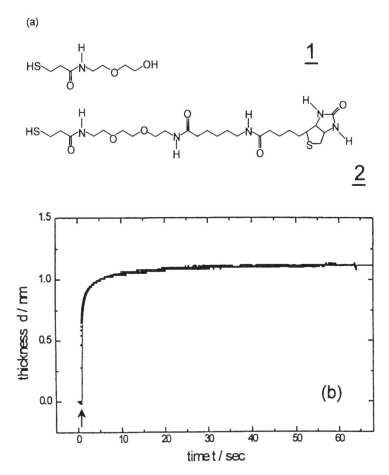

Figure 8 (a) Structure formula of water-soluble thiol molecules with different end group functionalities, i.e., –OH and –biotin, respectively. (b) Thickness increase as a function of time of a binary mixed monolayer composed of thiols 1 and 2, see (a). Arrow indicates addition of thiol solution.

This is examplified for a thin film of polystyrene prepared on an Ag surface by spin casting. This film was first measured in air ($n_1 = 1$, $\Delta\theta_1 = 10.47°$), then in methanol ($n_2 = 1.327$, $\Delta\theta_2 = 8.78°$), in 2-propanol ($n_3 = 1.383$, $\Delta\theta_3 = 7.83°$), and in ethylenegycol ($n_4 = 1.429$, $\Delta\theta_4 = 6.62°$). The n, d curves compatible with these measured angular shifts are all plotted in Figure 9. Even if we take into account some error in the angle determination of $\delta\Delta\theta_i = 0.3°$, it is obvious that there is not only one single intersection but rather a range of intersections. Possible explanations are (1) nonideality of the reference scans and the substrate samples, e.g., a thin oxide or sulfide coating of the Ag substrate before the polymer deposition (when measured in different solvents — or solvents of different refractive index — this "pre-coating" would result in a nonlinear shift relative to bare Ag) and (2) the whole approach is valid only if the film properties, e.g., n and d are not changing in going from one solvent to another. In particular, any swelling, i.e., uptake of only traces of solvent, e.g., in voids, would change this assumption.[41]

Nevertheless, this approach allows at least for a partial decoupling of n and d, narrowing the range of possible values considerably. For our sample, the index of refraction is between $n = 1.562$ and $n = 1.572$ at a thickness of $d = 37$ to 41 nm. In particular, if measured in one of the liquids, the assumption of a bulk refractive index for polystyrene, $n = 1.591$, would have resulted in a substantial underestimation of the true thickness. On the other hand, the curve derived from the measurement in air, i.e., at the highest contrast, is relatively insensitive to wrong assumptions concerning the refractive index.

7. MONITORING REFRACTIVE INDEX CHANGES IN THIN FILMS

So far, we have been concerned with the determination of thin film thicknesses or their time-dependent variation, e.g., in an ablation experiment or during film formation by adsorption from solution. In this

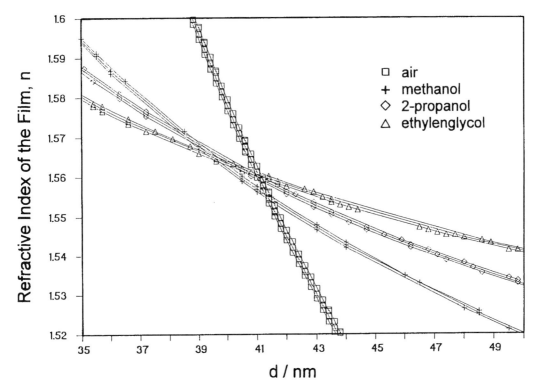

Figure 9   Refractive index contrast variation experiment: plotted are the pairs of refractive index n and thickness d data compatible with the angular shifts of the surface plasmon resonance $\Delta\theta_i$ measured for a polystyrene film in different solvents, as indicated, and in air.

Figure 10   (a) Structure formula of a copolymer system with phenylbenzoate and azobenzene side groups. (b) Surface plasmon spectroscopy experimental arrangement in the ATR-Kretschmann setup. The probe is a 632.8-nm He-Ne laser beam, and the reflectivity of the sample is recorded as a function of the incidence angle. The photoactive (pump) beam direction of propagation is perpendicular to the plane of the sample. (c) Surface plasmon resonance curves before (-●-) and under steady-state illumination with 360 nm light (-△-) in two layers of the copolymer, see (a). The reflectivity change at $\theta = 44.3°$ is monitored in a kinetic measurement. (d) Time-dependent refractive index change n_z calculated from PSPS measurements in a system of Cr/Au/eight layers polyglutamate/two layers of the copolymer.

section we present some results obtained with surface plasmon spectroscopy for thin films the thicknesses of which remained constant while the refractive index was manipulated externally.

The first example refers to the structural and optical rearrangements induced in thin multilayer assemblies of azobenzene-functionalized liquid-crystalline polymers by photoisomerizing the chromophores from the trans- to the cis-conformer, and vice versa.[39] Figure 10a gives the structure formula of the employed copolymer system. Sample preparation was performed according to the LBK technique. First, eight layers of an optically inert system of polyglutamate was deposited onto the Ag-coated substrates in order to separate the following chromophore layers from the metal and thus prevent transfer

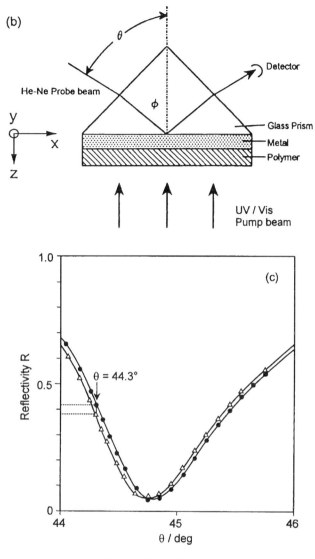

Figure 10 (continued)

of excitation energy from the azobenzene groups to the acceptor states in the Ag. On top of the spacer film two monolayers of the copolymer were then deposited. Figure 10c shows the reflectivity scan as obtained from the sample with its chromophores being mostly in the dark-adapted trans-conformer state as prepared (full circles). After the sample was illuminated with UV photons of $\lambda = (350 \pm 30)$ nm (power density P = 1.75 mW/cm²) for about 10 min (see the experimental setup given in Figure 10b), which was enough to reach a new photostationary state of the chromophore population, the resonance curve for excitation of surface plasmons was shifted to smaller angles (Figure 10c, open triangles) — a clear indication that the sample with most of its chromophores being in the cis-conformer was optically thinner. X-ray reflectometry measurements, on the other hand, had indicated that the geometric thickness as probed by the electron density profile normal to the film surface was nearly unchanged upon illumination of the sample. However, as indicated qualitatively, in Figure 1 the field components of the surface plasmon modes are dominated by E_z, which means that PSP light mostly probes the refractive index normal to the surface, n_z. The result of Figure 10c, hence, indicates that the anisotropy of the refractive index of the LBK double layer changed under the influence of the photoinduced isomerization reaction.

If the reflected intensity was monitored at constant angle ($\theta = 44.3°$, see arrow in Figure 10c) as a function of illumination time, the kinetics of this process could be followed directly. This is presented

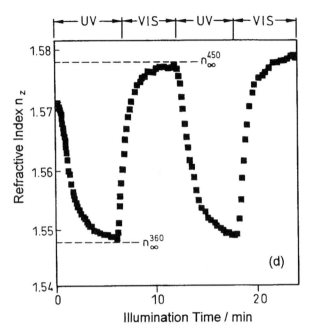

Figure 10 (continued)

in Figure 10d. Plotted are the time-dependent refractive index data, n_z, as obtained from the isomerization-induced shift of the resonance curve. During illumination of the sample with UV light (indicated in Figure 10d by |←UV→|), n_z drops within minutes significantly from $n_z = 1.572$ to $n_z = 1.548$. This process was fully reversed if the sample was illuminated with visible light [|←VIS→| in Figure 10d, $\lambda = (450 \pm 20)$ nm at a power density of $P = 1.25$ mW/cm^2], and the isomerization cycle could be repeated many times. The details of the kinetics of this structural change were rather complicated, with a broad range of relaxation times. We should note, however, that the sensitivity of surface plasmon spectroscopy — as judged from the signal-to-noise ratio observed in the data of Figure 10d — allowed us to follow relative index changes of better than $\Delta n/n = 10^{-3}$ for a polymer film sample of only 5 nm in thickness!

The second example concerns the measurement of index changes induced by the Pockels effect in a $\chi^{(2)}$-active chromophore.[42] The molecules used in these studies — an asymmetrically derivatized amphiphilic stilbene analogue — is shown in Figure 11a. Sample preparation was performed by organizing the dye molecules as a monomolecular layer at the water/air interface and then depositing onto an Ag-coated substrate two of these monolayers on the upstroke separated by a polyglutamate monolayer deposited at the downstroke. This way, a noncentrosymmetric architecture of the NLO-active chromophore layers needed for electrooptical measurements was ensured. The angular scan of this sample shows the excitation of a surface plasmon mode as given in Figure 11b, full squares.

In order to be able to apply an electric DC field across this ultrathin film sample a top electrode was required in addition to the bottom electrode (the latter carrying the surface plasmon mode). Since direct evaporation of a metal layer onto this thin LBK assembly was not possible without shortcuts, another strategy was taken. The top electrode was prepared by evaporation onto a separate glass slide which was then mechanically pressed against the NLO-active film using a 3-μm-thick mylar film as spacer.

Since the changes of the refractive index induced by an applied electric DC field were too small to be measured directly, a lock-in technique had to be used. To this aim the applied voltage was modulated at f = 2.23 kHz with an amplitude of U = 30 V (peak to peak).

This generated a slight intensity modulation of the reflected light which could be monitored with a phase- and frequency-sensitive detector locked to 2.23 kHz. The amplitude of this modulation is a measure of the electric field–induced refractive index change in the thin film and hence is proportional to the electro-optical coefficients.[43] The obtained data are plotted in Figure 11b and indicate that by this technique relative reflectivity changes of better than $\Delta R/R = 10^{-6}$ can be detected. The quantitative evaluation of the electro-optical coefficients is not exactly possible, mostly because the electric field strength cannot be determined sufficiently accurately. A rough estimate, however, gave a Pockels response of this material of $\chi^{(2)}_{33} = 1.6$ pm/V.

(a)

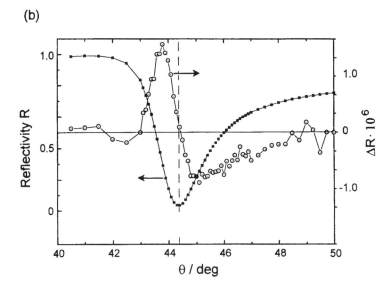

(b)

Figure 11 (a) Structure formula of the amphiphilic NLO-active stilbene derivative. (b) R vs. θ, -■-, and ΔR vs. θ scan, -⊙- for a double layer of the stilbene chromophore.

8. SURFACE PLASMON MICROSCOPY

Up to now, we have discussed the use of surface plasmon spectroscopy for the characterization of thin films with respect to their optical properties varying only along the z-coordinate; the lateral in-plane variations, either as fluctuating heterogeneties or as purposefully designed pattern, were totally ignored. All data measured and analyzed were averaged over the spot size of the laser probe beam.

Recently, it was established, however, that surface plasmons as specific "surface light" also couple to lateral index or thickness variations like normal photons do to bulk index variations.[44] These modes then couple back out through the prism with an angular intensity distribution that corresponds to the Fourier transform of this spatial (lateral) index variation. It is then a straightforward step to Fourier-back-convert this light from k-space into real space by means of a lens, thus generating an image of the interface, in particular, of its lateral index (or thickness) variations.[15,16] This is completely equivalent to the procedure in any other microscopy which uses plane waves of, e.g., light or electrons. In our case, it just means that we "illuminate" the sample with surface-bound light rather than with "normal" photons like it is done, e.g., in ellipsometric microscopy of thin films.[8] The schematics of the corresponding setup is given in Figure 12a. Any PSP spectrometer can be easily converted to a microscope by simply adding a lens to the detection side and monitoring the obtained image on a screen or with the help of a TV (CCD) camera. In the latter case we can store these pictures on magnetic tape and use image analysis computer routines to analyze quantitatively any stored information.[20] With a typical frame transfer time of ~20 ms, we also can follow kinetic property changes or processes at interfaces or in thin films.

If one takes images in this microscopic mode at a given incident angle, then only those areas that are at resonance for this angle appear dark in reflection while all other areas reflect some of the laser intensity because they are more or less off-resonance.

An example is given in Figure 12 for a sample that was laterally structured according to the scheme depicted in Figure 12b: the azo-initiator system introduced in Figure 4a was activated in the presence

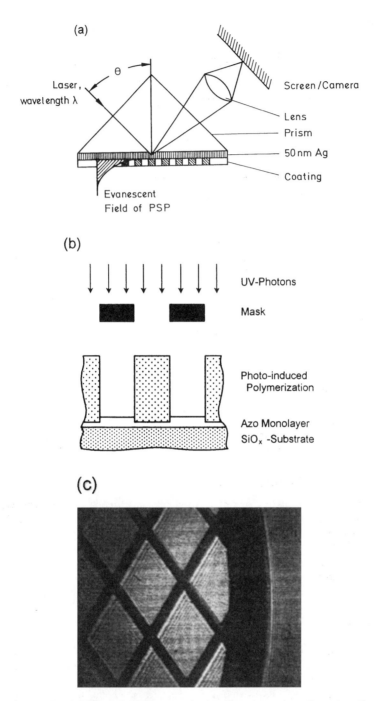

Figure 12 (a) Schematic of SPM and OWM setup in the Kretschmann configuration. The scattered and outcoupled guided light is Fourier-transformed by a lens to give an image of the interface in real space. (b) Schematic description of the procedure used for patterning of covalently bonded polymer films by photopolymerization. (c) SPM images of a polystyrene thin film grown by photopolymerization though a mask. The angle of incidence, $\theta = 48.7°$, was chosen such as to excite surface plasmons in the uncoated areas (which hence appear dark). The stripes are 48 μm wide. (d) Same as (c) but after tuning to $\theta = 57.7°$ where the coated areas are in resonance.

of styrene monomers by illuminating the functionalized surface with UV light through an electron microscopy copper grid used as the structuring mask. The photoinduced polymerization thus generated a polystyrene coating only in the illuminated areas. The corresponding surface plasmon microscopy

(d)

Figure 12 (continued)

(SPM) image taken at $\theta = 48.7°$ (Figure 12c) finds the unexposed areas in resonance, i.e., dark in reflection, while the polymer-covered illuminated areas are bright. (The periodic intensity variation at the left side of the bright squares originates from the interference of transmitted PSP light and the incident plane waves).[45] By increasing the incident angle to $\theta = 57.7°$ the polymer-coated areas gradually tune into resonance while the unexposed areas are again highly reflecting. The result is an identical image but with reversed contrast (Figure 12d).[46]

The obtainable contrast, i.e., the difference between the reflectivities from different areas at a given angle, is determined by the slope of the resonance curve for PSP excitation and the thickness- or index-dependent shift of its angular position. For example, for Ag at $\lambda = 633$ nm, an increase of the coating thickness (with n = 1.50, measured in air) of only 0.1 nm accounts for an intensity change of up to 4%, which is enough contrast to generate an image.

The highest possible contrast would be obtainable for resonance curves with the narrowest half-width. A narrow half-width, on the other hand, means a resonator with a high Q-value or with a low damping. This could be achieved by using IR light for PSP excitation because (noble) metals are more ideal in the IR, i.e., have a smaller ε'' which determines the dissipative losses and, hence, the PSP damping. As we have seen above, however, this damping also controls the propagation length of surface plasmons (through k'') which means that IR surface plasmons propagate farther than those with energies in the visible. For example, the propagation length, at $\lambda = 456$ nm for a surface plasmon at an Ag/air interface, is $L_x \approx 4$ μm, whereas at $\lambda = 1152$ nm one expects $L_x \approx 250$ μm. Along their propagation at the interface PSP waves integrate over any lateral heterogeneity and, hence, give an averaged information of the optical architecture.

As a consequence, in SPM like in any other microscopy, contrast and lateral resolution cannot be optimized simultaneously. For samples with lateral structures in the range of a few micrometers, the necessary compromise, nevertheless, allows for the imaging of thickness variations of only a few tenths of a nanometer. Two further aspects should be noted. (1) By changing the operational wavelength from $\lambda = 456$ nm to $\lambda = 1152$ nm (which is less than a factor of 2.5), one can change the lateral resolution for PSPs at a silver surface by more than a factor of 50. (2) The PSP mode damping and, hence, the lateral resolution are a function of the dielectric constant of the employed material and its wavelength dependence. While for Ag, e.g., at $\lambda = 633$ nm, the propagation length is ~50 μm, the interband transition occurring for Au at ~520 nm increases already at $\lambda = 633$ nm the imaginary part of ε_{Au} substantially, with the result that the propagation length drops to $L_x = 4$ μm. The practical consequence is that with Au substrates high-resolution SPM is possible even when working with an HeNe laser as the light source of $\lambda = 633$ nm.

In passing we should note the possibility for a quantitative analysis of SPM pictures.[12] If a series of images is taken at different angles of incidence, an image analysis computer routine can take little frames of preselected size of the images and calculate the average intensity of the corresponding pixel field. The obtained gray value is a measure of the intensity reflected from this (local) area. If one now plots these intensities as a function of the respective angle of incidence (for the various areas), one obtains a curve which is equivalent to the angular reflectivity scans discussed above. Again, a Fresnel fit calculation can yield the optical thickness of the different areas. For the example presented in Figure 12, the angular

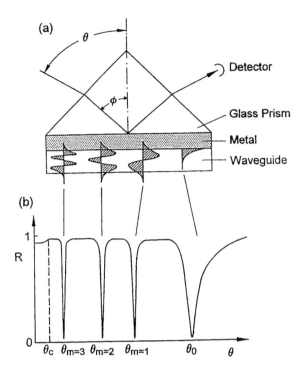

Figure 13 (a) In addition to the PSP wave, various waveguide modes can be excited by this configuration, provided the dielectric thin film (waveguide) structure is thick enough. (b) Excitation of these modes can also be seen in the reflectivity curve as very sharp dips as one measures the reflectivity R as a function of the angle of incidence θ. Modes are indexed according to the number of nodes of their field distribution in the waveguide structure. The latter is schematically sketched in the waveguide slap, see (a).

shift of $\Delta\theta = 9.0°$ corresponds to a thickness difference of $\Delta d = 21.5$ nm and, once again, gives an estimate of the sensitivity of the technique. Other than in the usual attenuated total reflection (ATR) scan, however, this time also a high lateral resolution was obtained.

9. OPTICAL WAVEGUIDE SPECTROSCOPY

The concept of surface plasmon spectroscopy and SPM can be directly extended to waveguide spectros-copy if thicker samples are to be characterized.[19,44] Then, in addition to PSP waves, guided optical modes can be excited. These new resonances can be seen in the angular scan of the reflected intensity. The extension of the concept of bound waves from evanescent surface plasmons to guided optical waves is schematically depicted in Figure 13a. Generally, these modes can be excited if the light traveling inside such a thin slab configuration is totally reflected at the boundaries to the surrounding media and if they fulfill the well-known mode equation:[47]

$$2k_z d + 2\beta_0 + 2\beta_1 = 2\pi m \tag{19}$$

with k being the wave vector of the mode of order m, d the thickness of the waveguide structure, and $2\beta_i = r_i'' / r_i'$ with $r_i' + ir_i''$ the complex reflection coefficient at the interface between waveguide and metal and between waveguide and air, respectively. (For a more detailed description see, e.g., Reference 30.) The excitation of these modes of different order again can be seen if the reflected intensity is recorded as a function of the angle of incidence, θ (Figure 13b): narrow dips in the reflectivity curve indicate the existence of the various guided waves. Provided the refractive index of the coupling prism is sufficiently high, the resonance observed at the highest angle θ_0 is still the PSP excitation. If the film thickness exceeds the penetration depth of the plasmon, l_c, then its angular position is not dependent upon the sample thickness, but only monitors the out-of-plane index n_z. For an all-dielectric waveguide configu-ration (e.g., a glass substrate, a higher-index glass waveguide, and air as superstrate), this resonance would correspond to the m = 0 mode. The next resonance seen at smaller angles is the first guided optical wave, the m = 1 mode. The number of modes that can be excited for a given waveguide material depends only on the thickness of the guiding layer because this determines how many resonances can fulfill the mode equation. These modes are also nonradiative, i.e., they, too, need a prism (or grating) to couple to photons. The configuration shown in Figure 13a is somewhat special in that this geometry ensures that guided light is constantly coupled out again through the prism so that the propagation length L_x is also reduced to a few micrometers, necessary for the microscopic use of these modes. What makes guided optical waves a particularly valuable diagnostic tool is the fact that they can be excited with both TM and TE light, i.e., with p- and s-polarized photons. For polymeric materials with optical anisotropy this means that different components of the dielectric tensor of the thin film structure can be probed. The theoretical treatment for birefringent materials, however, can be rather complicated.

The first example that we present concerns the angular reflectivity spectrum obtained with a solid polyelectrolyte sample (structure formula given in Figure 14a) with ethylorange sulfonate as an NLO-active counterion prepared by spin casting from solution.[48] Figure 14b and c give the obtained waveguide resonances for s- and p-polarized light, respectively. The corresponding Fresnel fit calculation (full curves in Figure 14b and c) shows excellent agreement with the measured data. Since each of these modes depends on n (n_x, n_y, n_z) and d, a very accurate determination of the waveguide thickness and its anisotropic refractive index can be obtained. For the example presented the Fresnel fit yields d = 1372 nm for the thickness and an indicatrix that is characterized by two different refractive indices $n_x = n_y = 1.694$ and $n_z = 1.693$, thus indicating a slight uniaxial birefringent behavior of the film, presumably induced by the spin-casting process.

The full potential of waveguide spectroscopy becomes evident for the next example that was prepared by the LBK technique from monolayers of the azobenzene-functionalized polyglutamate $P_{2,10}$.[49] Its structure formula is given in Figure 15a. One hundred fifty-six monolayers were assembled into a waveguide structure of d = 0.37 μm so as to allow for the excitation of one s- and one p-polarized waveguide mode. The angular scan as obtained for the dark-adapted trans-state of the chromophore with p-polarization is given in Figure 15b, labeled dark.

Upon irradiation with unpolarized UV light, $\lambda = (360 \pm 30)$ nm, following the procedure explained above (see Figure 10b), trans⇒cis photoisomerization takes place, and the mode shifts its angular position to lower incidence angles (labeled UV in Figure 15b). The mode recovers exactly its initial angular position, after blue light, $\lambda = (450 \pm 30)$ nm, irradiation, with the subsequent cis⇒trans back-photo-reaction (see Figure 15b). The angular reflectivity scans for both TM and TE light were taken in the dark, i.e., with the pump beam off, and with the waveguide modes propagating successively parallel and perpendicular to the dipping direction of the LBK preparation by rotating the sample. For each switching step, the irradiating light (UV and blue) was kept onto the sample for 12 min to reach the photostationary state of the photoisomerization reaction. After the UV light irradiation, the angular positions of the guided modes remain unchanged during the reflectivity scan (in the dark), because the cis⇒trans thermal back-reaction needs more than 15 h to be completed at room temperature, and the cis state could be considered to be stable within the time scale of minutes.

From all measurements (with p-polarized light and with s-polarization with $k_x \perp$ dipping and ∥ dipping direction), one derives to the evolution of in-plane (n_x, n_y) and out-of-plane (n_z) refractive indices of the $P_{2,10}$ LBK structure, under successive UV and blue unpolarized light irradiation cycles as given in Figure 15c. The mean refractive index, $n = (n_x + n_y + n_z)/3$, is also depicted in this figure. In all the columns (labeled New, UV, and Blue, corresponding, respectively, to the LBK structure before any irradiation, after UV, and after blue light irradiation) a small and persistent in-plane anisotropy ($n_y - n_x$) can be noted between the dipping direction (y) and the direction x perpendicular to it. This is due to the LBK film deposition process where the flow orients the rods parallel to the transfer direction.

Figure 15c also shows in the columns labeled New and Blue, that the out-of-plane refractive index is much higher than the in-plane refractive indices ($n_z - n_{x,y} \approx 0.14$; where $n_{x,y}$ is the in-plane average refractive index), which means that the side chains with the azobenzene are highly oriented and point out perpendicularly to the plane of the substrate. In the New film this orientation is due to the structure of the monolayer at the air/water interface on the trough which is conserved by transfer and which is common to all azo-polyglutamates. When the LBK film is exposed to the UV light, n_z decreases significantly ($\Delta n_z \approx 0.1$), and n_x and n_y increase both by nearly the same amount (see UV columns in Figure 15c). This shows that the polarizibility of the azobenzene molecules decreases as a consequence of the change in their electronic and structural properties induced by the photoisomerization from a planar to a bend structure. This can be seen also at the value of n (the mean refractive index), which is

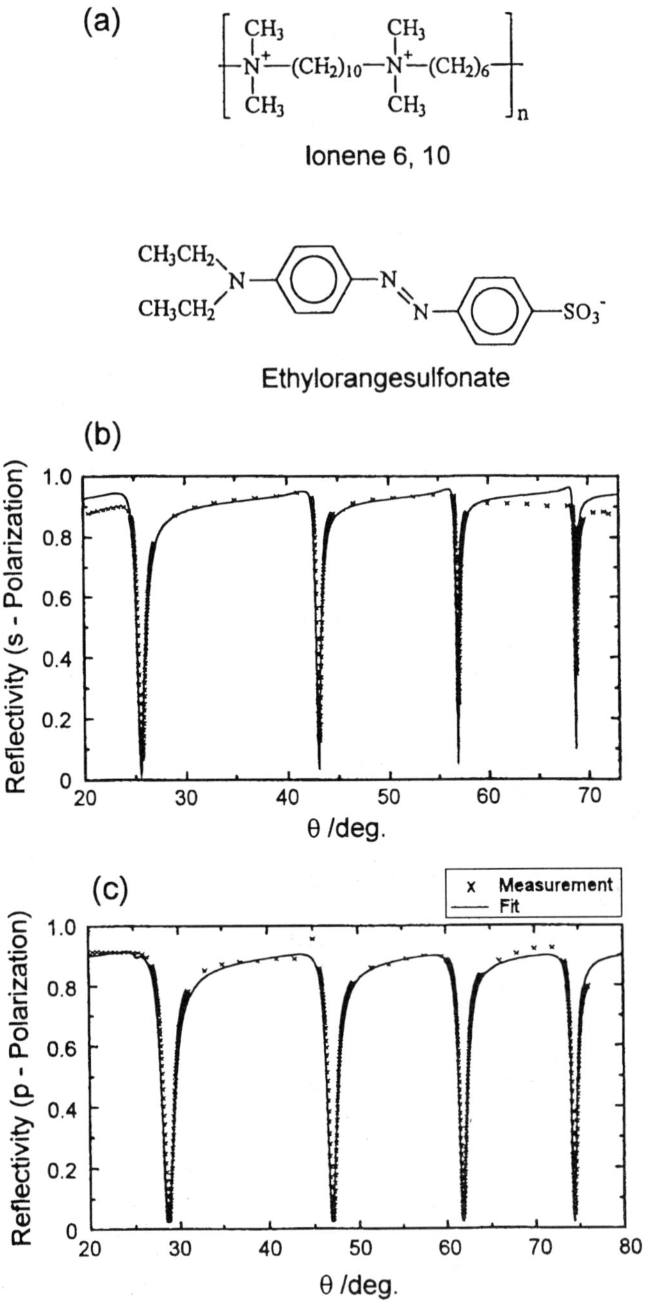

(a)

Ionene 6, 10

Ethylorangesulfonate

(b)

(c)

Figure 14 (a) Structure of the ionene 6, 10 which was doped (by ion exchange) with ethylorange sulfonate. (b) Waveguide mode pattern of a thin film of ionene 6, 10 containing ethylorange sulfonate with s-polarized light. (c) As (b), but taken with p-polarized light.

not conserved after the UV irradiation (see UV columns in Figure 15c). The thickness of the LBK film does not change under UV and blue light irradiation, as it was also confirmed by X-ray experiments. When the sample is exposed to the blue light, n_z increases, and n_x and n_y decrease to nearly their initial values before UV irradiation. This means that the packing of the end chains cannot be irreversibly distorted by the isomerization. The film structure is stable and the chromophores retain full memory of their initial orientation in the dark state prior to the UV irradiation.

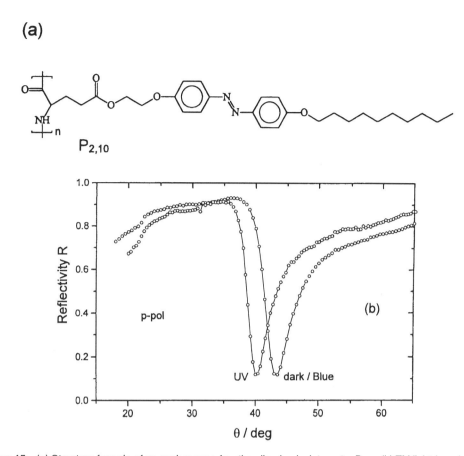

Figure 15 (a) Structure formula of an azobenzene-functionalized polyglutamate, $P_{2,10}$. (b) TM light (p-polarized) mode spectra guided into the LBK polymer film consisting of 156 $P_{2,10}$ monolayers. Dark, UV, and Blue refer, respectively, to the angular position of the mode before any irradiation, after a 360-nm UV light irradiation, and after 450-nm blue light irradiation. (c) Evolution of the indices of refraction in the plane of the LBK ($P_{2,10}$ 156 monolayers) structure (n_x, n_y), and in the perpendicular direction (n_z) under different conditions. The first column refers to the sample before any irradiation (New), and the columns labeled UV or Blue refer, respectively, to the sample after UV (360 nm) and blue (450 nm) light irradiations. The very high anisotropy shown in the columns labeled New or Blue indicates a highly optically anisotropic LBK structure. The columns labeled UV show a much less optically anisotropic structure. The evolution of the mean refractive index n under successive UV and blue light irradiation suggests that the azo-molecules are switched between the two conformations (e.g., cis and trans), without a photobleaching effect.

This example demonstrates the enormous potential of waveguide spectroscopy for a detailed analysis of optically anisotropic thin polymer films. Even subtle changes of their thickness or refractive index (anisotropies) can be monitored with high precision and, in principle, with good time resolution for transient phenomena.

10. OPTICAL WAVEGUIDE MICROSCOPY

As in the case of surface plasmons also guided optical waves can be used to illuminate the thin film sample in a microscopic mode.[44] The configuration sketched in Figure 13 is particularly suited because the guided modes excited at discrete angles are constantly coupled out through the prism and hence propagate for only a few micrometers. After Fourier-transforming this scattered light by a lens (in a setup that is identical to SPM), one therefore obtains images of the polymer film that can have a correspondingly high lateral resolution. This is demonstrated in Figure 16 for a sample that was prepared as a thin polymer film spin cast onto an Ag-coated substrate.[50] The structure formula of the employed photoreactive material is given in Figure 16a. This polymer had been shown to undergo a change in

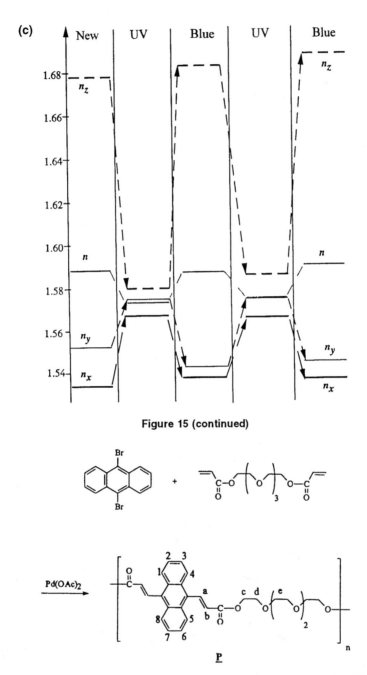

(c)

Figure 15 (continued)

Figure 16 (a) Synthetic route to polymer P. (b) OWM photograph taken at an angle of incidence $\theta = 53.6°$ from a film of P irradiated for 15 min in an Ar atmosphere through a grid. The bar corresponds to 100 μm.

thickness and in refractive index upon illumination with UV light of $\lambda = 345$ nm wavelength. The sample imaged in Figure 16b was photostructured by irradiating the thin film through a mask. The image was taken at $\theta = 53.6°$. At this angle waveguide modes are excited in the illuminated areas which hence appear dark. The pristine material is optically (and geometrically) thicker and therefore is still bright. Increasing the angle of incidence would result in a reversal of the contrast.

The significant advantage of optical waveguide microscopy (OWM) results from the fact that guided waves of different order and excited with light of different polarization can be used for the illumination and hence allow for the separate imaging of lateral heterogeneities of the thickness, the refractive index or its anisotropy. OWM is therefore a very powerful tool for the characterization of thin polymer films.

(b)

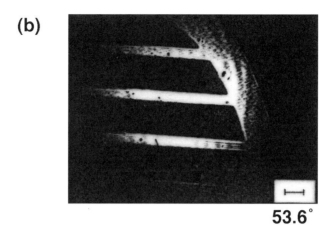

53.6°

Figure 16 (continued)

We should point out that the technique, by no means, is limited to the analysis of the linear optical properties and their lateral variations in thin films. Also NLO properties, e.g., lateral variations of the Pockels coefficient, can be imaged and hence allow for a very valuable characterization of optoboards where structures with dimensions in the range of a few micrometers are found.[48]

ACKNOWLEDGMENTS

It is my pleasure to acknowledge the discussion and collaboration with many colleagues, in particular, with R. Advincula, E. F. Aust, M. Büchel, W. Hickel, S. Ito, D. Kamp, R. Lawall, M. Liley, H. Menzel, W. Meyer, H. Motschmann, K. Müllen, H. Orendi, S. Paul, D. Piscevic, O. Prucker, R. Reiler, B. Rothenhäusler, J. Rühe, M. Sawodny, A. Schmidt, T. Seki, Z. Sekkat, J. Spinke, M. Stamm, G. I. Stegeman, S. Stein, J. D. Swalen, G. Tovar, G. Wegner, and S. Weiss. I thank J. D. Swalen for many helpful recommendations during the preparation of this manuscript.

REFERENCES

1. Bowden, M. G. J. (1988) Polymers for electronic and photonic applications, in: M. J. Bowden, S. R. Turner, Eds., *Electronic and Photonic Applications of Polymers,* Advances in Chemistry Series No. 218, Washington, D.C., American Chemical Society.
2. Seki, T., Sakuragi, M., Kawanishi, Y., Suzuki, Y., Tamaki, T., Fukuda, R., Ichimura, K. (1993) "Command surfaces" of Langmuir-Blodgett films, photoregulation of liquid crystal alignment by molecularly tailored surface azobenzene layers. *Langmuir* 9, 211–218.
3. Ikeda, T., Sasaki, T., Ichimura, K. (1993) Photochemical switching of polarization in ferroelectric lequid crystal films. *Nature* 361, 428–430.
4. Stegeman, G. I., Seaton, C. T., Zanoni, R. (1987) Organic films in nonlinear integrated optics structures. *Thin Solid Films* 152, 231–263.
5. Burroughes, J. H., Bradley, D. C., Brown, A. R., Marks, R. N., Mackay, K., Friend, R. H., Burn, P. L., Holmes, A. B. (1990) Light-emitting diodes based on conjugated polymers. *Nature* 347, 539–541.
6. Azzam, R. M. A., Bashara, N. M. (1977) *Ellipsometry and Polarized Light,* Elsevier, Amsterdam.
7. Motschmann, H., Reiter, R., Lawall, R., Duda, G., Stamm, M., Wegner, G., Knoll, W. (1991) Ellipsometric characterization of Langmuir monolayers of "hairy rod" polymers at the air-water interface. *Langmuir* 7, 2743–2747.
8. Reiter, R., Motschmann, H., Orendi, H., Nemetz, A., Knoll, W. (1992) Ellipsometric microscopy imaging monomolecular surfactant layers at the air-water interface. *Langmuir* 8, 1784–1788.
9. Laxhuber, L. A., Rothenhäusler, B., Schneider, G., Möhwald, H. (1986) Thermodesorption of ultrathin organic films studied by reflection. *Appl. Phys. A* 39, 173–181.
10. Gauglitz, G., Brecht, A., Kraus, G., Nahm, W. (1993) Chemical and biochemical sensors based on interferometry at thin (multi-) layers. *Sens. Actuators B* 11, 21–27.
11. Biegen, J. F., Smythe, R. A. (1988) *Proc. SPIE, Int. Soc. Opt. Engin.* 897, 207–219.
12. Knoll, W. (1991) Optical characterization of organic thin films and interfaces with evanescent waves. *MRS Bull.* 16, 29–39.
13. Burstein, E., Chen, W. P., Chen, Y. J., Hartstein, A. (1974) Surface polaritons-propagating electromagnetic model at interfaces. *J. Vac. Sci. Technol.* 2, 1004–1009.

14. Raether, H. (1988) *Surface Plasmons on Smooth and Rough Surfaces and on Gratings,* Springer Tracts in Modern Physics, Vol. 111, Berlin, Springer-Verlag.
15. Rothenhäusler, B., Knoll, W. (1988) Surface plasmon microscopy. *Nature* 332, 615–617.
16. Hickel, W., Kamp, D., Knoll, W. (1989) Surface plasmon microscopy. *Nature* 339, 186.
17. Hickel, W., Knoll, W. (1990) Surface plasmon optical characterization of lipid monolayers at 5 μm lateral resolution. *J. Appl. Phys.* 67, 3572–3575.
18. Swalen, J. D. (1979) Optical wave spectroscopy of molecules at surfaces. *J. Phys. Chem.* 83, 1438–1445.
19. Aust, E. F., Ito, S., Sawodny, M., Knoll, W. (1994) Investigation of polymer thin films using surface plasmon modes and optical waveguide modes. *Trends Polym. Sci.* 2, 313–323.
20. Page, R. H., Jurich, M. C., Reck, B., Sen, A., Twieg, R., Swalen, J. D., Bjorklund, G. C., Willson, C. G. (1990) Electrochromic and optical waveguide studies of corona-poled electro-optic polymer films. *J. Opt. Soc. Am B* 7, 1239–1246.
21. Hickel, W., Knoll, W. (1990) Optical waveguide microscopy. *Appl. Phys. Lett.* 57, 1286–1288.
22. Aust, E. F., Knoll, W. (1992) Electro-optical waveguide microscopy. *J. Appl. Phys.* 73, 2705–2708.
23. Ulman, A. (1991) *Ultrathin Organic Films,* Academic Press, San Diego.
24. Prucker, O., Rühe, J. (1993) Grafting of polymers to solid surfaces using immobilized azo-initiators. *Mater. Res. Soc. Symp. Proc.* 304, 167–174.
25. Rühe, J. (1994) Maßgeschneiderte oberflächen. *Nachr. Chem. Tech. Lab.* 42, 1237–1246.
26. Otto, A. (1968) Excitation of nonradiative surface plasma waves in silver by the method of frustrated total reflection. *Z. Phys.* 216, 398–403.
27. Kretschmann, E., Raether, H. (1968) Radiative decay of non-radiative surface plasmons excited by light. *Z. Naturforsch.* 23, 2135–2136.
28. Gordon, II, J. G., Swalen, J. D. (1977) The effect of thin organic films on the surface plasma resonance on gold. *Opt. Commun.* 22, 374–378.
29. Pockrand, I., Swalen, J. D., Gordon, II, J. G., Philpott, M. R. (1978) Surface plasmon spectroscopy of organic monolayer assemblies. *Surf. Sci.* 74, 237–244.
30. Swalen, J. D. (1986) Optical properties of LB-films. *J. Mol. Electron.* 2, 155–181.
31. Tovar, G., Paul, S., Knoll, W., Prucker, O., Rühe, J. (1995) Patterning molecularly thin films of polymers — new methods for photolithographic structuring of surfaces. *Supramol. Sci.* 2, 89–98.
32. Advincula, R., Aust, E., Meyer, W., Steffen, W., Knoll, W. (1996) *Polym. Adv. Technol.,* in press.
33. Wang, J., Meyer, W., Wegner, G. (1995) Synthesis and solid-state properties of comb-like ionenes. *Acta Polymerica* 46, 233–238.
34. Decher, G., Hong, J.-D. (1991) Buildup of ultrathin multilayer films by a self-assembly process: II. consecutive adsorption of anionic and cationic bipolar amphiphiles and polyelectrolytes on charged surfaces. *Ber. Bunsenges. Phys. Chem.* 95, 1430–1434.
35. Advincula, R., Aust, E. F., Meyer, W., Knoll, W. (1996) In-situ investigation of polymer self-assembly solution adsorption by surface plasmon spectroscopy. in preparation.
36. Sawodny, M., Schmidt, A., Stamm, M., Knoll, W., Urban, C., Ringsdorf, H. (1991) Photoreactive Langmuir-Blodgett-Kuhn multilayer assemblies from functionalized liquid crystalline side chain polymer, I. homopolymers containing azobenzene chromophores. *Polym. Adv. Technol.* 2, 127–136.
37. Hickel, W., Knoll, W. (1991) Time- and spatially resolved surface plasmon optical investigations of the photodesorption of Langmuir-Blodgett multilayer assemblies. *Thin Solid Films* 199, 367–373.
38. Piscevic, D. (1995) Ph.D. thesis, Johannes-Gutenberg-Universität Mainz, Mainz, Germany.
39. Sawodny, M., Schmidt, A., Urban, C., Ringsdorf, H., Knoll, W. (1992) Photoreactions in Langmuir-Blodgett-Kuhn multilayer assemblies of liquid crystalline azo-dye side-chain polymers. *Prog. Colloid Polym. Sci.* 89, 165–169.
40. Spinke, J., Liley, M., Piscevic, D., Knoll, W. (1996) in preparation.
41. Levy, Y., Jurich, M., Swalen, J. D. (1985) Optical properties of thin layers of SiO_x. *J. Appl. Phys.* 57, 2601–2605.
42. Aust, E. (1994) Ph.D. thesis, Johannes-Gutenberg-Universität Mainz, Mainz, Germany.
43. Dumont, M., Levy, Y., Morichère, D. (1990) in *Proc. Nato Conf. on Organic Molecules for Nonlinear Optics and Photonics,* J. Messier, Ed., ASI Series, Kluwer Acad. Publisher, La Rochelle.
44. Aust, E. F., Sawodny, M., Ito, S., Knoll, W. (1994), Surface plasmon and guided optical wave microscopies. *Scanning* 16, 353–361.
45. Rothenhäusler, B., Knoll, W. (1988) Surface plasmon interferometry in the visible. *Appl. Phys. Lett.* 52, 1554–1556.
46. Paul, S. (1995) Ph.D. thesis, Johannes-Gutenberg-Universität Mainz, Mainz, Germany.
47. Tien, P. K. (1977) Integrated optics and new wave phenomena in optical waveguides. *Rev. Mod. Phys.* 49, 361–420.
48. Weiss, S., Meyer, W. H., Aust, E. F., Knoll, W. (1996) Novel polymeric materials, architectures, and characterization techniques for electro-optics. *Mol. Cryst. Liq. Cryst.,* 280, 257–270.
49. Büchel, M., Sekkat, Z., Paul, S., Weichart, B., Menzel, H., Knoll, W. (1995) Langmuir-Blodgett-Kuhn multilayers of polyglutamates with azobenzene moieties: investigations of photoinduced changes in the optical properties and structure of the films,. *Langmuir* 11, 4460–4466.
50. Paul, S., Stein, S., Knoll, W., Müllen, K. (1994) Photostructuring of a polymer containing anthrylene-bisacrylate subunits. *Acta Polym.* 45, 235–243.

Chapter 14

Optical Properties of Thin Metal Films Interacting with Gases

P. Wißmann

Reviewed by U. Merkt

TABLE OF CONTENTS

1. INTRODUCTION

Only a few groups have systematically studied the influence of gas adsorption on the optical properties of metals up to now [Azzam and Bashara 1977, Bootsma et al. 1982, Abeles et al 1984, Watanabe 1987]. In many textbooks and handbook articles, this effect is only mentioned briefly [Hummel 1971, Wooten 1972] or not at all [Chopra 1969]. The variation of the optical properties during gas adsorption is usually considered to be very small and difficult to be detected. In addition, the metal surface has to be carefully cleaned by means of modern ultrahigh-vacuum (UHV) techniques before starting any adsorption experiment.

Nowadays, however, the preparation of such "clean" surfaces is standard procedure. Two general paths can be selected for doing so.

First, one can purify a bulk single crystal of the desired metal by heat treatment and/or ion bombardment, while the degree of outgassing and structure healing is controlled *in situ* by Auger electron spectroscopy (AES) and low-energy electron diffraction (LEED). This preparation procedure leads to clean surfaces with a well-defined structure, allowing a comprehensive check of even sophisticated adsorption theories by experiments. A huge experience has been accumulated on details of preparation parameters and their influence on the physics of the single crystals and on "tailoring" their surface properties. One can estimate that 90% of the published literature on adsorption phenomena deals with single-crystal surfaces; ellipsometric measurements have also been reported [Bootsma et al. 1982].

The method, however, also has big disadvantages. One is the difficulty of measuring absolute gas coverages, so that usually only the gas dosage is recorded, assuming that no interaction of the gas with the metal of the vacuum chamber takes place. The other is the need to fit various experimental conditions to an acceptable compromise. Among others, we should mention here the small surface area, large beam spots, the weight of the vacuum chamber or the ellipsometer effective during the adjustment procedure, local precision of the sample holder and the manipulator, window problems, and a heating facility of the chamber far away from the sensitive optical elements like polarizer and analyzer. Moreover, the calibration of the gas coverage by resistivity measurements is impossible in the case of bulk single crystals for obvious reasons.

Balancing all these difficulties we decided to follow the second path of preparing our clean metal surfaces by depositing spectrally pure metals under UHV conditions on suitable substrates. Then, the vacuum chamber can be made all of glass so that a more quantitative determination of coverage is possible. In complicated cases, a calibration of coverage scale by resistivity measurements is easily done. All adjustment problems of the sample can be solved to a high degree of accuracy, as is shown in Section 2. We have preferred the second path in spite of the fact that thin metal films usually exhibit a polycrystalline structure and a remaining roughness in the surface. Hence, it is necessary to carefully characterize the structure of the films by independent methods summarized in Section 3. The influence of film structure on the optical properties of the clean films is then described in Section 4.

Adsorption measurements on the films clearly show that the gas modifies the optical properties in many cases. Particularly drastic changes are observed if the gas penetrates into the interior of the film. As a consequence, oxidation processes can be optically studied by using standard ellipsometry [Azzam and Bashara 1977, Bootsma et al. 1982]. On the other hand, the modern technique of ellipsometrical measurement and data handling make possible the detection of even the smallest changes in the ellipsometric angles, down to 10^{-3} degrees. Hence, more or less all adsorption systems are accessible for optical studies nowadays. Various theoretical models have been developed to interpret the experimental evidence; details of these models are discussed in Section 5.

A motivating factor for optical studies on adsorption systems is to get insight in heterogeneous catalytic reactions at surfaces, for example. The advantage of light optics compared with electron spectroscopic characterization [Ertl and Küppers 1985] is the possibility of *in situ* application at high reactant pressures. Moreover, the gas/metal interaction plays an important role in corrosion protection, surface refinement, and sensorics. In all these cases the measurement of the optical constants can lead to useful information on the properties of the metal surfaces involved. Related problems are nondestructive reliability checks in microelectronics [Hummel 1989] and kinetic studies in solid state reactions. Note that the ellipsometer can usually detect changes in buried layers as far as a hundred Ångstroms from the surface even in the case of metals.

In Section 6, we will restrict ourselves to the example of thin copper films covered with oxygen, carbon monoxide, and carbon dioxide. These gases are advantageous as adsorbates because the characteristics of the chemical bond, like strength and geometry, are widely known because of their importance for the catalytic CO oxidation [Schlosser 1972]. Hence, a detailed check of the various model calculations should be possible. Oxygen is chemisorbed relatively strongly at 77 and 293 K, dissociation takes place at both temperatures, and there is a high reactivity with other adsorbed species. At higher temperatures the oxygen tends to enter the interior of the film and to start volume oxidation. Even here, the optical analysis is very promising because the penetration depth of the light corresponds well with the thickness range of interest. CO is adsorbed perpendicularly on the copper surface at 77 K, but at 293 K it is completely desorbed. Hence, the adsorption energy is medium, and no dissociation occurs. By repeating many adsorption/desorption cycles at 77 and 293 K, respectively, one can easily check the reproducibility of the measured phenomena. CO_2 is physisorbed very weakly on copper at 77 K; the adsorption energy has the order of magnitude of the condensation energy. Here, a multilayer adsorption becomes quite probable at higher gas pressures. On the other hand, the optical properties of the three gases are comparable because of their similar polarizability. Thus, it is quite obvious that a comparison of the three gases gives strong hints as to which theoretical models have to be favored for the interpretation of the optical data, and which models have to be replaced by making assumptions closer to reality.

The electronic band structure of copper is much easier to handle than that of catalytically important transition metals, which favors the interpretation of the optical data in the visible wavelength range. Last but not least, the structure of copper films has been intensively studied in the literature so that all the information on the physical status of the film surface necessary for a reliable interpretation is at hand to a wide extent.

A certain disadvantage of optical studies is that the lateral resolution is of the order of the wavelength of the light used and, hence, rather large as compared with atomic dimensions. Usually, an average over a millimeter scale is considered in such investigations. The evaluation is particularly simplified when the wavelength range is restricted to the near infrared, where the application of Drude theory for free electrons is justified. Therefore, the Drude range has been preferred for the quantitative study of the kinetics of copper oxidation. Details are reported in Section 7.

There are several open questions in surface physics which can be further elucidated by means of optical measurements on metal/gas systems. Among them, we shall refer to the density of free electrons in thin copper films or to the dependence of the optical constants on temperature. These special topics are briefly summarized in Section 8. We cannot include the treatment of interaction of gases with semiconductor surfaces into our chapter because of space limitations. Here, we have to refer to numerous handbook articles [Zemel 1975, Mahler et al. 1991, Mönch 1995] which reflect the great technical relevance for microelectronics. For the same reason, we have treated the famous bulk single-crystal measurements of Bootsma's group [1982] only relatively briefly when their results are directly correlated to our thin film measurements (refer to Section 8.4).

In order to include measurements on other adsorption systems to a certain extent, we have selected some representative and instructive examples. Characteristic statements are mentioned in Section 9, but it should be emphasized that the selection is rather individually motivated and by no means complete. A comprehensive specification of all the data published so far would certainly extend the framework of this article inadequately.

2. EXPERIMENTAL ARRANGEMENT

Ellipsometry is a particularly suitable tool for *in situ* studies of the change of the optical constants during gas adsorption [Azzam and Bashara 1977]. Figure 1 shows an approved cell made all of glass. In the figure, two windows Q of the UHV chamber allow the light beam to pass through. The windows are made of quartz glass to include the ultraviolet spectral range into investigation. They are molten to the Duran-50 cell by seven glass transitions in order to allow heating up to 350°C. During the pumping and heating process, the whole vacuum apparatus is moved out of the ellipsometer. Four electrical contacts F provide an *in situ* measurement of the resistivity by means of a compensation method [Merkt 1978].

The angle of incidence is kept constant at about 70° to guarantee a high detection sensitivity. The evaporation source E consists of a tungsten helix H with a pearl of specpure copper fixed on it. Thin copper films are deposited on the glass substrate S by direct current heating of the helix. The film thickness is estimated from the recorded resistivity. The exact value of the thickness of the film is independently determined after completion of the experiments by dissolving the film in nitric acid and performing quantitative analysis with the help of atomic absorption spectroscopy. The maximum error in the thickness determination was estimated to be about 5%.

The optical characterization is performed at the desired temperature which can be adjusted by introducing a heating source or liquid nitrogen into the reentrant cavity R of the sample holder (refer to Figure 1). The temperature is measured by means of a thermocouple fixed on a tungsten terminal which simultaneously serves as electrical contact for the resistivity measurements. Gas dosage is provided by magnetically breaking the seals of ampoules filled with the desired quantitites of chemically pure gas. Hence, the measurement is reduced to the detection of very small variations of the ellipsometric angles while the adjustment remains unchanged. Alternatively, the ellipsometer can be operated at a few selected wavelengths or with a continuous variation of wavelength (usually called spectroscopic ellipsometer).

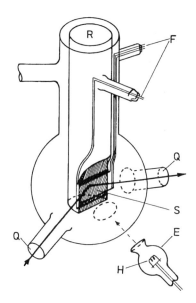

Figure 1 Glass cell for optical measurements on metal/gas systems. For details see text.

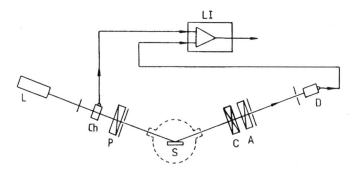

Figure 2 Beam path in a monochromatic ellipsometer in the PSCA mode. For details see text.

In the first case, an arrangement shown in Figure 2 has been proved to work successfully. A low-pressure mercury lamp serves as light source L, where the wavelength is adjusted with suitable filters. Arbitrarily, He/Ne lasers are used for 633- or 1152-nm wavelength. A germanium diode D detects the intensity I of the through-passing light, where the sensitivity of registration of the minimum intensity is increased using a tuning-fork chopper Ch and a lock-in amplifier LI. The ellipsometer is operated in the PSCA mode (Polarizer-Sample-Compensator-Analyzer), particularly successful in precision measurements [Merkt 1978]. In order to determine the exact position of the absolute minimum of intensity (x_0, y_0) we first estimate a preliminary minimum (x_0', y_0'), which serves as origin of a system of compensator positions x and analyzer positions y. Then an array of at least 3×3 compensator and analyzer positions is examined around the preliminary minimum where the step width S is chosen to be constant for both coordinates (Figure 3). The theoretically expected parabolic dependence

$$I(x, y) = ax^2 + by^2 + cxy + dx + ey + f \tag{1}$$

is evaluated with the help of a computer program which fits the experimental intensities to the paraboloid of Equation 1. In addition to the position of the absolute minimum

$$x_0 = \frac{ce - 2bd}{4ab - c^2} \qquad y_0 = \frac{cd - 2ae}{4ab - c^2} \tag{2}$$

one also obtains statements on the scattering width of the data.

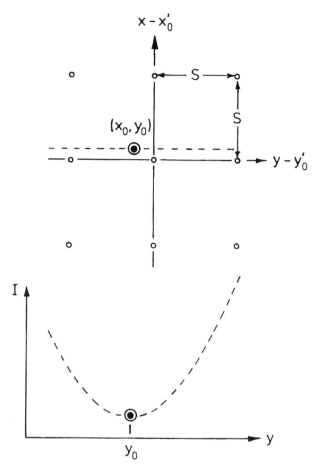

Figure 3 Quadratic mesh for the calculation of (x_o, y_o) and the parabolic dependence of transmitted intensity I on the analyzer azimuth y. For details see text.

An accuracy of a few thousandths of degrees can be achieved which is completely sufficient for a reliable detection of the optical constants at coverages in the submonolayer range. For kinetic measurements or a systematic variation of wavelength, however, such an experimental setup is not recommended because of the relatively large time consumption of 5 min for analyzing one single data point. In such a case, an automatic control of the ellipsometer is advantageous in spite of the fact that it is usually accompanied by a reduction of sensitivity by one order of magnitude.

Figure 4 shows a typical example of a so-called spectroscopic ellipsometer working in the wavelength range 400 to 900 nm [Wölfel et al. 1993]. The light source L is now a broadband tungsten lamp in connection with a monochromator M. The polarizer and the analyzer azimuths are modulated periodically by means of a magneto-optical rotation of the plane of polarization (Faraday cells F). The signal is detected by a photomultiplier PM using a two-phase lock-in technique where an on-line computer optimizes the adjustment of polarizer P and analyzer A, respectively. Simultaneously, the computer ensures that the Soleil–Babinet compensator C remains quarter-wave during wavelength variation. Here, the time consumption for a complete wavelength scan is about 15 min.

The evaluation method will be explained on the basis of Figure 5. The angular position of the main axis β of the elliptically polarized light of the reflected beam is measured as well as the ellipticity η. The incoming light is linearly polarized with an azimuth of 45° (PP in Figure 5) against the incident plane IP and the film plane FP. The angle of incidence is again about 70°. In the position of the absolute minimum, the compensator creates linearly polarized light the angular position of which is determined by the analyzer. From β and η we recalculate the phase shift Δ and the amplitude ratio ψ using the transformation equations

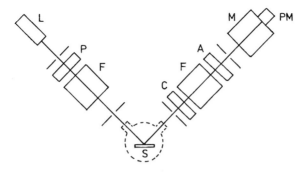

Figure 4 Beam path in a spectroscopic ellipsometer in the PSCA mode. For details see text.

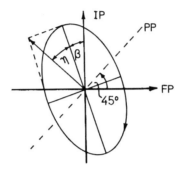

Figure 5 Quantities characterizing the position of the ellipse (for details see text).

$$\tan \Delta = \frac{\tan(2\eta)}{\sin(2\beta)} \quad \sin(2\psi) = \frac{\sin(2\eta)}{\sin \Delta} \qquad (3)$$

On the other hand, Δ and ψ define the ratio of the reflected components of the electrical field vector parallel and perpendicular to the incident plane, i.e., $R^{\parallel}$ and $R^{\perp}$, by the basic ellipsometric equation

$$\frac{R^{\parallel}}{R^{\perp}} = e^{-i\Delta} \tan \psi \qquad (4)$$

Therefore, Δ and ψ can be calculated in dependence on the optical constants n' and k of the metal film by applying suitable layer models. (The notation n has been reserved for gas coverage.) An iteration procedure as described in Section 4.1 is essentially used for the evaluation. In order to improve the justification of the assumption of a plane-parallel layer system, we first will discuss the real microstructure of the films. It is shown that the measured optical data sensitively reflect the peculiarities of the film structure, which is less ordered than that of bulk metals.

3. FILM STRUCTURE

Grain boundary density and surface roughness are enhanced in the case of thin films as compared with the bulk metal. We will concentrate our discussion on about 30-nm-thick copper films deposited at 77 or 293 K on glass substrates and subsequently annealed at higher temperatures. The films were evaporated from a tungsten helix by direct current heating. The residual gas pressure during deposition was in the 10^{-10}-mbar range, the deposition rate was about 1 nm/min.

The structure of the copper films investigated here has been extensively described by Buck et al. [1988]. The structure is polycrystalline, and the mean extension of the crystallites in the film plane approximately equals to film thickness. Note, however, that a strong variation of the crystallite size around a mean value is observed. The extension of the crystallites in the direction perpendicular to the surface is limited by film

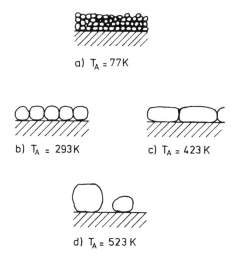

a) $T_A = 77K$

b) $T_A = 293K$ c) $T_A = 423K$

d) $T_A = 523K$

Figure 6   Crystallite structure in dependence on annealing temperature T_A for a film of medium thickness of 30 nm (schematically).

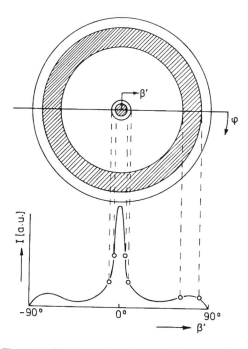

Figure 7   Pole figure for a 30-nm-thick copper film.

thickness, as the quantitative evaluation of the line width of X-ray diffraction peaks show [Gebhardt et al. 1991]. A comparison of the mean crystallite size obtained for the various directions in the film provides the real physical basis of the crystallite shapes schematically presented in Figure 6.

The formation of a (111) fiber texture is concluded from the absence of peaks of higher indices in the X-ray diffraction spectra. More-direct information is available from a texture analysis and the resulting pole figures which directly reflect the properties of the reciprocal lattice [Buck et al. 1988]. Figure 7 shows such a pole figure for the (111) reflection of a 30-nm-thick copper film. β' is the inclination angle against the film normal, and φ is the azimuth. One easily recognizes the maxima of intensity at $\beta' = 0°$ and $\beta' = 70.5°$, i.e., the crystallites are preferably oriented with their (111) planes parallel to the glass surface, and no azimuthal alignment occurs.

408

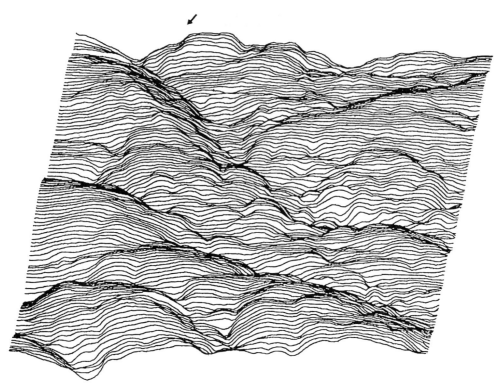

Figure 8 STM micrograph of a 30-nm-thick copper film. The area of the micrograph is 100 × 100 nm, the step height indicated by an arrow is 1 nm. We would like to thank Dr. D. Schumacher, Düsseldorf, for performing the STM investigation on our copper samples. For details see text.

A scanning tunneling micrograph (STM) of a 30-nm-thick copper film may illustrate these findings (Figure 8). The film was protected against atmospheric oxygen by additionally depositing a few angstroms of gold on top of the copper before insertion into the sample holder of the microscope. One can easily see the remaining roughness in the surface for such a film annealed only at room temperature. Again, the lateral extension of the crystallites roughly corresponds to film thickness. The step height indicated by an arrow is 1 nm.

Work function measurements provide additional information on the surface structure and orientation. Since the work function $e\phi$ is higher for smooth and densely packed surfaces and lower for rough and open structures [Somorjai 1994], the comparison of film values with $e\phi$ values published for (111) oriented single-crystal copper surfaces is useful. Figure 9 shows the dependence of ϕ on the annealing temperature T_A of the films along with the (111) copper bulk value [Hölzl and Schulte 1985]. The deviations are evident, indicating that even at high annealing temperatures the films still exhibit a certain surface roughness. The same conclusion holds when roughness factors f_R are derived from resistivity measurements [Wedler and Fouad 1963]. Figure 10 shows the dependence of f_R on the annealing temperature for CO-covered copper films. The roughness decreases steadily, but no saturation behavior is observed for $T < 373$ K. For the sake of completeness we should mention that contrary to early assumptions no thickness dependence of surface roughness can be detected (Figure 11).

Summarizing, we may state that the copper films prepared in the described manner exhibit various film structures depending on film thickness, as well as on details of the annealing treatment. A schematic survey was presented in Figure 6. After deposition at 77 K, a porous structure develops. The films consist of many small, ball-shaped crystals which are separated by pores and cavities between each other (Figure 6a). The crystallites grow with increasing annealing temperature so that a homogeneous film structure is formed. A more or less flat surface and a rather small density of crystallite boundaries is obtained for annealing temperatures of 423 K. At still higher annealing temperatures the films may crack and form an island structure, as schematically shown in Figure 6d.

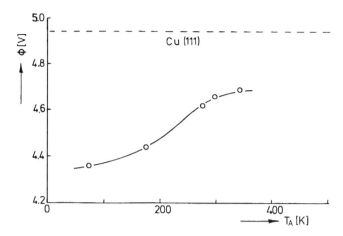

Figure 9 Work function φ in dependence on the annealing temperature T_A for 30-nm-thick copper films. (From Wißmann, P., in *Springer Tracts in Modern Physics,* Vol. 77, Springer, Berlin, 1975. With permission.)

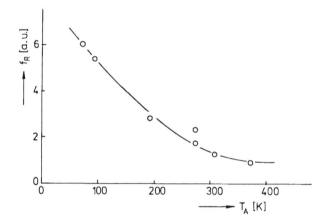

Figure 10 Roughness factor f_R derived from resistivity measurements in the Cu/CO system for differently annealed films. (From Wißmann, P., in *Springer Tracts in Modern Physics,* Vol. 77, Springer, Berlin, 1975. With permission.)

4. OPTICAL PROPERTIES OF PURE COPPER FILMS

Many groups have measured the dielectric function $\hat{\varepsilon} = \varepsilon' - i\varepsilon''$ of copper [Pells and Shiga 1969, Hagemann et al. 1974, Johnson and Christy 1975, Hanekamp et al. 1982]. The literature data on ε' and ε'', however, scatter considerably. The reason is that optical measurements are very sensitive to the surface conditions because of the limited penetration depth of the light. Hence, surface roughness as well as chemical compound formation can modify the results. Both effects are discussed in detail in the following sections. The starting point for each theoretical treatment is the so-called two-layer model, which represents the most simple application of Fresnel's formulae. Later, the model is extended by introducing further layers with the desired properties.

4.1. Two-Layer Model

To start with, the layers are assumed to be plane parallel without any interaction between film and substrate (Figure 12). The structure should be homogeneous and isotropic, only characterized by one uniform (complex) refractive index $\hat{n}$. On the basis of Fresnel's formulae we calculate the intensity of the reflected light with a component parallel to the incident plane ($R^{\parallel}$) and perpendicular to the incident plane ($R^{\perp}$) by

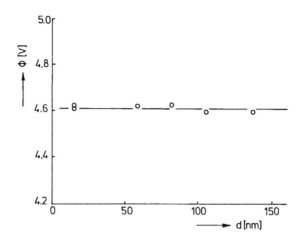

Figure 11 Thickness dependence of the work function ϕ for copper films annealed at room temperature.

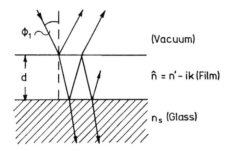

Figure 12 Two-layer model for the calculation of the optical constants n' and k of thin copperfilms deposited on glass substrates.

$$R^{\parallel} = \frac{r_{12}^{\parallel} + r_{23}^{\parallel} e^{\vartheta}}{1 + r_{12}^{\parallel} \, r_{23}^{\parallel} \, e^{\vartheta}} \tag{5a}$$

$$R^{\perp} = \frac{r_{12}^{\perp} + r_{23}^{\perp} e^{\vartheta}}{1 + r_{12}^{\perp} \, r_{23}^{\perp} \, e^{\vartheta}} \tag{5b}$$

with

$$\vartheta = 4\pi i n_2 \frac{\cos\phi_2 \cdot d}{\lambda} \tag{6}$$

Here, r_{12} are the reflection coefficients at the transition vacuum/film, and r_{23} the reflection coefficients at the transition film/substrate, respectively. n_1, n_2, and n_3 are the (complex) refraction indices of the first, second, and third layer, where n_1 is set equal to one. With Fresnel's formulae we obtain

$$r_{12}^{\parallel} = \frac{n_2 \cos\phi_1 - \cos\phi_2}{n_2 \cos\phi_1 + \cos\phi_2} \tag{7a}$$

$$r_{12}^{\perp} = \frac{\cos\phi_1 - n_2 \cos\phi_2}{\cos\phi_1 + n_2 \cos\phi_2} \tag{7b}$$

$$r_{23}^{\parallel} = \frac{n_3 \cos\phi_2 - n_2 \cos\phi_3}{n_3 \cos\phi_2 + n_2 \cos\phi_3} \tag{8a}$$

$$r_{23}^{\perp} = \frac{n_2 \cos\phi_2 - n_3 \cos\phi_3}{n_2 \cos\phi_2 + n_3 \cos\phi_3} \tag{8b}$$

and from Snell's law

$$\cos\phi_2 = \left[1 - \left(\frac{1}{n_2}\sin\phi_1\right)^2\right]^{1/2} \tag{9a}$$

and

$$\cos\phi_3 = \left[1 - \left(\frac{n_2}{n_3}\sin\phi_2\right)^2\right]^{1/2} \tag{9b}$$

Summarizing Equations 4 through 9 and separating the real and the imaginary part of Equation 4 with $n_2 = n' - ik$, we obtain two equations

$$\Delta = \Delta\left(n', k, d, \lambda, \phi_1, n_s\right)$$
$$\psi = \psi\left(n', k, d, \lambda, \phi_1, n_s\right) \tag{10}$$

which contain the two unknown quantities n' and k to be determined. λ is the wavelength of the incident light, which can be measured independently along with the angle of incidence ϕ_1 and the refractive index n_s of the glass substrate. The thickness d is obtained from a quantitative analysis of the chemically dissolved copper (refer to Section 2). We solve these equations using an iteration procedure, which can be easily managed on a personal computer. Sometimes it is preferable to evaluate the dielectric function $\hat{\varepsilon} = \varepsilon' - i\varepsilon''$ instead of the refractive index $\hat{n} = n' - ik$. The transformation equations are

$$\hat{\varepsilon} = n'^2 - k^2 \quad \text{and} \quad \varepsilon'' = 2n'k \tag{11}$$

It should be mentioned that a two-layer model can also be used to describe adsorption phenomena. The influence of the substrate is then assumed to be negligibly small which should be valid for films with a sufficiently high thickness. The first layer is then attributed to the adsorbed species, the second layer to the metal film. The situation can be considered a special case of the three-layer model described in Section 5.1.

4.2. Three-Layer Model Including Surface Roughness

The logical extension of the two-layer model for gas-covered or rough films is the introduction of an additional layer which characterizes the surface roughness or a layer of different chemical composition at the surface. The calculation follows the lines described in the last section, but three new parameters n_{eff}, k_{eff}, and d_r are introduced (refer to Figure 13).

First, we will concentrate on the roughness problem. We may calculate n_{eff} and k_{eff} from the properties of the pure metal and the vacuum by averaging in a proper manner. Such a treatment was first described for the case of bulk crystal surfaces by Fenstermaker and McCrackin [1969]. Here, we will assume for the sake of simplicity that the film surface exhibits a meander-like structure (Figure 13). Then, the most important quantity describing the surface roughness is the so-called filling factor f which can be

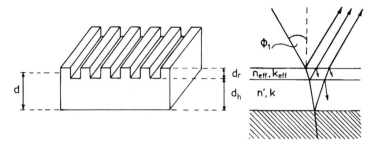

Figure 13 Three-layer model including a meander-shaped rough overlayer.

identified with the fraction of metal present in the rough overlayer. The exact lateral distribution of the metal, on the other hand, has no severe influence on the dielectric function.

In their pioneering paper, Fenstermaker and McCrackin [1969] used the Maxwell-Garnett theory for averaging. Recently, however, it was shown that an effective medium approximation (EMA) leads to more-reliable data just for filling factors around 0.5 [Theiß 1994]. We therefore preferred the EMA equation

$$0 = f \frac{\varepsilon_m - \varepsilon}{\varepsilon_m + 2\varepsilon} + (1 - f) \frac{1 - \varepsilon}{1 + 2\varepsilon} \tag{12}$$

for our simulations where ε_m is the (complex) dielectric function of the metal. The thickness of the rough overlayer d_r has been transferred to a roughness factor f_r by

$$f_r = \frac{d_r}{2d} \tag{13}$$

where d is the analytically determined film thickness. Such a roughness factor is characteristic for the volume fraction of the metal in the rough overlayer, but not for the lateral extension of the meander steps, contrary to the factor f_R plotted in Figure 10. For details of the definition of roughness factors we refer to the paper of Dayal et al. [1987]. On the basis of Equation 12 we can easily simulate the influence of roughness on the dielectric function. The result is shown in Figure 14 for a typical example. We have plotted the dielectric constant ε' vs. ε'', with the roughness factor f_r defined by Equation 13 as a parameter. The calculation was performed on the basis of $f = 0.5$, $d = 30$ nm, and $d_r = 4$ nm. Obviously, surface roughness causes ε' and ε'' to decrease.

It should be mentioned that the kind of averaging can be a certain problem when the thickness of the layer under investigation becomes too large. In the case of rough or gas-covered surfaces, however, the thickness of the surface layer is usually extremely small, and, hence, the differences due to the various averaging procedures are not significant. This is considered to be fortunate because detailed information about the structure of the surface layers is usually missing. For example, only little is known about the lateral distribution of the metal in the surface, and certainly the assumption of a meander-like profile of the surface layer is a very rough approximation. In the case of adsorbed gases, the formation of surface superstructures and the interaction between adsorbed molecules can play an important role [Somorjai 1994]. All in all, the application of Equation 12 seems to be a good compromise in the present case; for continued discussion we refer the reader to a comprehensive handbook article by Theiß [1994].

4.3. Experimental Results

A typical wavelength dependence of the dielectric constants measured with the experimental setup described in Figure 2b is shown in Figure 15. Several films annealed at 378 K were compared; the reproducibility was satisfactory [Rauh 1991]. The best agreement was found with the published values of Hanekamp et al. [1982] in spite of the fact that these authors used (110) oriented copper single crystals as adsorbent which represent a relatively open surface. The agreement with the data of Pells and Shiga [1969], Hagemann et al. [1974], and Johnson and Christy [1975] came out to be worse.

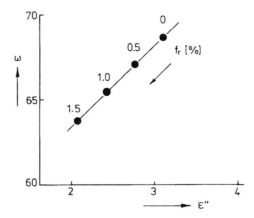

Figure 14 Plot of ε' vs. ε'' in order to show the calculated influence of the roughness factor f_r (refer to Equation 13) on the dielectric properties of copper.

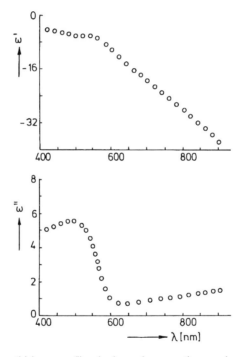

Figure 15 ε' and ε'' for 30-nm-thick copper films in dependence on the wavelength λ. (From Rauh, M., Thesis, Univ. Erlangen-Nürnberg, 1963. With permission.)

Generally, it is difficult to define a suitable standard for optical measurements. For example, our own data may contain certain restrictions due to the window effect of the measuring cell [Azzam and Bashara 1977] and due to the limited accuracy of thickness determination. We estimate our mean error in d to be about 5% (refer to Section 2). In each case, the ε' vs. λ curves clearly show the strong decrease at 550 nm which usually is attributed to a characteristic interband transition in copper at this wavelength [Hummel 1971]. At higher wavelengths the typical Drude behavior is observed, which is reflected in a proportionality of ε' to λ^2, and ε'' to λ^3 (Figure 16). Obviously, the optical properties of the copper films are directly comparable to bulk data; no features specific for films can be detected.

An instructive example of the limited reproducibility of literature values is shown for $\lambda = 1152$ nm in Figure 17. This wavelength is easily accessible using He/Ne lasers. The marked scattering is evidently due to a variation of microstructure and chemical composition of the copper surfaces under investigation. The bulk copper data (crosses in Figure 17) seem to be most reliable. On the other hand, vacuum

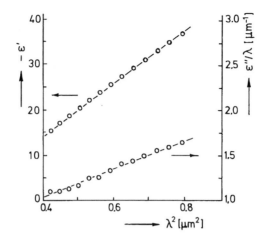

Figure 16 Check of the validity of Drude's theory (Equation 16) for the data of Figure 15.

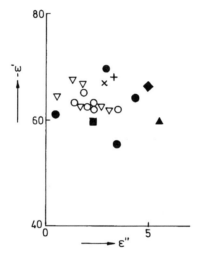

Figure 17 Dielectric constants measured at 1152 nm in a plot ε' vs. ε''. Bulk Cu values: × [Roberts 1960], + [Stoll 1969]. Film values: ∇, ○ [Rauh 1991], ● [Schmidt 1988], ◆ [Watanabe and Wißmann 1984], ▲ [Johnson and Christy 1975], ■ [Bispinck 1970].

conditions of 10^{-4} Pa are not sufficient to guarantee a clean surface. Hence, the ε' values of Johnson and Christy [1975], for example, come out to be too small (triangles in Figure 17).

Even if one tries to reproduce several measurements under well-defined UHV conditions with an optimum constancy of all preparation parameters, a remaining scattering of the optical properties cannot be totally avoided. The open symbols in Figure 17 represent a typical example [Rauh 1991], where the films with lower absolute ε' values seem to have rougher surfaces (refer to Figure 14). On the other hand, the strong change in ε'' values cannot be explained on the basis of roughness effects satisfactorily. Hence, the reason for the variation is not quite clear.

5. LAYER MODELS FOR THE DESCRIPTION OF ADSORPTION PHENOMENA

Adsorbed gases form a thin dielectric layer at the metal surface, thus modifying the optical properties. Therefore, a three-layer model just treated in Section 4.2 is a reasonable approach for the description of adsorption phenomena on thin films. Bootsma et al. [1982] have promoted such an interpretation and have called it "extrapolated macroscopic theory."

5.1. Extrapolated Macroscopic Theory

The word "extrapolated" may emphasize that the gas is considered to be a homogeneous, nonabsorbing layer even in the submonomolecular range. It is assumed that one effective refractive index n_g and one

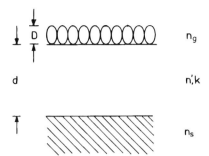

Figure 18 Three-layer model including adsorbed gases. D is the van der Waals diameter.

mean layer thickness D is sufficient to characterize the optical behavior. Figure 18 recalls the components of a three-layer model adapted to the present problem. Bootsma et al. [1982] calculated the refractive index n_g of the adsorbed layer on the basis of the Clausius–Mosotti equation

$$\frac{n_g^2 - 1}{n_g^2 + 2} = \frac{4}{3} \pi \frac{\alpha}{D} n \tag{14}$$

where α is a mean polarizability, D the van der Waals diameter of the adsorbed gas particles, and n the gas coverage in molecules per square centimeter of the geometric surface. In principle, we can also use other procedures of averaging to determine n_g. The problems involved were already discussed intensively in Section 4.2.

With literature values for the parameters in Equation 14 ($\alpha = 1.95$ Å^3 and $D = 3.8$ Å) we calculate the dependence of the change in the ellipsometrical angles $\delta\Delta$ and $\delta\psi$ on the degree θ of CO coverage ($\theta = n/n_{max}$). For $\lambda = 366$ nm, $n_s = 1.5$, $d = 30$ nm, and $\phi_1 = 70°$, we obtain the dependence shown in Figure 19a [Merkt 1978]. Here, $\delta\Delta$ and $\delta\psi$ are defined as proposed by Bootsma et al. [1982]

$$\delta\Delta = \Delta(n = 0) - \Delta(n = n_{max})$$
$$\delta\psi = \psi(n = 0) - \psi(n = n_{max}) \tag{15}$$

One easily recognizes that positive $\delta\Delta$ values and weakly negative $\delta\psi$ values are typical for a pure dielectric adlayer. Numerical values, however, depend on the system under investigation, i.e., on the size and polarizability of the adatoms (see Equation 14). Moreover, wavelength λ (refer to Figure 19b) and film thickness d (refer to Figure 20) will influence the results of the calculation. If the thickness approaches the penetration depth of the light in the metal or becomes smaller, more-drastic changes in the optical behavior occur, and $\delta\Delta$ may even become negative. This case, however, is of no severe physical interest since the films usually crack at these thicknesses to form an island structure which is accompanied by a drastic change in the optical properties. Therefore, we have concentrated our investigation on thicknesses in the range of 25 to 35 nm, where the thickness dependence of the ellipsometrical angles is negligibly small (refer to Figure 20).

Many other models have been discussed in the literature to explain adsorption phenomena, but the results were not at all convincing. In this context, the conductivity model and the Newns–Anderson model should be mentioned. On the other hand, the three-layer model can be refined by introducing a fourth layer which takes into account roughness or demetallization effects. All these models are discussed more in detail in the following sections.

Before doing so, we should mention a general criticism of the application of layer models on thin layers with thicknesses in the monolayer or submonolayer range. In a microscopic theory, nonlocal effects and the anisotropy of the dielectric function of the adsorbed layer should also be taken into consideration. A very instructive survey of the problems involved was given by Watanabe [1987]. There is, however, a lack of quantitative handling of copper surfaces in the literature. Therefore, a final discussion of necessary corrections of the layer models is an interesting area of future research. Note that the corrections are normally considered to be small for wavelengths that are large compared with the plasma wavelength of copper, i.e., in the visible and infrared wavelength regime [Watanabe 1987].

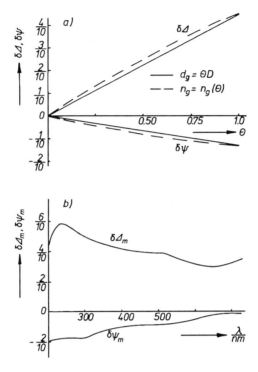

Figure 19 (a) Change in the ellipsometric angles δΔ and δψ in dependence on the degree of CO coverage θ. The curves have been calculated on the basis of Equation 14, assuming a coverage dependence of n_g (---) or of d_g (——). For calculation parameters see text. (b) Wavelength dependence of the maximum change in the ellipsometric angles $δΔ_m$ and $δψ_m$ calculated for a three-layer model with the help of Equation 14. (From Merkt, U., Thesis, Univ. Erlangen-Nürnberg, 1978. With permission.)

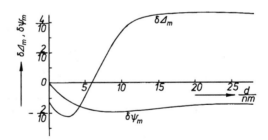

Figure 20 Thickness dependence of the maximum change in the ellipsometric angles $δΔ_m$ and $δψ_m$ calculated for a three-layer model with the help of Equation 14. For calculation parameters see text. (From Merkt, U., Thesis, Univ. Erlangen-Nürnberg, 1978. With permission.)

5.2. Conductivity Model

A further possibility for describing the influence of gas adsorption on the optical constants of thin metal films is based on the change in the conductivity σ during gas adsorption. It is advisable to look at the dielectric function $\hat{ε} = ε' - iε''$ since only $ε''$, but not $ε'$, is expected to be changed according to the Drude theory valid in the near infrared [Hummel 1971].

$$ε' = 1 - \frac{ω_p^2}{ω^2} \qquad ε'' = \frac{ω_p^2 ω_o}{ω^3} \tag{16}$$

with

$$ω_o = \frac{N'e^2}{mσ} \tag{17}$$

and

$$\omega_p = \left(\frac{4\pi N' e^2}{m}\right)^{1/2} \tag{18}$$

Here, σ is the conductivity of the film, e and m are the charge and optical mass of an electron, and N' is the electron density. ω_o and ω_p are called collision and plasma frequency, respectively; ω is the frequency of the incident light. With the electron density for bulk copper ($N' = 6.3 \cdot 10^{22}$ cm^{-3}) we obtain $\omega_p = 9.3$ eV which is in the far ultra violet. Hence, we have to expect strongly negative ε' values in the visible and near-infrared regime according to Equation 16. In principle, the free-electron density in thin films can be derived from the ε' values if the optical electron mass is known [Hummel 1971]. In practical cases, however, one must be very cautious in performing such an evaluation because of roughness effects (refer to Section 8.1).

Since the conductivity decreases with increasing gas coverage [Dayal et al. 1987], only ε'' should decrease according to Equation 16. We will show in Section 6.4, however, that a substantial correlation between optical and electrical data does not exist, contrary to some literature statements [Lin et al. 1994]. In all cases, ε'' as well as ε' are influenced by gas adsorption.

5.3. Bennett Model

An important consequence of gas adsorption on thin films is the diffuse scattering of the conduction electrons at the additional scattering centers created near the surface, as was mentioned in the last section. The effect is often described in terms of Fuchs's specularity p which represents the portion of the electrons specularly reflected at the surface. p is usually reduced during gas adsorption, leading to a lower conductivity. Bennett and Bennett [1966] have shown that the influence of p on the optical properties is particularly large for the regime of the so-called anomalous skin effect. The p-factor may modify ε'' even in the visible range, but again ε' should be constant because of the validity of Equation 16.

5.4. Newns–Anderson Model

Another consequence of the adsorption of gases on metal surfaces may be seen in the creation of new empty electron states near the Fermi level. These states can be partially filled by illuminating the surface by light. Hence, the absorption index is enhanced as compared with the pure metal. A typical example is the adsorption of carbon monoxide on copper, where the unoccupied $2\pi^*$ orbital is broadened and energetically lowered by the backdonation of metallic d-electrons [Watanabe and Wißmann 1984]. Figure 21 shows a schematic representation of the band structure for this metal/gas system. It is easily recognized that the adsorbed CO exhibits quasi-metallic properties; i.e., all energetic states up to the Fermi level are occupied, and above the Fermi level new empty states are available for the conduction electrons. In general, this behavior can be described by a dielectric function with negative ε' values and a marked complex portion ε'' [Watanabe and Wißmann 1984]. The resulting changes in $\delta\Delta$ and $\delta\psi$ are plotted in Figure 22 vs. gas coverage (dashed curve). The calculation is again based on the three-layer model of Figure 18. On easily recognizes that the $\delta\Delta$ values, as well as the $\delta\psi$ values, are enhanced as compared with a pure dielectric adlayer (dotted curve). Note that the Newns–Anderson model has also been applied to interpret the changes in resistivity during gas adsorption [Persson 1991]. The charge transfer z (refer to Figure 21) was considered to be the most important quantity in this connection [Rauh and Wißmann 1995]. Complications arise, however, when the σ-bond and the $2\pi^*$ backdonation result in charge transfers of opposite sign which partially compensate each other. Unfortunately, this is the case for the Cu/CO system [Xu et al. 1992].

5.5. Four-Layer Model Including Roughness Effects

A four-layer model is recommended for the description of adsorption effects if the gas is adsorbed on rough film surfaces. It can be derived from the three-layer model of Figure 18 by introducing an additional rough intermediate layer between the gas and the metal. Two further unknown parameters, however, arise from this procedure. The thickness as well as the dielectric function ε of the intermediate layer have to be fitted to the experimental data, taking into account that the gas will not form a closed adsorption layer but will first fill up successively the voids in the intermediate layer. Therefore, the EMA equation (Equation 12) now contains three terms with different filling factors characterizing metal, gas, and vacuum, respectively:

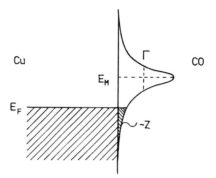

Figure 21 Shift and broadening of the unoccupied $2\pi^*$ orbital in the Cu/CO system (schematically). Γ and E_M are the width and the energetic position of the peak maximum, respectively.

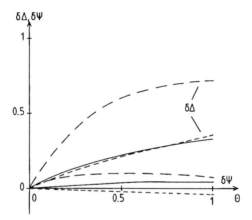

Figure 22 Calculated ellipsometric angles $\delta\Delta$ and $\delta\psi$ in dependence on the degree of coverage $\theta = n/n_{max}$. (----) $\varepsilon_g = 2$, (- -) $\varepsilon_g = -2 - 4i$, (——) $\varepsilon_g = +2 - 4i$. Numerical values: $\varepsilon_m = -64.9 - 2.5\ i$, $d = 34$ nm, $d_r = 0.4$ nm, $\lambda = 1152$ nm. (From Schmidt, R., Thesis, Erlangen-Nürnberg, 1988. With permission.)

$$O = f_m \frac{\varepsilon_m - \varepsilon}{\varepsilon_m + 2\varepsilon} + f_g \frac{\varepsilon_g - \varepsilon}{\varepsilon_g + 2\varepsilon} + f_v \frac{1 - \varepsilon}{1 + 2\varepsilon} \tag{19}$$

The sum of the filling factors is again 1. The numerical solution of this equation leads to the dependence shown in Figure 23 for three typical values of the dielectric function of the gas ($\varepsilon_g = 2$; $\varepsilon_g = 2 - 4i$; $\varepsilon_g = -2 - 4i$). A positive $\delta\psi$ is calculated while $\delta\Delta$ tends to be negative. The exact amount depends on the filling factors f_m under investigation (Figure 24).

The model of Dignam and Moskovits [1973] postulates the formation of sphere-shaped metallic clusters with the diameter L for the calculation of the dielectric properties. The authors have tested their model, however, only for thin silver films, where plasma oscillations can be created at the film surface [Wooten 1972; Forstmann and Gerhardts 1982]. Thus, the interpretation of silver data leads to additional complications. Contrary to the case of silver, plasma oscillations are fortunately of no significance for copper in the visible or infrared wavelength range.

5.6. Four-Layer Model Including a Transition Layer

Habraken et al. [1980] have also considered a strong chemical interaction between metal and adsorbed gas. They introduced a transition layer where the optical properties are modified because of the interaction. The resulting four-layer model is illustrated in Figure 25. With suitable assumptions of the thickness d_T and the optical constants n_T and k_T of the transition layer we calculate the adsorption effect

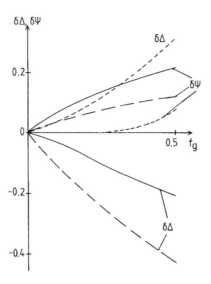

Figure 23 Calculated ellipsometric angles $\delta\Delta$ and $\delta\psi$ in dependence of the filling factor f_g of gas which is also a quantitative measure for coverage. f_m in Equation 19 was kept constant ($f_m = 0.5$); all other parameters correspond to the data of Figure 22. (----) $\varepsilon_g = 2$, (- -) $\varepsilon_g = -2 - 4i$, (——) $\varepsilon_g = +2 - 4i$. (From Schmidt, R., Thesis, Univ. Erlangen-Nürnberg, 1988. With permission.)

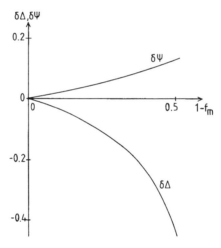

Figure 24 Calculated influence of a variation of f_m on the results of Figure 23. ε_g was kept constant ($\varepsilon_g = -2 - 4i$). (From Schmidt, R., Thesis, Univ. Erlangen-Nürnberg, 1988. With permission.)

shown in Figure 26. Here the maximum value $\delta\psi_{max}$ is plotted vs. the maximum change in phase shift $\delta\Delta_{max}$. The thickness and refraction index of the transition layer were chosen to be $d_T = 1$ nm and $n_T = 0.9$ (○ in the figure) or 1.2 (●), respectively. The absorption index k_T was varied systematically in the range 0.2 to 3.8. It is easily seen from Figure 26 that loop-shaped curves are obtained. As a consequence, it is impossible to attribute a simple well-defined couple of optical constants n_T and k_T of the transition layer to a measured couple of $\delta\Delta_m$ and $\delta\psi_m$ values. A successful application of this model is only to be expected when additional information on the nature and chemistry of the transition layer is available.

Similar statements result for five-layer models taking into account the chemical interaction between gas and metal as well as surface roughness. In special cases, even more layers have been considered, for example, to provide a description of complicated roughness profiles [Fenstermaker and McCrackin 1969]. In conclusion, however, one should be cautious not to enhance the number of layers too much. Due to the increase of fit parameters with each additional layer, a modeling of the measured data is

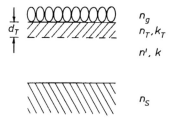

Figure 25 Four-layer model including a transition layer of thickness d_T.

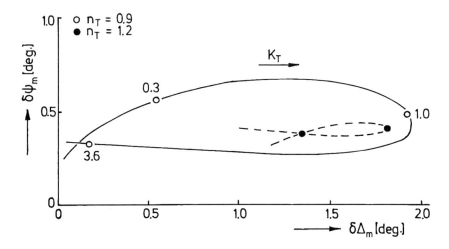

Figure 26 Influence of a transition layer of the thickness $d_T = 1$ nm on the ellipsometric angles $\delta\Delta_m$ and $\delta\psi_m$. Cu/Co, $T_M = 77$ k, $\lambda = 633$ nm, $d = 30$ nm. (From Merkt U., Thesis, Univ. Erlangen-Nürnberg, 1988. With permission.)

often accessible but physically ambiguous. Hence, it is recommended that one keep the number of layers under consideration as small as possible.

6. ADSORPTION OF O₂, CO, AND CO₂ ON COPPER SURFACES

The gases O_2, CO, and CO_2 are particularly suitable for an experimental check of the model calculations described above, because much information on the adsorption properties on copper is available [Engel and Ertl 1982; Campuzano 1990]. The interaction of copper with oxygen leads to the formation of a strong chemisorption bond which results in a partial or complete dissociation of O_2 in the temperature range between 77 and 293 K. On single-crystal copper surfaces, various characteristic superstructures are detected with LEED [Ertl and Küppers 1985], indicating that many adsorption states exist. At higher temperatures, the oxygen may penetrate into the interior of the copper but only after filling up all adsorption sites at the surface [Rauh et al. 1993].

Carbon monoxide does not dissociate at 77 K and is adsorbed more weakly than oxygen. At 293 K total desorption occurs. The molecule is bound perpendicular to the surface with the oxygen on the vacuum side. The resistivity increase is five times as large as compared with the Cu/O system, while the change in the optical properties is comparable in both systems (refer to Section 6.4).

The CO_2 bond on copper is very weak; even at 77 K practically no resistivity increase is observed [Rauh and Wißmann 1995]. Therefore, only a few data on this system are available in the literature. The molecule lies flatly on the surface, and no dissociation occurs at 77 K. The heat of adsorption is of the order of the condensation energy [Rauh et al. 1995], and, again, the molecules totally desorb during the warming up of the film to room temperature.

In order to compare the three systems mentioned above, we have kept all preparation parameters of the pure copper films as constant as possible. If not mentioned separately in the text, the conditions were $\lambda = 1152$ nm, $d \approx 30$ nm, angle of incidence $\phi_1 \approx 70°$; refractive index of the substrate $n_s = 1.47$; residual

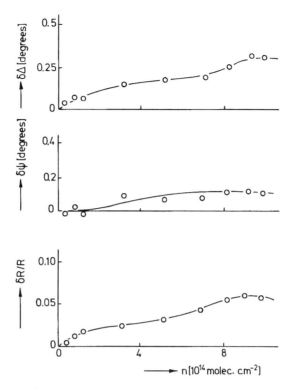

Figure 27 Change in the ellipsometric angles and the film resistance R with oxygen coverage n. $T_M = 77$ K, $\lambda = 1152$ nm, $d = 46$ nm. (From Schmidt, R., Thesis, Univ. Erlangen-Nürnberg, 1988. With permission.)

gas pressure during deposition $p = 3 \cdot 10^{-10}$ mbar; deposition rate $v = 1$ nm/min, deposition temperature $T_D = 293$ K, annealing temperature $T_A = 373$ K, and measuring temperature $T_M = 77$ K. All optical values are more or less pronounced equilibrium values. In reality, first a very quick kinetics is observed after breaking magnetically the seals of ampoules filled with a certain amount of the desired gas. This initial kinetics is attributed to the gas streaming into the cell [Rauh and Wißmann 1995]. It cannot be optically recorded because of the relatively small time resolution of our monocromatic ellipsometer described in Section 2. Therefore, at least 5 min after gas inlet is necessary before a kinetic measurement can be started. As a result, only very slow kinetic processes are studied with such a technique; a typical example is discussed in Section 7 dealing with the volume oxidation of the copper films.

In all cases, we have additionally recorded the change in the electrical resistivity during gas adsorption. In cases where the coverage dependence of the resistivity is known from precision measurements with spherical glass cells [Dayal et al. 1987], we have recalibrated the coverage scale for our optical measurements and have plotted $\delta\Delta$ and $\delta\psi$ vs. coverage n. A corresponding plot is shown in Figures 27 and 28 for the Cu/O and Cu/CO systems along with the resistance data. For the Cu/CO$_2$ system, such a recalibration seemed to be too ambiguous because of the small resistivity changes, and so we have plotted the ellipsometrical angles only versus the amount N of admitted molecules in this case. Note that N may be a quantitative measure of n for small coverages, while at higher coverages the deviations become larger. The physical reason for this effect is the rather undefined adsorbing area of the cell shown in Figure 1 [Dayal et al. 1987].

6.1. Cu/O System

A typical experimental result is shown in Figure 27. As mentioned above, we have plotted the change in the phase $\delta\Delta$, in the amplitude ratio $\delta\psi$, and in the resistance $\delta R/R$ as a function of the oxygen coverage n. The relatively complicated shape of the curves seems to indicate that several adsorption states are involved in the chemisorption bond. Comparing with Figure 22, we have to state that the assumption of a pure dielectric oxygen monolayer with a dissappearing complex part of the dielectric function is not a good approach for the real situation. On the contrary, a complex dielectric function should be postulated to interpret the curves of Figure 27 (particularly with respect to the positive $\delta\psi$ changes).

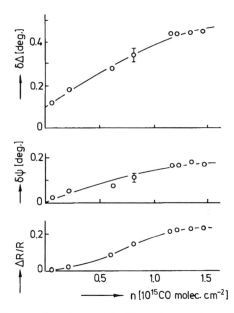

Figure 28 Change in the ellipsometric angles and the film resistance R with coverage n of carbon monoxide. $T_M = 77$ K, $\lambda = 1152$ nm, $d = 40$ nm. (From Watanabe, M. and Wißmann, P., *Surf. Sci.*, 138, 95, 1984. With permission.)

Problems also arise from the fact that oxygen is partially dissociated at the copper surface at 77 K [Buck et al. 1988] and may even enter the outer copper layers to form a subsurface species. This interpretation is supported by the observation that the oxygen uptake does not saturate at the monolayer coverage, which is assumed to be 1.5×10^{15} oxygen atoms/cm^2 [Wedler 1976]. On the contrary, $\delta\Delta$ can be further increased up to $0.8°$ by applying oxygen pressures higher than 5×10^{-4} mbar to the film. Volume oxidation starts if the measurements are performed at 378 K, but only after completion of the monolayer coverage [Rauh 1991]. Details of the oxidation process are reported in Section 7.

6.2. Cu/CO System

Again $\delta\Delta$, $\delta\psi$, and $\delta R/R$ are plotted vs. the gas coverage n at 77 K (Figure 28). The resulting change in the ellipsometrical angles is similar to the case of Cu/O in spite of the fact that CO does not dissociate at 77 K but is molecularly adsorbed in a perpendicular position. If we restrict ourselves to the evaluation of the maximum change in the ellipsometrical angles $\delta\Delta_m$ and $\Delta\psi_m$, we obtain the dependence on the wavelength of the incident light shown in Figure 29. The agreement with the theoretical curves of Figure 19b is rather poor, indicating once more that a simple three-layer model is insufficient to explain the experimental data. In particular, the sign of $\delta\psi$ changes from negative to positive at about 380 nm. This result is obviously correlated with the fact that the Drude range and the interband transition range must be discussed separately. In the following, we will concentrate on the Drude range at 1152-nm wavelength. Thus, we can circumvent many complications associated with surface states, surface anisotropy, etc. [Ertl and Küppers 1985].

The change in the ellipsometrical angles also depends on film thickness, as can be derived from Figure 30. Here the maximum values $\delta\Delta_m$ and $\delta\psi_m$ are plotted vs. film thickness d for eight different films. Obviously, very thin films crack after annealing at room temperature and form an island structure. The consequence is negative $\delta\Delta$ values at small thicknesses in Figure 30. A saturation behavior only develops for higher thicknesses of about 20 nm. Hence, optical data on gas adsorption can serve, among others, as a criterion for the homogeneity of the films [Merkt 1978].

6.3. Cu/CO$_2$ System

The dependence of the change in the ellipsometrical angles on the amount of admitted CO$_2$ molecules N is shown in Figures 31 and 32. N can only be considered as a qualitative measure of coverage at small coverages because cooled portions of the substrate holder shown in Figure 1 can also adsorb CO$_2$. Hence, the sample area cannot be identified with the area effective for the calculation of coverage for higher coverages as mentioned above.

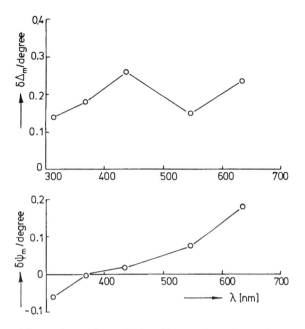

Figure 29 Dependence of the maximum change in the ellipsometric angles on the wavelength of incident light. Cu/CO system, $T_M = 77$ K, $d = 25$ nm. (From Merkt, U., Thesis, Univ. Erlangen-Nürnberg, 1978. With permission.)

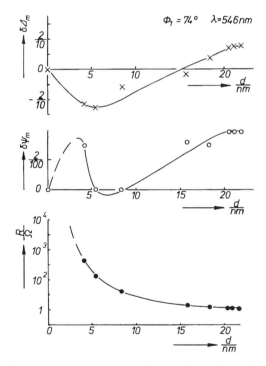

Figure 30 Dependence of the maximum change in the ellipsometric angles on film thickness d. Cu/CO system, $T_M = 77$ K, $\lambda = 546$ nm, R is the simultaneously recorded resistance of the film. (From Merkt, U., Thesis, Univ. Erlangen-Nürnberg, 1978. With permission.)

Two typical kinds of curves are detected; both were reproduced several times. Obviously, the curves with the saturation values (Figure 31) can be attributed to the formation of a monolayer of CO_2 which is completed by all molecules aligned flatly on the surface. No further CO_2 is then adsorbed even at higher CO_2 dosages. Note that the resistivity change is extremely small for this system and can even become negative [Rauh and Wißmann 1995]. On the other hand, occasionally the laser beam used as

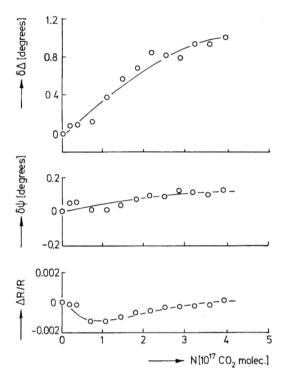

Figure 31 Change in the ellipsometric angles and the film resistance R with the amount N of admitted CO_2 molecules (pure adsorption). $T_M = 77$ K, $\lambda = 1152$ nm, $d = 30$ nm. (From Rauh, M. and Wißmann, P., *Fresenius Z. Anal. Chem.*, 1995. With permission.)

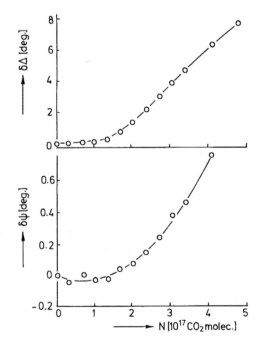

Figure 32 Change in the ellipsometric angles with the amount N of admitted CO_2 (condensation). For details see text.

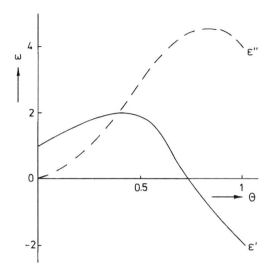

Figure 33 Change in the dielectric constants ε' and ε'' during oxygen adsorption calculated for the example $\varepsilon_g = -2 - 4i$ and numerical values of Figure 22. (From Schmidt, R., Thesis, Univ. Erlangen-Nürnberg, 1988. With permission.)

the light source meets special centers of crystallization where condensation occurs. Perhaps these centers are at a slightly lower temperature than the surrounding substrate area since small inhomogeneities in the temperature distribution cannot be avoided for the substrate of the cell of Figure 1. Then the ellipsometrical angles may increase by 8° and more (Figure 32). No saturation behavior is detected, and the total change in $\delta\Delta$ and $\delta\psi$ is a direct measure of the overlayer thickness [Watanabe et al. 1985].

6.4. Comparing Discussion

Comparing the three adsorption systems discussed above, we indeed can draw some general conclusions. First, we note that the three systems differ strongly in their adsorption energy and in the change in resistivity during gas adsorption. The binding energy is large for the Cu/O system, medium for the Cu/CO system, and small for the Cu/CO$_2$ system. The resistivity increase is large for the Cu/CO system, medium for the Cu/O system, and negligibly small for the Cu/CO$_2$ system. On the other hand, the changes in the ellipsometrical angles $\delta\Delta$ and $\delta\psi$ are comparable if one takes into account the usual scattering width of the experimental values. Hence, it is immediately concluded that the conductivity model, as well as the Newns–Anderson model, cannot explain the observed behavior. Both models predict a very large change in the optical properties for the CO adsorption and minor changes for the CO$_2$ adsorption, contrary to the experimental evidence.

A further argument against the conductivity model and/or the anomalous skin-effect model can be seen in the fact that these models predict only a change in ε'' while ε' should not be influenced. In reality, however, strong changes in ε' are observed (see, for example, Figure 33). With respect to the interpretation of resistivity data we refer to the literature [Dayal et al. 1987] because of space limitations. We can conclude from the above-mentioned results, however, that there is no strong correlation between optical and electrical properties in the present case.

The comparison of Figures 27, 28, and 31 clearly indicates that slightly positive $\delta\psi$ values are always detected. Hence, a simple three-layer model with a pure dielectric gas is not sufficient to explain the experimental data, and a four-layer model should be applied. A transition layer due to chemical interactions between gas and metal cannot play a decisive role because the interaction should be strong for the Cu/O system and very weak for the Cu/CO$_2$ system. Thus, it is concluded that the surface roughness of our copper films is responsible for the slightly positive $\delta\psi$ values.

To finally check the validity of this interpretation we plan measurements on single-crystal copper surfaces at 1152 nm adsorbing the same gases oxygen and carbon dioxide. Results are expected in the near future. For the discussion of corresponding measurements of Bootsma et al. [1982] on copper single crystals, we refer to Section 8.4. The role of surface roughness, however, has also been improved by evaluating measurements on other adsorption systems. A selected choice of such measurements is separately presented in Section 9.

The main physical quantity to be derived from the optical measurements on the adsorption systems is evidently the mean polarizability α of the gas. From Equation 14 we obtain, for example, $\varepsilon = 2.1$ for CO on copper. A comparison with literature data is difficult because of the anisotropy of the adsorption bond and the fixed angle of incidence of about 70°. For the sake of completeness we should mention that the approximative formula [Bootsma et al. 1982]

$$\delta\Delta = c\alpha \frac{n}{\lambda} \qquad \delta\psi \leq o \qquad (20)$$

does not fit the measured wavelength dependence very well. This becomes immediately evident by evaluating Figure 19b. (n is the gas coverage in Equation 20 and c a characteristic constant of proportionality.) Moreover, the positive changes in $\delta\psi$ shown in Figures 27, 28, and 31 contradict Equation 20.

Another check of the interpretation presented above can be seen in optical measurements on the Cu/Xe system. Such measurements are just now being performed by our group [Walter 1996]. If the roughness produces positive changes in $\delta\psi$, we should observe such an effect also in the Cu/Xe system, contrary to the Au/Xe system described in Section 9.4.

7. OXYGEN ABSORPTION AND KINETICS OF OXIDATION

In the preceding section we have shown that oxygen dissociates during adsorption on copper surfaces at 77 K and at 293 K. A penetration of oxygen into the interior of the film is only observed for higher temperatures. At 465 K, for example, the values of Figure 34 were measured at a time interval of 15 min after gas dosage and, hence, represent more or less equilibrium values. The resistivity vs. coverage curve runs through a characteristic maximum and thereafter through a minimum. The minimum can be approximately attributed to a monolayer coverage, and up to this point oxygen is only adsorbed at the copper surface [Dayal et al. 1987]. At a higher dosage, however, absorption begins to become effective, and resistivity as well as the ellipsometric angles clearly increase, indicating that the formation of an oxide phase in the copper volume begins. A similar behavior has also been observed for the Pd/H system at 77 K, where an α-hydride phase is formed. Because of space limitations, however, we have to refer to the literature for details [Watanabe et al. 1985].

A temperature of 378 K is sufficient to start volume oxidation of the copper films, but the time necessary to reach saturation values is now much larger. Therefore, we have kept constant the oxygen pressure in these experiments. Figure 35 shows the measured time dependence of $\delta\Delta$ and $\delta\psi$ in the wavelength range 400 to 900 nm measured with the experimental setup described in Section 2. The kinetics is interpreted on the basis of the simple layer model shown in Figure 36, where the condition

$$d_{ox} = V(d_o - d_t) \qquad (21)$$

holds. The thickness of the oxide layer d_{ox} is directly coupled to the thickness of the copper film at the time t (d_t) and at the beginning of the oxidation process (d_o) by a stoichiometric factor V which amounts to 1.7, to a good approximation for copper oxides [Rauh et al. 1993]. The evaluation of Figure 35 on the basis of the layer model of Figure 36 leads to the dielectric constants of the oxide layer shown in Figure 37 for a typical example of the oxidation of a 64-nm-thick copper film. The comparison with literature data [Rauh et al. 1993] makes evident that the oxide layer mainly consists of Cu_2O. A more detailed analysis, however, leads to the conclusion that the model should be further refined. For example, the maximum or shoulder in the ε'' vs. λ curve at about 600 nm can be only explained if the roughness at the Cu/Cu_2O interface and/or copper inclusions in the oxide layer are taken into consideration. The marked thickness dependence of the calculated ε values points to deviations from the stoichiometry; probably a CuO layer is formed on top of the layer system for thicker oxide films [Karlsson et al. 1982]. Nevertheless, Cu_2O is the main component dominating the structural properties of the oxide layer. This can be easily derived from X-ray diffraction investigations of the films. In addition to the expected copper reflections, one only detects the strongly broadened $Cu_2O(111)$ peak as shown in Figure 38

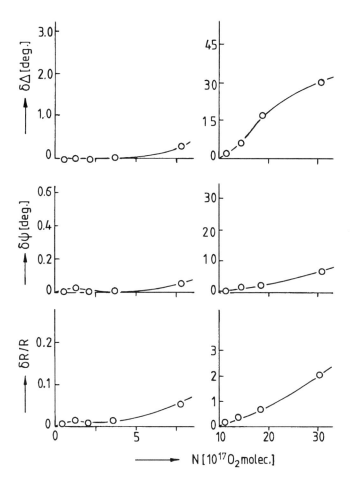

Figure 34 Change in the ellipsometric angles and the film resistance R with the amount N of admitted oxygen molecules. T_M = 465 K, λ = 1152 nm, d = 45 nm. (From Schmidt, R., Thesis, Univ. Erlangen-Nürnberg, 1988. With permission.)

[Gebhardt et al. 1991]. One reason for the missing CuO peaks may be that X-ray diffraction is not very sensitive to surface phenomena.

The behavior is particularly easily understandable when the wavelength is kept constant at 1152 nm. Here $\varepsilon'' \approx 0$ is valid to a good approximation for the oxide layer. As a consequence, $\delta\Delta$ varies proportional to the oxidation time (Figure 39) which corresponds to a linear time law:

$$d_{ox} = k_1 t \tag{22}$$

Such a time law is characteristic for a surface-controlled reaction where the volume diffusion is no longer the rate-determining step. The oxidation is activated; from an Arrhenius plot an activation energy E_a = 64 kJ mol^{-1} is derived, which corresponds well to published data on surface-controlled oxidation reactions [Rauh 1991]. The reaction mechanism can be further specified by analyzing the pressure dependence of the rate constant k_1. If the adsorption of a molecular oxygen species at the copper surface is involved in the rate-determining step, then the pressure dependence can be expressed according to Ritchie and Hunt [1969] on the basis of a Langmuir-type adsorption isotherm by

$$k_1 = k_o \, \frac{N_m \cdot p}{K' + p} \tag{23}$$

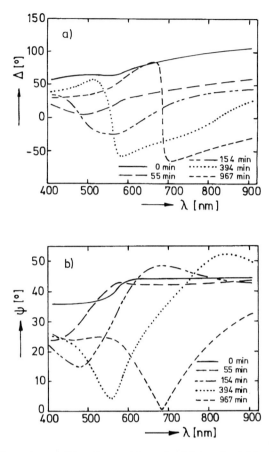

Figure 35 Wavelength dependence of the ellipsometric angles Δ (a) and ψ (b) for different times of oxidation. d_o = 64 nm, T_M = 378 K, P_{ox} = 31 Pa, ϕ_1 = 78°. (From Rauh, M., Thesis, Univ. Erlangen-Nürnberg, 1991. With permission.)

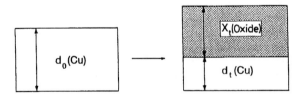

Figure 36 Schematic representation of the layer model applied to copper oxidation. d_{ox} is the thickness of the oxide layer.

where k_o, K' and N_m are characteristic constants. The evaluation of Figure 39 leads to the pressure dependence of k_1 shown in Figure 40. Obviously, Equation 23 can describe the results with sufficient accuracy (solid curve in Figure 40). This result agrees with the statements of other authors. For example, Hardel and Schön [1972] reported for copper foils oxidized at 800 to 1000°C at pressures between 5 and 93 Pa that the dissociation of the adsorbed oxygen molecules at the interphase gas/Cu$_2$O is the decisive step dominating the kinetics of oxidation. Similar conclusions were drawn by Stotz [1966].

Deviations from the linear time law are evident in Figure 39 at small pressures. They can be attributed to a volume diffusion, a field-aided oxide growth, or the influence of space charges. For a comprehensive survey of kinetic data including a parabolic, a logarithmic, or a power time law, we refer to various monographs [Leidheiser 1971, Dürrschnabel and Voßhühler 1976].

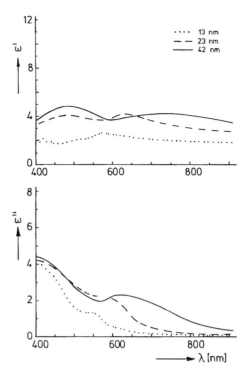

Figure 37 ε' and ε'' for the oxide layer calculated from Figure 35 on the basis of the layer model of Figure 36. Parameter in the thickness of the oxide layer d_{ox}.

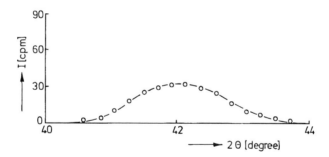

Figure 38 The $Cu_2O(111)$ peak for a 150-nm-thick copper film oxidized at $T_M = 423$ K for 3 h at an oxygen pressure of 25 mbar. (From Gebhardt, R. et al., *Fresenius Z. Anal. Chem.*, 341, 332, 1991. With permission.)

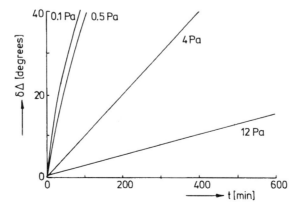

Figure 39 $\delta\Delta$ as a function of time t for various oxygen pressures (parameter). $d_o = 53$ nm, $T = 378$ K, $\lambda = 1152$ nm. (From Rauh, M., Thesis, Univ. Erlangen-Nürnberg, 1991. With permission.)

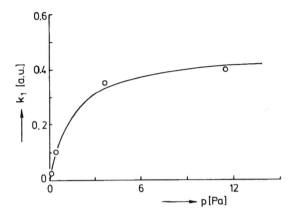

Figure 40 Plot of the data of Figure 39 to check the validitiy of Equation 23 (solid curve).

8. SPECIAL TOPICS

Up to now we have investigated the influence of adsorbed and absorbed gases on the optical properties of evaporated copper films. There are, however, some other open questions which are strongly related to adsorption and absorption phenomena. Among those, we will first check the constancy of free-electron density in copper films. In order to say it more accurately, we will answer the questions whether or not the free-electron density is reduced in thin films and whether there is a remarkable change during gas adsorption experiments or not. Then, we will look at the temperature dependence of the optical constants and study the possible influence of adsorption/desorption cycles in the case of poor vacuum conditions. Another area of interest is the properties of porous films which can be easily produced by depositing at 77 K without subsequent annealing. These films show many drastic peculiarities in their optical properties, and the interaction with gases exhibits features of adsorption and absorption as well. Last but not least, we will discuss adsorption experiments on bulk copper single crystals. There are some literature data on the Cu/O system available, and they allow one to independently check the concepts developed here for the interpretation of thin film data. Only such single-crystal surfaces are mentioned that have been prepared in accordance to the well-known surface-cleaning procedures, and whose surface properties have been checked *in situ* under ultrahigh-vacuum conditions using LEED and Auger [Ertl and Küppers 1985]. The main advantage of experiments on single-crystal surfaces is that roughness effects should be strongly reduced. We will discuss this crucial point in Section 8.4.

8.1. Free-Electron Density

A constancy of free-electron density independent of film thickness was an implicit presupposition for the model calculations presented in Figure 20. Since the justification of this assumption is still disputed in the literature [Bispinck 1970; Hoffmann 1983], we will add some remarks on this topic in the following.

Hall-effect measurements are a suitable method to determine the free-electron density in thin films. Figure 41 shows a typical example of measurements of the Hall constant R_H for evaporated copper films [Wedler and Wiebauer 1975]. The Hall constant R_H corresponds well to the bulk value of free-electron density for copper films of medium thickness. Only for ultrathin films a certain enhancement of R_H is evident. This enhancement should not, however, be attributed to size effects as predicted theoretically by Sondheimer [Chopra 1969]; more likely an influence of surface roughness is effective. The effect of CO adsorption on the Hall constant is rather small (Figure 41) and cannot at all be interpreted by a reduction of the electron density [Wedler and Wiebauer 1975].

A further powerful method for the determination of the free-electron density N' in thin films is the measurement of the real part ε' of the dielectric function in the Drude range. According to Equation 18, a proportionality between ε' and N' is to be expected in the infrared wavelength region, where the constant of proportionality is a material-specific factor only depending on the optical mass m_o and light wavelength λ. We have plotted the measured ε' values for copper films of various thicknesses at $\lambda = 1152$ nm in Figure 17 along with bulk copper data. No systematic deviations between film and bulk data are detected, the obvious scattering of the measured points can be easily traced back to roughness effects (refer to Section 4.2).

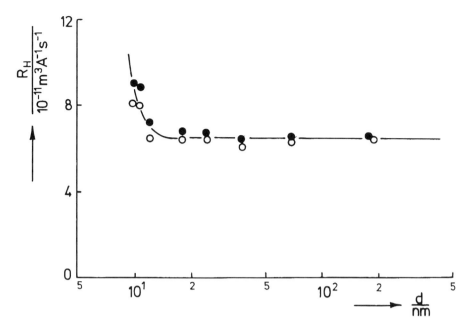

Figure 41 Thickness dependence of the Hall constant of pure (●) and CO-covered (○) copper films. $T_A = 298$ K, $T_M = 77$ K. (From Wedler, G. and Wiebauer, W., *Thin Solid Films*, 28, 65, 1975. With permission.)

In conclusion, we have to state that a thickness dependence of the free-electron density can be excluded for the films discussed here. The N' values of the films approach the corresponding bulk values satisfactorily. This statement remains valid even if the influence of adsorbed gases is taken into account. Contrary to early assumptions [Merkt et al. 1983], the gas can modify the electron density only in the upmost surface layer, which would lead to very small changes in ε'. This effect calculated on the basis of Equation 16 is usually overcompensated by the influence of the dielectric gas layer (refer to Section 5.1).

8.2. Temperature Dependence of the Dielectric Constants

One must be very cautious in performing measurements of the temperature dependence of the optical constants in the range 77 to 293 K. Excellent UHV conditions are a prerequisite for such investigations. Otherwise, the effect of residual gases adsorbed on the surface cannot be ruled out. Particularly, carbon monoxide, always present in an appreciable amount in the residual gas, is adsorbed by Cu at 77 K, but is totally desorbed while warming up the metal to room temperature.

Hence, the possibility of the superposition of an adsorption–desorption cycle during the measurement of the temperature dependence of the optical constants must be taken into consideration while discussing the data published for copper surfaces under relatively bad vacuum conditions. On the other hand, the UHV cell of Figure 1 enables us to measure the optical constants *in situ* immediately after annealing at 293 K, and at 77 K as well. In order to get insight into the temperature dependence the following quantities were calculated:

$$\frac{\Delta\varepsilon'}{\Delta T} = \frac{\varepsilon'(293\text{ K}) - \varepsilon'(77\text{ K})}{293\text{ K} - 77\text{ K}} \quad \text{and} \quad \frac{\Delta\varepsilon''}{\Delta T} = \frac{\varepsilon''(293\text{ K}) - \varepsilon''(77\text{ K})}{293\text{ K} - 77\text{ K}} \quad (24)$$

The experimental results are shown in Figure 42. The changes are very small, especially in the medium-wavelength range reported here.

For comparison, numerical values obtained by other authors for $\lambda = 633$ nm have been plotted in Figure 43 vs. the year of publication. The inserted lines seem to indicate that the amount of the temperature dependence becomes smaller with the increasing refinement of vacuum techniques. The smallest values are observed for films deposited under UHV conditions [Merkt 1978], which, again, points to the modification of the results by residual gas adsorption at 77 K.

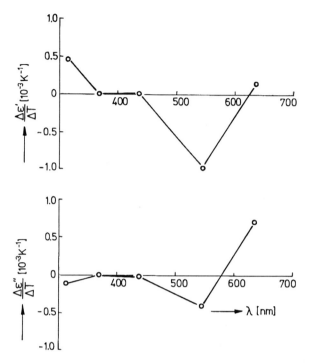

Figure 42 $\Delta\varepsilon'/\Delta T$ and $\Delta\varepsilon''/\Delta T$ (refer to Equation 24) in dependence on light wavelength λ. (From Merkt, U., Thesis, Univ. Erlangen-Nürnberg, 1978. With permission.)

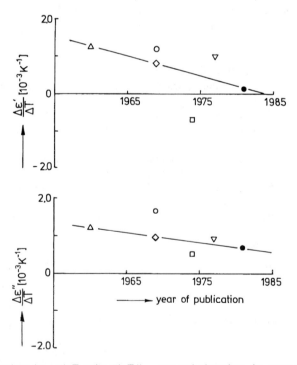

Figure 43 Literature values for $\Delta\varepsilon'/\Delta T$ and $\Delta\varepsilon''/\Delta T$ (λ = 633 nm) plotted vs. the year of publication in order to illustrate the effect of developing vacuum technique. (Δ) Roberts [1960], (○) Pells and Shiga [1969], ($\Diamond$) Stoll [1969], (□) Johnson and Christy [1974], (∇) Hollstein et al. [1977], (●) Merkt [1978].

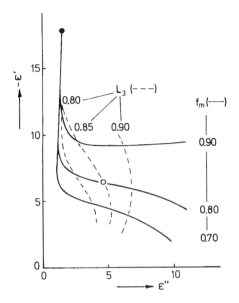

Figure 44 ε' vs. ε'' for porous copper films. ($\circ$) measured value for $T_M = 77$ K, $\lambda = 633$ nm, $d = 40$ mm. Curves calculated on the basis of Equation 25 for $\varepsilon_m = -18 - 1.6i$ ($\bullet$). f_m and L_3 are varied systematically. (From Schmidt, R. and Wißmann, P., *Fresenius Z. Anal. Chem.*, 333, 410, 1989. With permission.)

Smaller discrepancies may arise from the fact that, in principle, the thermal expansion of the metal lattice also affects the temperature dependence of optical constants [Hummel 1971]. In the case of evaporated films, thermally induced strains can develop which are caused by the difference in thermal expansion coefficients of glass substrate and metal film. Hence, no complete agreement between thin film and bulk data can be expected even at higher temperatures. Moreover, phonon excitation can play a certain role in this connection [Hummel 1971].

8.3. Porous Copper Films

Porous metal films differ strongly in their optical properties from the usual behavior discussed so far. Such films are prepared by depositing copper on glass substrates at 77 K without subsequent annealing. In order to interpret the data, we use a generalized EMA equation

$$O = \sum_{i=1}^{3} \left[f_m \frac{\varepsilon_m - \varepsilon}{\varepsilon + L_i(\varepsilon_m - \varepsilon)} + f_v \frac{\varepsilon_v - \varepsilon}{\varepsilon + L_i'(\varepsilon_v - \varepsilon)} \right] \tag{25}$$

where f_m and ε_m are the volume fraction and the dielectric function of the metal, respectively. f_v and ε_v are the corresponding quantities of the voids with $\varepsilon_v = 1$ and $f_m + f_v = 1$. ε is the measured dielectric function, and L_i and L_i' are depolarization factors in the x, y, and z direction. L_i can be calculated for idealized ellipsoidal particles or cavities [Kittel 1980]. We assume $L_1 = L_2 < L_3$ for geometric reasons, with $\sum L_i = 1$. Equation 25 is solved with the help of a suitable computer program for various real couples of parameters f_m and L_3. ε_m is kept constant and replaced by $\varepsilon_m = -18 - 1.6i$ [Schmidt and Wißmann 1989]. The result of the calculation is shown in Figure 44 in a plot ε' vs. ε''. The measured value is additionally included in the figure (open circle). The main result is that a coincidence of theoretical and experimental values can only be achieved for a strongly anisotropic depolarization of the metal. For the isotropic case $L_1 = L_2 = L_3$ a fit comes out to be impossible.

The experimental value is theoretically represented by $f_m = 0.80$ and $L_3 = 0.85$ (Figure 44). The solution is unique, all other solutions obtained from Equation 25 show negative ε'' values and, hence, are unqualified for a meaningful physical interpretation. Thus, the void fraction in the films amounts to 20%, in good agreement with the qualitative picture of Figure 6a. The relatively high L_3 value can be traced back either to the (111) fiber texture of the porous films [Granquist and Hunderi 1977], or to a

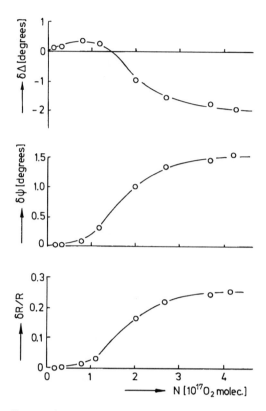

Figure 45 Change in the ellipsometric angles $\delta\Delta$ and $\delta\psi$ with the amount N of admitted oxygen molecules for the film described in Figure 44. (From Schmidt, R. and Wißmann, P., *Fresenius Z. Anal. Chem.*, 333, 410, 1989. With permission.)

plate-like shape of the crystallites [Kittel 1980]. For annealed and homogeneous (111) textured films, however, no depolarization factors are necessary for the theoretical description of the experimental values. Hence, the shape of the crystallites seems to be the most important factor for the optical behavior of the porous films.

We can also try to go a step further by describing the interaction between gas and film by an extension of Equation 25:

$$0 = \sum_{i=1}^{3} \left[f_m \frac{\varepsilon_m - \varepsilon}{\varepsilon + L_i(\varepsilon_m - \varepsilon)} + f_v \frac{\varepsilon_v - \varepsilon}{\varepsilon + L_i'(\varepsilon_v - \varepsilon)} + f_g \frac{\varepsilon_g - \varepsilon}{\varepsilon + L_i''(\varepsilon_g - \varepsilon)} \right] \qquad (26)$$

where f_g and ε_g are the volume fraction and dielectric function of the gas. The measured curves in Figure 45 can be interpreted on the basis of $\varepsilon_g = 1.1 - 1.6i$, assuming that all voids in a surface layer of 0.4 nm thickness are empty before adsorption and filled with oxygen after gas admittance [Schmidt and Wißmann 1989]. The initial increase in $\delta\Delta$ at small gas doses is probably due to a pure surface adsorption (refer to Figure 19) at this stage of interaction. Note, however, the strong negative $\delta\Delta$ values at high gas doses typical for inhomogeneous film structures (refer to Figure 20). These high negative $\delta\Delta$ values are practically independent of the depolarization factors L_i of the surface layer as model calculations on the basis of Equation 26 point out [Schmidt et al. 1989].

Summarizing, we may conclude that optical measurements on porous metal films are a suitable tool to characterize the structure and to quantify the void concentration. Oxygen tends to penetrate only in the near-surface portion of the voids.

8.4. Measurements on Copper Single Crystals

The usual procedure for preparing clean metal surfaces is the alternative heating and ion bombardment of single crystals in the UHV as mentioned in the introduction. The success of the cleaning process can

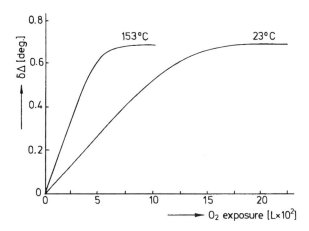

Figure 46 Change in $\delta\Delta$ for a Cu(111) single crystal in dependence on oxygen exposure $P_{ox} = 10^{-6}$ to 10^{-5} torr, T_M is parameter. (From Bootsma, G. A. et al., in *Springer Series in Chemical Physics,* Vol. 20, *Chemistry and Physics of Solid Surfaces IV,* Vanselow, R. and Howe, R., Eds., Springer-Verlag, Berlin, 1982, 77. With permission.)

easily be checked by a LEED/Auger analysis [Ertl and Küppers 1985]. Roughness should be small for such surfaces to a first approximation, and anisotropy effects, i.e., the influence of crystallic orientation on the optical properties, should be accessible to experiments.

Optical measurements on copper single crystals have been mainly reported by Bootsma et al. [1982]. A typical experimental curve is shown in Figure 46 where $\delta\Delta$ is plotted vs. oxygen dosage in Langmuir units for oxygen adsorbed on Cu(111). One can easily recognize the increase of $\delta\Delta$ at small dosages and the flattening to a saturation value at higher dosages. The curves widely correspond to Figure 27. The change in $\delta\psi$ is positive and reaches with $\Delta\psi_{max} = 0.2$ a remarkably large saturation value, which is attributed by the authors to the formation of a transition layer. Taking into account the discussion presented in Section 6.4, however, it seems possible that even on the rather closed (111) surface roughness can influence the results. The tolerance of the cut of the surface was only about 2°, which is rather bad. Hence, a high step density is to be expected at the surface, and, unfortunately, no LEED characterization of the pure copper surface was presented in the paper of Habraken et al. [1979] to specify this problem.

On the other hand, the accuracy is sufficient to verify effects of anisotropy in the surface. Figure 47 shows the dependence of $\delta\Delta$ on the azimuthal orientation of the incident beam of the ellipsometer. Here a (110) crystal is studied where particularly large anisotropic effects are to be expected because of the groove-shaped surface structure. For an incidence parallel to the grooves, rather large $\delta\Delta$ and rather small $\delta\psi$ values are found. We have to conclude, therefore, that in experiments on single-crystal surfaces generally the azimuthal direction of the incident beam should be noticed.

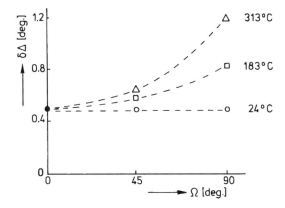

Figure 47 Change in $\delta\Delta$ for oxygen on Cu(110) at a coverage of $\theta = 0.5$. Ω is the angle between the [1$\bar{1}$0] plane and the plane of incidence. (From Habraken, F. H. P. M. et al., *Surf. Sci.,* 96, 482, 1980. With permission.)

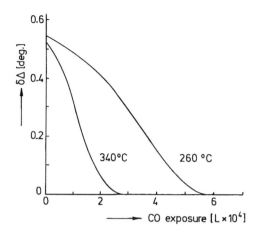

Figure 48 Change in $\delta\Delta$ for a Cu(111) surface precovered with oxygen in dependence on CO exposure. P_{CO} = 10^{-5} to 10^{-4} torr, T_M is parameter. (From Habraken, F. H. P. M. et al., *Surf. Sci.*, 83, 45, 1979. With permission.)

If we adsorb on a Cu(111) surface, the effect of anisotropy is smaller, as expected. Moreover, the influence of the light wavelength is not very marked [Bootsma et al. 1982]. The curves agree widely with the Cu/CO system (Figure 29), but disagree with the theoretical calculation on the basis of a simple three-layer model (refer to Figure 19b).

Habraken et al. [1979] also present an impressive example of a chemical reaction where the kinetics has been studied by ellipsometry (Figure 48). A certain dosage of oxygen is preadsorbed at 340°C on a Cu(111) single crystal. If CO is now admitted, the initial increase of $\delta\Delta$ is partially reduced. Obviously, the CO is oxidized at the surface, and the CO_2 molecules formed desorb at 340°C quantitatively into the gas phase, where they can be detected with the help of a mass spectrometer. The evaluation leads to the conclusion that a Langmuir–Hinshelwood mechanism is effective in this case; i.e., both species are to be adsorbed before the reaction takes place. Similar results were reported also for CO oxidation on Pd [Engel et al. 1982].

Finally, we should mention that adsorbed oxygen often induces very large changes in the optical constants. It is preferred as a test gas by many research groups, therefore. Results on the oxygen interaction with copper single crystals have additionally been published by Lin et al. [1994].

9. OTHER SELECTED METAL/GAS SYSTEMS

In the preceding sections we have concentrated on the adsorption of O_2, CO, and CO_2 on copper. Other systems have only been mentioned marginally. There are some results obtained for other systems, however, which should finally be discussed because they elucidate certain peculiarities not mentioned so far. First, the adsorption of hydrogen on iron films is of interest in this context. The polarizability of hydrogen is extremely small, and hence only minor changes in the optical parameters are observed. Moreover, iron serves as a catalyst in the Fischer–Tropsch synthesis, which can be studied by ellipsometry *in situ* under reaction conditions. Second, we deal with oxygen adsorption on thin silver films. These films tend to coagulate even at relatively moderate annealing temperatures of 200°C. Thus, this system is particularly suitable for studying the effect of island formation on the optical properties. Third, the CO adsorption on Ru (11$\bar{2}$0) is reported as an example of a gas/metal system which can be satisfactorily described by a simple two-layer model at a fixed wavelength of 546 nm. Fourth, we mention the system Xe/Au where a noble gas is adsorbed on a noble metal. Since, here, the interaction energy is extremely small and since the gold films of 70 nm used as adsorbent present extremely flat surfaces [Geiger et al. 1985] and a sufficiently large thickness, Xe/Au can serve as an ideal model system for the applicability of a two-layer model for the whole wavelength range investigated here (400 to 900 nm).

9.1. Hydrogen Adsorption on Thin Iron Films

Iron is a well-known hydrogenation catalyst. From this point of view the study of the hydrogen adsorption on thin iron films is of special interest. Contrary to the case of oxygen, dissociated hydrogen has a very small polarizability which leads to the expectation of very small changes in $\delta\Delta$ and $\delta\psi$. On the other hand, hydrogen tends to penetrate into the interior of the films to form hydride phases.

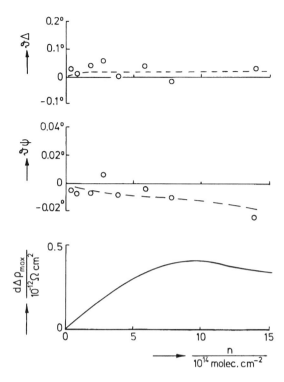

Figure 49 Change in the ellipsometric angles and the film resistance R of a 30-nm-thick iron film in dependence on hydrogen coverage n. $T_M = 293$ K, $\lambda = 1152$ nm. (From Schmidt, R. et al., *Vacuum*, 41, 1590, 1990. With permission.)

Figures 49 and 50 show typical measured curves for the dependence of $\delta\Delta$ and $\delta\psi$ on hydrogen coverage at 293 and 77 K, respectively. Indeed, the measured changes are very small; the order of magnitude is a few hundredths of a degree. The main difference between both temperatures is the sign of $\delta\psi$, which varies from negative (293 K) to positive (77 K). Taking into account that the resistivity change, work function, and heat of adsorption do not show marked differences [Schmidt et al. 1990], we assume that the small hydrogen atoms are stronger dipped into the iron surface at 77 K, thus squeezing the surrounding iron atoms and forming a reconstructed surface.

In spite of the minor changes, the accuracy of the ellipsometer used is sufficient to check the applicability of the method and to control *in situ* the reaction parameters under the conditions of the Fischer–Tropsch synthesis [Dry 1981]. Figure 51 shows the time dependence of $\delta\Delta$ and $\delta\psi$ for a hydrogenation reaction of a carbon species on a thick Fe foil at 400°C. Hydrogen is dosed in the time range 10 to 20 s up to an equilibrium pressure of 100 torr in the gas phase. A very rapid and large decrease in $\delta\Delta$ is seen due to an elimination of the carbon layer and a penetration of hydrogen into the bulk to form a subsurface species. Simultaneously, CH_4 molecules were detected in the gas phase with the help of a quadrupole mass spectrometer [Watanabe and Wißmann 1991]. Obviously, ellipsometry is a very suitable tool to study the kinetics and the formation of reactive species under realistic catalytic conditions, which is a strong advantage as compared with most of the other surface-sensitive methods of investigation [Somorjai 1994].

9.2. Oxygen Adsorption on Thin Inhomogeneous Silver Films

Silver films tend to coagulate at higher annealing temperatures. A typical transmission electron (TEM) micrograph of a 30-nm-thick silver film deposited at room temperature on glass and annealed for 1 h at 473 K is shown in Figure 52. Several small silver islands separated from each other can be recognized. Accordingly, the ε' values are much smaller than those of bulk silver [Wittmann 1984].

A strongly negative $\delta\Delta$ and a slightly positive $\delta\psi$ are observed when such a cracked silver film is covered with oxygen at 473 K (Figure 53). The theoretical prediction of Figure 23 is therefore fulfilled by the experiments; $\delta\Delta$ amounts to several degrees. It should be mentioned, however, that a marked kinetics is observed particularly at small coverages. Since the kinetics is effective on the scale of some

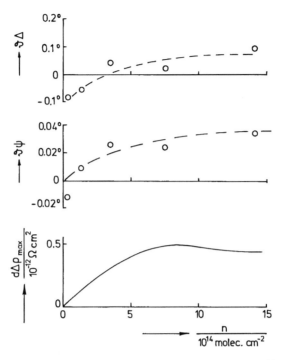

Figure 50 The same plot as in Figure 49 for hydrogen adsorption on iron at 77 K, d = 30 nm, λ = 1152 nm. (From Schmidt, R. et al., *Vacuum*, 41, 1590, 1990. With permission.)

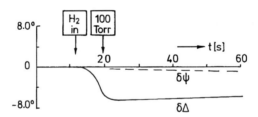

Figure 51 Time dependence of the change in the ellipsometric angles $\delta\Delta$ and $\delta\psi$ for the hydrogenation of carbon on a thick Fe foil at 400°C. For details see text. (From Watanabe, M. and Wißmann, P., *Catal. Lett.,* 7, 15, 1991. With permission.)

minutes, we have waited for 15 min after each gas dosage in order to get an equilibrium value for the optical constants. As a consequence, it is very difficult to study the interaction of oxygen with ethylene on such films at 473 K, which would be of great interest to simulate the conditions of technical ethylene oxidation. The result is that the change in $\delta\Delta$ and $\delta\psi$ due to the interaction with ethylene overlaps with the change due to the unstable film structure. The principal information available from such experiments becomes evident, however, even if ethylene is dosed at room temperature. Figure 54 shows that $\delta\Delta$ induced by a precoverage of oxygen is partially reduced by the ethylene. As discussed above (refer to Section 8.4), carbon dioxide is formed and completely desorbed at room temperature.

9.3. Interaction of Ru(11$\bar{2}$0) Single Crystals with CO

The investigation of CO adsorption on Ru(11$\bar{2}$0) is an impressive example for adsorption experiments on bulk single-crystal surfaces. Carroll et al. [1980] checked the surface structure with LEED and found a (1 × 2) superstructure during the adsorption process. Figure 55 shows the change in Δ and ψ with increasing CO exposure, measured in Langmuirs. In spite of the fact that the wavelength of 546 nm is in the visible region, the authors observe a clear increase of $\delta\Delta$ and a small decrease of $\delta\psi$. The changes are proportional to exposure at small coverages and approach to a saturation value of higher coverages. Hence, the theoretical prediction of Figure 19b is realized to a large extent.

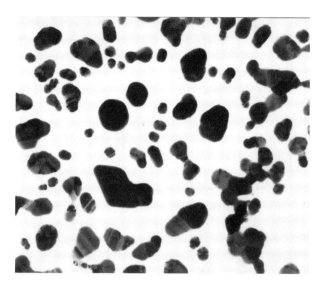

Figure 52   TEM micrograph of a 30-nm-thick silver film annealed at 473 K for 1 h. The width of the micrograph corresponds to 2.2 μm. (From Wittmann, E., Thesis, Univ. Erlangen-Nürnberg, 1984. With permission.)

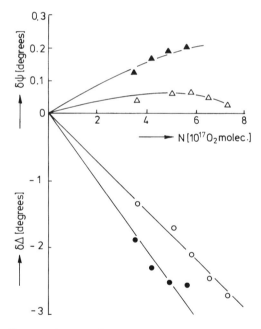

Figure 53   Change in the ellipsometric angles δΔ and δψ of two different inhomogeneous silver films (Δ, ▲) in dependence on the amount N of admitted oxygen molecules. T_A = 473 K, TM = 473 K, d = 30 nm, λ = 1152 nm. (From Wittmann, E., Thesis, Univ. Erlangen-Nürnberg, 1984. With permission.)

9.4. System Xe/Au

Finally, we will discuss the system Xe/Au. Here, a noble gas is adsorbed on a noble metal; i.e., the interaction between the gas atoms adsorbed, as well as the interaction of the gas with the metal surface, is extremely small. Moreover, the gold films can be prepared with a rather flat surface by annealing at 100°C [Geiger et al. 1985] so that roughness phenomena can be totally neglected. Hence, this system seems an ideal candidate to check the theoretical predictions of simple models. In fact, the first xenon layer is packed slightly more densely than the second because of the binding to the gold surface. With the third monolayer the packing density of bulk xenon is practically reached. As a consequence, the first monolayers are filled layer-by-layer with increasing xenon dosage, and the curves δΔ vs. dosage show characteristic steps (Figure 56), which can be attributed to the various monolayers [Wölfel et al. 1993].

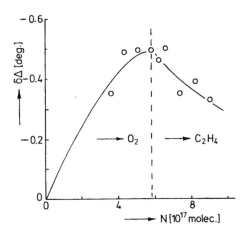

Figure 54 Change in the phase shift δΔ for a 30-nm-thick silver film annealed at 473 K and precovered with oxygen at 293 K in dependence on ethylene adsorption. T_M = 293 K, λ = 1152 nm. (From Wittmann, E., Thesis, Univ. Erlangen-Nürnberg, 1984. With permission.)

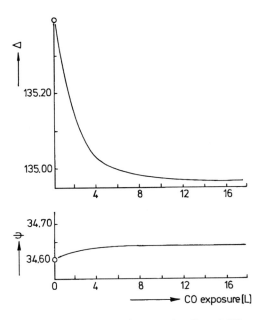

Figure 55 Change in the ellipsometric angles Δ and ψ as a function of CO exposure for (11$\bar{2}$0) ruthenium single crystals. T_M = 293 K, λ = 546.1 nm, ϕ_1 = 70. (From Carrol, J. J. et al., *Surf. Sci.*, 96, 508, 1980. With permission.)

The thickness of the gold film (d = 70 nm) is sufficiently large to assume that the adsorption of the first monolayer can be described with a two-layer model. For d_{Xe} = 0.354 nm and ε = 2.25, we calculate the dotted curve plotted in Figure 57 for the wavelength dependence of the ellipsometrical angles. The agreement between experimental and theoretical values is rather acceptable for curve 1 and worse for curves 2 and 3. Here, a better agreement is achieved if ε = 2.15 and ε = 2.10 are used for the second and the third xenon monolayer, respectively.

All in all, this system offers very clear and understandable conditions. δΔ is positive, δψ is slightly negative, both with a wavelength dependence predicted by the theory. (Note that exceptionally Δ–Δ$_o$ and ψ–ψ$_o$ are plotted in Figure 57, contrary to the definition of Equation 15.) A two-layer model in its most simple form can be applied, and with that we return to the starting point of our discussion of the various models (refer to the considerations at the end of Section 4.1). Once again, it becomes evident that the dielectric properties of the adsorbed gases are the essential physical quantity for the explanation of the measured phenomena.

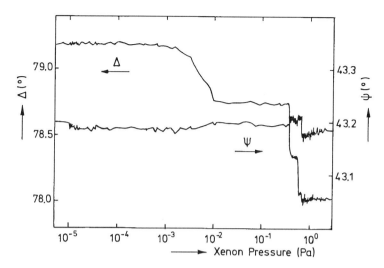

Figure 56 Change in the ellipsometric angles of a 70-nm-thick gold film in dependence on xenon pressure. T_M = 77 K, λ = 650 nm. (From Wölfel, M. et al., *Fresenius Z. Anal. Chem.*, 346, 362, 1993. With permission.)

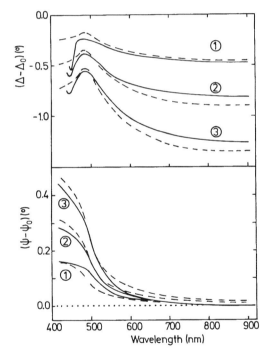

Figure 57 Change in the ellipsometric angles on the film of Figure 56 in dependence on light wavelength λ. The numbers denote the monolayers adsorbed. (From Wölfel, M. et al., *Fresenius Z. Anal. Chem.*, 346, 362, 1993. With permission.)

10. CONCLUSIONS AND OUTLOOK

The consequent application of modern UHV technique makes possible the direct measurement of the influence of adsorbed and absorbed gases on the optical constants of thin metal films. The high sensitivity and resolution of PSCA ellipsometers allow the investigation of even very weakly bound gases (physisorption systems), as is shown for the example Xe/Au. Main emphasis, however, is put on the study of the interaction of O_2, CO, and CO_2 with polycrystalline copper films. The interpretation is based on suitable layer models which work satisfactorily in the Drude range, where the optical properties are governed by the free electrons. In the interband transition range, on the other hand, many questions on quantitative interpretation remain open. A critical review is given on all models available up to now; the systematic variation of the angle of incidence is expected to supply us with further information in the near future, particularly with respect to anisotropic phenomena.

The results show that the optical measurements are a fruitful method to study adsorption states as well as the properties of the clean films. A quantitative interpretation of the data obtained for technical catalysts, however, has been up to now difficult because of a lack of theoretical information on the role of surface roughness and film inhomogeneities in optical measurements. Here, we expect another key point of future work.

ACKNOWLEDGMENTS

The author would like to thank Prof. Dr. U. Merkt, Hamburg, for many helpful comments and for critically reading the manuscript. Financial support of the Fonds der Chemischen Industrie and the Deutsche Forschungsgemeinschaft is gratefully acknowledged.

REFERENCES

Abeles, F., Borensztein, Y., and Lopez-Rios, T. (1984). Optical properties of discontinuous thin films and rough surfaces of silver, in *Festkörperprobleme*, Vol. 24, Vieweg, Braunschweig, p. 93.

Azzam, R. M. A., and Bashara, N. M. (1977). *Ellipsometry and Polarized Light*. North-Holland, Amsterdam.

Bennett, H. E., and Bennett, J. M. (1966). *Optical Properties and Electronic Structure of Metals and Alloys*, ed. F. Abeles, North-Holland, Amsterdam.

Bispinck, H. (1970). Der Einfluß von Gitterdefekten auf die optischen Konstanten von Kupfer. *Z. Naturforsch.* 25a:70.

Bootsma, G. A., Hanekamp, L. J., and Gijzeman, O. L. J. (1982). Chemisorption investigated by ellipsometry, in *Springer Series in Chemical Physics,* Vol. 20, *Chemistry and Physics of Solid Surfaces IV*, eds. R. Vanselow and R. Howe, Springer-Verlag, Berlin, p. 77.

Buck, H., Schmidt, R., and Wißmann, P. (1988). Widerstandsmessungen an dünnen Kupferschichten bei Wechselwirkung mit Sauerstoff (in German), in *Nichtmetalle in Metallen,* ed. D. Hirschfeld, Verlag DGM-Informationsges, Wiesbaden, p. 219.

Campuzano, J. C. (1990). The adsorption of carbon monoxide by the transition metals, in *Chemical Physics of Solid Surfaces and Heterogeneous Catalysis*, Vol. 3A, eds. D. A. King and D. P. Woodruff, Elsevier, Amsterdam, p. 389.

Carroll, J. J., Madey, T. E., Melmed, A. J., and Sandstrom, D. R. (1980). The room temperature adsorption of oxygen, hydrogen and carbon monoxide on (1120) ruthenium: an ellipsometry-LEED-characterization, *Surf. Sci.* 96:508.

Chopra, K. L. (1969). *Thin Film Phenomena,* McGraw-Hill, New York.

Dayal, D., Finzel, H.-U., and Wißmann, P. (1987). Resistivity measurements on pure and gas covered silver films, in *Studies in Surface Science and Catalysis,* Vol. 32, *Thin Metal Films and Gas Chemisorption*, ed. P. Wißmann, Elsevier, Amsterdam, p. 53.

Dignam, M. J., and Moskovits, M. (1973). Optical properties of sub-monolayer molecular films, *Trans. Faraday Soc.* II 69:56.

Dry, M. E. (1981). The Fischer-Tropsch synthesis, in *Catalysis,* Vol. 1, eds. J. R. Anderson and M. Boudart, Springer, Berlin, p. 159.

Dürrschnabel, W., and Voßhühler, H. (1976). Data on the Cu/O system, in *Gase und Kohlenstoff in Metallen,* eds. E. Fromm and E. Gebhardt, Springer, Berlin, p. 657

Engel, T., and Ertl, G. (1982). Oxidation of carbon monoxide, in *The Chemical Physics of Solid Surfaces and Heterogeneous Catalysis,* Vol. 4, eds. D. A. King and D. P. Woodruff, Elsevier, Amsterdam, p. 73.

Ertl, G., and Küppers, J. (1985). *Low Energy Electrons and Surface Chemistry,* VCH, Weinheim.

Fenstermaker, C. A., and McCrackin, F. L. (1969). Errors arising from surface roughness in ellipsometric measurements, *Surf. Sci.* 16:85.

Forstmann, F., and Gerhardts, R. R. (1982). Metal optics near the plasma frequency, in *Festkörperprobleme,* Vol. 22, ed. P. Grosse, Vieweg, Braunschweig, p. 291.

Gebhardt, R., Rauh, M., and Wißmann, P. (1991). Structure investigations of pure and partially oxidized copper films, *Fresenius Z. Anal. Chem.* 341:332.

Geiger, H., Häupl, K., Wißmann, P., and Wittmann, E. (1985). Structure investigations on gold films with flat surfaces, *Vakuum-Technik* 34:135.

Granquist, G., and Hunderi, O. (1977). Optical properties of ultra-fine gold particles, *Phys. Rev. B* 16:3513.

Habraken, F. H. P. M., Kieffer, E. Ph., and Bootsma, G. A. (1979). A study of the kinetics of the interaction of O_2 and NO_2 with Cu(111), *Surf. Sci.* 83:45.

Habraken, F. H. P. M., Gijzeman, O. L. J., and Bootsma, G. A. (1980). Ellipsometry of clean surfaces, submonolayer and monolayer films, *Surf. Sci.* 96:482.

Hagemann, H. H., Gudat, W., and Kunz, C. (1974). *Handbook of Optical Constants of Solids*, ed. E. D. Palik, Academic Press, Orlando, FL.

Hanekamp, L. J., Lisowski, W., and Bootsma, G. A. (1982). Spectroscopic ellipsometric investigation of clean and oxygen covered copper single-crystal surfaces, *Surf. Sci.* 118:1.

Hardel, K., and Schön, D. (1972). Phasengrenzreaktionen bei der Oxidation von Metallen: das System $Cu/Cu_2O/O_2$, *Z. Phys. Chem.* (Frankfurt) 77:293.

Hoffmann, H. (1983). Free electrons in polycrystalline metal films, in *Proc. 9th Vacuum Congress,* ed. J. L. de Segovia, Madrid, ASEVA Publ., p. 351.

Hollstein, T., Kreibig, N., and Leis, F. (1977). Optical properties of Cu and Ag in the intermediate region between pure drude and interband absorption, *Phys. Status Solidi B* 82:545.

Hölzl, J., and Schulte, F. K. (1979). Work function of metals, in *Springer Tracts in Modern Physics,* Vol. 85, ed. G. Höhler, Springer, Berlin.

Hummel, R. E. (1971). *Optische Eigenschaften von Metallen und Legierungen.* Springer. Berlin.

Hummel, R. E. (1989). A new look at the reliability of thin film metallizations for microelectronic devices, in *Festkörperprobleme,* Vol. 29, ed. V. Rössler, Vieweg, Braunschweig, p. 251.

Johnson, P. B., and Christy, R. W. (1975). Optical constants of noble metals, *Phys. Rev. B* 11:1315.

Karlsson, B., Ribbing, C. G., Roos, A., Valkonen, E., and Karlson, T. (1982). Optical properties of some metal oxides in solar absorbers, *Phys. Scripta* 25:826.

Kittel, Ch. (1980). *Einführung in die Festkörperphysik,* R. Oldenbourg, München, p. 434.

Leidheiser, H., Jr. (1971). *The Corrosion of Copper, Tin, and Their Alloys,* The Corrosion Monograph Series, John Wiley, New York.

Lin, K. C., Tobin, R. G., and Dumas, P. (1994). Adsorbate-induced changes in the broadband infrared reflectance of oxygen on Cu(100), *Phys. Rev. B* 49:17273.

Mahler, G., Körner, H., and Teich, W. (1991). Optical properties of quasi-molecular structures, from single atoms to quantum dots, in *Festkörperprobleme,* Vol. 31, ed. V. Rössler, Vieweg, Braunschweig.

Merkt, U. (1978). Ellipsometrische Untersuchungen an reinen und mit CO belegten Metallschichten. Thesis, Univ. Erlangen-Nürnberg (in German). See also: Merkt, U., and Wißmann, P. (1979). *Thin Solid Films* 57:65; Merkt, U., and Wißmann, P. (1980). *Surf. Sci.* 96:529; Merkt, U. (1981). *Appl. Opt.* 20:307; Merkt, U., and Wißmann, P. (1983). *Z. Phys. Chem. (Frankfurt)* 135:227.

Mönch, W. (1995). *Semiconductor Surfaces and Interfaces,* Springer, Berlin.

Pells, G. P., and Shiga, M. (1969). The optical properties of copper and gold as a function of temperature, *J. Phys. C* 2:1835.

Persson, B. N. J. (1991). Surface resistivity and vibrational damping in adsorbed layers, *Phys. Rev. B* 44:3277.

Rauh, M. (1991). Ellipsometrische Untersuchungen zur Oxidation dünner Kupferschichten, Thesis, Univ. Erlangen-Nürnberg (in German). See also: Rauh, M., and Wißmann, P. (1993). *Thin Solid Films* 228:121; Rauh, M., Wißmann, P., and Wölfel, M. (1993). *Thin Solid Films* 233:289; Rauh, M., and Wißmann, P. (1995). *Fresenius Z. Anal. Chem.*, 353:769.

Ritchie, J. M., and Hunt, G. L (1969). The kinetics and pressure dependence of surface controlled metal oxidation reactions, *Surf. Sci.* 15:524.

Roberts, S. (1960). Optical properties of Cu, *Phys. Rev.* 118:1513.

Schlosser, G. (1972). *Heterogene Katalyse,* VCH, Weinheim.

Schmidt, R. (1988). Ellipsometrische Untersuchungen in den Systemen Cu/O_2 und Fe/H_2. Thesis, Univ. Erlangen-Nürnberg (in German). See also: Schmidt, R., and Wißmann, P. (1988). *Surf. Interface Anal.* 12:407; Schmidt, R., and Wißmann, P. (1989). *Fresenius Z. Anal. Chem.* 333:410; Schmidt, R., Wedler, G., and Wißmann, P. (1990). *Vacuum* 41:1590.

Somorjai, G. (1994). *Surface Chemistry and Catalysis,* John Wiley, New York.

Stoll, M. P. (1969). Optical constants of copper measured with a polarimetric method, *J. Appl. Phys.* 40:4533.

Stotz, S. (1966). Untersuchungen über die Geschwindigkeit der Reaktion $O_2(g) \rightarrow 2O(ads)$ an der Oberfläche von Cu_2O und NiO, *Ber. Bunsenges. Phys. Chem.* 70:769.

Theiß, W. (1994). The use of effective medium theories in optical spectroscopy, in *Festkörperprobleme,* Vol. 33, ed. R. Helbig, Vieweg, Braunschweig.

Walter, T. (1996). Ellipsometrical studies of xenon adsorption on Ag and Cu films, Thesis, Univ. Erlangen-Nürnberg.

Watanabe, M., and Wißmann, P. (1984). Dielectric constants of adsorbed CO on Cu and Ag, *Surf. Sci.* 138:95.

Watanabe, M., Wedler, G., and Wißmann, P. (1985). Ellipsometric response to hydrogen absorbed on Pd films, *Surf. Sci.* 154:L207.

Watanabe, M. (1987). Optical analysis of adsorbed gases on metals and metal films, in *Studies in Surface Science and Catalysis,* Vol. 32, *Thin Metal Films and Gas Chemisorption*, ed P. Wißmann, Elsevier, Amsterdam, p. 389.

Watanabe, M., and Wißmann, P. (1991). States of surface hydrogen under reaction conditions of the Fischer-Tropsch synthesis, *Catal. Lett.* 7:15.

Wedler, G., and Fouad, M. (1963). Die Schichtdicken- und Temperaturabhängigkeit der Adsorption von Kohlenmonoxid an aufgedampften Metallfilmen, *Z. Phys. Chem.* (Frankfurt) 40:12.

Wedler, G., and Wiebauer, W. (1975). Resistivity and Hall effect of copper films before and after adsorption of carbon monoxide, *Thin Solid Films* 28:65.

Wedler, G. (1976). *Chemisorption,* Butterworths, London.

Wißmann, P. (1975). The electrical resistivity of pure and gas covered metal films, in *Springer Tracts in Modern Physics,* Vol. 77, Springer, Berlin.

Wißmann, P. (1994). The interaction of gases with thin metal films, in *Growth and Applications of Thin Films*, eds. L. Eckertova and T. Ruzicka, Prometheus, Prague, p. 25.

Wittmann, E. (1984). Ellipsometrische Untersuchungen zur Adsorption und Koadsorption von Sauerstoff und Ethylen an aufgedampften Silber- und Goldschichten. Thesis, Univ. Erlangen-Nürnberg (in German), see also: Schmiedl, E., Wißmann, P., and Wittmann, E. (1983). *Surf. Sci.* 135:341; Wißmann, P., and Wittmann, E. (1985). *Surf. Sci.* 152:638; Finzel, H.-U., Wißmann, P., and Wittmann, E. (1986). *Surf. Sci.* 166:L126.

Wölfel, M., Rauh, M., and Wißmann, P. (1993). Spectroscopic ellipsometry on xenon monolayers adsorbed on gold films, *Fresenius Z. Anal. Chem.* 346:362.

Wolter, H. (1956). Optik dünner Schichten, in *Handbuch der Physik*, Vol. 24, ed. S. Flügge, Springer, Berlin.

Wooten, F. (1972). *Optical Properties of Solids*, Academic Press, San Diego.

Xu., X., Wang, N., and Zhang, O. (1992). Ab initio calculations for the CO/Cu chemisorption system, *Surf. Sci.* 274:386.

Zemel, J. N. (1975). Gas effects on IV-VI semiconductor films, in *Surface Physics of Phosphors and Semiconductors*, eds. C. G. Scott and C. E. Reed, Academic Press, London.

Chapter 15

Optical Properties of High-T$_c$ Superconducting Surfaces

Rajiv K. Singh

Reviewed by J. Narayan

TABLE OF CONTENTS

1. INTRODUCTION

The discovery of high-critical-temperature (high-T$_c$) superconductivity above 30 K in copper oxide–based perovskites [Bednorz and Muller, 1986] led to tremendous research efforts to discover new materials that exhibit superconductivity phenomena at higher T$_c$ values (>77 K) and to understand the nature of superconductivity in these materials. Subsequent to the discovery by Bednorz and Muller, several new materials systems have been discovered, including Y–Ba–Cu–O, Bi–Sr–Ca–Cu–O, Tl–Ba–Ca–Cu–O, etc., all of which exhibit superconductivity above liquid nitrogen temperatures [Wu et al., 1987; Maeda et al., 1988; Sheng et al., 1988]. Of all these materials systems, YBa$_2$Cu$_3$O$_7$ is perhaps the most widely studied materials system in terms of understanding both the superconducting phenomenon and its use in many potential applications [Singh and Narayan, 1991].

There have been several experiments conducted to understand the nature of superconductivity phenomena in YBa$_2$Cu$_3$O$_7$. One of the most fundamental aspects in superconductivity is the effect of lattice vibrations on the superconducting phenomena [Ramirez et al., 1987]. The phonon density of states has been determined by neutron scattering [Renker et al., 1987], while Raman and infrared spectroscopy yields information on the phonon frequencies, the microstructure of the superconducting phase, and the dynamics of the lattice vibrations in the superconducting unit cell [Fiele, 1989]. At present, very few techniques are available to determine the exact location of light atoms, like oxygen, in the superconducting unit cell. Polarization experiments in Raman spectroscopy combined with lattice dynamics calculations can be applied to identify the location of the oxygen signal from the superconducting unit cell. Raman spectroscopy also provides detailed information on the temperature dependence of the lattice vibrations in the superconducting material and its dependence on the oxygen content, the nature of the superconducting gap, and determination of the isotope effect in superconductivity [Osipyan et al., 1988; Fiele, 1989].

In this review, we plan to summarize Raman investigations of YBa$_2$Cu$_3$O$_7$ surfaces. The penetration depth of the 514.5 nm Raman radiation in YBa$_2$Cu$_3$O$_7$ is approximately 1200 Å; thus, essentially the near surface of the material is investigated in the Raman probe [Singh et al., 1991]. Single-crystal, thin film, and bulk polycrystalline YBa$_2$Cu$_3$O$_7$ surfaces have been probed by various researchers. All these surfaces have unique aspects which will be discussed in detail in the following sections.

2. STRUCTURE AND STRUCTURE–PROPERTY CORRELATION IN YBa$_2$Cu$_3$O$_7$

To understand the optical properties of YBa$_2$Cu$_3$O$_7$, the nature of the superconducting unit cell should first be examined. Figure 1A shows the position of all the atoms of a fully oxygenated YBa$_2$Cu$_3$O$_{7-\delta}$

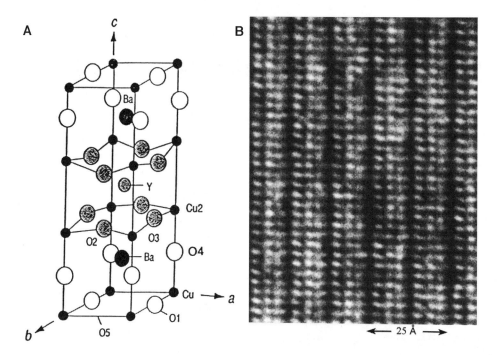

Figure 1 (A) Schematic diagram of a fully oxygenated superconducting $YBa_2Cu_3O_7$ unit cell and (B) HRTEM showing the the unit cell symmetry. The zone axis is [100] for the micrograph.

($\delta = 0$) superconducting unit cell [Jorgenson et al., 1987; Singh and Narayan, 1991]. A corresponding high-resolution transmission electron micrograph (HRTEM) with (100) zone axis is shown in Figure 1B. The unit cell has an orthorhombic structure with the c-axis (11.77 Å) almost three times the "a" (3.82 Å) and "b" (3.88 Å) axes. Thus, the unit cell can be defined as a tripled perovskite unit cell structure with oxygen vacancies at the yttrium and the Cu–O(5) planes. Figure 1B clearly shows a stacked planar structure consisting of barium–oxygen and yttrium planes separating the copper–oxygen planes and chains. The HRTEM micrograph also shows the layered structure and the corresponding symmetry of the superconducting unit cell.

A unique characteristic of this structure is the tetragonal to orthorhombic phase transition which occurs at high temperature and is accompanied by a corresponding increase in the oxygen content ($\delta \leq 0.5$). At high temperatures, the unit cell structure is oxygen deficient ($\delta > 0.5$) and possesses a tetragonal symmetry (space group = D_{2h} P4/mmm), while a fully oxygenated superconductor ($\delta = 0.0$) is transformed into an orthorhombic structure (space group = D_{4h} Pmmmm) [Jorgenson et al., 1987]. The dependence of T_c on the oxygen content is influenced by the preparation conditions. Reducing the oxygen content from 7.0 reduces the superconducting transition temperature continuously, and superconducting vanishes at $\delta = 0.5$. Cava et al. [1987] have shown that annealing at lower temperature in argon atmospheres leads to oxygen gettering which gives pronounced plateaus of T_c at 60 K for $\delta = 0.4$ to $\delta = 0.2$ due to oxygen vacancies at the Cu–O chains. These samples do not show superconductive behavior for $\delta > 0.5$.

Another interesting characteristic of $YBa_2Cu_3O_7$ is the appearance and disappearance of copper–oxygen chains [Jorgenson et al., 1987; Liu et al., 1988; Cava et al, 1990; Gupta and Gupta, 1991; Jorgenson, 1991]. For $\delta = 1$, the $YBa_2Cu_3O_7$ is nonmetallic and nonsuperconducting, while for $\delta < 0.1$, a maximum T_c of 93 K is observed. The transition from the superconducting to the nonsuperconducting state is accompanied with the onset of antiferromagnetism. During this transformation a vacancy is created in the O(5) position of the unit cell which leads to the formation of Cu(1)–O(1) chains. There has been considerable debate on the contribution of chain oxygen atoms to the superconducting mechanism. The disappearance of superconductivity with concomitant disordering of the Cu(1)–O(5) chains in oxygen-deficient samples led to initial speculation that the ordered Cu(1)–O(1) chains were responsible for superconductivity. However, with the discovery of the copper–oxygen chain-free bismuth-based superconductors [Maeda et al., 1988], researchers have argued that copper–oxygen planes and not the chains were primarily responsible for high-temperature superconductivity. Although the exact mechanisms for the occurrence of superconductivity are still not clear, it is widely believed via indirect observations that the superconductivity in these compounds arises only as a result of the two-dimensional $Cu–O_2$ planes [Jorgenson, 1991]. A direct correlation has been established between the hole concentration in the $Cu–O_2$ planes and T_c. In an insulating $YBa_2Cu_3O_6$ structure, the lack of oxygen atoms at the chains isolated the CuO_2 planes from the chains, rendering the sample nonsuperconducting, whereas in a fully oxygenated structure, the chains act as an electron reservoir for electron transfer from the planes to the chains [Cava et al., 1990].

The orthorhombic phase exhibits 39 translational modes: 3 acoustic, 21 IR active, and 15 Raman active [Fiele, 1989]. In the tetragonal phase three fewer modes occur because of the missing oxygen atom. Among these modes several are doubly degenerate modes which reduce the number of observable lines in the spectra. It should also be noted that deviation from the ideal composition can change the symmetry selection rules, thus causing forbidden lines to appear in the Raman spectra.

3. SINGLE-CRYSTAL INVESTIGATIONS

Several Raman studies have been conducted on polycrystalline ceramic single crystals and thin film materials to understand the nature of lattice vibrations. Most of the initial studies were conducted on polycrystalline sintered pellets which provide the Raman signal from a large number of small grains of unknown orientation [Hemley and Mao, 1987; Narayan et al., 1987]. This complexity initially combined with the presence of second phases (Y_2BaCuO_5, barium- or copper-rich phases) led to incorrect assignment of some of the lattice vibrations observed in the superconducting spectra [Rosen et al., 1988]. The dominant Raman-active vibrations were found to occur at the following frequencies: ~116, ~145, 230, ~440, ~340, ~500, and 600 cm^{-1}. However, in some samples some of these peaks were found to be missing from the Raman spectra [Fiele et al., 1988].

With the availability of single crystals of $YBa_2Cu_3O_{7-\delta}$, consistent assignment of the Raman peaks to various lattice vibrations was first established. Initial experiments were conducted on small single crystals which were twinned [Hemley and Mao, 1987; Krol et al., 1987; 1988; Thomsen et al., 1988]. Due to the twinned structure, the "a" and "b" axes of the superconductor cannot not be differentiated. Krol et al. [1988] obtained results from small $YBa_2Cu_3O_7$ and $GdBa_2Cu_3O_7$ crystals. In their experiments, the lines at 140, 330, 435, and 500 cm^{-1} were observed under different polarization conditions. These Raman peaks were assigned to A_g vibrations which occur along the c-axis of the $YBa_2Cu_3O_7$ superconductor. These modes were assigned to the symmetric stretching of the Ba planes, Cu(2)axial symmetry stretching, O(2)/)(3) out-of-phase and in-phase bending, and O(1) oxygen symmetric bond vibrations. The peaks at 500 cm^{-1} were assumed to arise from the disorder scattering from the nonstoichiometry which allows $q \neq 0$ scattering to be observed in Raman measurements as local defect modes. This peak was found to shift to lower frequencies because of the oxygen content in the sample, which will be considered in detail in the next section. Krol et al., [1988]. Cooper et al. [1988] in their single-crystal experiments observed lines at ~116, ~150, ~340, ~440, and ~500 cm^{-1} and identified them to have the A_g symmetry. Table 1 lists the various identified A_g modes observed in single-crystal experiments.

Table 1 Raman shifts for twin free $YBa_2Cu_3O_7$ crystals

Vibration mode	Polarization	Raman frequencies				
		Experimental				
A_g	α_{xx}	116	149	336	493	
A_g	α_{yy}	114	149	335	493	
A_g	α_{zz}	119	149	435	498	
B_{2g}	α_{zx}	70	142	210	579	
B_{3g}	α_{xy}	83	140	303	526	
		Calculated				
A_g		116	157	355	378	508

Note: Calculated values are adapted from Kress et al. [1988]. α_{xx} refers to the component of the polarizability tensor permitted by the polarization geometry.

Adapted from McCarty, K. F. et al., *Phys Rev. B*, 41, 8792, 1990.

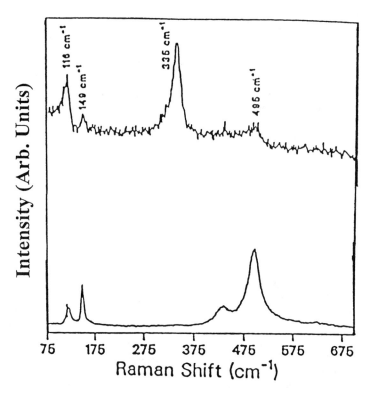

Figure 2 Raman spectra of twin free $YBa_2Cu_3O_7$ obtained with the incident laser beam propagating (a) parallel to the "c" axis and (b) perpendicular to the "c" axis. (Adapted from McCarty et al. 1990.)

Perhaps the most definitive study on untwinned $YBa_2Cu_3O_7$ single crystals was conducted by McCarty et al. [1990]. The majority of the earlier investigations were conducted for "single crystals" containing high density of twins which interchange the "a" and "b" directions in the crystal. Because of the "a–b" anisotropy, a complete analysis of the Raman–active modes was not possible using twinned crystals. In particular, the assignments of the B_{2g} and B_{3g} modes cannot be made using untwinned crystals.

McCarty et al. confirmed that peaks at 340, 440, and 500 cm^{-1} in the Raman spectra were due to the oxygen vibrations, while the two other lines at 116 and 145 cm^{-1} represent the Cu and Ba modes, respectively. The A_g modes for two different polarization conditions are shown in Figure 2. The Raman peak at 145 cm^{-1} has been assigned to vibrations due to symmetric stretching of barium planes, while the 116-cm^{-1} peak arises because of Cu(2) vibrations along the "c" axis in the CuO_2 planes of the superconducting unit cell. The Ba vibration at 116 cm^{-1} has been observed to have an asymmetric line shape caused by the interference of the scattering of the discrete phonon with the scattering of the electronic continuum. The interference is constructive on the high-frequency side of the phonon and destructive on the low-frequency side. The destructive interference causes the intensity to dip below the base line on the high-frequency side. The higher-frequency Raman modes have been assigned to different oxygen locations. The Raman peak at 340 cm^{-1} has been unambiguously assigned to the in-phase bond bending of the O(2) and O(3) atoms in the CuO_2 planes, while the peak at 500 cm^{-1} has been assigned to the axial symmetric stretching of the Cu(1)–O(4) bonds as shown in Figure 1. The assignment of the 500-cm^{-1} peak has been confirmed by varying the peak position as a function of oxygen composition in $YBa_2Cu_3O_7$.

In contrast to the A_g modes, the intensity of the B_{2g} and B_{3g} modes obtained from untwinned single-crystal studies was found to be extremely weak and nearly 100 times less intense than the A_g modes [McCarty et al. 1990]. In some investigations it was found that imperfect polarization or polarization leakage resulted in contamination of the B_{2g} and B_{3g} spectrum with the A_g modes [McCarty et al., 1990]. Table 1 lists the frequencies of the observed A_g, B_{2g}, and B_{3g} modes in untwinned $YBa_2Cu_3O_7$ along with the lattice dynamic calculations, which are discussed in detail in Section 4 [Fiele, 1989]. The B_{2g} geometry contained peaks at 70, 142, 210, and 579 cm^{-1}, while the B_{3g} spectrum exhibited peaks at 83, 140, 303, and 526 cm^{-1}. The B_{2g} and B_{3g} modes involve essentially the same vibrations with the B_{2g}

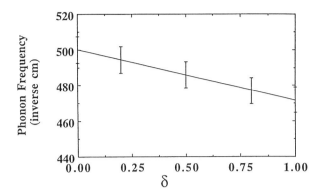

Figure 3 Variation in the Raman shift with oxygen deficiency of the fully oxygenated ~500-cm⁻¹ peak.

modes involving atomic motion along the "a" axis and the B_{3g} modes involving motion along the "b" axis. However, due to the Cu(1)–O(1) chains that run along the "b" direction, the B_{2g} and B_{3g} modes can differ in frequency, particularly for modes involving atoms adjacent to the Cu(1)–O(1) chains, i.e., Y and O(4) atoms.

4. INFLUENCE OF OXYGEN STOICHIOMETRY

The understanding of the oxygen stoichiometry in $YBa_2Cu_3O_7$ is of critical importance for the understanding of the superconducting phenomena. Several groups have investigated the influence of the oxygen content on the Raman spectra [Yamanaka et al., 1987; Cardona et al., 1987; Krol, 1988; Burns et al., 1988; Bhadra et al., 1988]. The Raman vibration at 500 cm⁻¹ was found to shift considerably with change in oxygen content of the superconducting unit cell. Various researchers have provided Raman data with differing concentrations of oxygen and concomitant correlation with the Raman shift of the ~500 cm⁻¹ vibration mode. Data by several researchers (Yamanaka et al., Cardona et al., Krol et al.) have been near the stability limits, while the data from Burns et al. have been near the orthorhombic–tetragonal transition which occurs at $\delta = 0.5$. Based on the data presented by various authors, and analysis presented by other researchers, a linear relationship of the peak shift with oxygen content has been determined [Fiele, 1989]. Figure 3 shows the calculated phonon frequency of the ~500-cm⁻¹ peak as a function of the oxygen content in the $YBa_2Cu_3O_7$. The error bars show the upper and lower limits of the experimental data observed by various researchers. The equation of the Raman frequency, ω, as a function of the oxygen deficiency can be given by

$$\omega = 500 - 28\delta$$

where ω corresponds to the frequency of the ~500-cm⁻¹ peak and δ corresponds to the oxygen deviation from the maximum oxygen content in the superconducting unit cell. The large change in the values of the 500-cm⁻¹ peak with the oxygen content can thus be used to estimate the oxygen content in the superconducting thin films. Besides the phonon peak at 500 cm⁻¹, significant peak shifts with oxygen content were not observed. Although some researchers have shown the pressure effects on the variation in the Raman peak shifts [Syassen et al., 1988], adequate explanation of the variation for peak shifts with oxygen content is still unclear.

5. SUPERCONDUCTING THIN FILMS

Raman spectroscopy of superconducting thin films is particularly suited because of the flat surface which gives less elastically scattered light than a ceramic sintered surface. In addition, like single crystals, single orientation of the films can be obtained. Epitaxial superconducting thin films have been fabricated on a wide variety of substrates with excellent superconducting properties [Singh et al. 1989; Poppe et al., 1989]. Raman measurements have been conducted on thin films to determine the texture, presence of impurity phases, oxygen stoichiometry, and the epitaxial nature of the film. A detailed investigation of the thin film superconductors was first conducted by Fiele et al. [1988]. They observed all the allowable Raman vibrations expected from a polycrystalline film.

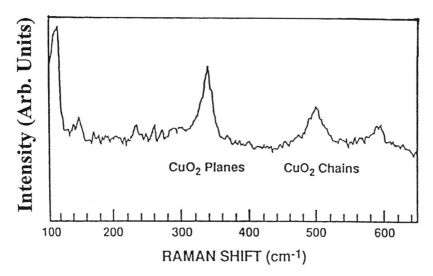

Figure 4   Raman spectra obtained from an epitaxial superconducting $YBa_2Cu_3O_7$ film on an yttria-stabilized zirconia substrate.

Berberich et al. [1988] determined the effect of the preparation conditions on the superconducting properties of polycrystalline evaporated thin films. The 500-cm^{-1} line shifted because of oxygen deficit in lower-quality films, which is consistent with the results discussed in the previous section. In some earlier investigations, films with a sharper superconducting transition indicated the presence of $BaCuO_2$ [Berberich et al., 1988]. This suggested that the presence of the $BaCuO_2$ may serve as a flux for growth of better-quality films, as in the case of single crystal. However, with the sophistication in the growth process, very high quality films have been grown since then which do not exhibit substantial amounts of $BaCuO_2$ phase [Singh et al., 1989].

The best films have been grown on lattice-matched substrates, like $SrTiO_3$, $LaAlO_3$, yttria-stabilized zirconia, using *in situ* deposition methods like pulsed laser deposition (PLD), off-axis sputtering, etc. In these films the three axes are aligned to the substrates and possess very high critical current densities ($>5 \times 10^6$ A/cm^2) at 77 K and zero magnetic field. The large critical current densities are due to lack of large-angle grain boundaries in the film which significantly improves the electrical transport.

A typical Raman spectra on the high-quality superconducting thin film on yttria-stabilized zirconia substrate is shown in Figure 4 [Singh et al., 1995]. The figure shows the typical A_g peaks expected from a c-axis-oriented superconducting thin film. All the A_g peaks are present except the peak at 435 cm^{-1}, which is generally observed in single crystals when the Raman axis is parallel to the "a–b" planes. Another interesting feature observed in these films is the lattice vibration at ~500 cm^{-1}. The scattering efficiency of the Cu(1)–O(4) oxygen vibrations at 500 cm^{-1} has been attributed to the Cu(1)–O(1) chains. Single-crystal experiments conducted by McCarty et al., (discussed in earlier section) showed that the intensity of the 500-cm^{-1} peak is the strongest when the Raman beam is perpendicular to the c-axis. However, this geometry is not achievable in thin films because of the thin dimensions in those directions. Singh et al. have attributed the ~500-cm^{-1} peak in the epitaxial film to the "polarization leakage" effect which arises because of slight misorientation in the experimental setup.

Singh et al. [1995] have conducted a series of investigations using Raman spectroscopy to understand the nonequilibrium disordering of the the Cu–O chains in the superconducting unit cell. By employing extremely rapid thermal treatments, such as pulsed nanosecond lasers, they have been able to selectively disorder the copper–oxygen chains without affecting the copper–oxygen planes in the superconducting unit cell. Figure 5 shows the Raman spectra of epitaxial $YBa_2Cu_3O_7$ film on $LaAlO_3$ irradiated with 25-ns KrF laser beams at different energy densities (1) 0 J/cm^2, (2) 90 mJ/cm^2, (3) 150 mJ/cm^2, (4) 250 mJ/cm^2, and (5) 400 mJ/cm^2. Upon laser irradiation of the $YBa_2Cu_3O_7$ films at different energy densities, some distinct changes are observed. The crystallinity of the films shows distinct improvement until an energy density of 150 mJ/cm^2. This is shown by stronger and cleaner low-frequency peaks at 145, 230, and 340 cm^{-1}. The first two peaks correspond to the vibrations from the copper and barium atoms, while the stronger peak at 340 cm^{-1} shows an increased crystallinity of the oxygen atoms in the copper–oxygen planes. However, there is deterioration and broadening of the high-frequency Raman

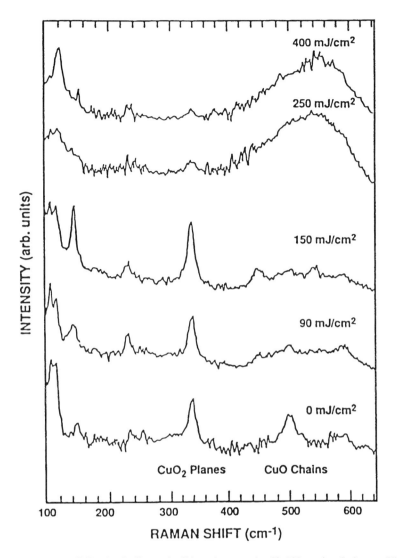

Figure 5 Raman spectra of $YBa_2Cu_3O_7$ film on $LaAlO_3$ substrate after XeCl laser irradiation at different energy densities. Increased ordering of the $Cu-O_2$ planes coincides with disordering in Cu–O chains for increasing energy density up to the melt threshold.

peak at 500-cm^{-1}, which suggests disordering of the chain oxygen atoms with increasing energy density. Singh et al. [1995] have further correlated these results with the superconducting properties and determined than the chain oxygen atoms are not critical for superconductivity.

6. LATTICE DYNAMICS

Lattice dynamic calculations have been performed on $YBa_2Cu_3O_7$ to calculate the electron–phonon interactions and to assign the observed vibrational excitations. For $YBa_2Cu_3O_7$ the calculation of the phonon frequencies on the basis of the force constant model was first reported by Stavola et al. [1987]. In this model the structure of the unit cell and the force constant for the short-range interactions were used. The force constants were then adjusted in order to match with observed Raman spectra. Subsequent to this work, several researchers adopted different values of force constants so that an adequate prediction of the frequencies can be estimated. Fiele et al [1988] also developed a lattice dynamic model with first and second nearest-neighbor force constants. However, the Raman frequencies obtained for the Raman-active frequencies were too high because of misinterpretation of the disorder active IR active mode at 600 cm^{-1}. The phonon frequencies obtained with modified force constants were on the order of 500 cm^{-1}.

The main problem in the calculations mentioned above was that a direct correlation between the calculated and observed Raman vibrational modes was rather vague, primarily because a pure force constant model which may be applied to covalent crystals is probably too simple for $YBa_2Cu_3O_7$. It was noted that the long-range and short-range coulombic interactions in the unit cell should also be accounted for [Bruesch and Buhrer, 1988].

Kress et al. [1988] developed a different route to calculate the phonon frequencies in $YBa_2Cu_3O_7$. The short-range forces were represented by a repulsive Born–Mayer potential V_{ij} and is given by

$$V_{ij} = a_{ij} \exp(-b_{ij} r)$$

where r corresponds to the interatomic distance and a_{ij} and b_{ij} are constants. This formula gives rise to different force constants for the same atomic pairs at different distances, thus reducing the number of parameters for short-range interactions. From the phonon-dispersion curves of $YBa_2Cu_3O_7$, the coulombic forces between the ions were also determined. Calculated values shown in Table 1 were found to be in good agreement with the experimental results. In addition, the calculated phonon density of states reproduced the overall features of neutron scattering experiments. Thus, these calculations combined with the experimental data can provide insight into the nature of lattice dynamics in these superconducting materials.

7. CONCLUSIONS

A summary of the Raman spectroscopy of the high-T_c superconducting surfaces from thin films, single crystals, and sintered pellets of $YBa_2Cu_3O_7$ has been presented in this chapter. Detailed polarization studies on $YBa_2Cu_3O_7$ single crystals provide insights into the nature of lattice vibration in $YBa_2Cu_3O_7$. The dominant Raman-active modes were found to be of the A_g type, in which the vibrations occur along the c-axis of the superconducting unit cell. Significant shifts in the ~500 cm^{-1} peak were observed with change in the oxygen content of the sample. Raman studies on thin films also provide information on the nature of the microstructure (presence of impurity phases, texture, etc.). Lattice dynamics calculations based on short- and long-range coulombic interactions were found to be in good agreement with the experimental results.

REFERENCES

Bednorz, J. G. and K. A. Muller, *Z. Phys. B* 64, 189, 1986.

Berberich, P., W. Dietsche, H. Kinder, J. Tate, C. Thomsen, and B. Scherzer, *Physica C*, 153–155, 1451, 1988.

Bhadra, R., T. O. Brun, M. A. Beno, B. Dabroski, and D. G. Hinks, *Phys. Rev B* 37, 5142, 1987.

Bruesch, P. and W. Buhrer, *Z. Phys. B* 70, 70, 1988.

Burns, G., F. H. Dacol, and M. W. Shafer, *Solid State Commun.* 62, 687, 1988.

Cava, R. J., B. Batlogg, C. H. Chen, E. A. Reitman, S. M. Zahurak, and D. Werder, *Phy. Rev. B* 36, 5719, 1989.

Cava, R. J., A. W. Hewat, E. A. Hewat, B. Batlogg, M. Marizio, K. M. Rabe, J. J. Kajeswski, W. F. Peek, and L. W. Rupp, *Physcia C*, 165, 419, 1990.

Cooper, S. L., M. V. Klein, B. G. Pazol, J. P. Rice, and D. M Ginsberg, *Phys. Rev B* 37, 5920, 1987.

Cordona, M., L. Genzel, R. Liu, A. Wittlin, H. Mattausch, F. Garcia Alvardo, and E. Garcia Gonzalez, *Solid State Commun.* 64, 727, 1987.

Fiele, R., U. Schmitt, and P. Leiderer, *Physica C* 153–155, 292, 1988.

Fiele, R., *Physica C* 159, 1, 1989.

Hemley, R. J. and H. K. Mao, *Phys. Rev. Lett.* 58, 2340, 1987.

Jorgenson, J. D., M. A. Beno, D. G. Hinks, L. Soderholm, K. J. Volin, R. L. Herman, J. D. Grace, I. V. Schuller, C. U. Segre, K. Zhang, and M. S. Kleesfisch, *Phys. Rev. B* 36, 3608, 1987.

Jorgenson, J. D., *Physics Today*, 31, 1991.

Kress, W., U. Schroder, J. Prade, A. Kulkarni, and F. W. deWette, *Phys. Rev. B* 39, 2906, 1988.

Krol, D. M., M. Syavola, L. F. Schneemeyer, J. V. Waszczak, and W. Weber, in *High Temperature Superconductors*, M. B. Brodsky, R. C. Dynes, K. Kitazawa, and H. L. Tuller, Eds., *MRS Proc.* No. 99, MRS, Pittsburgh, PA, 1988, 781.

Krol, D. M., M. Stavola, W. Wever, L. F. Schneemeyer, J. V. Waszcak, S. M. Zahurak, and S. G. Kosinski, *Phys. Rev. B* 36, 8325, 1987.

Maeda, H., H. Tanaka, M. Fukotomi, and T. Asano, *Jpn. J. Appl. Phys.* 27, L209, 1988.

McCarty, K. F., J. Z. Liu, R. N Shelton, and H. B. Radoisky, *Phys. Rev. B* 41, 8792, 1990.

Narayan, J., V. N. Shukla, S. J. Lukasiewicz, N. Biunno, and R. K. Singh, A. F. Schreiner, and S. J. Pennycook, *Appl. Phys. Lett.* 51, 940, 1987.

Osipyan, Y. A., V. B. Timofeev, and I. F. Sclegolev, *Physica C* 153–155, 1133, 1988.

Poppe, U., P. Priesto, J. Scubert, H. Soltner, and K. Urban, *Solid State Commun.,* 71, 569, 1989.

Ramirez, A. P., B. Batlogg, G. Aeppli, R. J. Cava, E. Rietman, A. Goldman, and G. Shirane, *Phys. Rev. B* 35, 8833, 1987.

Renker, R., F. Gompf, E. Gering, N. Nucker, D. Ewert, W. Reichardt, and H. Rietschel, *Z. Phys. B* 67, 15, 1987.

Rosen, H. J., R. D. Macfarlane, E. M. Engler, V. Y. Lee, and R. D. Jacowitz, *Phys. Rev. B* 73, 695, 1988.

Sheng, Z. Z., A. M. Hermann, A. Elali, C. Almasan, J. Estrada, T. Datta, and R. J. Matson, *Phys. Rev. Lett.* 60, 937, 1988.

Singh, R. K., J. Narayan, A. K. Singh, and J. Krishnaswamy, *Appl. Phys. Lett.* 54, 2271, 1989.

Singh, R. K. and J. Narayan, *Phys. Rev. B* 41, 8843, 1990.

Singh, R. K. and J. Narayan, *J. Met.* 43, 13, 1991.

Singh, R. K., D. Bhattacharya, S. Harkness, J. Narayan, C. Jahncke, and M. Paesler, *Appl. Phys. Lett.* 59, 1380, 1991.

Singh, R. K., S. Harkness, P. Tiwari, J. Narayan, C. Jahncke, and M. Paesler, *Phys. Rev. B* 51, 9155, 1995.

Stavola, M., D. M. Krol, W. Weber, S. Sunshine, A. Jayaraman, G. Kouroulis, R. J. Cava, and E. A. Rierman, *Phys. Rev. B* 37, 850, 1987.

Syassen, K., M. Hanfland, and K. Strosser, *Physica C* 153–155, 264, 1988.

Wu, M. K., J. R. Ashburn, C. J. Torng, P. H. Hor, R. L. Meng, L. Gao, Z. J. Huang, Y. Q. Wang, and C. W. Chu, *Phys. Rev. Lett.* 58, 908, 1987.

Yamanaka, A., F. Minami, K. Watanabe, K. Inoue, S. Takekawa, and N. Iyi, *Jpn. J. Appl. Phys. 2* 26, 1404, 1987.

INDEX